W0107167

Science and Technology of Thin Film Superconductors

Science and Technology of Thin Film Superconductors

Edited by
Robert D. McConnell

Solar Energy Research Institute
Golden, Colorado

and
Stuart A. Wolf

Naval Research Laboratory
Washington, D.C.

Springer Science+Business Media, LLC

Library of Congress Cataloging in Publication Data

Conference on the Science and Technology of Thin Film Superconductors (1988: Colorado Springs, Colo.)
  Science and Technology of Thin Film Superconductors / edited by Robert D. Mc-Conell and Stuart A. Wolf.
     p.   cm.
  "Proceedings of the Conference on the Science and Technology of Thin Film Super-conductors, held November 14–18, 1988, in Colorado Springs, Colorado" — T.p. verso.
  Bibliography: p.
  Includes index.
  ISBN 978-1-4684-5660-8   ISBN 978-1-4684-5658-5 (eBook)
  DOI 10.1007/978-1-4684-5658-5
  1. Superconductors — Congresses. 2. Thin Film devices — Congresses. 1. McConnell, Robert D. II. Wolf, Stuart A. III. Title.
  TK7872.S8C67 1988   89-33432
  621.3 — dc20   CIP

Proceedings of the Conference on the Science and Technology
of Thin Film Superconductors, held November 14–18, 1988,
in Colorado Springs, Colorado

© 1989 Springer Science+Business Media New York
Originally published by Plenum Press, New York in 1989
Softcover reprint of the hardcover 1st edition 1989

All rights reserved

No part of this book may be reproduced, stored in a retrieval system, or transmitted in any form or by any means, electronic, mechanical, photocopying, microfilming, recording, or otherwise, without written permission from the Publisher

PROGRAM COMMITTEE

R. McConnell, Solar Energy Research Institute
R. Berdahl, Lawrence Berkeley Laboratory
R. Blaugher, Intermagnetics General Corp.
A. DasGupta, U.S. Department of Energy
R. Hammond, Stanford University
R. Harris, National Institute of Standards and Technology
A. Hermann, University of Arkansas
R. Simon, TRW, Inc.
K. Wasa, Matsushita Electric Industrial Co., Ltd.
S. Wolf, Naval Research Laboratory

CONFERENCE ORGANIZERS

U.S. Department of Energy
Solar Energy Research Institute
National Institute of Standards and Technology
Naval Research Laboratory
Lawrence Berkeley Laboratory

CONFERENCE COORDINATOR

D. Christodaro, Solar Energy Research Institute

PROCEEDINGS

R. D. McConnell, Solar Energy Research Institute
S. A. Wolf, Naval Research Laboratory

PREFACE

The Conference on the Science and Technology of Thin Film Superconductors was conceived in the early part of 1988 as a forum for the specialist in thin film superconductivity. The conference was held on November 14-18, 1988, in Colorado Springs, Colorado. Although many excellent superconductivity conferences had been convened in the wake of the 1986-1987 discoveries in high temperature superconductivity, thin film topics were often dispersed among the sessions of a more general conference agenda. The response to the Conference on the Science and Technology of Thin Film Superconductors confirmed the need for an extended conference devoted to thin film superconductors. These proceedings are a major contribution to the technnology of thin film superconductivity because of the breadth and quality of the articles provided by leaders in the field. The proceedings are divided into articles on laser deposition, sputtering, evaporation, metal organic chemical vapor deposition, thick film, substrate studies, characterization, patterning and applications, and general properties. Most of the articles discuss scientific issues for high temperature thin film superconductors, although the conference was to be a forum for technology and scientific questions for both low and high temperature superconductivity.

For the first day of the 5 day conference, Lawrence Berkeley Laboratory had organized an excellent set of short courses in superconducting thin film devices. For the rest of the conference, the program committee had invited speakers for about 25 of the 80 presentations; the remaining contributions appeared either as poster presentations or as additional oral presentations. The program committee was pleased with the quality of the contributed and invited presentations. Ultimately the conference hosted 210 researchers from 13 countries.

This conferences was more than a set of technical presentations. It provided an important opportunity for private communications, for the purpose of understanding the latest research and identifying new research topics. Both the ambiance of the Red Lion Inn and other conference activities enhanced this opportunity. The poster session, on the second night of the conference, included a wine and cheese reception. Allen Hermann, whose trombone is as well known in jazz clubs throughout the U.S. as is his work on the thallium compounds, organized the superb Wednesday night entertainment entitled "Dixieland Music Scene." Thursday night's banquet provided an opportunity to reflect on what had been learned and to hear an industry perspective, presented by Donald Hicks, Group Vice President for Ball Aerospace Systems Group, on what needs to be done to realize the commercial potential of high temperature superconductivity. Friday morning was devoted to the Colorado Superconductivity Shoot Out, an opportunity to compare the thallium compounds, bismuth compounds, and 123 compounds. Randy Simon's presentation was a highlight during the Shoot Out because of his succinct comparison of these compounds using a Consumer Report's format. The conference included visits to the extensive thin film facilities at the Solar Energy Research Institute and the well known superconductivity facilities at Boulder's National Institute of Standards and Technology.

There are many people and organizations responsible for the success of a conference and we thank the Solar Energy Research Institute and the Office of Energy Storage and Distribution within the U.S. Department of Energy for supporting us in organizing the conference. The Naval Research Laboratory, the National Institute of Standards and Technology in Boulder, and Lawrence Berkeley Laboratory provided important contributions. We thank the program committee who enticed so many excellent researchers into making presentations. Diane Christodaro, from the Solar Energy Research Institute, skillfully coordinated the preparations for the conference. Her help was appreciated by all attendees as she and her group resolved individual problems while ensuring the smooth flow of scheduled activities. We thank the many contributors who took the time and effort out of busy schedules to prepare thoughtful, quality, descriptions of their research activities. Finally, we thank Kristine Weber-McConnell and Iris Wolf for their understanding and support while we have pursued this exciting and important new area of research.

Robert D. McConnell and Stuart A. Wolf

CONTENTS

LASER DEPOSITION

EVAPORATION

SPUTTERING

THICK FILM

SUBSTRATE STUDIES

CHARACTERIZATION

PATTERNING and APPLICATIONS

GENERAL PROPERTIES

SUPERCONDUCTING PROPERTIES OF

ULTRA-THIN FILMS OF $Y_1Ba_2Cu_3O_{7-x}$

T. Venkatesan, X.D. Wu[a], B. Dutta[b], A. Inam[a], M. S. Hegde[c],
D. M. Hwang, C. C. Chang, L. Nazar and B. Wilkens

Bellcore, Red Bank, NJ 07701
(a) Physics Department, Rutgers University, Piscataway, NJ 08854
(b) Physics Department, Middlebury College, Middlebury, VT 05753
(c) Solid State Structural Chemistry Unit, Indian Institute of Science
Bangalore, India, 560012, and
Center for Ceramics Research, Rutgers University, Piscataway, NJ 08854

ABSTRACT

We have grown ultra-thin films of $Y_1Ba_2Cu_3O_{7-x}$ *in situ* on (001) $SrTiO_3$ by pulsed laser deposition. The zero resistance transition temperature (T_{co}) is > 90 K for films > 300 Å thick. The critical current density (J_c at 77 K) is 0.8×10^6 A/cm^2 for a 300 Å film and $4\text{-}5 \times 10^6$ A/cm^2 for a 1000 Å film. The T_{co} and J_c deteriorate rapidly below 300 Å, reaching values of 82 K and 300 A/cm^2 at 77 K, respectively, for a 100 Å film. Films only 50 Å thick exhibit metallic behavior and possible evidence of superconductivity without showing zero resistance to 10 K. These results are understood on the basis of the defects formed at the film-substrate interface, the density of which rapidly decreases over a thickness of 100 Å. We have studied these defects by ion channeling measurements and cross section transmission electron microscopy. Our results suggest that the superconducting transport in these films is likely to be two dimensional in nature, consistent with the short coherence length along the c-axis of the crystals.

INTRODUCTION

In situ epitaxial growth of $Y_1Ba_2Cu_3O_{7-x}$ (123) films at substrate holder temperatures of 600-700 C has been demonstrated using a number of different techniques[1-6]. On (001) $SrTiO_3$ or MgO single crystal substrates highly c-axis oriented films have been grown. Recently, an oriented film as thin as 100 Å has been demonstrated on (001) $SrTiO_3$ by Bando et al[6] with a zero resistance temperature (T_{co}) of 82 K. While it is technologically interesting that such thin films could be made, from the point of view of basic principles this raises some important questions. For example, in the 123 system the superconductivity is assumed to take place in the Cu-O planes, with essentially a 2-dimensional transport in the planes and negligible interaction across the planes owing to the short coherence length across the planes[7]. If one could synthesize very thin films, on the order of one or two unit cells thick, one may be able to understand the coupling across the planes and study the proximity effect between planes (eg., is more than one plane necessary for high T_c ?). In this paper we will describe the results of our study in the synthesis of ultrathin superconducting film by a pulsed laser deposition technique[2].

EXPERIMENT

The preparation of films by pulsed laser deposition has been extensively described in earlier publications[2,8]. In this work, we performed the deposition in oxygen pressures of 100 mTorr instead of a few mTorr as in our earlier work[2,8]. We have shown by optical spectroscopic techniques that deposition under such high oxygen pressures leads to the formation of oxides of the three cations which enhances the quality and reproducibility of the films[9]. Subsequent to deposition the films were allowed to cool to room temperature in about 200 Torr of oxygen. Transport measurements were made by painting silver ink contacts on the film. To measure critical current densities a constriction was produced using a diamond scribe. The width of the constriction (typically 50 μm) was measured with an optical microscope. The crystallinity (or conversely, the disorder) in the film as a function of depth was measured by using a 2 MeV He[+] ion backscattering technique with an ion beam spot size of 1 mm diameter and a detector placed in the backscattering geometry[10-11]. The thickness of the films was also measured using a mechanical stylus (Alfa Step). The accuracy of the film thickness measurement is better than 10%. Transmission electron microscopy (TEM) of cross-sectioned samples were performed with a JEOL 4000 FX 400 kev microscope.

RESULTS

In Fig. 1 are shown the resistance vs. temperature values for films of different thicknesses. Several features can be noted in this figure. The 1000 Å film shows a normal state resistance which extrapolates to the origin. However, as the thickness decreases, the normal state resistance shows a deviation from linearity, the onset of the transition becomes more gradual and the transition width broadens. Below 300 Å the T_{co} deteriorates rapidly, with the 100 Å film showing a T_{co} of 82 K and the 50 Å film showing evidence for a superconducting transition (the dip at 90 K). For the 50 Å film the drop in the resistivity at lower temperatures may be due to the increased resistance of the contacts. In Fig. 2 is shown the temperature dependence of the critical current density (defined at 10 μV); the lower temperature bound for each data set is determined by the largest measurable current in our system (measurements below 76 K were possible only for 300 Å and 100 Å films). The critical current density for a 1000 Å film extrapolates to 4--5 x 10^6 A/cm^2 at 77 K, while even for a 300 Å film at 77 K the critical current density is as high as 0.8 x 10^6 A/cm^2. However, for a 100 Å film the J_c drops to 300 A/cm^2 at 77 K though at 10 K the value is a respectable 10^5 A/cm^2. Since the 50 Å film did not show full onset of superconductivity, a critical current density could not be measured. The thickness dependences of the transport properties indicate a critical thickness of approximate 100 Å, above and below which the film properties are dramatically different, with deterioration at lower thicknesses and improvement at larger values. What is the origin of this thickness effect ? In order to study this we explored the depth dependent disorder in the film.

In Fig. 3 are shown the random and ion channeling spectra for a 100 Å and a 1000 Å film. Only the Ba spectrum is easy to interpret unlike those of Y and Cu which are more difficult owing to their overlape with the substrate elemental spectra. However, the Ba spectrum should be quite representitive of the crystalline order in the film. The channeling minimum yield (χ_{min}) for the 100 A film is 35 % implying that 67 % of the atoms are in the proper lattice sites. The channeling spectra of the 1000 Å film is more interesting. The minimum yield near the surface is 5 % (by comparison, that for a bulk single crystal is 3.5 % [10]), implying > 98 % of the atoms are in the right lattice sites while at the interface the scattering yield is significantly larger. This implies that the film-substrate interface has residual disorder which rapidly decreases as one moves away from the interface towards the surface. The T_{co} as a function of disorder (defined as 1- χ_{min} - 0.02), is plotted in Fig. 4. As the film thickness decreases to 50 A, the disorder increases rapidly. Indeed the thickness of this disordered interface (as measured in the thick films) is of the order of 100-200 Å, which is consistent with the transition thickness observed by the transport measurements. In x-ray diffraction, the films exhibited predominantly (00L) (L=integer) reflections for all the films measured implying c-axis orientation normal to the substrate surface even for the thinnest (50 Å) films. In Fig. 5 are shown for comparison the (001) reflection obtained in a 2Θ scan for a 100 and a 1000 Å film. There is significant broading of the peak for the 100 Å film, which would be due both to the size effect as well as the modulation of the plane orientation.

In order to understand the nature of the defect at the interface, TEM cross sectional specimens were prepared and examined in a high resolution microscope. The details of the preparation of the samples have been described elsewhere[12-13]. In Figs. 6a and b are shown the lattice images of a 2000 Å film on SrTiO$_3$ giving an idea of the interface. In Fig. 6a one sees the film-substrate interface and the transition is abrupt over one atomic spacing. There is no interdiffusion at the interface. Further, the epitaxy of the (110) planes (vertical) with the substrate is perfect though the c-plane stacking (horizontal) is imperfect. Within a 50 Å distance from the substrate one does not see completely connected 123 layers horizontally though the cells exhibit perfect order vertically.

2

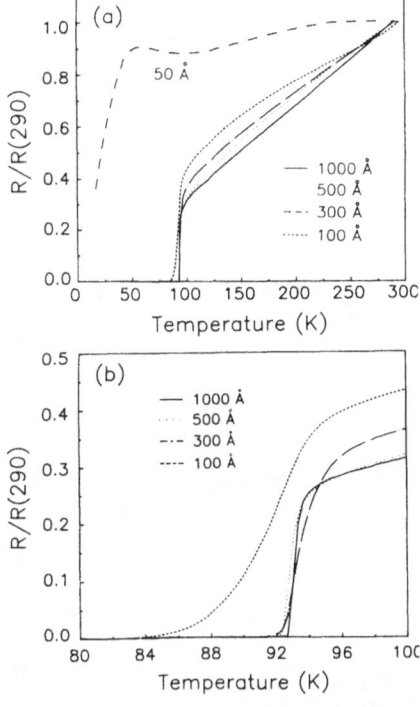

Fig. 1. Resistance vs. temperature curves for different film thicknesses: (a) reduced scale, and (b) expanded scale.

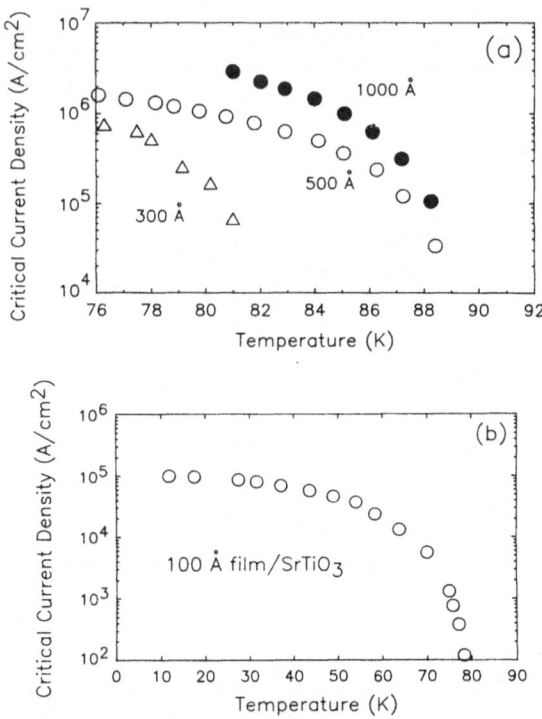

Fig. 2. J_c vs. temperature for films of different thicknesses. (a) 300, 500, and 1000 Å, and (b) 100 Å. Note the change of scales in the two figures.

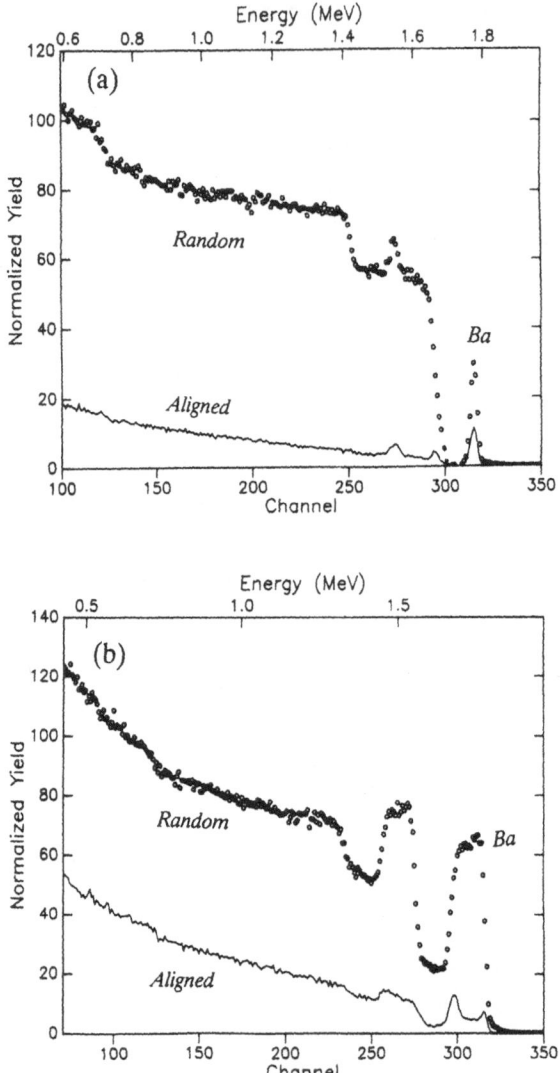

Fig. 3. Random and channeling spectra for $Y_1Ba_2Cu_3O_{7-x}$ films on $SrTiO_3$, for two different film thicknesses: (a) 100 Å and (b) 1000 Å.

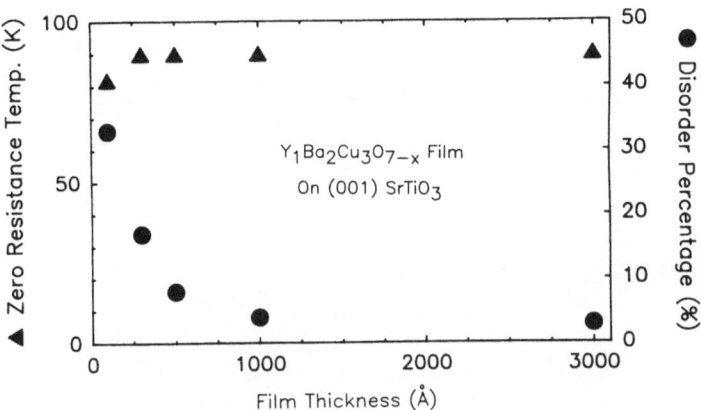

Fig. 4. T_{co} and disorder as a function of film thickness.

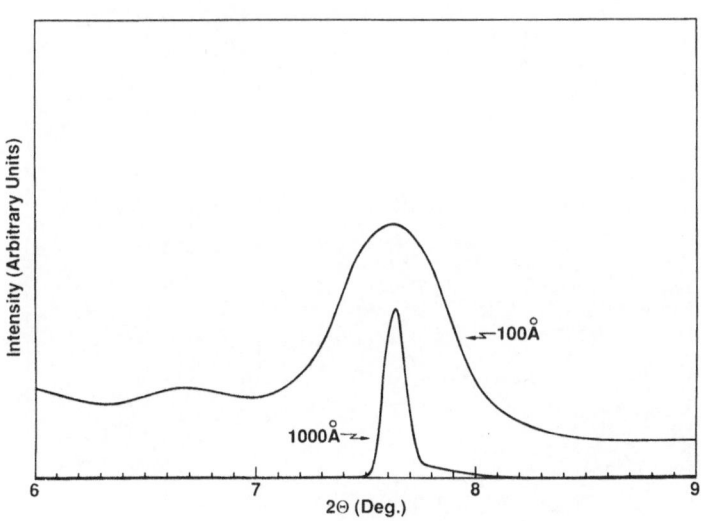

Fig. 5. Comparison of (001) reflection in the X-ray 2Θ
Scan for a 100 and 1000 Å films of $Y_1Ba_2Cu_3O_{7-x}$.

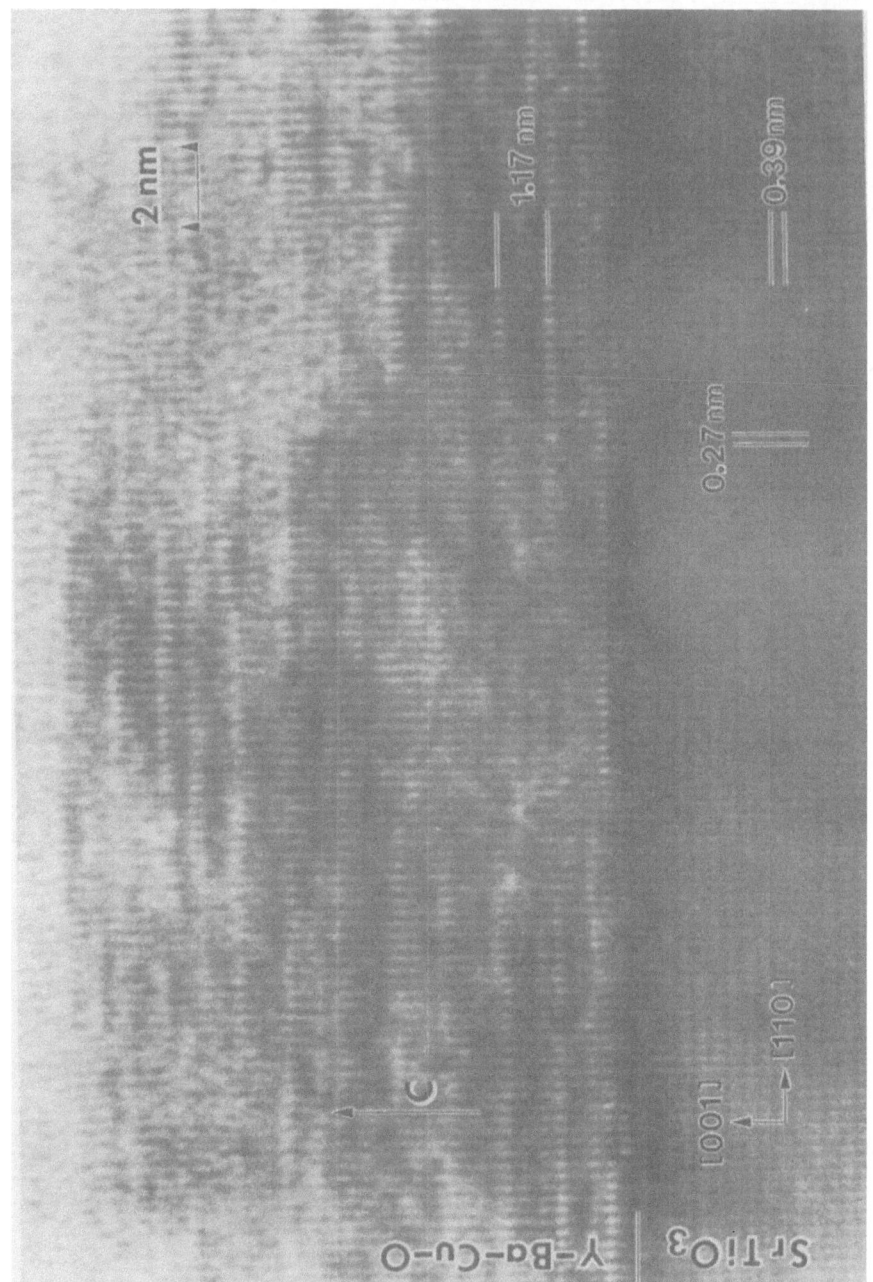

Fig. 6a. TEM cross sectional lattice images of a 2000 Å epitaxial 123 film on (001) SrTiO$_3$ showing a portion of the film close to the interface region.

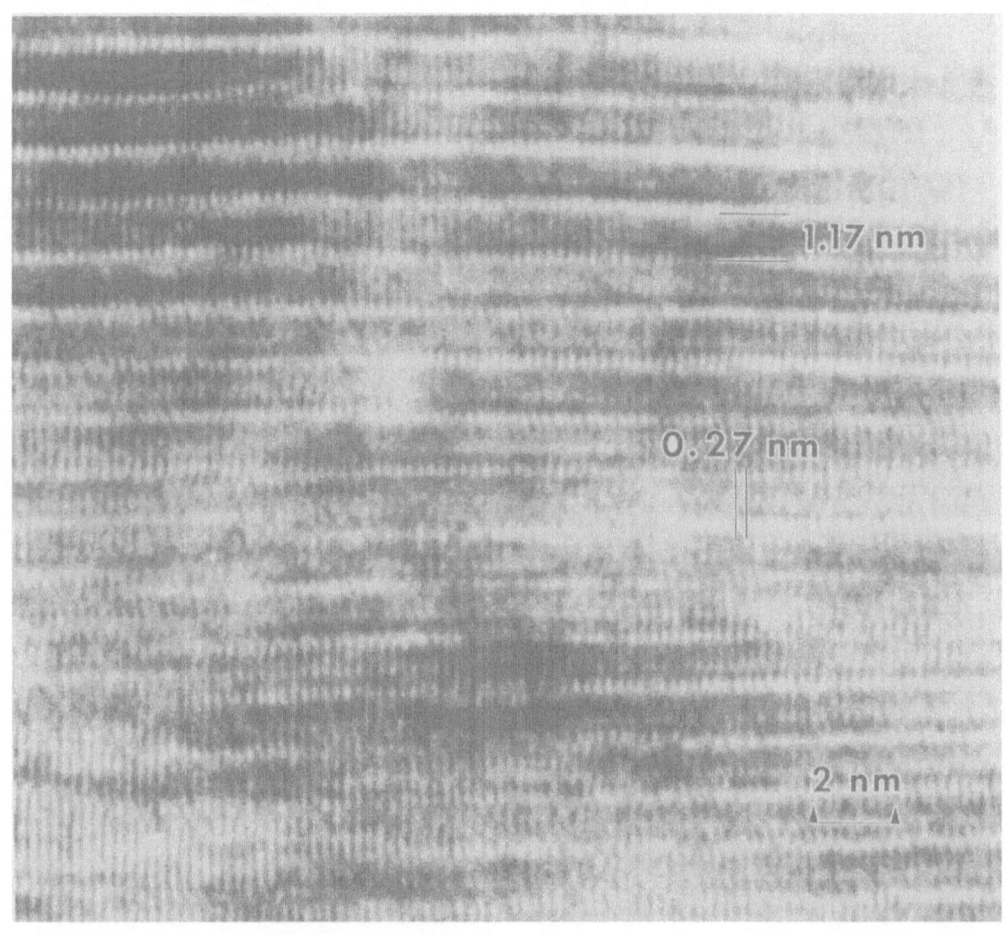

Fig. 6b. TEM cross sectional lattice images of a 2000 Å epitaxial 123 film on (001) SrTiO$_3$ showing the area several 100 Å away from the interface.

But as we move away from the interface the lateral coherence of the layers improves as shown in Fig. 6b, the horizontal continuity has increased significantly, though the vertical coherence has deteriorated.

We now discuss our ideas about the origin of this transition region. The (001) $SrTiO_3$ substrate has a unit cell whose cross section is a square. The c-axis oriented orthorhombic 123 phase has a rectangular cross section; the b-axis matches closely the side of the cubic $SrTiO_3$ unit cell (0.5 % lattice mismatch). However, the a-axis has a misfit of about 2 %. Epitaxial growth on a substrate with such a misfit would introduce a significant number of defects at the interface, which however with the further growth of the 123 phase would present a surface layer which is closer to being orthorhombic. Hence, the 123 phase immediately on top of the $SrTiO_3$ will not only be closer to a tetragonal phase, but because of the lattice mismatch may either have misfit dislocations or be strained. From our experimental data it appears that after a growth of a film on the order of a 100 Å, the crystallinity as well as the orthorhombicity of the film may be improving, thereby improving the transport properties.

CONCLUSION

The results are exciting from another point of view. For a 100 Å thick film, there are about 8 unit cells of the 123 stacked on top of each other. The bottom few unit cells are probably not superconducting because of their proximity to the substrate (they look very different from the rest of the film in the TEM micrographs) while the top one unit cell has generally been found to be not of the right superconducting phase.[11] As a result, the superconducting transport is probably occuring in the middle few unit cells, which is consistent with the idea of a two dimensional superconducting system. Even these layers must be highly defective since the critical current density has decreased by more than two orders of magnitude with respect to the thick films. If these defects could be eliminated, in principle a film only a unit cell thick may be superconducting. Recent measurements[14] of extremely high J_c's in sub-micron wires made on these films suggest further that the pulsed laser deposited films could be processed with little deterioration of the film properties. Thus attempts to produce such ultra-thin films with ultra-fine geometries may not only enable us to optimize the deposition process, but also enable us to study new physics in these small dimensions.

ACKNOWLEDGMENTS

The authors are grateful to J. M. Tarascon, P. Barboux, E. W. Chase, J. M. Rowell, M. Giroud, W. L. McLean and J. B. Wachtman for various help and fruitful discussions. One of the authors (MSH) would also like to acknowlege financial support by the National Science Foundation Grant No. NSF-MSM-87-16317 1/1.

REFERENCES

1.   H. Adachi, K. Hirochi, K. Setsune, M.Kitabake, and K. Wasa, Appl. Phys. Lett., **51**, 2263 (1987).

2.   A. Inam, M.S. Hegde, X. D. Wu, T. Venkatesan, T. Venkatesan, P. England, P. F. Miceli, E. W. Chase, C. C. Chang, J. M. Tarascon and J. Wachtman, Appl. Phys. Lett., **53**, 908 (1988).

3.   H. C. Li, G. Linker, F. Ratzel, R. Smithey, and J. Geerk, Appl. Phys. Lett., **52**, 1098 (1988).

4.   R. M. Silver, A. B. Berezin, M. Wendman, and A. de Lozanne, Appl. Phys. Lett., **52**, 2174 (1988).

5.   D. K. Lathrop, S. E. Russek, and R. A. Burhman, Appl. Phys. Lett., **51**, 2263 (1987).

6.   Y. Bando, T. Terashima, K. Iijima, K. Yamamoto and H. Mazaki, Physica **C 153-155** (1988) 810-811.

7.   W. J. Gallagher, T. W. Worthington, T. R. Dinger, F. Holtberg, D. L. Kalser, and R. L. Sandstrom, in *Superconductivity in Highly Correlated Fermion Systems*, edited by M. Tachiki, Y. Muto, and S, Mackawa (North-Holland, Amsterdam, 1987) pp. 228-232.

8.   X. D. Wu, A. Inam, T. Venkatesan, C. C. Chang, E. W. Chase, P. Barboux, J. M. Tarascon and B. Wilkens, Appl. Phys. Lett., **52**, 754 (1988).

9.   X. D. Wu, B. Dutta, M. S. Hegde, A. Inam, T. Venkatesan, E. W. Chase, C. C. Chang and R. Howard, Appl. Phys. Lett. (to be published).

10.   N. G. Stoffel, P. A. Morris, W. A. Bonner, and B. J. Wilkens, Phys. Rev. **B37**, 2297 (1988).

11.   X. D. Wu, A. Inam, M. S. Hegde, T. Venkatesan, C. C. Chang, E. W. Chase, B. Wilkens and J. M. Tarascon, Phys. Rev. **B38**, 9307(1988).

12.   L. Nazar, Proc. 46th Ann. EMSA Meeting, ed. by G. W. Bailey (San Francisco Press, San Francisco, 1988), p. 862.

13.   D. M. Hwang, L. Nazar, T. Venkatesan, and X. D. Wu, Appl. Phys. Lett. **52**, 1834(1988).

14.   P. England (unpublished).

PLASMA-ASSISTED LASER DEPOSITION OF SUPERCONDUCTING THIN FILMS - A BASIC STUDY

H.S. Kwok, J.P. Zheng, Z.Q. Huang, Q.Y. Ying,
S. Witanachchi and D.T. Shaw

Institute on Superconductivity
State University of New York at Buffalo
Buffalo, NY 14260

ABSTRACT

An experimental study of the process of plasma-assisted laser deposition was described. It was found that atomic beams of Y,Ba,Cu, and O were inportant for the in-situ growth of superconducting films. These atomic beams could be modelled very well by a supersonic expansion mechanism. The atomic kinetic energies and their spatial dependence were measured and correlated to the thin film properties.

1. INTRODUCTION

Since the early days of high power laser-target interaction studies, it was realized that copious amounts of atoms and ions could be liberated from the target[1,2]. Various attempts were made to form thin films from these atoms. It was generally believed that that laser evaporation was similar to normal e-beam evaporation or ion sputtering in the sense that it was just another way of providing concentrated energy to the target in order to knock atoms out of it[2]. For various reasons, laser evaporation was not considered practical and serious attempts of applying it to thin film formation were scarce[3]. Hanabusa et al reported in 1981 that amorphorous silicon could be produced by pulsed Nd:YAG laser reactive evaporation of silicon in a hydrogen atmosphere[4].

One nagging criticism of the pulsed laser deposition technique was that the films formed were always marred with large particulates. This is generally due to large particles expelled from the target during the often violent laser-target interaction[5]. Cheung et al later discovered that large particle ejection could be avoided by keeping the target at a constant high temperature[6]. Smooth and high quality films have been obtained for dielectric as well as semiconducting films[7-9].

With the discovery of the new high temperature superconductors (HTSC), the pulsed laser deposition technique was applied to deposit thin films of these materials. Early results obtained by Venkatesan et al were quite impressive in producing stoichiometric films[10]. However, post- annealing was required to form the proper superconducting phase. More recently, much attention was devoted to the formation of in-situ films without high temperature post annealing. In this paper, we shall discuss the physics of the laser deposition of in-situ superconducting films. It will be demonstrated that in order to obtain as-deposited superconducting films, one has to ensure (1) the proper coverage rate of the various atoms on the substrate, and (2) the introduction of extra oxygen to form the proper superconducting phase.

The condition on coverage rates arises because laser evaporation is a pulsed process, as opposed to sputtering or MBE which are continuous processes. There-fore, the laser produced Ba, Cu, Y, O atoms may travel at different speeds and arrive at the substrate at different times. Obviously this is not inducive to epitaxial film for-mation. The condition on extra oxygen is related to the fact that oxygen does not stick to the substrate with unity efficiency, therefore extra oxygen is needed, besides those from the target, to form the correct superconducting phase of the high T_c material.

Several methods have been introduced to provide additional oxygen to the deposited film. Reactive sputtering in an O_2 atmosphere was used to produce in-situ HTSC films[11,12]. Similar approach was also adopted in pulsed laser deposition to achieve good quality film at a substrate temperature of 600°C.[13] By incorporating a d.c. oxygen plasma discharge, we showed that in-situ superconducting films could be grown at a substrate temperature of 400°C.[14] This plasma-assisted laser deposition (PLD) technique has been successfully applied to a variety of substrates.

In this paper, we report a basic study of the laser target interaction mechanism during pulsed laser deposition. Besides understanding the physics of the supercon-ducting film formation process, the eventual goal is to optimize PLD even further. The question of coverage rates and the role played by the activated oxygen atoms will be examined. It will be shown that as-deposited films can be produced at an optimal target- substrate separation of 7 cm, and that the oxygen atoms are highly activated in the sense that their kinetic energies are in the ~ 8 eV range. Such activation is es-sential for surface mobility and epitaxial film formation.

2. SUMMARY OF PLD RESULTS

The technique of PLD has been successfully applied to a variety of substrates. Generally, it was found that 400°C was the lowest limit for the substrate tempera-ture. The superconducting properties (T_c and J_c) will be better for films deposited at higher temperatures, such as 500°C. The substrate temperature was measured on the substrate surface itself. The temperature on the heater block was generally higher by about 50°C. Silver paste was always used to enhance thermal contact to the heater black.

Table I shows a summary of results of deposition by PLD on various substrates. The substrates can be divided into three catagories for different application pos-sibilities, including (1) dielectrics (SrTiO3, ZrO2, Al2O3 and MgO), (2) semiconduc-tors (Si and GaAs) and (3) metals (Ni, Cu and stainless steel). Negative results have been obtained so far for Cu and GaAs. As-deposited films on the remaining sub-

Table 1. Summary of Results on Low Temperature PLD Films

Substrate	Buffer Layer	T(on set)/Tc (R=0) and Jc	Comments
$SrTiO_3$ (110)	None	$90°K/85°K$ $\sim10^5 A/cm$ (80°K) $\sim5\times10 \text{ Å}/cm$ (4.2°K)	Best J_c, T_c
ZrO_2 (100)	"	$89°K/85°K$ $\sim0.7\times10^5 A/cm^2$ (75°K)	Standard Sample
MgO	"		Potentially Best Diffusion Barrier
Sapphire (1012)	MgO None	$88°K/78°K$; $4\times10^3 A/cm^2$ (40°K) $89°K/76°K$; $9\times10^3 A/cm^2$ (40°K)	Excellent Film
GaAs (100)	MgO None		Not Superconducting yet
Si (100)	None MgO	$89°K/45°K$; $>50 A/cm^2$ (30°K) $88°K/70°K$; $3\times10^3 A/cm^2$ (30°K)	Thermal Matching Surface Preparation
Ni (Poly. Cryst.)	None	$85°K/40°K$	
Cu	Ni		Not Superconducting
Stainless Steel	None	$89°K/79°K$ $\sim10^4 A/cm^2$ (40°K) $\sim2\times10^3 A/cm^2$ (75°K)	Very Good J_c, T_c

strates show superconducting transitions as shown in the table. These results are rather encouraging. Obviously further optimization is possible by adjusting the deposition parameters which include (1) substrate temperature and conditioning (such as ion etch cleaning), (2) Oxygen flow rate and partial pressure, (3) d.c. discharge voltage and current, (4) laser fluence and (5) possibility of ion-assist by a separate ion source. In the following, we shall describe the experiment to study the physics of PLD.

3. ATOMIC BEAMS FORMATION

The deposition and film formation mechanism of PLD is rather complicated. Because of the ArF laser used, direct photoionization and photodissociations of O_2 is possible:

$$O_2 + h\nu \rightarrow O_2^+ + e$$

$$O_2 + h\nu \rightarrow O + O$$

The electrons and ions are affected by the d.c. discharge. In addition, electrons, Cu, Y, Ba, O atoms and ions are produced from the target. Electron impact ionization of various atoms and recombinations can also play an important role:

$$e + A \rightarrow 2e + A^+$$

$$A^+ + e + M \rightarrow A + M$$

where M is any particle that enables the conservation of momentum in these reactions. Regardless of the detailed plasma chemistry occuring within the laser plume and the d.c. discharge, the final result is that a pulse or pulses of the constituent atoms and ions, plus electrons, oxygen atoms and ions is formed, impinging on the substrate. We performed an optical time-of-flight experiment to examine these atomic beams. The laser plume with and without the separate oxygen gas source were examined separately. In this section, we shall discuss the Y,Ba,Cu,O atomic beams formed in a vacuum.

The apparatus for the optical time-of-flight (TOF) measurement is shown in Fig. 1. The optical emission from the laser plume was imaged into the 100 μm slit of a 0.6 m spectrometer using two lenses with F1 = 100 mm and F2 = 200 mm. The spatial resolution was 100 μm. The target was a standard superconducting pellet of Y-Ba-Cu-O. Thus there were many different species of atoms and ions present in the laser generated plasma plume in addition to the electrons. The spectrometer could be tuned to the various emission lines of the Cu, Ba, Y and O atoms and ions, thus allowing examination of each individual species[15,16]. The emission was time-resolved by a fast photomultiphier (10 ns resolution) and digitized for further analysis. The F1 lens/mirror assembly could be translated continuously to examine different parts of the laser plume at different distances from the target.

The laser was an ArF laser emitting at 193 nm, focused to a 0.5 mm x 1 mm spot on the target. The laser fluence could be varied continuously. In this experiment, fluences of 1-7 J/cm^2 were used because this range encompassed the best

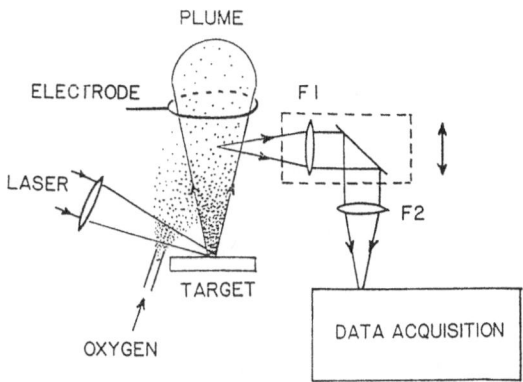

PLUME

ELECTRODE   F I

LASER

                                  F2

TARGET

OXYGEN   DATA ACQUISITION

Fig. 1 Experimental arrangement. The oxygen gas jet intersects the laser beam and the
plasma plume. The ring electrode has no effect on the shape of the plume.
F1 = 100 mm, F2 = 200 mm.

deposition condition of the HTSC films. The vacuum chamber was evacuated to 10^{-5}
torr typically.

Fig. 2 (a),(b),(c) show the time-of-flight spectra of Cu, Y and Ba atoms respectively at a distance of 7.2 cm from the target at a fluence of 5.3 J/cm^2. For each
atomic species, it was ascertained that the TOF spectrum was independent of the
emission line monitored, thus assuring that fluorescence lifetimes do not affect the
results. Typically, 200 laser shots were averaged to obtain a single TOF spectrum.
The solid lines in Fig. 2 are theoretical fits based on the theory of isentropic supersonic expansions. Such expansions lead to the formation of supersonic molecular
beams, which are described by the following velocity distribution function[17]:

$$f(v) = Av^3 \exp[-m(v-v_0)^2/2kT_s] \tag{1}$$

where A is a normalization constant, v_0 is the so-called stream velocity, m is the mass
of the atom or ion under consideration and T_s is a temperature parameter describing
the velocity spread. The expansion Mach number M is a measure of the randomness
of the particle motion and is related to v_0 and T_s by

$$\gamma M^2 kT_s = mv_0^2 \tag{2}$$

where γ is the usual specific heat ratio. A value of 1.3 was used in the calculation to
account for the degrees of freedom associated with ionization and excitation[18]. For
the case of $v_0 = M = 0$, eq. (1) reduces to the ordinary Maxwell-Boltzman distribution
and T_s becomes the usual translational temperature. The most probable speed v_p is
given by

$$v_p = v_0/2 + (v_0^2/4 + 3 kT_s/m)^{1/2} \tag{3}$$

From Fig. 2, it can be seen that the theory of supersonic expansion works very
well for the laser generated atomic beams. Table 2 summarizes the results of fitting
the experimental time-of-flight spectra for all the atoms present in the plume at a
laser fluence of 5.3 J/cm^2. Several interesting observations can be made: (1) In all

15

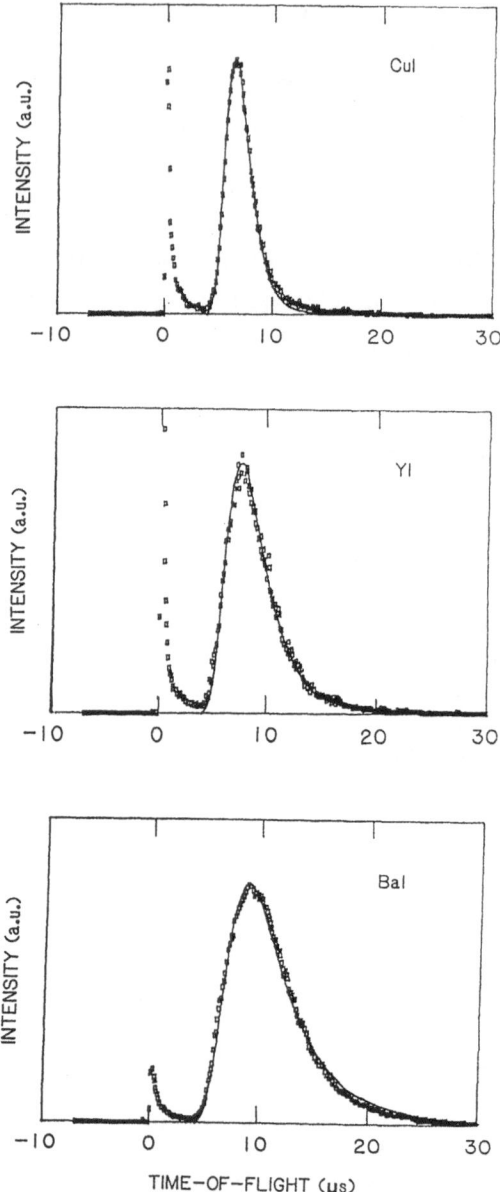

Fig. 2 TOF spectrum of CuI, YI and BaI at 7.2 cm from the target. The solid lines are theoretical fits using eq. (1). The initial spike is due to scattering and fluorescence due to the laser, and can be used as a time marker.

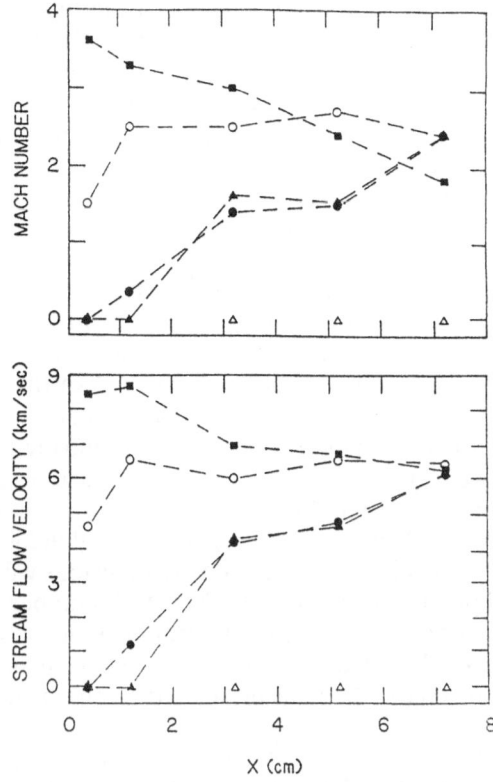

Fig. 3 Dependence of the Mach number and v_0 on distance from the target for various species: ■ CuI; ○ :YI; ▲ :BaI; ● :OI; △ :BaII.

cases, the final velocities of the particles are more or less the same $\sim 10^6$ cm/s, which correspond to kinetic energies of tens of eV. It is believed that the constant velocity of the various constituents is important in achieving good films on the substrate. (2) The Mach number is generally larger for the lighter atoms. This is consistent with the notion that there are many longitudinal elastic collisions among all the species, and the lighter atoms conform to the mean velocity better. (3) The ion velocity distributions are approximately Maxwellian with $M = 0$. This may be due to the long range Coulomb forces between ions and electrons present in the plasma, causing the collisions to be not so efficient in equalizing the velocities. (4) Finally, it was found that the atomic oxygen emission was very faint and could not be time-resolved.

4. THE OXYGEN JET AND OXYGEN ATOMIC BEAM

The oxygen gas jet as shown in Fig. 1 was very critical for the formation of as-deposited superconducting films. It was found that if the tip of the oxygen jet was close to the target with the excimer laser beam passing through the gas stream, a red glow could be observed emitting from the jet. It was also found that the laser triggered red glow could be enhanced and actually changed to a yellowish color upon the application of the d.c. voltage. It is the behavior of the oxygen jet and its effect on the various atomic beams that we wish to characterize in this paper.

The laser generated atomic beams of Y,Ba,Cu and O travel towards the substrate and form the as-deposited Y-Ba-Cu-O film. Experimentally, it was found that the O-atom emission was weak and could not be time- resolved if the chamber was at 10^{-5} torr. When the gas jet was turned on, with a steady state pressure of 5 mtorr as measured near the base plate, the O-atom emission became very strong.

Under this real deposition condition, the spatial development of the various atomic beams was measured as a function of distance from the target. Fig. 3 shows the dependence of the Mach number and v_0 on the distance. It reveals several interesting features: (1) The Cu atoms travel much faster than the other atoms initially. However, as the various beams develop, their velocities equilibrate to $\sim 0.6 \times 10^6$ cm/s at a distance of 7 cm. This exchange of speeds is due mostly to the many collisions between the fast and slow atoms. (2) The oxygen and barium beams start out at very slow speed, and increases to $\sim 0.6 \times 10^6$ cm/s. The O atoms come from the dissociation of oxygen molecules which have an average thermal velocity of merely 0.5×10^5 cm/s. In both cases the increase in speed is due to collisions with the more energetic Y and Cu atoms, and possibly fast ions. (3) The Mach number of the various beams also converge as they propagates. The equilibrium is reached at 7 cm. (4) The ion beam is always Boltzman-like with $v_0 = M = 0$. However, its peak velocity v_p is also $\sim 0.5 \times 10^6$ cm/s. From eq.(3), it indicates that the ions have higher translational temperatures.

The above observations collaborate an experimental fact that we discovered in laser deposition of superconducting films. It was found that in order to obtain good quality as-deposited films, the target- substrate distance has to be at least 7 cm at 5 mtorr O_2 background pressure. For closer separation, the deposition rate may be higher, but the film quality is not as good. We believe that this is related to the velocity distribution of the various species. It can be seen from Fig. 3 that at distances longer than 7 cm, the velocity distributions of Y,Ba,Cu,O are almost the same. This implies that the densities of the constituents on the surface will be uniform and the coverage rates are the same. Therefore there is a better chance the atoms to rearrange among themselves to form a stoichiometric epitaxial film.

At shorter distances, the density of the various constituents will not be uniform. Specifically, following each laser pulse, the Cu atoms will arrive first, followed by the Y, Ba and lastly the O atoms. This situation is not inducive to epitaxial film formation and is actually similar to sequential evaporation of various constituent metals. Therefore, for as- deposited films, it is important that the various atoms travel at the same speed and cover the substrate at the same rate.

The results in Fig. 3 also indicates that the various atomic beams exchanges energy as they travel towards the substrate. This implies that the kinetic energy is not constant, which is quite different from a simple supersonic expansion for the formation of molecular beams. This is due to the higher background pressure used in the present study. Because of the dependence of velocity on distance, it should be pointed out that the velocities quoted in the discussions above are all averaged over the distance traveled. The decovolution to obtain the true instantaneous velocities is quite complicated and will not be considered here.

The emission intensity of the O atom beam became stronger as the middle electrode was turned on. However the velocity distribution was not affected significantly by the presence of a voltage. Currently, we are trying to vary the d.c. discharge conditions and correlate the superconducting film properties to the atomic beam parameters (M and v_0). Such a correlation is important for the optimization of the laser deposition technique.

At present, the role played by diatomics and larger radicals in the film formation is not known. The TOF of such species are harder to measure because of their much broader emission spectra and inadvertent overlap with strong atomic lines[15]. Techniques such as laser-induced- fluorescence may be necessary. However, from the relative emission intensity, we believe that only a small fraction of the laser plume is in the form of diatomics such as YO, BaO, CuO and O_2. Because of the large kinetic energies of the atoms and ions, it is doubtful that larger molecules will be formed in significant quantities.

5. CONCLUSIONS

In this paper, we have described the present experimental results on the as-deposited superconducting films using the PLD technique. This novel deposition method is capable of depositing in-situ films at a low substrate temperature of 400°C, and presents a significant improvement over normal reactive laser deposition without a d.c. plasma discharge. In that method, the substrate is typically at a temperature of 600°C[13].

The physics of the deposition process was studied by examining the atomic beams of various species liberated from the target. It was found that the atomic beams could be described very well by a supersonic expansion mechanism. An interesting change in the velocity distribution of the various atomic beams as a function of distance from the target was also observed. This change could be correlated to the deposited film property and the optical target/substrate separation. The oxygen gas jet produces a strong oxygen atomic beam with a velocity of $\sim 0.6 \times 10^6$ cm/s. It is believed that this energetic oxygen atomic beam is responsible for the low temperature as- deposited superconducting film formation.

ACKNOWLEDGEMENT

This research was supported by the New York State Institute on Superconductivity. H.S.K. is a National Science Foundation Presidential Young Investigator.

REFERENCES

1. L.P. Levine, J.F. Ready and E. Bernal, IEEE J. Quant. Elect. QE-4, 18 (1968).
2. C. Cali, V. Daneu, A. Orioli and S. Riva-Sanseverino, Appl. Opt. 15, 1327 (1976).
3. R.F. Bunshah et al, Deposition Technologies for Films and Coatings, Noyes Publ., New Jersey 1982.
4. M. Hanabusa and M. Suzuki, Appl. Phys. Lett. 39, 431 (1981).

5. D. Lubben, S.A. Barnett, K. Suzuki and J.E. Greene in <u>Laser Controlled Chemical Processing of Surfaces</u>, p.359, ed. by A.W. Johnson, D.J. Ehrlich and H.R. Schlossberg, North-Holland, 1984.

6. J.T. Cheung and D.T. Cheung, J. Vac. Sci. Tech. <u>21</u>, 182 (1982).

7. J.T. Cheung, G. Niizawa, J. Moyle, N.P. Ong, B.M. Paine and T. Vreeland, J. Vac. Sci. Tech. <u>A4</u>, 2086 (1986).

8. H. Sankur in <u>Laser Controlled Chemical Processing of Surfaces</u>, p.373, ed. by A.W. Johnson, D.J. Ehrlich and H.R. Schlossberg, North- Holland, 1984.

9. J.I. Dubowski, P. Norman, P.B. Sewell, D.F. Williams, F. Krolicki and M. Lewicki, Thin Solid Films, <u>147</u>, L51 (1987).

10. D. Dijkkamp, T. Venkatesan, X.D. Wu, S.A. Schaheen, N. Tisrawi, Y.H. Ming-Lee, W.L. McLean and M. Croft, Appl. Phys. Lett. <u>51</u>, 619 (1987).

11. D.K. Lanthrop, S.E. Russek and R.A. Buhrman, Appl. Phys. Lett. <u>51</u>, 1554 (1987).

12. K. Kar, A.D. Kent, A. Kapitulnik, M.R. Beasley and T.H. Geballe, Appl. Phys. Lett. <u>51</u>, 1370 (1987).

13. A. Inam, M.S. Hedge, X.D. Wu, T. Venkatesan, P. England, P.F. Miceli, E.W. Chase, C.C. Chang, J.M. Tarascon and J.B. Wachtman, Appl. Phys. Lett. <u>53</u>, 897 (1988).

14. S. Witanachchi, H.S. Kwok, X.W. Wang and D.T. Shaw, Appl. Phys. Lett. <u>53</u>, 234 (1988).

15. Q.Y. Ying, D.T. Shaw and H.S. Kwok, Appl. Phys. Lett. <u>18</u>, 1762 (1988).

16. P.E. Dyer, R.D. Greenough, A. Issa and P.H. Key, Appl. Phys. Lett. <u>53</u>, 534 (1988).

17. J.B. Anderson, R.P. Andres and J.B. Fenn, Adv. in Chem. Phys. <u>10</u>, 275 (1966).

18. Ya B. Zeldovich and Yu P. Raiser, <u>Physics of Shock Wave and High Temperature Hydrodynamic Phenomena</u>, vol. 2, p.573, Academic Press, London 1968.

FORMATION OF Bi(Pb)-Sr-Ca-Cu-O THIN FILMS AT 500°C BY A SUCCESSIVE DEPOSITION METHOD USING EXCIMER LASER

Tomoji Kawai, Masaki Kanai, Hitoshi Tabata[1]
and Shichio Kawai

The Institute of Scientific and Industrial Research, Osaka
University, Mihogaoka, Ibaraki, Osaka 567 Japan
1)Technical Institute, Kawasaki Heavy Industries, Ltd
Kawasakicho, Akashi, Hyogo 673 Japan

1. Introduction

Thin films of high Tc superconductors including La-Sr-Cu-O, Y-Ba-Cu-O, Bi-Sr-Ca-Cu-O and Tl-Ba-Ca-Cu-O have been prepared by various methods. For a device application, as-grown superconducting films prepared with low temperature process are needed. In the present study we have prepared thin films of Bi-Sr-Ca-Cu-O (BSCCO) and Bi(Pb)-Sr-Ca-Cu-O(BPSCCO) by a laser ablation method, and have succeeded in the formation of a crystallized as-grown thin film at the substrate temperature as low as 500°C.

The films were prepared by "successive deposition method"(Fig.1) using multi-targets. This method may be suitable for the film formation of the compounds having two dimensional layer structures, and has advantages to control the film structure artificially. The BSCCO[1] or BPSCCO[2] have two dimensional layer structure, in which the 110K phase contains three $Cu-O_2$ layers[3], 80K phase two $Cu-O_2$ layers[4][5] and semiconducting phase single $Cu-O_2$ layer.[4][5] The application of the successive deposition method to control the number of layers in BSCCO film was first tried with a multi-target magnetron sputtering method[2]. We have tried to control the structure of these Bi based superconducting films by the successive deposition method at low substrate temperatures with laser ablation in combination with activated oxygen. Our results showed, with regard to low temperature

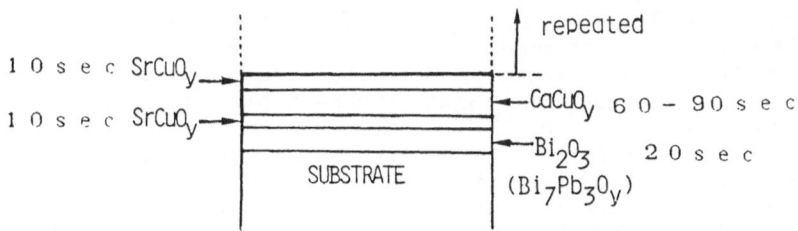

Fig.1 A schematic representation of multi-target successive deposition for Bi(Pb)-Sr-Ca-Cu-O film.

processing, a sufficient crystallization is not enough under oxygen gas flow, but a treatment with stronger oxidizing gas is necessary for the sufficient crystallization of the oxide to form as-grown films at low temperature. A N_2O gas is easily dissociated and release oxygen atoms above 300°C under atmospheric pressure, so that N_2O becomes strong oxidizing agent. Thus as-grown films were prepared under N_2O gas flow and its effect was examined. Under this condition, that is, laser ablation and N_2O gas flow in combination with successive deposition method, we have successfully formed as-grown Bi(Pb)-Sr-Ca-Cu-O film around 500°C, controlling the numbers of $Cu-O_2$ layers artificially.

2. Experimental

The films were prepared by pulses of ArF excimer laser(193nm) focused on the target placed in a vacuum chamber.[1,2] Emitted atoms and molecules from the target were accumulated on a substrate placed at the opposite side to form a film. Laser intensity was 1 - 10J/cm² pulse on the target after focusing. The substrate used was MgO(100) single crystal. Targets were synthesized from Bi_2O_3, PbO, $SrCO_3$, $CaCO_3$ and CuO. The Bi_2O_3, $SrCuO_y$ and $CaCuO_y$ pellets sintered in air were used as targets for BSCCO film, and $Bi_7Pb_3O_y$, $SrCuO_y$ and $CaCuO_y$ for BPSCCO films. The Bi_2O_3 and $Bi_7Pb_3O_y$ targets were calcined at 650°C for 3 hours, and $SrCO_3$ and $CaCuO_y$ targets were calcined at 950°C for 15 hours in air. The films were deposited successively from Bi_2O_3 or $Bi_7Pb_3O_y$ for 10 sec, $SrCuO_y$ for 10sec, $CaCuO_y$ for appropriate second to be changed and $SrCuO_y$ for 10 sec in one cycle. The deposition time for $CaCuO_y$, was changed, between 60 sec and 90 sec.(Fig.1) for different films to control the number of $Cu-O_2$ layers. The samples were prepared with 20 - 30 cycles under oxygen or N_2O gas flow (the order of 10^{-1} to 1 torr). The substrate temperature was from 400°C to 600°C as measured by chromel-alumel thermocouple attached to the substrate surface. The repetition rate of laser pulses was changed from 7 to 20Hz and the film thickness was mainly controlled by this procedure to form 600Å - 2000Å thickness.

3. Results and Discussion

3.1 Successive deposition under O_2 atmosphere

The results on the successively deposited films under O_2 gas flow were summarized as follows. As-grown films of BSCCO with 2000Å thickness made by 20 cycle deposition, had very weak and broad X-ray diffraction peaks when the substrate temperature was 600°C. At lower substrate temperature, the diffraction patterns of the films were amorphous like one. These results showed that the crystallization of the as-grown film did not occur or hardly proceeded in that temperature region under oxygen atmosphere. At higher substrate temperature, diffraction peaks of impurity phase, which could be assigned as $Bi(Sr,Ca)O_y$, grew and peaks of BSCCO phase did not become higher. Therefore, it is not easy to get a crystallized as-grown film in that condition. The weak and wide diffraction peaks at 2θ= 4 - 6 corresponding to (002) of BSCCO system, however, were shifted to lower angle with the increase of the deposition time of $CaCuO_y$ in the cycle. This shift of the peak position indicates that the precursor of BSCCO phase was formed in the as-grown film and the lattice constant of the c-axis of that precursor became longer with increase of the deposition time of $CaCuO_y$. Then, the films were sintered to obtain more prominent diffraction peaks. Fig.2 shows the X-ray diffraction patterns of the films sintered at 790°C for 3 hours in air for which the deposition time of CaCuO$_y$ was 60 sec and 80 sec. The film with shorter deposition time for CaCuO$_y$ had both semiconducting phase ($2\theta = 7.2°$) with single Cu-O$_2$ layer and 80K phase ($2\theta = 5.8°$) with double Cu-O$_2$ layers.(Fig.2 a) On the other hand, the peaks of semiconducting phase disappeared in the film with longer deposition time, and only 80K phase remained.(Fig.2 b) This change of the diffraction pattern indicates that it is possible to control the number of Cu-O$_2$ planes by this laser ablation method, but O$_2$ atmosphere during deposition is not enough to form a crystallized as-grown film below 600°C. The sintering process around 800°C is needed to control the numbers of Cu-O$_2$ layers. The same results were obtained in the BSCCO film for the thickness of 600Å – 800Å with 25 – 30 cycle deposition, and also in these as-grown films, sufficient crystallization was not obtained at 600°C.

3.2 Successive deposition under N$_2$O atmosphere

We desire to obtain crystallized as-grown films at low temperature with controlling the number of the films. The N$_2$O gas, which is much stronger oxidizing agent, was used instead of O$_2$ gas in order to form a crystallized as-grown film around 500°C. Under the N$_2$O gas flow, the BSCCO film (700 Å) having diffraction peaks of 80K phase was actually obtained at substrate temperature of 520°C, though it contained Bi(Sr,Ca)O$_y$.

As Pb doping for this system is known to make the formation of high Tc phase easier[7], BPSCCO films were prepared by the same successive deposition method using Bi$_7$Pb$_3$O$_y$, SrCuO$_y$, CaCuO$_y$ targets at 520°C. Fig.3 shows the change of the X-ray diffraction patterns of these as-grown BPSCCO films against the changes of the deposition time of CaCuO$_y$. These films had thickness of 500Å – 800Å after 20 – 30 cycle deposition. The diffraction patterns in Fig.3 indicate following results; When the deposition time of CaCuO$_y$ was short, 60 sec, the film had diffraction peaks due to 80K phase, whose lattice constant c is 30.6Å.(Fig.3 a) As the deposition time of CaCuO$_y$ was increased, the lattice constant c became longer and 110K phase appeared whose c-axis length is 37.3Å.(Fig.3 b) The phase having larger lattice constant than 37Å was obtained in the region of longer deposition time of CaCuO$_y$.(Fig.5 c,d) The lattice constants c of these phases calculated based on the index shown in Fig.3 (c,d) were 43.5Å and 50.0Å. These

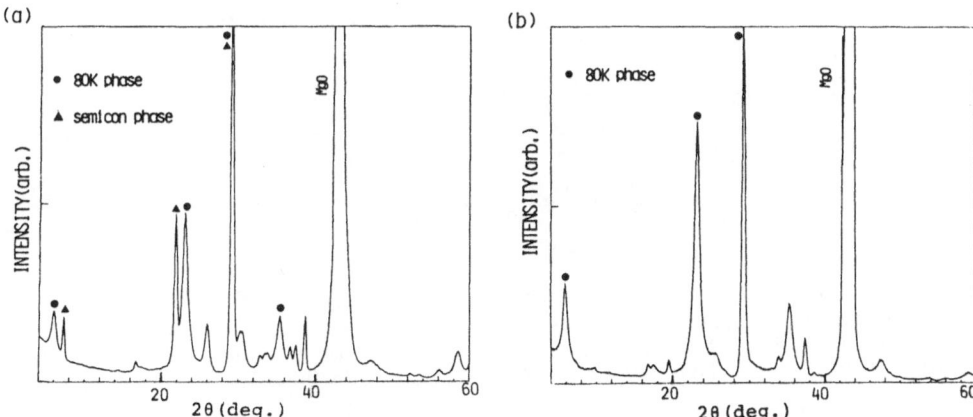

Fig.2 The X-ray diffraction patterns of sintered Bi-Sr-Ca-Cu-O thin films, which were prepared by multi-target successive deposition. The deposition time of CaCuO$_y$ in 1 cycle was (a) 60 sec. (b) 30 sec.

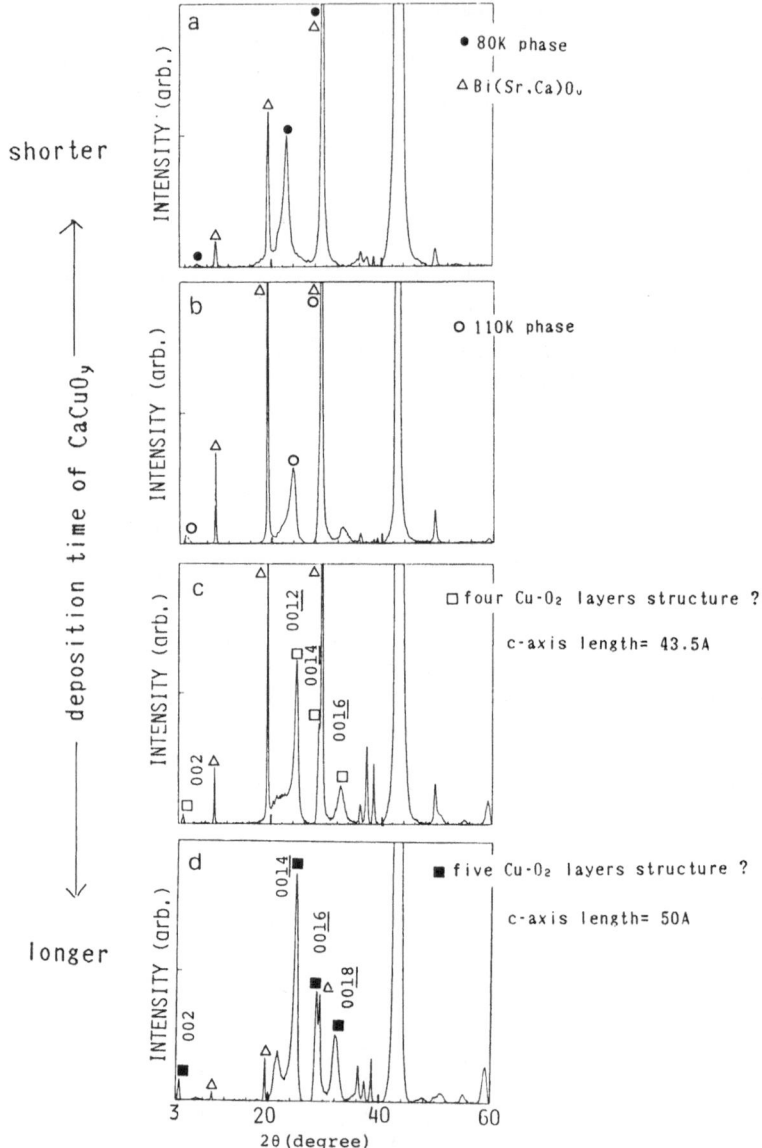

Fig.3 The change of the X-ray diffraction patterns of Bi-Pb-Sr-Ca-Cu-O
as-grown films at 520°C under N₂O gas flow with the change of the
deposition time of CaCuO$_y$ in a cycle. The deposition time for CaCuO$_y$
was (a) 60sec (b) 70 sec (c) 75 sec (d) 90 sec

values correspond to the lattice constant of the structures having four $Cu-O_2$ layers and five $Cu-O_2$ layers between adjacent Bi_2O_2 layers. Thus, we believe that the four and five $Cu-O_2$ layer structures with the c-axis oriented perpendicular to the substrate surface was achieved in these films. These results show that it is possible to control the number of $Cu-O_2$ layers even in the as-grown films made at 520°C. The diffraction peaks of these phases, however, were not very strong, that is, the crystallization was not sufficient, and the films contained impurity phase, $Bi(Sr,Ca)O_y$. The film formation at lower temperature was effective to reduce the impurity phase as shown in the X-ray diffraction pattern of the film prepared at substrate temperature of 480°C (Fig.4) Though the peaks of BPSCCO became a little bit weaker, the $Bi(Sr,Ca)O_y$ was not seen in this diffraction pattern.

The films having four or five $Cu-O_2$ layer shown in Fig.3 were sintered at 790°C for 3 hours in air. The X-ray diffraction patterns were shown in Fig.5. The sharp peaks due to long lattice constant, 43 or 50Å, disappeared and, instead, peaks of 80K phase dominantly appeared. At the same time, broad peaks due to large lattice constant appeared, and the lower angle contribution of the peaks became larger with the longer deposition of $CaCuO_y$.(Fig.5) This shows that the structure in the as-grown state affected the structures even after sintering, similar to the results of sintered BSCCO films shown in Fig.2. The changes of the diffraction patterns after sintering indicate that four or five $Cu-O_2$ layer structure is not stable in such a high temperature, which induces transformation to 80K and 110K phase being stable thermodynamically. The reason for the growth of relatively unstable phases, i.e. four and fine layer structure, in the as-grown films would be that these films are prepared at low substrate temperature. The successive deposition force the film to make four or five $Cu-O_2$ layer structure in these circumstances.

4. Conclusion

1) The numbers of $Cu-O_2$ layers in Bi-(Pb)-Sr-Ca-Cu-O compound can be controlled by a laser ablation combined with successive deposition method using multi targets. In BPSCCO as-grown films, the

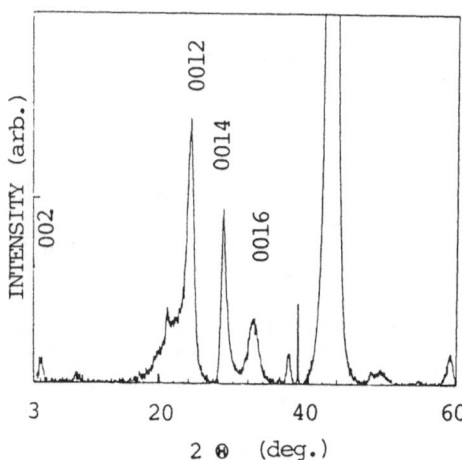

Fig.4 The X-ray diffraction pattern of Bi(Pb)-Sr-Ca-Cu-O as-grown film by successive deposition at substrate temperature as low as 480°C. The lattice constant c is 43.5Å corresponding to four $Cu-O_2$ layer structure.

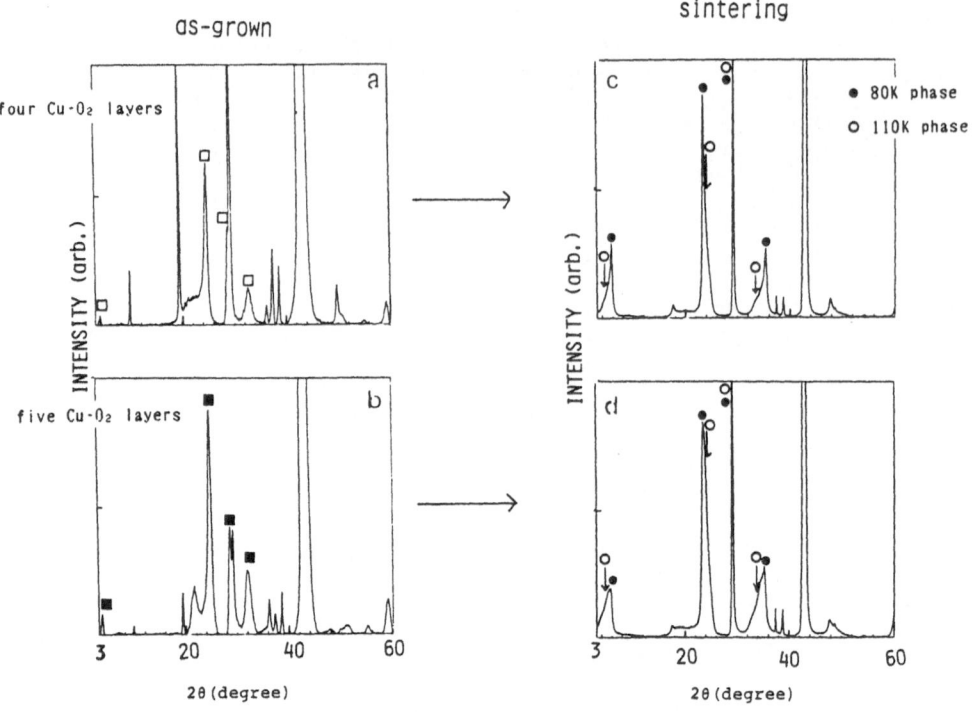

Fig.5 The X-ray diffraction patterns of sintered Bi-Pb-Sr-Ca-Cu-O films which had contained four and five Cu-O$_2$ layers in the as-grown state. The as-grown film containing four Cu-O$_2$ layers(a) and the sintered one at 790°C. The as-grown film containing five Cu-O$_2$ layers(c) and the sintered one at 790°C.

structure having large lattice constant of c-axis corresponding to four and five Cu-O$_2$ layer structure(c= 43A, 50A) can be formed even at 500 C by this method.

2) The introduction of dinitrogen monoxide(N$_2$O) gas during the laser ablation definitely enhances the crystallization of the as-grown films to let X-ray diffraction peaks of BPSCCO phase seen in as-grown films even at the substrate temperature of 480 C.

References

1) M.Kanai, T.Kawai, M.Kawai and S.Kawai: Jpn.J.Appl.Phys. 27 (1988) L1293
2) H.Tabata, T.Kawai, M.Kanai, O.Murata and S.Kawai: to be published
3) H.Nobumasa, K.Shimizu, Y.Kitano and T.Kawai: Jpn.J.Appl.Phys. 27 (1988) L846
4) E.Takayama-Muromachi, Y.Uchida, A.Ono, F.Izumi, M.Onoda, Y.Matsui, K.Kosuda, S.Takekawa and K.Kato: Jpn.J.Appl.Phys. 27 (1988) L365
5) M.A.Subramanian, C.C.Torardi, J.C.Calabrese, J.Gopalakrishnan, K.J.Morrssey, T.R.Askew, R.B.Flippen, V.Chowdahri and A.W.Sleight: Science 239 (1988) 1015
6) H.Adachi, S.Kohiki, K.Setsune, T.Mithuyu and K.Wasa: Jpn.J.Appl.Phys, in press (1988)
7) S.A.Sunshine, T.Siegrist, L.F.Schneemeyer, D.W.Murphy, R.J.Cava, B.Batlogg, R.B.van Dover, R.F.Fleming, S.H.Glarum, S.Nakahara, P.Farrow, J.J.Krajewski, S.M.Zahurak, J.V.Waszczak, J.H.Marshall, P.Marsh, L.W.Rupp,Jr. and W.F.Peck: Phys.Rev.B 38 (1988) 893

THE EFFECT OF ANNEALING CONDITIONS ON LASER DEPOSITED

SUPERCONDUCTING Bi-Sr-Ca-Cu-O THIN FILMS

B. F. Kim, J. Bohandy, E. Agostinelli[*], T. E. Phillips,
W. J. Green, F. J. Adrian and K. Moorjani

The Milton S. Eisenhower Research Center
The Johns Hopkins University Applied Physics Laboratory
Laurel, Maryland 20707

INTRODUCTION

The method of laser ablation processing (LAP) for deposition of high T_c superconducting thin films has been demonstrated to be relatively simple in practice when compared to some other techniques [1,2]. Thus far, however, additional processing is necessary in order to produce electrically continuous, superconducting films. Generally, heated substrates and/or post deposition annealing at elevated temperatures in the presence of oxygen are required. A wide range of substrate temperatures, post deposition annealing temperatures, and annealing times are possible, and a particular set of processing parameters produces films of optimal quality. The optimum processing parameters for a given film/substrate structure is of obvious interest. The range of these parameters which can produce films of near optimum quality is also important since this indicates how critical the processing is.

Films of BSCCO have been deposited by the LAP method on a number of substrates including crystalline quartz[3], $SrTiO_3$ [4], MgO [5,6], and cubic zirconia.[6,7] Although the bulk material has two prominent phases with $T_c \sim 110$ K and ~ 80 K,[8] films prepared by this method thus far exhibit predominantly the low temperature phase. It has been noted in previous reports [4-6] that BSCCO films require annealing treatment which must be optimized. There has been no report, however, on how critical the annealing process is for production of good quality films. In this report, we present preliminary results on films of Bi-Sr-Ca-Cu-O (BSCCO) on substrates of cubic zirconia and crystalline quartz with thin cubic zirconia buffer layers and find that short post deposition annealing times, of the order of a few minutes, are required to produce optimum films. This could be of practical importance because short annealing times will reduce damage owing to thermal diffusion and may produce less thermal distress on the film/substrate structure. On the other hand, the processing is very critical in that small deviations from the optimum processing parameters markedly reduce film quality.

---

* Permanent affiliation: Istituto di Teoria e Struttura Elettronica
  e Comportamento Elettrochimico dei Composti di Coordinazione-Consiglio
  Nazionale delle Ricerche (ITSE-CNR), Area Della Ricerca di Roma, 00016
  Monterotondo Scalo, Italy.

The results of a study of the relation between film quality and processing parameter depends upon the criteria for quality and the methods used to characterize the films. For the purpose of this study, we have taken the transition temperature, T_c, and the phase purity of the film as indicators of film quality. The methods used to characterize the films are magnetically modulated microwave absorption (MAMMA)[1,9] and dc resistance. The MAMMA technique provides both phase purity and T_c, which with this method is an average T_c for the entire sample. Resistance measurements, on the other hand, provide the T_c for the first superconducting path in the sample, and does not, in general, provide information on the phase purity for the entire sample. In this regard, it has been shown that films prepared under different annealing conditions with correspondingly different degrees of phase purity and average T_c as measured by the MAMMA method can exhibit virtually identical resistance vs temperature curves.[7] The resistance measurements are, nevertheless, of interest in this study because they show the behavior of resistance above T_c.

EXPERIMENT

Thin films of BSCCO were deposited on substrates of ZrO_2 stabilized with 9.5 Mol % Y_2O_3, and on crystalline quartz with a thin buffer layer of ZrO_2 interposed between the substrate and the film. The films and ZrO_2 buffer layers were both deposited by the LAP method, an important step toward formation of multilayer structures by a single processing method. An ArF excimer laser operating at 150 mJoules per pulse and pulse repetition rate of 10 pps was used for this purpose. The laser beam was focused onto the target pellet with a focal size of 0.5 mm². Depositions were done in a vacuum cell at 10^{-3} mm Hg pressure. Details of the preparation of the target material with molar ratios of Bi:Sr:Ca:Cu: = 1:1:1:2 was described in a separate report.[7]

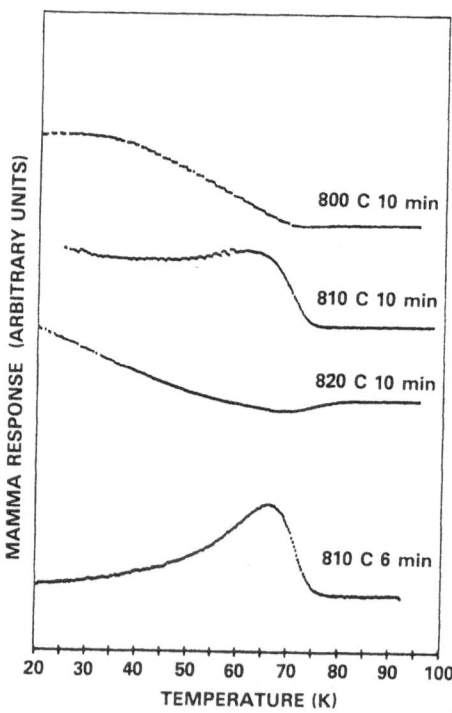

Fig. 1. MAMMA response vs. temperature for films deposited on unheated ZrO_2 substrates for different annealing temperatures and annealing times

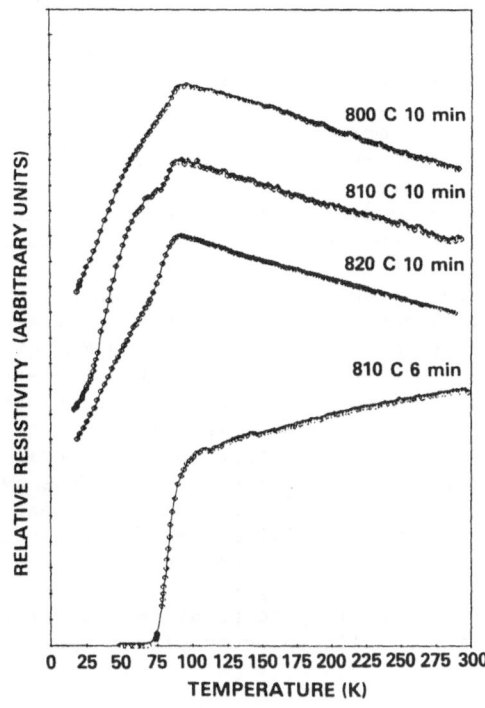

Fig. 2. dc resistance vs. temperature
for films in fig. 1

Post deposition annealing was done in air in a tubular furnace. The
films were inserted into a furnace with temperature maintained at the
desired annealing temperature, and withdrawn after a specified time. The
measured annealing times were taken to be the time during which the films
were at the annealing temperature. The time required to raise the temperature of the film to the annealing temperature was typically 30 seconds and 10-20 seconds was the approximate cool down time. Resistance
measurements were performed using the four point probe technique. The
MAMMA method, recently developed in this laboratory,[9] was used with a
30 G bias magnetic field.

RESULTS

We consider first the effect of annealing temperature on two sets of
BSCCO films on ZrO_2 substrates. One set was deposited with the substrates
at room temperature, while in the second set the substrates were heated
to 300°C. Figure 1 shows the MAMMA response vs temperature for the first
set of films (deposited with substrates at room temperature), and figure
2 shows the corresponding resistance measurements. The first three
curves in each figure show the results of annealing for 10 minutes at
800°C, 810°C, and 820°C. The MAMMA results in figure 1 show that the
film deposition at 810°C has the highest average T_c (determined by the
position of the peak) and the best phase purity (indicated by the width
of the peak). It is clear from this figure that the production of an
optimum film in this set occurs for a temperature range of only a few
degrees. While a complete study of the effect of annealing time is still
in progress, the fourth curve in figures 1 and 2 which illustrates the
result of annealing at the optimum temperature 810°C for 6 minutes, shows
that the resulting film is superior to the film annealed at 810°C for 10
minutes and suggests that optimum films can be produced for only a narrow
range of annealing times.

The corresponding resistance vs temperature curves shown in figure 2 are particularly interesting in the temperature region above T_C. The films annealed for 10 minutes exhibit a temperature dependence characteristic of semiconductors, the resistance increasing with decreasing temperature. The film annealed at 810°C for 6 minutes, on the other hand, exhibits a response more characteristic of metals. The semiconductor type response is interpreted to be due to granularity in the films which results in localization of carriers at trapping sites. Thus, the annealing process not only affects the superconducting phases in the films, but also seems to reduce the degree of granularity in the films.

The results for the second set of films, deposited on heated ZrO_2 substrates (300°C) and annealed for 10 minutes are shown in figures 3 and 4. The optimum film, according to the MAMMA data in figure 3, is the film annealed at 790°C, which is lower than the annealing temperature required to produce an optimum film deposited on an unheated substrate. In addition, the range of annealing temperatures which can produce films of near optimum quality is greater than for the previous set of films. Finally, the resistance curves in figure 4 show less evidence of granularity than for the previous set of films. (Note that there is little evidence of differences in these films from their resistance measurements.) Thus, one of the effects of heating the substrate during deposition is to deposit films which are less granular before annealing which allows a slightly lower annealing temperature and perhaps more importantly, a less critical annealing process with respect to the required temperature.

A third set of films deposited on crystalline quartz with a thin ZrO_2 buffer layer illustrates the effect of annealing time on film quality. These films were deposited on films heated to 300°C and annealed at 810 °C for 4, 6, 8 and 10 minutes. The MAMMA results, shown

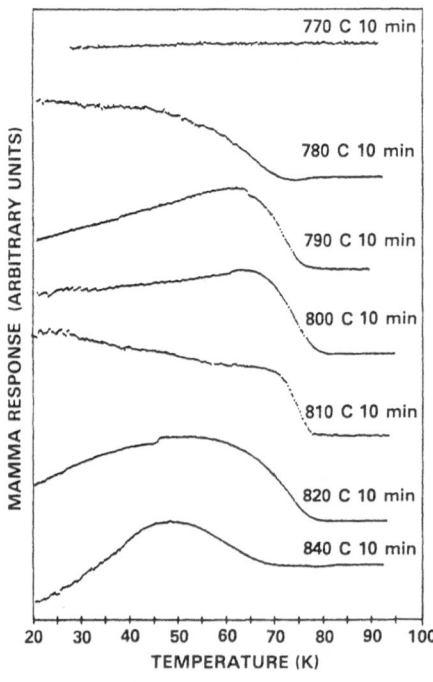

Fig. 3. MAMMA response vs. temperature for films deposited on ZrO_2 substrates at 300°C and annealed for 10 minutes at various temperatures

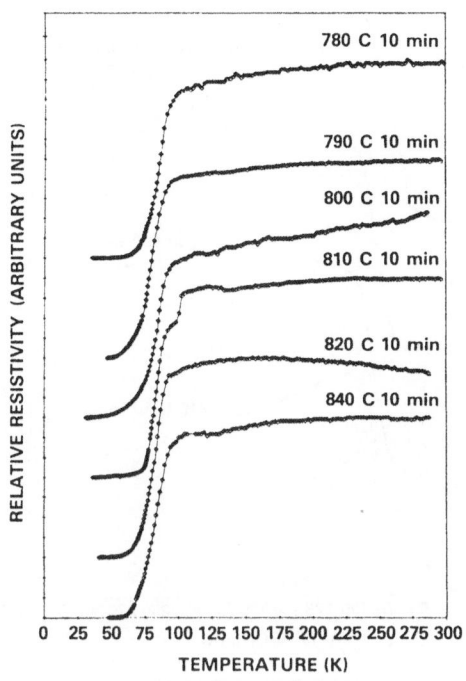

Fig. 4.   dc resistance vs. temperature
for films in fig. 3

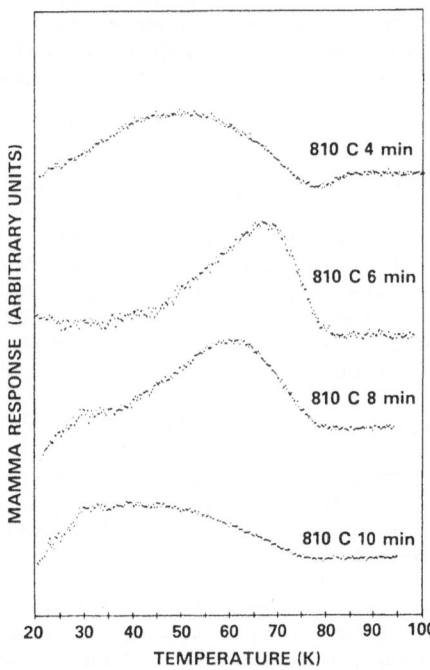

Fig. 5. MAMMA response vs. temperature for films
deposited on crystalline quartz substrates
with ZrO_2 buffer layers at 300 °C and annealed
for various times at 810°C

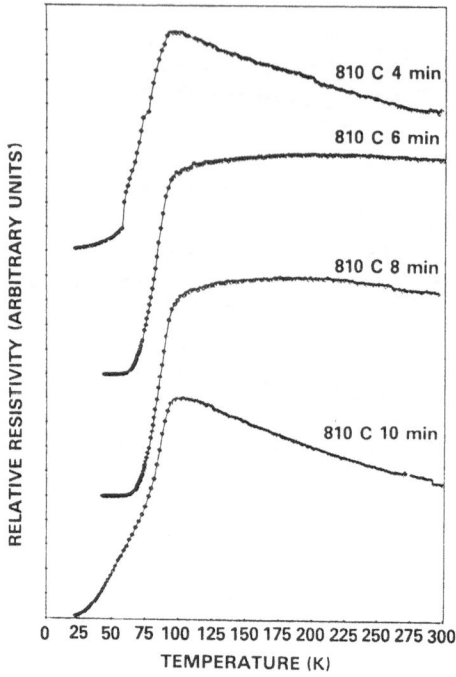

Fig. 6. dc resistance vs. temperature
for films in fig. 5

in figure 5, indicate that the film annealed for 6 minutes is the optimum
film, having the highest average T_c and sharpest peak. They also show
that optimum films are produced by a surprisingly small range of annealing times. The corresponding resistance measurements in figure 6 show
that the optimum film, determined by the MAMMA data, exhibits the least
granularity, which is consistent with the previous results.

CONCLUSION

The results of this study show that post deposition annealing conditions can be critical for producing good films by the LAP method. This
study also illustrates the importance of using a global measurement of
superconductivity, such as the MAMMA method, for characterizing thin
films. Resistance measurements, which were also used in this study, do
not resolve superconducting film quality to the extent of the MAMMA
method, but provide interesting information on the granularity of films
by their resistance characteristics above T_c.

The results in this report show that optimum quality films of BSCCO
on cubic zirconia are produced by a narrow range of post deposition
annealing temperature and time. The effect of heating the substrate
during deposition is to allow a slightly lower annealing temperature for
producing an optimum film, and a larger range of temperatures for producing near optimum films. We also reported here, for the first time, the
fabrication of thin films of BSCCO on crystalline quartz with a thin
buffer layer of cubic zirconia. The properties of these films, which
will be described in detail in a separate report,[3] are generally similar
to that of films deposited on cubic zirconia. These films were used to
observe the effect of annealing time on the quality of the films. The
required annealing time for producing an optimum film was only 6 minutes,
and deviations of only 2 minutes from this optimum time produced films
which were measurably inferior to the optimum film.

ACKNOWLEDGMENT

This work was supported by the Space and Naval Warfare Systems Command under Contract No. N00039-87-C-5301. One of us (E. A. is indebted to NATO-CNR organization for providing the fellowship (Grant No. 106701/00/8700746).

REFERENCES

1. K. Moorjani, J. Bohandy, F. J. Adrian, B. F. Kim, R. D. Shull, C. K. Chiang, L. J. Swartzendruber and L. H. Bennett, Phys. Rev. B 36:4036 (1987).
2. D. Dijkkamp, T. Venkatesan, X. D. Wu, S. A. Shaheen, N. Jisrawi, Y. H. Min-Lee, W. L. McLean and M. Croft, Appl. Phys. Lett. 51:619 (1987).
3. J. Bohandy, E. Agostinelli, T. E. Philips, B. F. Kim, W. J. Green, F. J. Adrian, and K. Moorjani (to be published).
4. D. K. Fork, J. B. Boyce, F. A. Ponce, R. I. Johnson, G. B. Anderson, G. A. N. Connell, C. B. Eom, and T. H. Geballe, Appl. Phys. Lett. 53:337 (1988).
5. C. R. Guornieri, R. A. Roy, K. L. Saenger, S. A. Shivashankar, D. S. Yee, and J. J. Cuomo, Appl. Phys. Lett. 53:532 (1988).
6. J. Perriere, E. Fogarassy, G. Hauchecorne, X. Z. Wang, C. Fuchs, F. Rochet, I. Rosenman, C. Simon, R. M. Defourneau, F. Kerherve, J. P. Enard, and A. Laurent, Solid State Commun. 67:345 (1988).
7. B. F. Kim, J. Bohandy, T. E. Phillips, W. J. Green, E. Agostinelli, F. J. Adrian, K. Moorjani, L. J. Swartzendruber, R. D. Shull, L. H. Bennett, and J. S. Wallace, Appl. Phys. Lett. 53:321 (1988).
8. H. H. Maeda, Y. Tanaka, M. Fukutomi, and T. Asano, Jpn. J. Appl. Phys. 27:L209 (1988).
9. B. F. Kim, J. Bohandy, K. Moorjani, F. J. Adrian, J. Appl. Phys. 63:2029 (1988).

PULSED LASER EVAPORATION OF Tl-Ba-Ca-Cu-O FILMS

S.H. Liou
Department of Physics and Astronomy
University of Nebraska-Lincoln
Lincoln, Nebraska 68588-0111

N.J. Ianno, B. Johs, D. Thompson, D. Meyer, and John A. Woollam
Department of Electrical Engineering
University of Nebraska - Lincoln
Lincoln, Nebraska 68588-0511

ABSTRACT

Pulsed Laser Evaporation (PLE) has been shown to produce superconducting films of excellent quality. We will be discussing the results obtained from the PLE of Tl-Ba-Ca-Cu-O using a frequency doubled Nd:YAG laser operating at 532 nm. Films were deposited on $SrTiO_3$, MgO, yttrium stabilized ZrO_2, and polycrystalline Al_2O_3. Nearly single phase films of $Tl_2Ba_2Ca_2Cu_3O_{10}$ on MgO were routinely obtained. The best films exhibited a superconducting transition onset temperature of about 125K and zero resistance at 110K. The films had a c-axis orientation perpendicular to the substrates. X-ray microprobe fluorescence measurements indicate that a typical composition of the films is $Tl_{0.66}Ba_{1.77}Ca_{1.46}Cu_3O_x$, which is low in Tl compared to that expected for the 2:2:2:3 phase. The typical grain size is greater than 10 μm as revealed by scanning electron microscopy.

INTRODUCTION

Pulsed Laser Evaporation (PLE) is a powerful technique capable of depositing a wide variety of materials in thin film form[1-7]. Recently this technique has been used to deposit thin films of (Rare Earth) $Ba_2Cu_3O_x$ and Bi-Sr-Ca-Cu-O high temperature superconductors[8-13]. It has been shown that stoichiometric films can be produced from stoichiometric compound targets at wavelengths from 193 nm to 308 nm. This may result from the fact that the evaporation process is primarily photochemical at these short wavelengths[2]. However, good quality $YBa_2Cu_3O_x$ films have been obtained at other wavelengths, specifically 1.06 μm and 532 nm from a Nd:YAG laser and 10.6 μm from a CO_2 Laser[14-16].

Recently, T_c's up to 125K have been achieved in bulk samples of $Tl_2Ba_2Ca_2Cu_3O_{10}$ $(2:2:2:3)$[17]. Tl-based films have been prepared by electron-beam evaporation[18] and single-target[19-22] or multi-target[2,24] sputtering. D.S. Ginley et al.[18] reported a transport J_c above 2.4×10^5 A/cm^2 at 77K for unoriented 2:2:2:3 phase films, with little magnetic field dependence observed. W.Y. Lee et al. have achieved T_c's of up to 120K in oriented multi-phase films[9]. Critical current densities as high as 1×10^5 A/cm^2 at 100K in a zero magnetic field have been reported by M. Hong et al.[22] on a SrTiO$_3$ substrate with the oriented 2:2:2:3 phase.

In this paper, we are reporting the results of PLE of Tl-Ba-Ca-Cu-O thin films from a composite stoichiometric $Tl_2Ba_2Ca_2Cu_3O_{10}$ target using a frequency doubled Nd:YAG laser operating at 532 μm. We have deposited films on MgO (100), SrTiO$_3$ (100), yttrium stabilized ZrO$_2$ (100), and polycrystalline Al$_2$O$_3$. We show that virtually single phase superconducting films can be achieved under proper annealing conditions.

EXPERIMENTAL APPARATUS AND PROCESSING CONDITIONS

The deposition system, as shown in figure 1, consists of a stainless steel hexagonal chamber, pumped from below by an oil diffusion pump to a base pressure of 1×10^{-5} torr. The target is mounted on the end of a stainless steel rod, which is rotated at 10 rpm. The substrate is mounted on a heated stainless steel block directly in front of the target at a distance of approximately 3 cm.

The output of a Nd:YAG laser is frequency doubled to 532 nm and passed through a prism to separate it from the remaining 1.06 μm radiation, which is subsequently dumped into a beam block. The 532 nm radiation is focused, passed through a window in the deposition chamber, and made incident on the target.

Typical deposition conditions are: a substrate temperature of 200°C, deposition time of 30 minutes, laser pulse rate of 4 Hz, laser pulse width of 10 nsec, and an energy density of 1.2 J/cm^2. The laser beam spot size was determined by placing a glass slide coated with liquid graphite in front of the target and exposing it to a few laser pulses. Dividing the measured area of the spot by the energy of the laser pulse (as measured by a calorimetric power meter) yields the energy density. However, due to spatial non-uniformities in the laser beam and measurement uncertainties, the error in the reported energy density is on the order of 25%. In spite of this error, this method allows reproducible experimental conditions to be established.

The target was a composite of Tl-Ba-Ca-Cu-O made by sintering a mixture of Tl$_2$O$_3$, BaO, CaO, and CuO with a metal cation ratio of 2:2:2:3.

The as-deposited films, about 1 μm thick, were not conducting, and a post annealing step at 840°C-870°C was required to make them superconduct. Post annealing of films was carried out under 1 atm of O$_2$ or air in a sealed quartz tube. The procedure was similar to that used by Lee et al.[9], but the annealing temperature of our best films was lower than theirs by about 30°C.

The film compositions were determined by x-ray fluorescence microprobe spectroscopy. A typical film composition after annealing was $Tl_{0.66}Ba_{1.77}Ca_{1.46}Cu_3O_x$. The Tl content was found to vary spatially by 20% about the film surface.

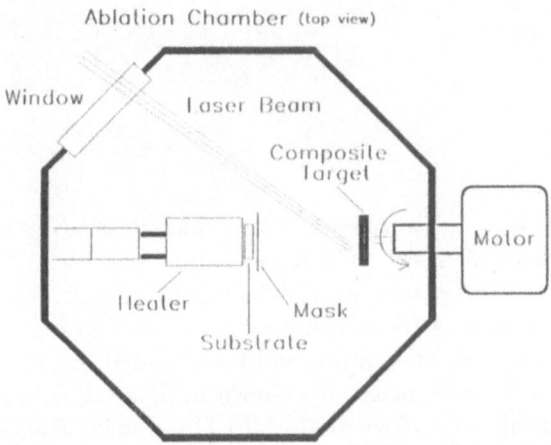

Laser Ablation System

Optical Set Up

Nd:YAG Laser

10 ns pulse
4 pulse/s
200 mJ/pulse

Harmonic
Generator

Prism

Visible (532nm)

IR

Beam Block

Lens

Ablation Chamber

Ablation Chamber (top view)

Window

Laser Beam

Composite
Target

Motor

Heater

Mask

Substrate

Figure 1. Pulsed Laser Evaporation apparatus for high T_c superconductors

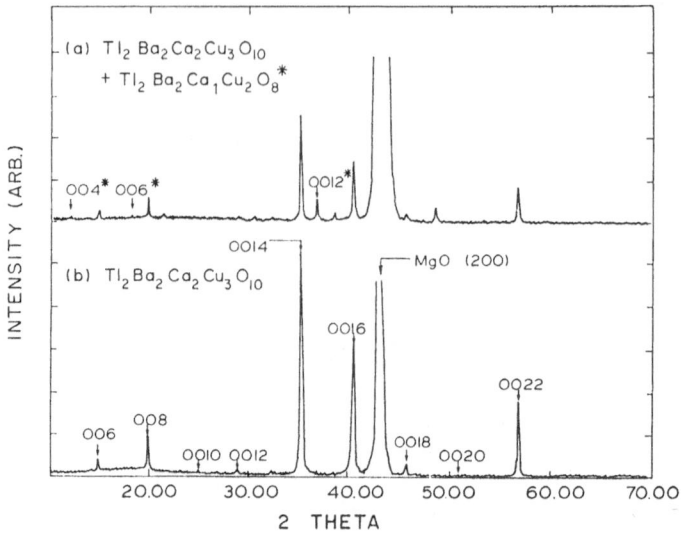

Figure 2. X-ray diffraction pattern of Tl-Ba-Ca-Cu-O films on MgO substrates (a) annealed at 870°C for 5 min. which contains mostly 2:2:2:3 and 2:2:1:2 phases. Some (00l) peaks of 2:2:1:2 phase are indexed. (b) annealed at 870°C for 10 min. which contains nearly pure 2:2:2:3 phase. (00l) peaks of 2:2:2:3 phase are indexed.

STRUCTURE

The crystal structure of each annealed film was characterized by x-ray diffraction. A Rigaku θ-2θ diffractometer with Cu K$_\alpha$ radiation was used. From systematic studies, we found that the phases formed in the films depend strongly on heat treatment conditions, such as the temperture and duration of the reaction time. For the films grown on MgO substrates, and heat-treated less than 10 min. at or below 870°C, the phases were mixed. As shown in figure 2(a), a film annealed at 870° for 5 min. typically contained both 2:2:1:2 and 2:2:2:3 phases. The major peaks can be assigned to the diffraction from the c plane with lattice constants of c = 29.3 Å and 35.6 Å respectively. This mixed phase film is highly oriented, with the c-axis perpendicular to the film plane. For the films grown on MgO (100) and annealed at 870°C for 10 - 15 min., the 2:2:2:3 phase becomes the primary phase. In figure 2(b) sharp periodic peaks were observed in the pattern. This indicates that the film is highly oriented. All peaks can be assigned to the (001) peak of the 2:2:2:3 phase.

We also found that the phases formed in the films depend strongly on the substrate. Figure 3 illustrates the different x-ray diffraction patterns for films grown on Y-ZrO$_2$, SrTiO$_3$, and poly-Al$_2$O$_3$ and annealed at 870°C for 10 min. For the film grown on Y-ZrO$_2$, the 2:2:2:3 phase was the primary phase, and the 2:2:1:2 phase was also observed. The film has a strong c-axis orientation as evidenced by the x-ray diffraction pattern shown in figure 3(a). The (00l) peaks of 2:2:2:3 phase are labeled. For the film grown on SrTiO$_3$, both 2:2:1:2 and 2:2:0:1 phases were observed, as shown in figure 3(b). Only the 2:2:0:1 phase has been labeled in this figure. This film was also c-axis oriented, although the (00l) peaks are much broader than films

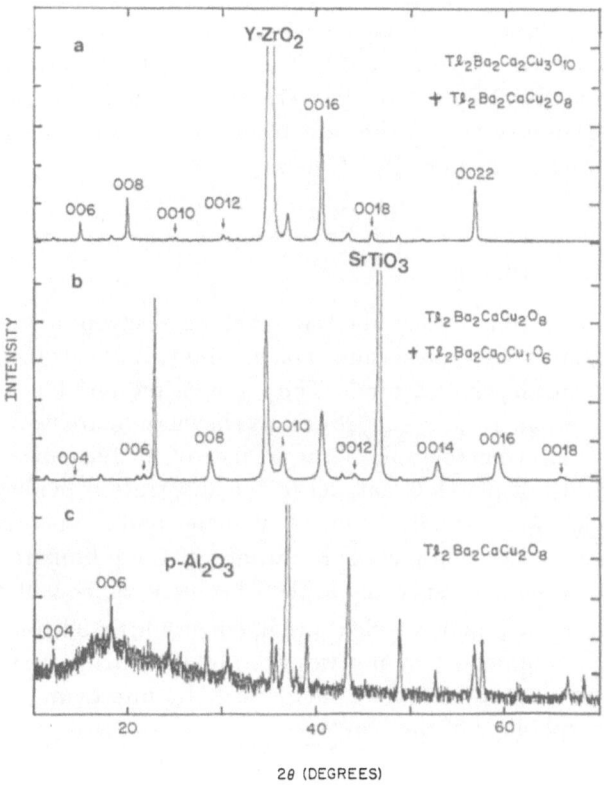

Figure 3. X-ray diffraction patterns of Tl-Ba-Ca-Cu-O films annealed at 870°C for 10 min. (a) Y-ZrO$_2$ substrate, (b) SrTiO$_3$ substrate, (c) poly-Al$_2$O$_3$ substrate.

grown on the other substrates, indicating smaller grains or more defects. The x-ray diffraction pattern for the film grown on polycrystatlline Al$_2$O$_3$ and annealed at 870°C for 10 min. is shown in figure 3(c). The 2:2:1:2 phase was clearly seen in the pattern. However, the peak intensities are very weak and embedded in a broad hump. This indicates that the film and substrate may have strong interactions, and grains of the 2:2:1:2 phase may be randomly oriented.

TRANSPORT PROPERTIES

Transport properties were measured using the standard four-point measurement technique using DC currents, where the polarization of the current was switched during the measurements. The room temperature resistivities were estimated, using the van der Pauw method, to be between 500 $\mu\Omega$-cm and 1500 $\mu\Omega$-cm. These values are considerably higher than those for $YBa_2Cu_3O_x$ and Bi-based superconductors. However, if we take into account the porosity and roughness of these films, the actual resistivity of Tl-based superconductors may be much smaller (the evidence of porosity and roughness of the films is discussed later). Figure 4 shows the resistivity versus temperature characteristics observed for the annealed sample with nearly pure 2:2:2:3 phase. The superconducting transition for this film starts around 125K, and the zero-resistance state is below 110K. The transport critical current density of this film was evaluated to be 10^4 A/cm^2 at 77K in zero magnetic field. This value of the critical current is lower than the best results that have been reported[18,22]. This may be partly due to the porosity of the samples.

SEM MICROSTRUCTURE

The morphology of these films was studied by scanning electron microscopy. Scanning electron micrographs of a typical sample on an MgO (100) substrate before and after heat treatment are shown in figure 5. The as-deposited film has a grain size of about 1 μm and the surface is somewhat rough, as shown in figure 5(a). Figure 5(b) shows a scanning electron micrograph of the surface of a film annealed at 870°C for 10 min. The growth of platelets parallel to the substrate is evident in the micrograph. The platelets are typically 10 μm in diameter and are poorly connected. Energy-dispersive x-ray microanalysis of the surface of these films revealed broad compositional inhomogeneities, especially in the Tl content which varied from point to point. There were many pinholes which contained very little Tl-compound. This indicates that the Tl-compound may not wet the MgO substrate very well. Based on these observations, we may be able to increase J_c by improving the wetting of the substrate and morphology of the film.

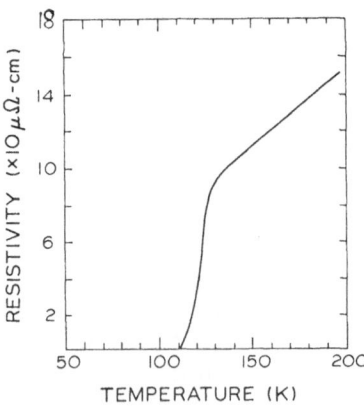

Figure 4.   Resistance versus temperature for a Tl-Ba-Ca-Cu-O film on MgO annealed at 870°C for 10 min.

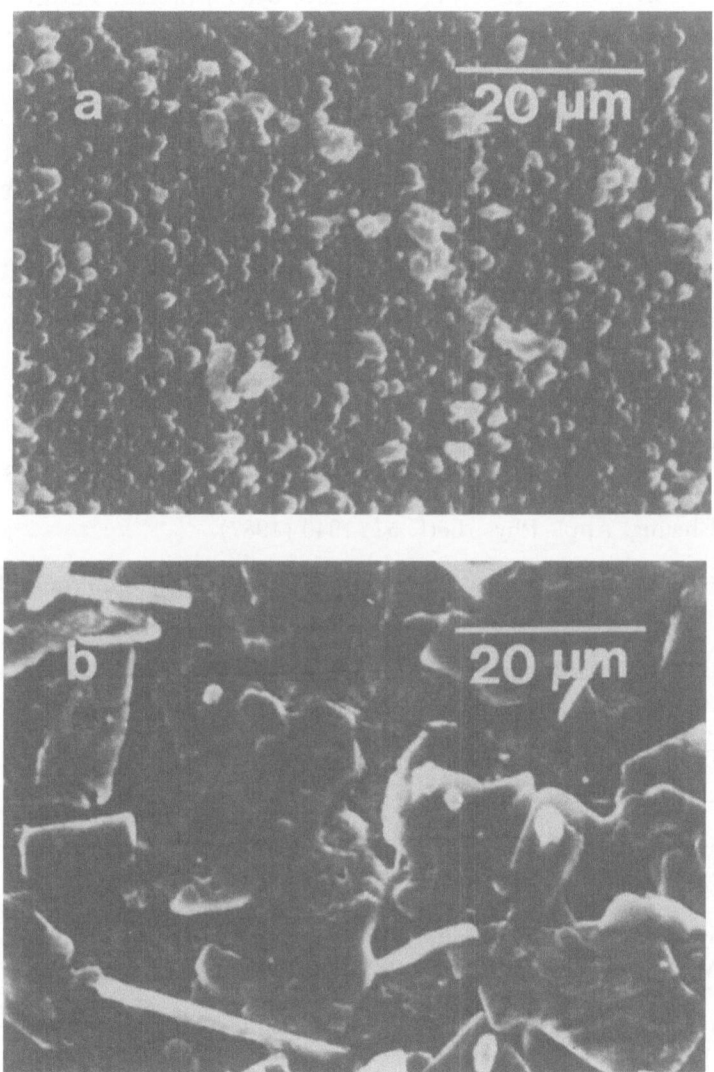

Figure 5. Scanning electron micrograph of the same films, as in figure 2, (a) as deposited, and (b) after annealing at 870°C for 10 min.

SUMMARY

In summary, we have prepared Tl-Ba-Ca-Cu-O thin films on MgO (100), Y-ZrO$_2$, SrTiO$_3$, and poly Al$_2$O$_3$ substrates using pulsed laser evaporation. The microstructure of the films are dependent on the post annealing conditions and substrates. Nearly single phase films of Tl$_2$Ba$_2$Ca$_2$Cu$_3$O$_{10}$ on MgO substrates are routinely obtained by keeping the same post annealing conditions. Superconducting films with T$_c$ onset at 125K and T$_c$ (R=O) of 110K have been achieved. Future work will be concentrated on improving the morphology and producing films with a high J$_c$.

ACKNOWLEDGMENT

This work is supported in part by NASA Lewis Grant NAG 3-866.

REFERENCES

[1] G. Gorodetsky, T.G. Kazyaka, R.L. Melcher, and R. Srinivasan, Appl. Phys. Lett. **46**, 828 (1985).

[2] R. Srinivasan, in Laser Processing and Diagnostics, edited by D. Baurle, Springer Series in Chemical Physics Vol. 39, (Springer-Berlin, 1984) p. 343.

[3] J.T. Cheung, Appl. Phys. Lett. **51**, 1940 (1987).

[4] J.T. Cheung, G. Niizawa, J. Moyle, N.P. Ong, B.M. Paine, and T. Vreeland, J. Vac. Sci. Technol. **A4**, 2086 (1986).

[5] M.I. Baleva, M.H. Maksimov, S.M. Metev, and M.S. Sendova, J. Mat. Sci. Lett. **5**, 533 (1986).

[6] J.J. Dubowski, P. Norman, P.B. Sewell, D.F. Williams, F. Krolicki, and M. Lewicki, Thin Solid Films **147**, L51 (1987).

[7] H.S. Kwok, P. Mattocks, D.T. Shaw, L. Shi, X.W. Wang, S. Witanachchi, Q.Y. Ying J.P. Zheng and P. Bush, Mat. Res. Soc. Symp. Proc. Vol99. 273(1988).

[8] D. Dijkkamp, T. Venkatesan, X.D. Wu, S.A. Shaheen, N. Jisrawi, Y.H. Min-Lee, W.L. McLean, and M. Croft, Appl. Phys. Lett. **51**, 619 (1987).

[9] S. Witanachchi, H.S. Kwok, X.W. Wang and D.T. Shaw, Appl. Phys. Lett. **53**, 234 (1988).

[10] A. Inam, M.S. Hegde, X.D. Wu, T. Venkatesan, P. England, P.F. Miceli, E.W. Chase, C.C. Chang, J.M. Tarascon and J.B. Wachtman, Appl. Phys. Lett. **53**, 908 (1988).

[11] D.K. Fork, J.B. Boyce, F.A. Ponce, R.I. Johnson, G.B. Anderson, G.A.N. Connell, C.B. Eom and T.H. Geballe, Appl. Phys. Lett. **53**, 337 (1988).

[12] C.R. Guarnieri, R.A. Roy, K.L. Saenger, S.A. Shivashankar, D.S. Yee, and J.J. Cuomo, Appl. Phys. Lett. **53**, 532 (1988).

[13] R.A. Neifeld, S. Gunapala, C. Liang, S.A. Shaheen, M. Croft, J. Price, D. Simons and W.T. Hill III, Appl. Phys. Lett. **53**, 703 (1988).

[14] L. Lynds, B.R. Weinberger, G.G. Peterson, and H.A. Krasinski, Appl. Phys. Lett. **52**, 320 (1988).

[15] S. Mazuk and M.S. Leung, Spring Meeting Materials Research Society, Abstract K3.10, April, 1988.

[16] S. Miura, T. Yoshitake, T. Satoh, Y. Miyasaka, and N. Shohata, Appl. Phys. Lett. **52**, 1008 (1988).

[17] S.S.P. Parkin, V.Y. Lee, E.M. Engler, A.I. Nazzal, T.C. Huang, G. Gorman, R. Savoy, and R. Beyers, Phys. Rev. Lett. **60**, 2539 (1988).

[18] D.S. Ginley, J.F. Kwak, R.P. Hellmer, R.J. Baughman, E.L. Venturini, and B. Morosin, Appl. Phys. Lett. **53**, 406 (1988).

[19] W.Y. Lee, V.Y. Lee, J. Salem, T.C. Huang, R. Savoy, D.C. Bullock, and S.S.P. Parkin, Appl. Phys. Lett. **53**, 329 (1988).

[20] M. Nakao, R. Yuasa, M. Nemoto, H. Kuwahara, H. Mukaida, and A. Mizukami, Jpn. J. Appl. Phys. **27**, L849 (1988).

[21] Yo. Ichikawa, H. Adachi, K. Setsune, S. Hatta, K. Hirochi, and K. Wasa, Appl. Phys. Lett. **53**, 919 (1988).

[22] M. Hong, S.H. Liou, D.D. Bacon, G.S. Grader, J. Kwo, A.R. Kortan, and B.A. Davidson, Appl. Phys. Lett. **53**, 2102 (1988).

[23] J.H. Kang, R.T. Kampwirth, and K.E. Gray, Phys. Lett. **A131**, 208 (1988).

[24] D.H. Chen, R.L. Sabatini, S.L. Qiu, D. DiMarzio, S.M. Heald, and H. Wiesmann, (unpublished).

PREPARATION OF HIGH TEMPERATURE SUPERCONDUCTOR THIN FILMS ON Si BASED SUBSTRATES BY EXCIMER LASER ABLATION, FILM-SUBSTRATE INTERACTIONS, AND PLASMA PROCESSING EFFECTS ON CRYSTALLINITY

R. A. Neifeld *, A. Hansen; Harry Diamond Laboratories, Adelphi, MD, 20783; R. Pfeffer, E. Potenziani, W. Wilbur, D. Basarab, R. Lareau, J. Shappirio, Electronic Technologies and Devices Laboratory, Fort Monmouth NJ; C. Wrenn, L. Calderon, Vitronics Corp., Eatontown NJ; W. Savin, NJIT, Newark NJ; A. Tauber, SCEEE; G. Liang, S. Gunapala, M. Croft, Department of Physics and Astronomy, Rutgers University, Piscataway, NJ; J. Price, US Navy - Naval Surface Warfare Center, Silver Spring, MD; D. Simons, Catholic University, Washington DC; W. T. Hill, III, B. Turner, A. Pinkas, J. Zhu; Institute for Physical Science and Technology, University of Maryland, College Park, MD; D. Ginley, Sandia National Laboratories, Albuquerque, NM

(* Currently at Electronic Technologies and Devices Laboratory)

ABSTRACT: Thin films were prepared by excimer laser ablation from Y-Ba-Cu-O and Tl-Ba-Ca-Cu-O targets. The films were prepared upon amorphous quartz and crystalline Si (111) substrates in partial pressures of oxygen. The substrate-film interaction, crystallization, composition, and microstructure of these films versus the deposition condition variables were studied including, substrate temperature, oxygen partial pressure, laser fluence and repetition rate. Additional deposition variables include the substrate/substrate holder and plasma generation loop DC and RF bias conditions and the target which can be DC biased. With the various biasing/plasma conditions we have studied to date, we have been unable to lower the substrate deposition temperature required for crystallization of Y-Ba-Cu-O films by more than 50 degrees C. Films of Y-Ba-Cu-O produced upon Si (111) are shown to be superconducting.

Preparation of high temperature superconductor (HTSC) films at low substrate temperatures is desirable in order to minimize film-substrate interdiffusion and thermal mismatch effects upon thermal cycling, and to be compatible with other microelectronics process steps. Development of low temperature processes to prepare such films upon semiconductor substrates is required for successful development of a hybrid high speed technology.

Much interest has developed in laser ablation since it has shown the ability to prepare HTSC films at reasonable substrate temperatures and allows considerable flexibility in process conditions.[1,2,3,4,5] In previous work on laser ablation, we have investigated the systematics of film properties prepared by laser ablation from several HTSC materials.[6,7] To investigate various processing schemes, we have constructed a chamber for reactive and plasma processing using laser ablation for deposition. In figure 1 is a schematic of this chamber. The chamber is equipped with a radiant heating source and thermocouple for substrate heating and control. The substrate holder is DC and RF biasable. A 2.5 inch diameter copper loop is 1 inch away from the the substrate surface and is also DC and RF biasable. The thermocouple is contacted by the back side of the substrate when the substrate is slid into position. Due to the poor thermal contact between the thermocouple and substrate, the numbers we quote for temperature which have small relative error may have a large absolute error. (However our film crystallization temperatures agree with those from other groups using similar process conditions.[2,3]) A plasma compatible thickness monitor can be slid into position in order to monitor depostion rate. Oxygen gas is admitted through a needle valve until the desired pressure is established. The target rotates at 5 rpm and can be DC biased. The laser pulse enters through a quartz window and impinges upon the rotating target. We use the laser at a wavelength of 248nm with pulse energies at the target of 0.1 to 0.2 joules. The laser repetition rate is variable from 0-150 hertz. The laser pulse is focused by a quartz lense, and the laser fluence (joules per square centimeter) is fixed for each sample.

Previous work in our lab on Tl superconductor based films indicated substantial loss of Tl in films prepared upon hot substrates. We have therefore used a Tl rich target with stoichiometry of $Tl_5Ba_2Ca_2Cu_3O_x$ for this study. We have prepared a series of films from this target for which only the substrate temperature was varied. The fixed process

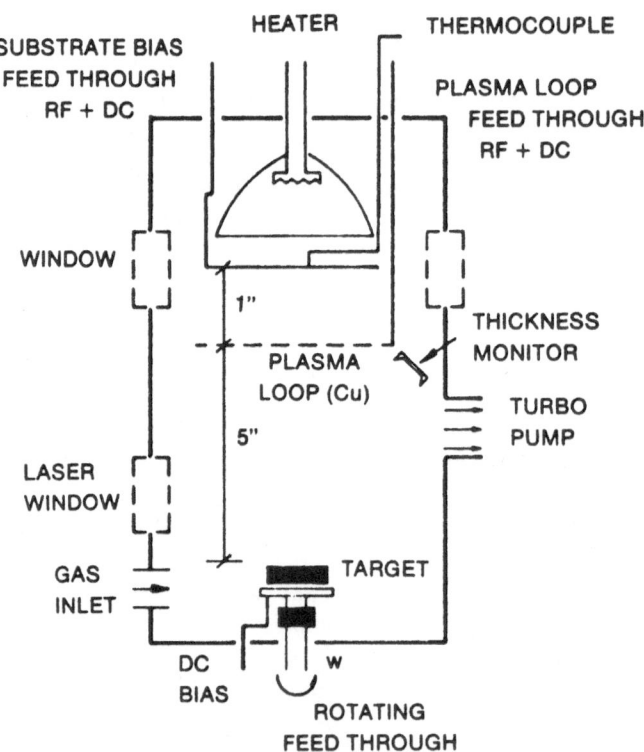

FIG. 1. Schematic of process chamber used for laser ablation. During operation a laser pulse enters the window and is intercepted by the target. A plume of material is ejected from the target, passes through the copper plasma generating loop, and is collected upon the heated substrate.

conditions were a laser fluence of 1.4 joules per square centimeter, laser repetition rate of 30 hertz, background pressure of 5 millitorr of oxygen, and 125 watts of RF power (at 13.56 megahertz) supplied to the copper loop. A glow discharge was initiated and continuously maintained by the RF power on the loop. Depositions lasted 30 minutes with film growth rates of 0.6 to 0.9 angstroms per second. These films were grown on quartz substrates. Results of film composition, as determined by RBS, and electrical nature are presented in figure 2. In figure 2 the Cu atom ratio is fixed to 3, since Cu_3 is the starting composition of the target and Cu appears to be the least volatile element. The relative variation of the Tl, Ba, and Ca concentrations is plotted versus substrate temperature. The Ba/Cu ratio remains relatively constant at 2/3, which is the same as the target, indicating that Ba has the same sticking coefficient (which is probably near unity) on the film as Cu. The Ca/Cu and Tl/Cu ratios are well below those for the target. Particularly interesting is the behavior of the Tl concentration versus temperature. From 250 to 350

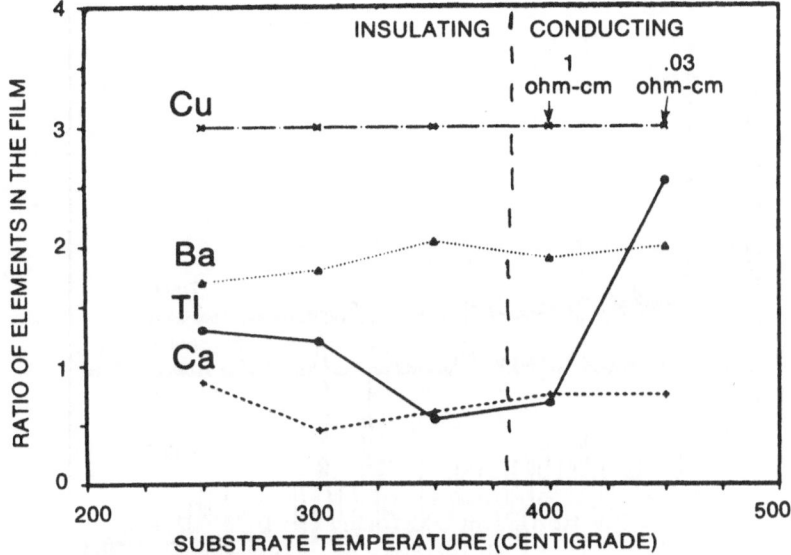

FIG. 2. Stoichiometry of the metal ions in the Tl-Ba-Ca-Cu-O films versus substrate temperature. Note the dividing line between conducting and insulating films.

centigrade the Tl concentration in the film decreases with temperature as expected due to high Tl vapor pressure. However above 350 centigrade the Tl concentration in the film increases rapidly. A possible explanation for the increase in Tl concentration is that above 350 centigrade, even though Tl has a very high partial pressure, the Tl rapidly diffuses into the substrate and is unavailable for evaporation from the surface of the film. In fact RBS of the film produced at 450 centigrade exhibited a surface layer deficient in Tl, and substantial diffusion. A SIMS profile of this film clearly indicates substantial diffusion of silicon from the substrate into the film, and almost to the front surface. While the films produced above 350 centigrade were conducting, x-ray diffraction indicated that they were amorphous.

Three films have been produced from our Tl target upon Si (111). These films were produced with a laser fluence of 0.5 joules per square centimeter with a laser repetition rate of 40 hertz in 2 millitorr of oxygen, and a deposition rate of roughly 4 angstroms per second, with no DC or RF biasing, but with the substrate and target floating. The substrate temperatures were 450, 500, and 550 centigrade. The film produced at 450 C was insulating while the other two were poor conductors with resistivities on the order of 1 ohm-centimeter.

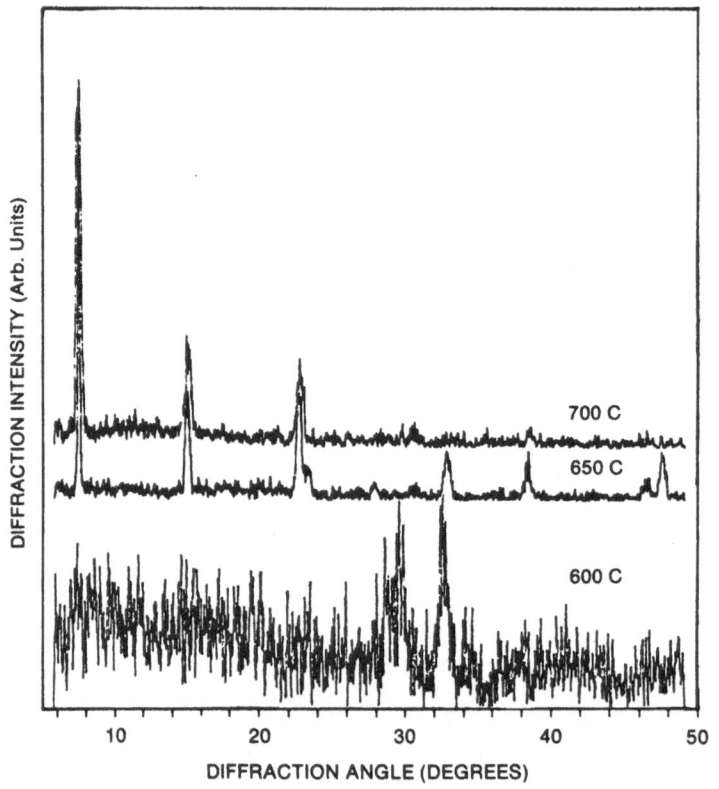

FIG. 3. X-ray diffraction patterns of Y-Ba-Cu-O films on quartz for substrate temperatures of 600, 650, and 700 centigrade.

Since more is known about $Y_1Ba_2Cu_3O_7$, we have tested the effect of various plasma processing conditions upon crystallization using this target material. We fixed the laser fluence at 2.4 joules per square centimeter and ran the laser at 30 hertz for 60 minutes for each deposition. The chamber pressure was fixed at 5 millitorr of oxygen. These films were deposited upon fused quartz with growth rates of 0.3 to 1.0 angstrom per second. After deposition was completed, the chamber oxygen pressure was increased to several torr, and the films were cooled at 50 degrees per minute. In figure 3 we present the x-ray diffraction patterns for films produced at 600, 650, and 700 C. No DC or RF power is supplied to these films. The film produced at 600 C is amorphous while those at 650 and 700 are crystalline with increasing degrees of c axis alignment. RBS indicates very little diffusion on the films prepared at 650 C and below and diffusion of Si all the way to the front edge of the film produced at 700 C. It is clear that the Y-Ba-Cu-O crystallization temperature under these conditions is approximately 650 C, in agreement with other studies.[1] We have prepared a series of films with the substrate temperature at either 600 or 550 C under the conditions listed above with the addition of various DC and RF biases to the substrate holder, target, and copper loop. The biasing conditions along with the laser induced plasma at the target surface usually resulted in a glow discharge observable near the substrate and copper loop region. Results upon these films are presented in table 1 along with results upon the film produced at 650 C discussed above for comparison. RBS spectra of these films were inspected to ensure film quality. All films examined by RBS had Y/Ba/Cu ratios within 1/1.5/2.1 to 1/2.4/3.6. The RBS determined ratio of Y/O ranged from 1/7 to 1/11. The RBS determined ratio of Y/O has considerable inaccuracy due to the small oxygen sensitivity and overlap of signal from oxygen in the film and the signal from the substrate. For a quantitative measure of crystallization, we present the number of observable x-ray peaks and the full width of the most narrow line between two theta of 5 and 50 degrees. Our results listing substrate temperature, bias on the target and loop and substrate holder, and x-ray results are presented in table 1, along with the data for the film produced at 650 C for comparison. Clearly none of the biasing techniques attempted so far have substantially lowered the crystallization temperature of the films in contrast to similar processing by others.[4,5]

Table 1. Process conditions and film crystallinity

Film No.	Substrate temperature (C)	Substrate holder bias dc volts dc amps rf watts	Copper loop bias dc volts dc amps rf watts	Target bias dc volts dc amps rf watts	X-ray data No. of peaks observed	Narrowest peak: index and FWHM (deg)
144	600	ground	ground	floating	1	(110) 0.6
149	650	floating	ground	floating	10	(001) 0.16
146	600	ground	125 W	floating	0	—
150	600	−160 V, 0.00 A	125 W	floating	3	(110) 0.5
151	600	125 W	+340 V, 0.05 A	floating	2	(110) 0.5
152	600	125 W	floating	+340 V	0	—
153	550	ground	125 W	+240 V, 0.01 A	0	—
155	600	ground	+340 V, 0.40 A	ground	2	—
162	550	ground	floating	+250 V, 0.20 A	0	—
163	550	ground	250 W	+220 V, 0.50 A	0	—
164	550	floating	250 W	+220 V, 0.50 A	0	—

Several films from the $Y_1Ba_2Cu_3O_7$ target have been prepared on Si (111). For all these films, we maintained the following deposition conditions: laser fluence of 1.5 joules per square centimeter, laser repetition rate of 40 hertz, target electrically floating, and substrate electrically floating. The film growth rates were roughly 1 angstrom per second. The depositions all started in vacuum with the substrate temperature at 750 centigrade (in an attempt to burn through the surface oxide and registrate with the Si

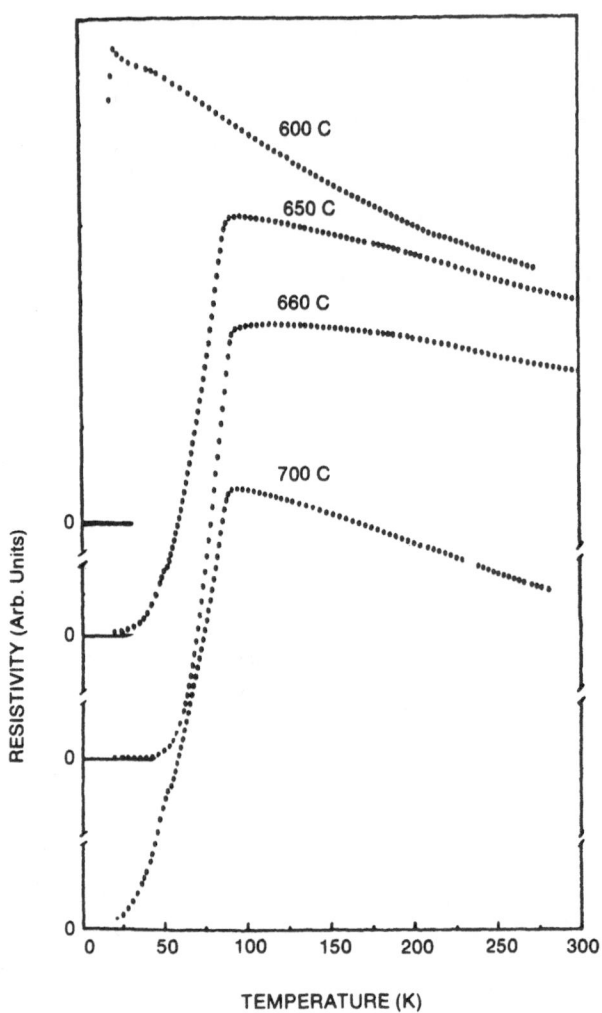

FIG. 4 Resistivity versus temperature for films prepared upon Si (111) at substrate temperatures of 600, 650, 660, and 700 centigrade.

surface) in vacuum. After 2 minutes the pressure was increased to 2 millitorr of oxygen and the substrate temperature which was then reduced for the remainder of the deposition was stabilized within 2 minutes. After the deposition was completed the laser was turned off, and the system pump was valved off. Oxygen continued to bleed into the chamber until the pressure was several torr. The films were cooled at 50 degrees per minute to below 150 centigrade and removed from the chamber. In figure 4, we present resistivity data upon some of these films. X-ray diffraction patterns of these films indicate rather poor crystal structure with little or no sign of c axis orientation. Several of the films are superconducting as seen in the resistivity data. Note that our best results occur for the film prepared at 660 centigrade with the resistivity drop onset at 88 Kelvin and zero resistance at 39 Kelvin. Presumably this is due to an increase of film contamination by interdiffusion with the substrate at higher substrate temperatures.

Acknowledgement

This work was supported in part by the National Science Foundation under grant No. PHY8451284.

REFERENCES

1. D. Dijkkamp, T. Venkatessan, X. D. Wu, S. A. Shaheen, N. Jisrawi, Y. H. Min-Lee, W. L. McLean, and M. Croft, Appl. Phys. Lett. 51, 861 (1987).

2. X. D. Wu, A. Inam, T. Venkatessan, C. C. Chang, E. W. Chase, P. Barboux, J. M. Tarason, and B. Wilkins, Appl. Phys. Lett. 52, 754 (1988).

3. T. Venkatessan, X. D. Wu, A. Inam, J. B. Wachtman, Appl. Phys. Lett. 52, 1193 (1988).

4. S. Watanachchi, H. S. Kwo, X. W. Wang, D. T. Shaw, Appl. Phys. Lett. 53, 234 (1988)

5. S. Witanachchi, H. S. Kwo, D. T. Shaw, submitted to Appl. Phys. Lett.

6. R. A. Neifeld, S. Gunapala, G. Liangf, S. A. Shaheen, M. Croft, J. Price, D. Simons, W. T. Hill, III, Appl. Phys. Lett. 53, 703 (1988)

7. R. Neifeld, et. al. in, "PROCESSING AND APPLICATIONS OF HIGH T_c SUPERCONDUCTORS: STATUS AND PROSPECTS." edited by W. Mayo (TMS, Warrendale, PA, 1988)

LASER ABLATION ANALYSIS OF 1-2-3 MATERIAL

R. Sega*, R. Lawconnell, and R. Motes

Frank J. Seiler Research Laboratory

T. Grycewicz

Department of Electrical Engineering

T. McNeil and J. McNally

Department of Physics
U.S. Air Force Academy
Colorado Springs, CO 80840

INTRODUCTION

Since the discovery of the Y–Ba–Cu–O material [1], several groups have investigated laser deposition techniques to produce superconducting thin films from the bulk 1–2–3 material [2–5]. We have performed laser ablation/evaporation research which supports the production of superconducting thin films. Control of the laser pulse and appropriate analysis of the vapor produced are necessary to understand the species that deposit on the substrate. Models for the laser–target interaction, the target plasma formation, and propagation to the substrate using a modified hydrodynamics code are presented.

Thin films of Y–Ba–Cu–O were deposited using a 40 psec width Nd:YAG laser operated at 10 pps which delivered approximately 100 mJ/pulse. The short pulse input to the target is modeled and the vapor/plasma is followed to the substrate in space and time. The vapor/plasma is being characterized experimentally for input to the model. Spectra of the target plasma obtained using an optical multichannel analyzer (OMA) for given laser deposition parameters are presented.

Using the short–pulsed laser, films were fabricated onto heated and unheated $SrTiO_3$ substrates in both an oxygen and an ozone background. The films were characterized using scanning electron microscopy (SEM) and energy dispersive spectroscopy (EDS). Results in this paper are for the Nd:YAG (1064 nm and 532 nm) laser. Optical spectra have been presented for these lines [6] with lower input power levels and longer pulse widths (10–100 nsec).

---

*R. Sega is currently with the Department of Electrical Engineering, University of Colorado at Colorado Springs.

THEORY

A hydrodynamics code is in development to model the vapor/plasma formed by a pulsed–laser incident on a superconducting bulk target. The complex nature of the ceramic superconductors has lead to a semi–empirical approach whereby the relevant species for the model are determined in the laboratory rather than by first principles. To simplify the calculation, it is assumed that the laser energy is deposited instantaneously onto the target because the experimental laser pulse widths are on the order of picoseconds — the hydrodynamic response of the target is on the order of microseconds. The vaporized target materials expansion is followed by conserving mass, momentum and energy in the hydro code.

A simulation for a single material target is presented to illustrate the technique. The parameters for this simulation are given in Table 1 with values approximately the same as the typical experimental conditions with the exception of the target–substrate separation which we have maintained between 3 and 4 cm in the experiment. In Figures 1–3, the target position is located at $Z=0$ cm and the substrate is located at $Z=0.94$ cm — cylindrical symmetry is assumed. The progression of the vapor/plasma is illustrated as one views an early time situation (1.5×10^{-5} sec) in Figure 2 and a later time view (3.9×10^{-5} sec) in Figure 3. The reaction of the species at the substrate will be dependent on vapor/plasma conditions along with substrate temperature, oxygen pressure, etc. For the correct species input to the model, experimental information is presently needed.

EXPERIMENT

The lasers used in the ablation/evaporation work have been Nd:YAG operated at the fundamental (1064 nm) and second harmonic (532 nm). The pulse width have varied from approximately 100 ns to 40 psec. The short–pulsed laser used a pulse rate of 10 pps with a nominal energy of 100 mJ/pulse. The target was a 0.75 in. diameter dish of stoichiometric Y–Ba–Cu–O rotated at 0.5 rps. The target substrate distance was variable from 3 to 4 cm. The vacuum system was cryo–pumped with base pressure of approximately 5×10^{-8} torr and O_2 flow to a typical backfill pressure of 10^{-4} torr. The substrate pre–heat was accomplished with a second laser.

The diagnostics for the vapor/plasma are an optical multichannel analyzer (OMA) and mass spectrometer. The Plasma/vapor species will be a function of laser wavelength, intensity, background environment, etc. Examples of spectral lines identified through use of an OMA are shown in Figures 4 and 5.

CONCLUSIONS

We have presented an approach and some preliminary results toward an understanding of the vapor formed by a laser pulse input to a superconducting bulk material target. The theoretical analysis should provide a predictive tool for vapor density, energy, and momentum in time and space as functions of incidence angle, power and wavelength of the laser. The use of a short pulse–width laser simplifies the modeling at the bulk target surface by limiting the interaction considerations both for the thermal penetration depth into the target as well as the vapor/plasma interaction with a longer laser pulse. Work is also in progress to determine the degradation of the target as a function of pulse width. The model presented will continue to be developed in conjunction with experimental results such as the species that are generated and target density. Finally, understanding of the plasma/vapor should provide an invaluable aide for control of the kinetics at the substrate surface required to produce high quality thin films.

Table I Parameters for Hydro Code Simulation

LASER SPOT SIZE RADIUS = 0.1524 cm

TIME LENGTH OF PULSE = 6 x 10^{-11} sec

INTENSITY OF PULSE = 1 X 10^{17} ergs/(cm^2 sec)

TARGET–SUBSTRATE DISTANCE = 0.94 cm

SUBSTRATE RADIUS = 0.635 cm

MESH = 37 x 37 CELLS

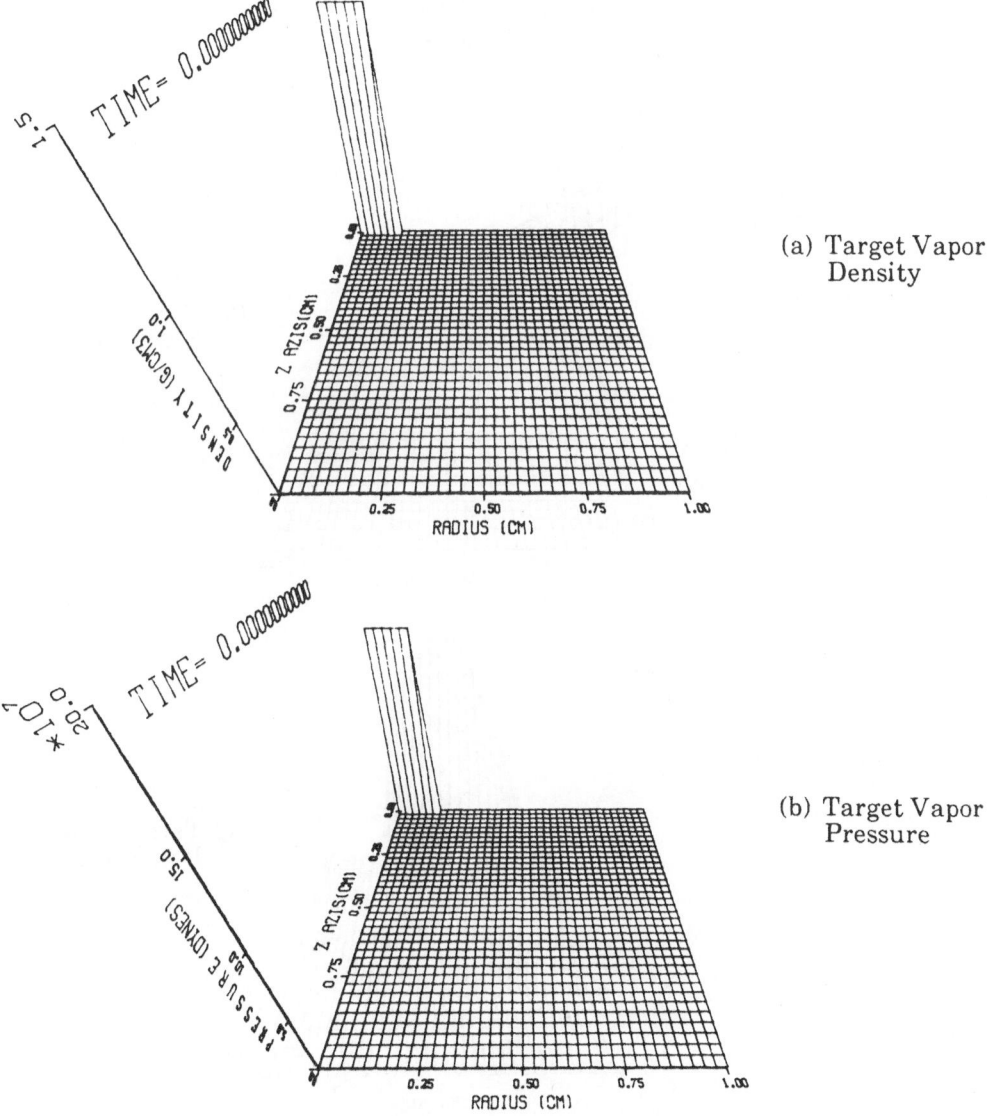

(a) Target Vapor Density

(b) Target Vapor Pressure

Figure 1. Initial Conditions for Simulation

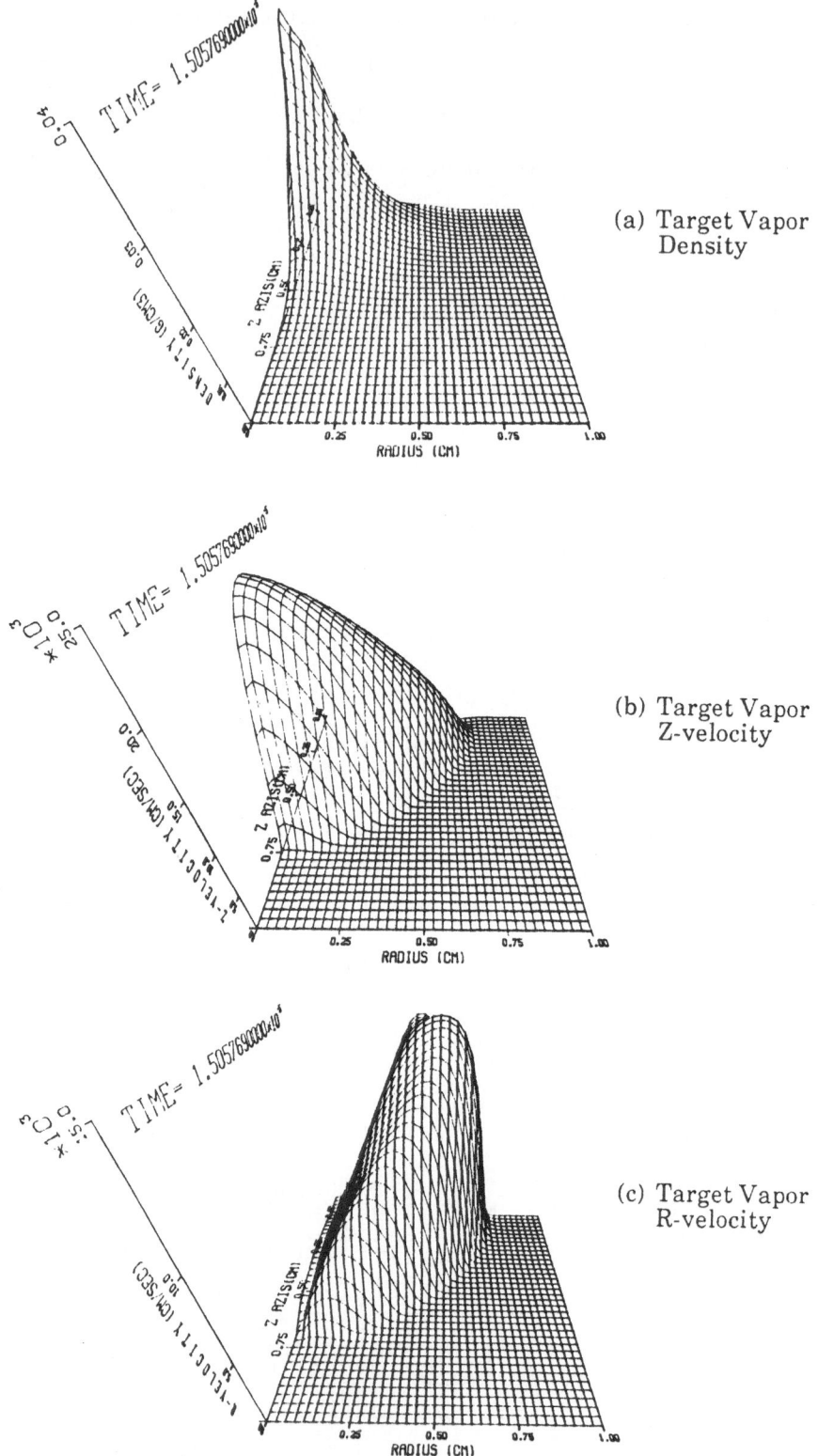

(a) Target Vapor Density

(b) Target Vapor Z-velocity

(c) Target Vapor R-velocity

Figure 2. Simulation Results at Early Time (1.5 x 10⁻⁵ sec)

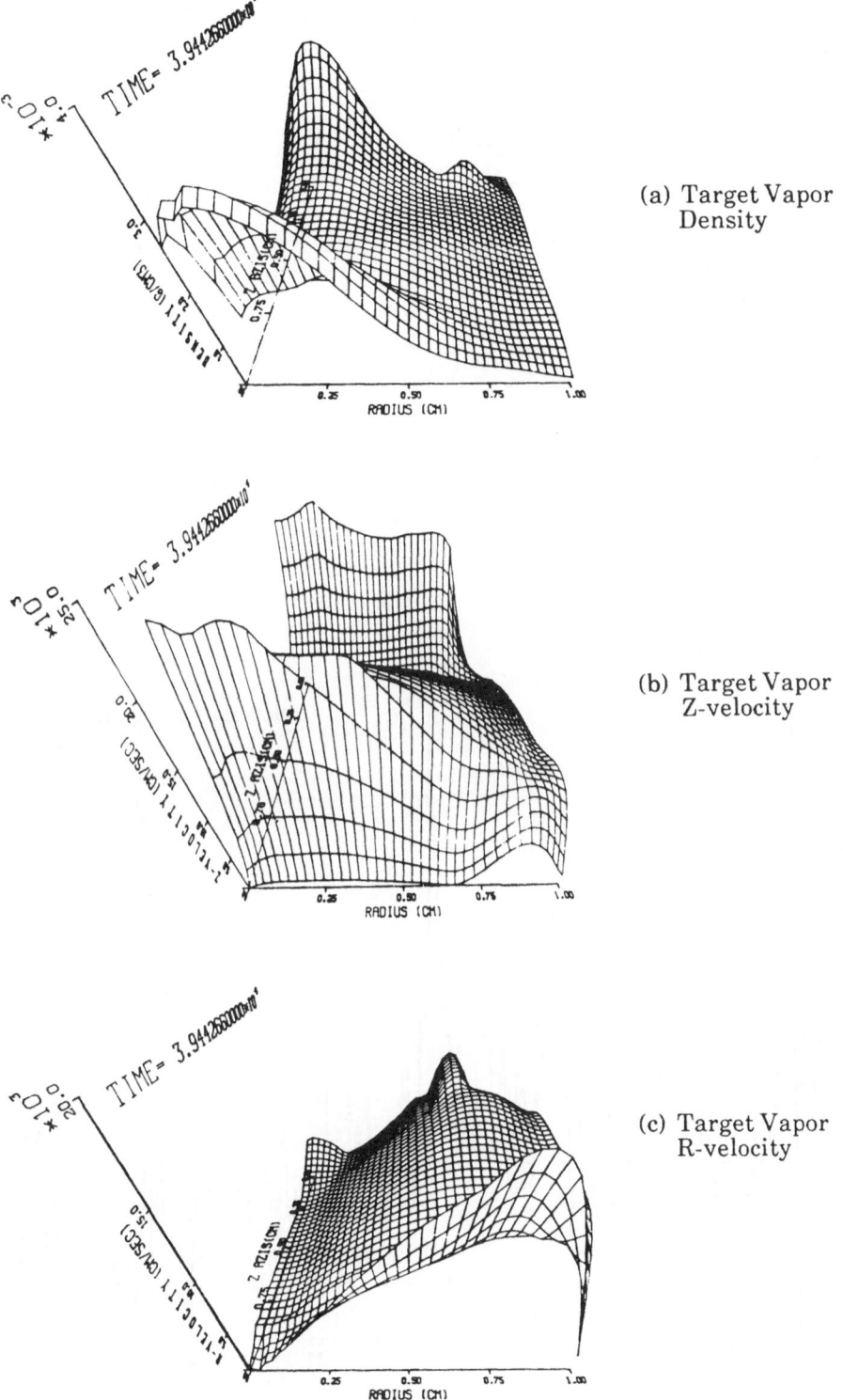

(a) Target Vapor
Density

(b) Target Vapor
Z-velocity

(c) Target Vapor
R-velocity

Figure 3. Simulation Results at Later Time (3.9 x 10⁻⁵ sec)

PLASMA SPECTROSCOPY (Y Ba$_2$Cu$_3$O$_x$) – 532 nm EXCITATION

LINE (OMA PEAK) (nm)	TENTATIVE IDENTIFICATION (nm)	COMMENTS
614.8	614.1 Ba II	STRONGEST
650.8	649.6 Ba II	NEXT STRONGEST
583.6	?	WEAKER
566.2	566.3 Y II	WEAKER

CALIBRATION LINES:

532 nm #400

632.8 nm #568

Figure 4. Spectral line identification from a 532 nm laser on 1–2–3 material

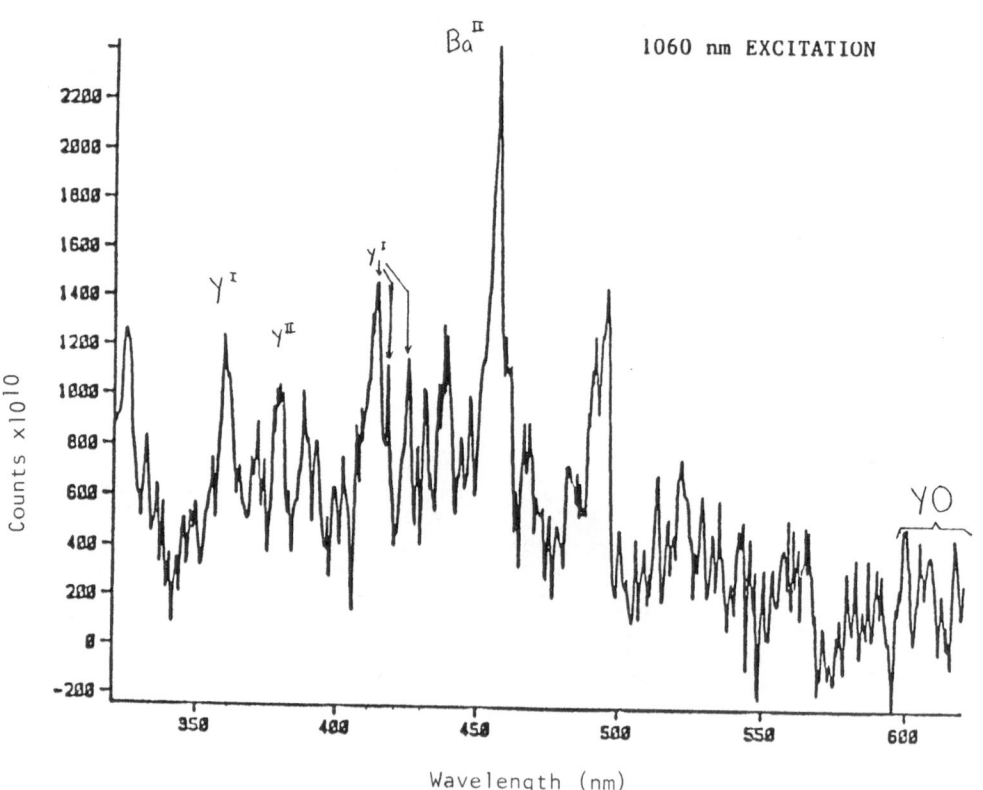

Figure 5. OMA spectra of 1064 nm laser illumination of a 1-2-3 target

REFERENCES

1. M. K. Wu, J. R. Ashburn, C. T. Torng, P. H. Hor, R. L. Meng, L. Gao, Z. H. Huang, Y. Q. Wang, and C. W. Chu, Phys. Rev. Lett. 58, 908 (1987)
2. D. Dijkkamp, T. Venkatesan, X. D. Wu, S. A. Shaheen, N. Jisrawi, Y. H. Min Lee, W. L. McLean, and M. Croft, Appl. Phys. Lett. 51, 619 (1987)
3. K. Moorjani, J. Bohandy, F. J. Adrian, B. F. Kim, R. D. Shull, C. K. Chiang, L. J. Swartzendruber, and L. H. Bennett, Phys. Rev. B36, 4036 (1987)
4. J. Narayan, N. Biunno, R. Singh, O. W. Holland, and O. Auciello, Appl. Phys. Lett. 51, 1845 (1987)
5. O. Auciello, S. Athavale, O. E. Hankins, M. Sito, A. F. Schreiner, and N. Biunno, Appl. Phys. Lett, 53, 72 (1988)
6. H. S. Kwok, P. Mattocks, D. T. Shaw, L. Shi, X. W. Wang, S. Witanachchi, Q. Y. Ying, J. P. Zheng, and P. Bush, MRS Symposium Proceedings (Materials Research Society, Pittsburgh, 1988) Vol. 99 p. 273

MICROSTRUCTURAL INVESTIGATION OF $YBa_2Cu_3O_{7-\delta}$ FILMS DEPOSITED BY

LASER ABLATION FROM $BaF_2/Y_2O_3/CuO$ TARGETS

R.L. Burton, C.U. Segre[*], H.O. Marcy[**] and C.R. Kannewurf[**]

Microwave and Electro-Optics Division
IIT Research Institute
Chicago, IL 60616

*) Department of Physics
Illinois Institute of Technology
Chicago, IL 60616

**) Department of Electrical Engineering and Computer Science
Northwestern University
Evanston, IL 60208

INTRODUCTION

Soon after the discovery of the high temperature superconductor $YBa_2Cu_3O_{7-\delta}$, it was realized that thin films of this compound could have significant technological applications. Within several months, many research groups had succeeded in fabricating good quality films by evaporation, sputtering, and laser ablation.[1] This last process holds much hope for commercial application and has received a great deal of attention. The first samples were ablated from a bulk $YBa_2Cu_3O_{7-\delta}$ target, and post-deposition annealing in oxygen at 900°C was used to produce the superconducting phase. The best films fabricated by this method exhibited good superconducting transitions with zero-resistance temperatures (T_{c0}) in the range of 80-88 K, but they consistently had low critical current densities (J_c).[2] These results were only achieved through accurate control of all aspects of the laser deposition process. The film composition, in particular, is strongly dependent on the uniformity and energy density of the laser beam. Heating the substrate to high temperatures (over 650°C) during the deposition has resulted in much improved values for both T_{c0} and J_c, but the inherent sensitivity to the deposition parameters is still the overriding consideration.[3-5]

A more recent development is a variation of the laser ablation method which uses targets consisting of the mixed but unreacted powders of BaF_2, Y_2O_3 and CuO. This technique has been reported[6] to yield consistently high quality films, in terms of T_{c0} and J_c, and is relatively insensitive to variations in the deposition process. Similar improvements in T_{c0} and reproducibility have been reported for other deposition schemes that used BaF_2 as a source material.[7-10]

The role of BaF_2 in the synthesis of these films has been much discussed, but it is still not well understood. It has been conjectured[6] that BaF_2 is deposited molecularly onto the substrate by the laser ablation process, and that by keeping the barium tightly bound to fluorine until the film is in the controlled environment of the annealing furnace prevents the formation of stable but undesirable Ba compounds, such as $BaCO_3$.

The BaF$_2$ derived films consistently show remarkable mixed-orientation epitaxial growth on SrTiO$_3$ substrates, features which have occasionally been observed with other deposition schemes. It is possible that the BaF$_2$ is helpful in obtaining such epitaxial growth. These films are often very porous, and this may affect their suitability for device applications. It is thus important to focus on the processing parameters in order to learn how to optimize the film microstructure. In this paper we present results that show how post-deposition annealing time and temperature affect the microstructure of films produced by this technique.

EXPERIMENTAL DETAILS

The laser ablation method of film deposition from targets containing BaF$_2$ will be referred to as the BaF$_2$ technique, whereas the earlier technique that features targets of the bulk superconductor will be called the YBCO technique. The BaF$_2$ technique has been quite well described by DeSantolo and coworkers.[6] Briefly, deposition targets (1" diameter) were prepared by mixing and grinding powders of BaF$_2$, Y$_2$O$_3$, and CuO in the metal ratios of Y:Ba:Cu=0.80:4.01:3.00. The resulting powder was pressed (20,000 psi) into pellets and sintered in dry oxygen at 800°C for 12 hours to improve mechanical stability. This sintering is inadequate to react the BaF$_2$. Unheated substrates of SrTiO$_3$ (100) or sapphire (unoriented) were mounted a few centimeters away from the target (typically 3 cm). The deposition chamber was evacuated to 5x10^{-6} torr. The excimer laser (Lumonics model TE-861) was operated at a wavelength of 248 nm. The pulse rate was 10 pps, with a 30 ns pulse width and energy of 20 mJ/pulse. The beam was focused on the target to a spot of approximately 1 mm x 2 mm. The output of the laser was monitored and held constant. During the deposition, ablated material coated the entrance window and absorbed significantly at the laser wavelength; over the course of a typical deposition, this resulted in a factor of two reduction in the laser energy at the target. The incidence angle of the laser beam on the target was 20° from normal and the substrate was centered in the plume of ejected material. The resulting deposition rate was 0.4 Å/pulse. For the films involved in this study, thicknesses ranged from 1.0-1.3 μm (determined by stylus profiler), unless otherwise noted.

After deposition the films were transferred to a quartz tube furnace for annealing. The heating schedule was typically 45 minutes from 25°C to peak temperature (PT), hold at PT, then ramp down to 440°C over two hours and furnace cooling from 440°C to 25°C (about 3 hr). The peak temperature was varied from 825°C to 875°C and the holding time at PT ranged from 30 minutes to 2 hours as will be discussed below. The heating and holding segments were carried out in the presence of flowing wet oxygen (250 cc/min); the cooling segments were performed in flowing dry oxygen (250 cc/min). The water vapor is essential to react the BaF$_2$ via the evolution of hydrogen fluoride. The wet oxygen was obtained by bubbling oxygen gas through an airstone (aquarium variety) submerged in 12 cm of distilled water (held at room temperature). It was learned that oxygen that was not sufficiently wet failed to react the BaF$_2$ completely.

The resistivity as a function of temperature was measured by the conventional 4-point method. Data were acquired and processed using a computer automated charge transport system described elsewhere.[11] Electrical contacts were made using gold paste and 25 μm gold wire. The microstructure of the films was examined by scanning electron microscopy (SEM) with an Hitachi model S-570LB. Samples were mounted with colloidal graphite and showed no evidence of charging. Structural measurements were performed on an automated GE XRD-5 X-ray diffractometer using Cu Ka radiation. Supplemental information on composition was obtained from energy dispersive X-ray analysis (EDX) and X-ray photoemission spectroscopy (XPS).

For comparison, some films were fabricated using the YBCO technique mentioned earlier. In this case the targets were bulk YBa$_2$Cu$_3$O$_{7-\delta}$ superconductor. The substrate temperature during the deposition was 400°C and the films were subsequently annealed in dry flowing oxygen at 900°C for 1 hr. The details of this method are described elsewhere.[12]

ELECTRICAL PROPERTIES

Figure 1 shows the resistivity vs. temperature for two films: (a) fabricated by the YBCO method, and (b) by the BaF_2 method. The BaF_2 film properties are superior in several important ways: (1) higher T_{c0}, 89 K vs. 70 K; (2) a narrower superconducting transition, 1 K vs. 15 K; and (3) a more metallic normal state resistivity, $(d\rho/dT) > 0$. In fact, the BaF_2 film strongly resembles the typical results obtained from good bulk or single crystal samples. We have not performed careful J_c measurements; however, by simply increasing the sample current during the resistivity measurement, we have established a lower limit for J_c of 2000 A/cm^2 (4.2 K) for the BaF_2 film. In contrast, the film from the YBCO method had a maximum J_c of 10 A/cm^2. Another important point is that the results shown in Figure 1 for the BaF_2 film are typical of every film produced by this method, while the other sample shown is the best of many samples prepared by the YBCO method. Furthermore, the BaF_2 films exhibited no degradation of T_{c0} after exposure to a typical indoor atmosphere for five weeks.

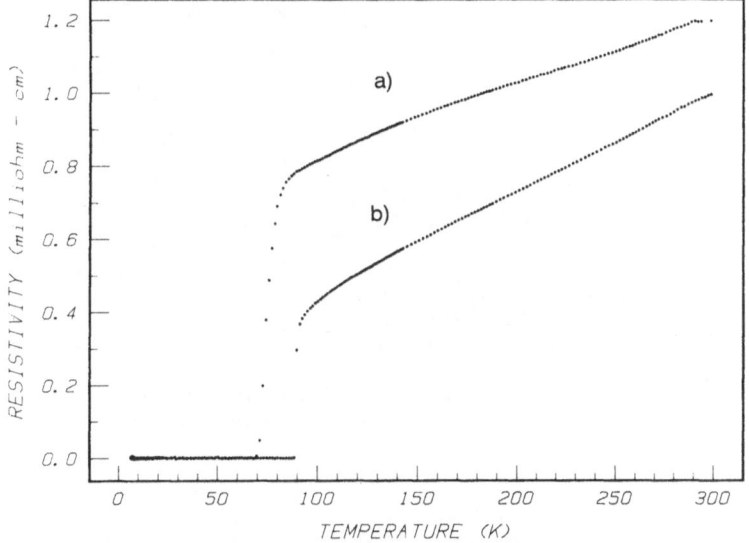

Fig. 1. Resistivity versus temperature for two films:
a) produced by the YBCO method, and b) by the BaF_2 method.

The BaF_2 technique is not limited to $SrTiO_3$ substrates. Several films were deposited by this method on sapphire, with the same annealing schedule as used for the $SrTiO_3$ substrates. It was found that T_{c0} ranged from 50 K to 75 K for these films. Their resistivity at room temperature is higher by a factor of five as compared to $SrTiO_3$, 5 milliohm-cm vs. 1 milliohm-cm. The minimum thickness for obtaining a superconducting film on sapphire by this technique is about 500 nm. Films less than this were semiconducting or insulating after annealing. A minimum thickness for films on $SrTiO_3$ substrates has not yet been determined, although it is certainly less than that for sapphire.

RESULTS AND DISCUSSION

All of the films discussed in this section were deposited on $SrTiO_3$ substrates. Perhaps the most dramatic result of this investigation is the remarkable difference in microstructure between films produced by the BaF_2 and YBCO methods as shown in Figures 2 and 3, respectively. These two films were both 1 μm thicK. The BaF_2 film is characterized by large (1 μm) interconnected crystallites of $YBa_2Cu_3O_{7-\delta}$ which are aligned perpendicularly to each other. These appear to be associated with a-orientation material. The structure echoes the twinning structure observed on the bare $SrTiO_3$ substrates. Under lower magnifications it is apparent that the entire film is covered with such structures, and they give it a distinctive "checkerboard" appearance, which has been much described in the literature for non-laser ablation techniques.[7-10] The crystallographic orientation of this film, as determined from XRD, has contributions from both c-axis and a-axis orientation, henceforth referred to as c-orientation and a-orientation.

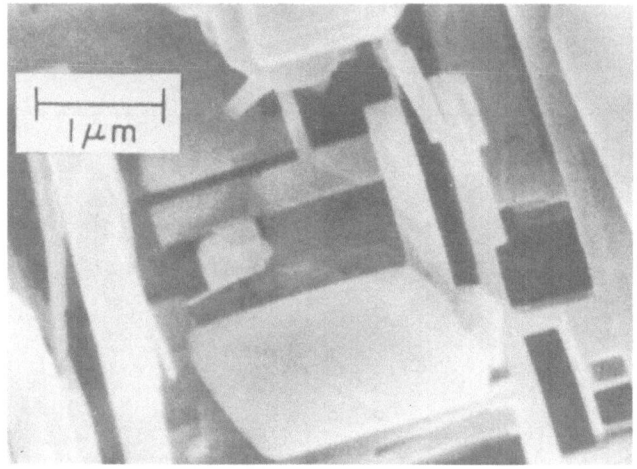

Fig. 2. $YBa_2Cu_3O_{7-\delta}$ film deposited on $SrTiO_3$ by the BaF_2 method; annealed at 850°C for 1 hr.

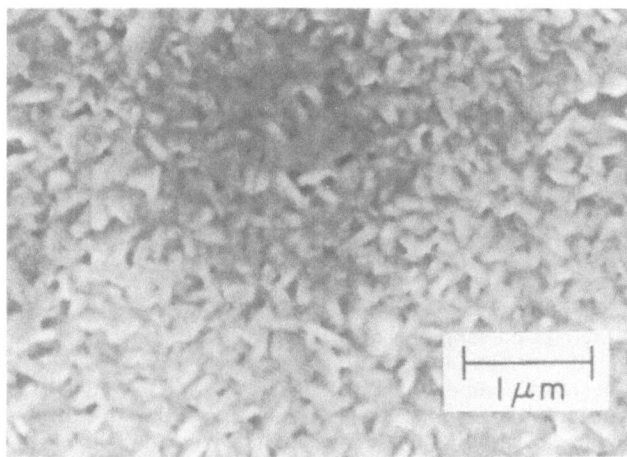

Fig. 3. $YBa_2Cu_3O_{7-\delta}$ film deposited on $SrTiO_3$ by the YBCO method; annealed at 900°C for 1 hr.

Fig. 4a. Film annealed at 825°C for 1 hr.

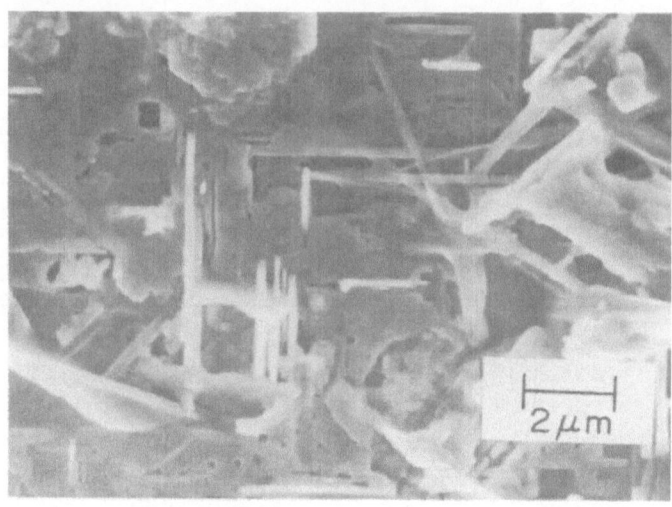

Fig. 4b. Film annealed at 875°C for 1 hr.

From Figure 3, the film produced by the YBCO method appears to contain a large number of small (100 nm) randomly oriented grains. Surprisingly, XRD reveals a significant quantity of a-orientation with only a small amount of random orientation and virtually no c-orientation. It is possible that the surface region shown by the SEM is not indicative of the entire thickness of the film. Alternatively, the ellipsoidal grains may naturally tend to form with the a and c axes in the plane of the surface, thus achieving some degree of preferential orientation. In view of the poor J_c for this film, it is likely that the grains are coated with insulating impurity phases that have segregated to the grain boundaries. This would constrain the intergrain supercurrent to tunneling mechanisms only, consistent with low J_c.

The rest of this discussion is concerned with films produced by the BaF_2 method. The first series of experiments varied the peak temperature of the annealing process. Films were annealed at 825°C, 850°C, and 875°C for 1 hr. Figures 4a and 4b are SEM micrographs of the 825°C and 875°C films, respectively. An example of a film annealed at 850°C for 1 hr was shown in Figure 2 (but with higher magnification). It is clear that the 875°C film exhibits a very disordered microstructure. The 825°C film, in contrast, features an emerging rectangular pattern of needle-like crystallites with uniformly gray material intervening, possibly material that has not yet fully crystallized. The 850°C film (Figure 2) exhibits the finest order of this series. These results indicate that 825°C may be insufficient to fully crystallize the superconducting phase in 1 hr, while the 850°C film is almost fully crystallized. Annealing at 875°C is detrimental to the well ordered surface microstructure, possibly by fostering the growth of other phases or orientations that inhibit the ordering of the crystallites.

The XRD patterns for these films are shown in figure 5. All of the samples show a mixture of c-axis and a-axis crystallographic orientation. The size of the (005) peak is useful for detecting c-orientation. It is much more difficult to assess the quantity of

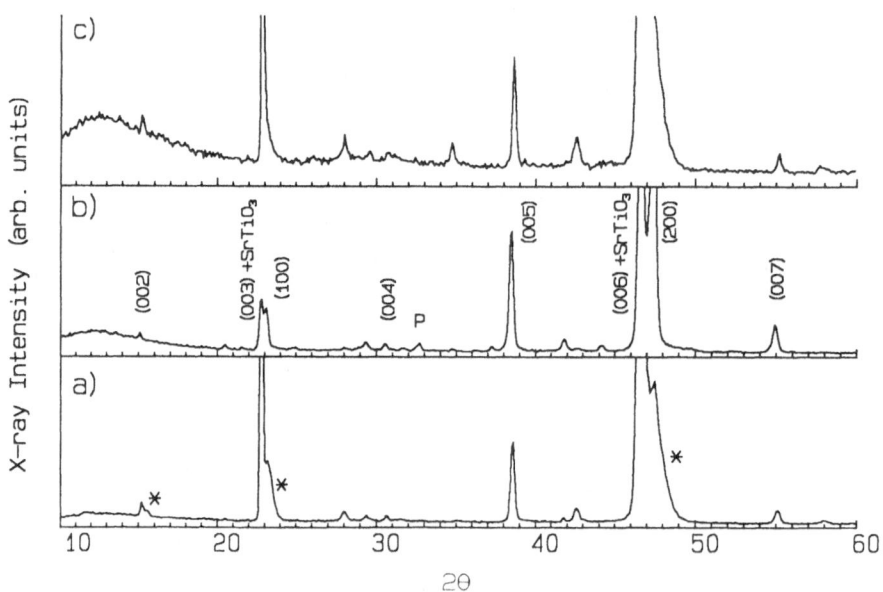

Fig. 5. X-ray diffraction patterns for three films annealed for 1 hr at: a) 825°C, b) 850°C, and c) 875°C. For clarity, the relevant $YBa_2Cu_3O_{7-\delta}$ peaks and substrate reflections are indicated only in the center pattern. The peak marked "P" indicates randomly oriented material, and the "*" indicates the shoulders described in the text.

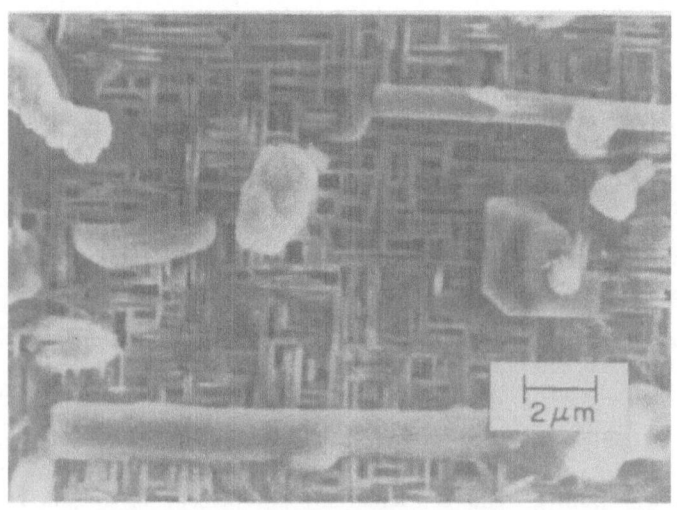

Fig. 6a. Film annealed at 850°C for 30 minutes.

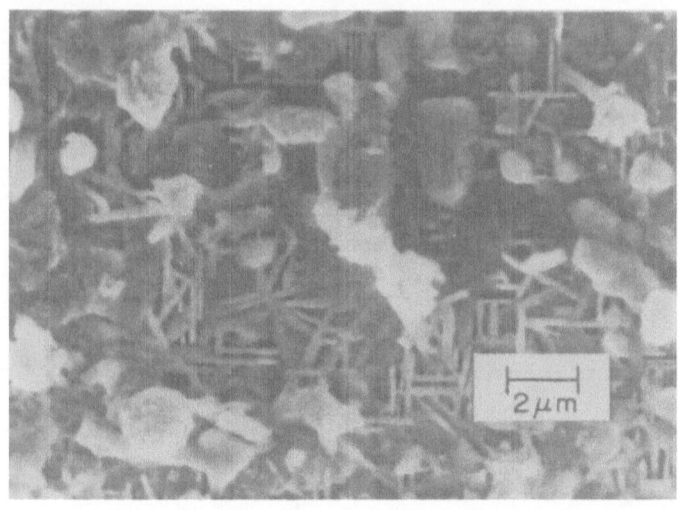

Fig. 6b. Film annealed at 850°C for 2 hr.

a-orientation as the (100) and (200) peaks are masked by the strong substrate peaks. The 825°C film shows a broad shoulder (indicated with a *) just above (in angle) the (002), (003)/(100)/(010), and the (006)/(200)/(020) reflections. These shoulders are significantly broader than the typical reflections from the film. The data from the other samples (850°C and 875°C) indicates that the higher annealing temperature results in nearly complete c-orientation for the film, even though some a-orientation crystallites still appear on the surface.

The second set of experiments varied the duration of the peak temperature anneal. The SEM micrographs of the surfaces of films annealed for 30 min and 2 hr at 850°C are shown in Figures 6a and 6b, respectively. Again, Figure 2 is representative of a 1 hr anneal at 850°C. The microstructure of the 30 min and 1 hr films is consistently well ordered, whereas the 2 hr film exhibits a disordered microstructure. Comparing the 2 hr film to the 875°C film, which was also disordered, it seems that the 2 hr film exhibits a higher density of needle-like crystallites. Examination of a large number of micrographs indicates that the 30 min anneal has slightly superior "checkerboard" order than does the 1 hr. The large rectangular blocks that grow from the surface of the 30 min film are impurity phases (discussed below) that seem to be interestingly aligned to the $YBa_2Cu_3O_{7-\delta}$ crystallites.

The XRD patterns for these films are presented in Figure 7. It is interesting to note that the 30 min film exhibits the same kind of broad shoulder structure as the low temperature annealed film (825°C). This feature may very well be associated with short annealing times and/or low annealing temperatures. We do not currently understand the presence of these features; however, such shoulders could indicate incomplete crystallization of $YBa_2Cu_3O_{7-\delta}$ since they disappear at longer annealing times and higher temperatures. Both a- and c-orientations are present in all the films. The shoulder structure makes it difficult to estimate the degree of a-orientation, so we only report orientation effects for the 1 hr and 2 hr films. The effect of the longer annealing times on the crystallographic orientation of the films is to reduce c-orientation, as evidenced by the reduction of the (005) peak compared to the (100).

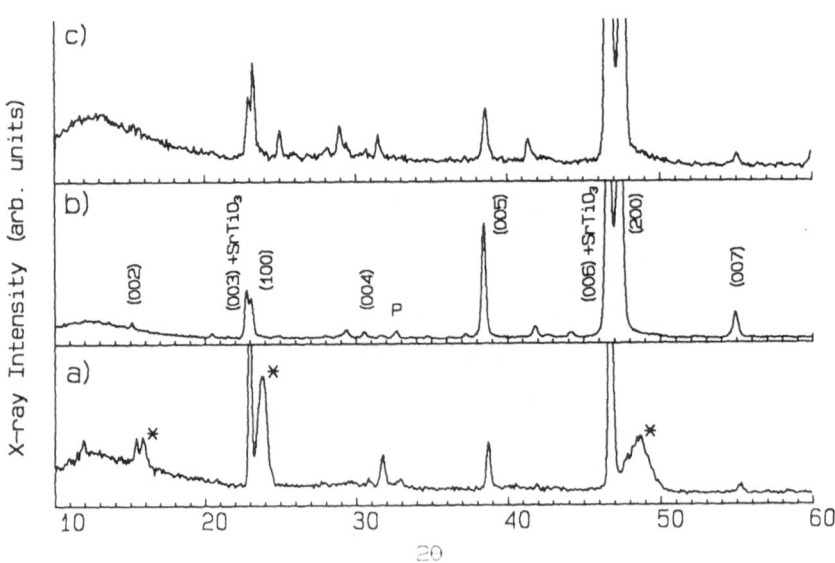

Fig. 7 X-ray diffraction patterns for three films annealed at 850°C for: a) 30 min, b) 60 min, c) 120 min. For clarity, the relevant $YBa_2Cu_3O_{7-\delta}$ peaks and the substrate reflections are indicated only in the center pattern. The peak marked "P" indicates a small amount of randomly oriented material, and the "*" indicates the shoulders described in the text.

All of the films shown in Figures 2,3,4, and 6 feature a substantial quantity of surface impurities. The overall terrain visible in micrographs can be classified into four types: (1) well ordered needle-like, grey crystals of $YBa_2Cu_3O_{7-\delta}$; (2) featureless grey regions; (3) localized white clumps (1-5 μm size); and (4) very large grey blocks (only seen on 30 min films). The EDX results indicate that the white clumps are very rich in barium and therefore could be $BaCO_3$, BaO, or unreacted BaF_2. Typical XPS results on these films indicate the presence of carbonate contamination on the film surface, which can be rapidly removed by sputtering. This would indicate BaO with a coating of $BaCO_3$ due to reaction with the atmosphere. Occasionally a fluorine peak is also seen in the XPS data, so the identity of this species is not completely clear. The large grey blocks only turn up in the short annealing durations. They are rich in both copper and barium, possibly $BaCuO_x$. There are several small impurity peaks in the XRD patterns. These do not correspond to any of the previously listed compounds and are currently unidentified. Since no combination of peak temperature and annealing time significantly reduced the observed surface impurities, it must be assumed that they are due to an imperfect starting composition in the target or are a consequence of this deposition technique.

From EDX, the featureless regions are roughly similar in composition to the $YBa_2Cu_3O_{7-\delta}$ regions. It has been reported[10] that the "checkerboard" pattern consists only of a-oriented material, whereas c-oriented material generally has no telltale microstructure observable by SEM[7]. Our data is consistent with this view: we observe both types of terrain, well ordered and featureless, and both a- and c-orientations in the XRD patterns.

CONCLUSIONS

Our results have shown that there is a tremendous difference in the surface microstructure of films produced by the BaF_2 technique and YBCO technique. There is also substantial improvement in the electrical properties of the films, as other groups have also reported.[6] We have found that the microstructure and crystallographic orientation of the 1 μm thick BaF_2 derived films is strongly dependent on the peak temperature of the annealing schedule and on the duration at peak temperature. It will be important to study the effects of this wide range of microstructures and orientations on the critical current densities of the films.

REFERENCES

1. Thin Film Processing and Characterization of High-T_c Superconductors: AIP Conference Proceedings No. 165, ed. by J.M.E. Harper, R.J. Colton, and L.C. Feldman (AIP, 1988), pp.1-465.
2. D. Dijkkamp, T. Venkatesan, X.D. Wu, S.A. Shaheen, N. Jisrawi, Y.H. Min-Lee, W.L. McLean, and M. Croft, Appl. Phys. Lett. 51, pp. 619-621 (1987).
3. X.D. Wu, A. Inam, T. Venkatesan, C.C. Chang, E.W. Chase, P. Barboux, J.M. Tarascon, and B. Wilkens, Appl. Phys. Lett. 52, pp. 754-756 (1988).
4. C.C. Chang, X.D. Wu, A. Inam, D.M. Hwang, T. Venkatesan, P. Barboux, and J.M. Tarascon, Appl. Phys. Lett 53, pp.517-519 (1988).
5. B. Roas, L. Schultz, and G. Endres, Appl. Phys. Lett. 53, pp. 1557-1559 (1988).
6. A.M. DeSantolo, M.L. Mandich, S. Sunshine, B.A. Davidson, R.M. Fleming, P. Marsh, and T.Y. Kometani, Appl. Phys. Lett. 52, pp. 1995-1997 (1988).
7. P.M. Mankiewich, J.H. Scofield, W.J. Skocpol, R.E. Howard, A.H. Dayem, and E. Good, Appl. Phys. Lett. 51, pp. 1753-1755 (1987).
8. A. Gupta, R. Jagannathan, E.I. Cooper, E.A. Giess, J.I. Landman, and B.W. Hussey, Appl. Phys. Lett. 52, pp. 2077-2079 (1988).
9. G.C. Hilton, E.B. Harris, and D.J. Van Harlingen, Appl. Phys. Lett. 53, pp. 1107-1109 (1988).

10. J.R. Gavaler, A.I. Braginski, J. Talvacchio, M.A. Janocko, M.G. Forrester, and J. Greggi, <u>High-Temperature Superconductors II</u>, ed. by D.W. Capone, W.H. Butler, B. Battlog, and C.W. Chu (MRS, 1988), pp. 193-196.
11. J.W. Lyding, H.O. Marcy, T.J. Marks, and C.R. Kannewurf, IEEE Trans. Instrum. Meas. **37**, p 76 (1988).
12. R.L. Burton, C.S. Batholomew, W.J. Wild, and J.L. Grieser, <u>Thin Film Processing and Characterization of High-Temperature Superconductors: AIP Conference Proceedings No. 165</u>, ed. by J.M.E. Harper, R.C. Colton, and L.C. Feldman (AIP, 1988) pp. 166-173.

PREPARATION OF THIN AND THICK FILM SUPERCONDUCTORS IN THE Tl-Ca-Ba-Cu-O SYSTEM

D. S. Ginley, J. F. Kwak, E. L. Venturini,
M. A. Mitchell, R. P. Hellmer, B. Morosin
and R. J. Baughman

Organization 1144
Sandia National Laboratories
Albuquerque, NM 87185

Introduction

The advent of high temperature superconductivity in Cu-O based materials has catalyzed a tremendous amount of activity in the synthesis, characterization and application of these materials [1]. The recent discovery of superconductivity above liquid nitrogen temperature in the Bi-Ca-Sr-Cu-O system [2] and Tl-Ca-Ba-Cu-O system [3] has led to hopes that these materials could overcome some of the problems experienced to date in the first system to superconduct above liquid nitrogen temperature, the rare earth $(Re)Ba_2Cu_3O_7$ [4]. The $(Re)Ba_2Cu_3O_7$ materials suffer from many problems including: weak links between the grains in polycrystalline aggregates [5], a complex equilibrium with oxygen, a phase change in the region where the material is processed and limited chemical stability. Further, the T_c of these materials near 90K is probably too close to 77K for many applications. These limitations mean substantial hurdles to the practical utilization of the $(Re)Ba_2Cu_3O_7$ materials.

The Bi-Ca-Sr-Cu-O system and Tl-Ca-Ba-Cu-O system offer considerable hope for obviating some of these difficulties. They have higher T_c's , up to 125K, and there are no phase changes in the region of interest. Oxygen appears to be much more stable in their structures, and they appear to be considerably more chemically stable than the $(Re)Ba_2Cu_3O_7$ materials. Very importantly, the Tl-Ca-Ba-Cu-O system has demonstrated strong intergranular links in films[6]. The Bi system has one phase, $Bi_2CaSr_2Cu_2O_8$ [7-8], which appears to be thermodynamically more stable than any of the others in this system, although $Bi_2Ca_2Sr_2Cu_3O_{10}$ can be stabilized by the addition of Pb [9]. The details of the structure of the Bi phases are complicated by an orthorhombic incommensurate superlattice.

On the other hand the Tl-Ca-Ba-Cu-O system has at least five stable, tetragonal phases all of which have comparable superconducting properties [7]. This multiplicity of phases produces a situation whereby strict phase purity may not be required for good superconducting properties. Table I lists the primary phases in these systems. Note that the a-axis parameters are not listed since they are nearly the same for all the phases in either of the systems, provided the subcells of the Bi materials are employed, and this allows epitaxial intergrowths between the phases. For corresponding phases, the Tl-Ca-Ba-Cu-O system has a T_c from 10 to 15K higher than the Bi-Ca-Sr-Cu-O system. This may be a consequence of a better lattice match between the CuO and TlO planes in the Tl system [10].

Table 1. Superconducting Phases in the Tl-Ca-Ba-Cu-O system and Bi-Ca-Sr-Cu-O system.

Phase	c-axis	Tc
$Tl_2CaBa_2Cu_2O_y$	29.4 Å	114K
$Tl_2Ca_2Ba_2Cu_3O_y$	35.8 Å	125K
$TlCaBa_2Cu_2O_y$	12.7Å	103K
$TlCa_2Ba_3Cu_4O_y$	19.8Å	117K
$TlCa_2Ba_2Cu_3O_y$	15.9Å	110K
$Bi_2CaSr_2Cu_2O_8$	30.3Å	80K
$Bi_2Ca_2Sr_2Cu_3O_{10}$ /Pb	36.4Å	110K

Our recent success with the Tl-based materials has led us to focus on these materials. In this paper we briefly summarize our results in this material system in bulk ceramic, thick film and thin film forms. The object of the paper is not to be all inclusive but to express some of the connected aspects of the processing, their relationship to the superconducting properties of all of these materials, and to speculate on future directions for processing.

Materials Preparation

All of the materials investigated are ceramic in nature. In these materials the grain boundaries will frequently dominate the structural and transport properties. The processing parameters must be carefully controlled to produce high quality intragrain material while optimizing the intergrain properties. In the Tl-Ca-Ba-Cu-O system the situation is further complicated by the volatility of the Tl species and the reversible equilibrium between the various Tl species as in equation 1.

$$Tl_2O_3 \leftrightarrow Tl_2O + O_2 \uparrow \leftrightarrow 2Tl + O_2 \uparrow \qquad (eq.\ 1)$$

All the above species are volatile under the processing conditions commonly employed for the Tl-Ca-Ba-Cu-O system. Below we briefly enumerate the synthesis procedures for our best materials to date in this system.

Bulk Ceramic

Bulk ceramics have been prepared by both a solid state sintering process and by melt processing. In all cases the best phase purity obtained has been around 90% by x-ray diffraction. Frequently the best samples from the perspective of transport have been multiphase and resulted from a stoichiometry that was off that for any particular phase.

The materials have all been synthesized from mixtures of the high purity oxides that are stored in an Ar inert atmosphere dry box. The oxides are mixed and ground in an agate mortar and pestle and then sieved to 30 μm. Approximately 12-15 gms. are pressed (5 kbar) as a 4 cm pellet. For the bulk sintered ceramic the pellets are then air sintered for 15 min. at 850 C. They are then reground, pressed and resintered and then oxygen annealed for 12 hours at 850 C with a 5 hour slow cool. A typical starting stoichiometry is $Tl_2Ca_2Ba_2Cu_3O_y$, resulting in predominantly the $Tl_2CaBa_2Cu_2O_y$ and $Tl_2Ca_2Ba_2Cu_3O_y$ phases. Considerable Tl is lost during the air sinter step (20-30 wt. %), while only moderate Tl loss is observed during the oxygen anneal (5 wt. %).

For "melt-processed" materials, the powders are directly loaded into a Pt crucible with a tight fitting Pt lid. The crucible is placed in a vertical tube furnace under oxygen. The temperature is rapidly raised to 950 C held there for 1 hour and then slowly cooled (to 700 C in 12.5 hours and then to 25 C in an other 3 hours). Portions of this melt are then oxygen annealed in the same way as described above for the bulk sintered ceramic. The best superconducting properties to date have resulted from initial stoichiometries of $TlCaBaCu_2O_x$, $Tl_2CaBa_2Cu_2O_y$ and $TlCa_{1.5}Ba_{0.5}Cu_2O_y$.

The morphologies of the sintered and "melt-processed" ceramics were similar and consisted of 10-100 μm grains. In both cases the habit is plate-like with the thin dimension of the plate along the c-axis. The plate-like character was more pronounced in the "melt- processed" materials and 1-3 mm single crystal plates as well as polycrystals could be isolated [7].

Thick Films

Thick films were prepared by screen printing techniques on flat substrates. Tl-rich bulk sintered ceramic of composition $Tl_3Ca_2Ba_2Cu_3O_y$ that had previously been processed as above was ground to a powder and sieved to 20 μm. The powder was mixed in an n-butanol and either screen printed or painted onto various substrates. These included $SrTiO_3$, $Y-ZrO_2$ and MgO. The organic binder was removed with an air presinter at 120 C for 10 min. The films were then air sintered (850 C, 10 min.) and oxygen annealed (750 C, 30 min., 1 atm followed by a furnace cool). Both the sintering and annealing were done in close proximity to bulk Tl-based ceramic of the same composition as the intended final phase to control the volatilization of the Tl from the ceramic thick film. The morphology was typically 10-20 μm grains exhibiting little evidence of melting during sintering.

Thin Films

The preparation of thin films consists of 3 main steps: deposition, sintering and annealing. Deposition is not critical as long as an approximately correct stoichiometry is obtained and the formation of intermetallic phases is prevented. Sintering is crucial, since phase and morphology are established here. In the anneal the superconducting properties are optimized, apparently to a large extent by improving grain boundary properties.

Deposition

To date films have been deposited by sequential electron beam evaporation. Under a slight oxygen overpressure (1-3 x 10^{-5} mbar) pure metals are evaporated onto the substrate of choice ($SrTiO_3$, $Y-ZrO_2$, MgO or sapphire with and without a coating of $Y-ZrO_2$). Evaporation is done in layers starting with Cu and ending with Cu to encapsulate the structure. Films have been prepared of 0.2 μm, 0.7 μm and 1.5 μm using 9, 25 and 49 layers respectively. The average layer thicknesses and order are Cu 159Å, Ba 675Å, Ca 450Å, and Tl 298Å. The substrate temperature was kept below 50 C. The resulting partially oxidized structures are very sensitive to oxygen and moisture and are consequently stored in a dry box before sintering and annealing.

Sintering

The sintering process is the most critical one to the formation of high quality films. In this process the film is fully oxidized and the Tl-based phases synthesized. The morphology of the films is also established at this time. The existence of liquid Tl phases, perhaps

Tl_2O, is crucial to the formation of high quality grain boundaries as reflected in transport measurements. To date sintering in pure oxygen has not produced films with superconducting properties as good as those that were air sintered, perhaps because of the suppression of liquid phase formation. For the 0.7 μm films the anneal time currently being employed is 7 min. at 850 C over bulk ceramic in a sealed Pt crucible . Figure 1 shows the transition temperature as a function of air sinter time. As the data show there is a window where reasonable T_c's are found but there is clearly considerable irreproducibility in the process. A significant part of this is the difficulty of reproducibly sintering for short times in conventional furnaces. To this end rapid thermal annealing is being investigated. In addition slight compositional or Tl vapor pressure differences can make a significant difference in the chemistry during sintering.

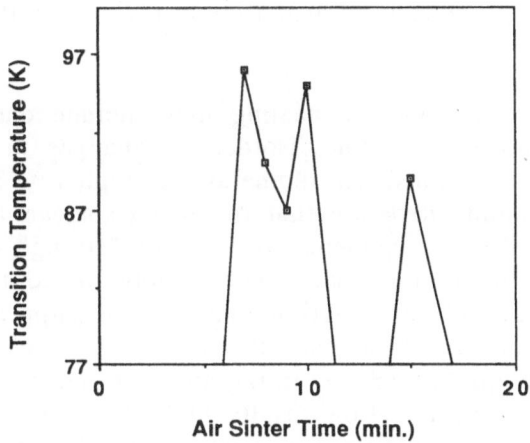

Figure 1. Illustrates the changes in transition temperature for various air sinter times for a 0.7 μm $Tl_2Ca_2Ba_2Cu_3O_y$ film on cubic zirconia. Only the points shown produced T_c greater than 77K, all other times produced lower T_c's.

The morphology of the films appears to be a function of the film thickness for this deposition technique. For very thin films (0.2 μm) the habit is polycrystalline highly oriented or epitaxial growth on $SrTiO_3$ substrates with 10-20 μm grains [11]. Films of this superconductors with current loops during static magnetization experiments much greater than 1 grain diameter [6]. The nature of the grain boundaries responsible for the strong links is currently not thickness are also oriented on other substrates but do not have comparable superconducting properties. Films 0.7 and 1.5 µm thick are randomly oriented polycrystalline aggregates [12]. Other films in the Tl-Ca-Ba-Cu-O system deposited by sputtering have shown orientation up to 1 µm thick [13-14]. This may be a factor of the nucleation process in the different films. The sputtered films may nucleate predominantly at the substrate or the film surface while the sequential e-beam films appear to nucleate at all of the multiple interfaces in the layer structure. While in general it is very difficult in this multiphase system to produce single phase materials, the thin films are typically 80% or higher phase pure. This is a consequence of the control of the stoichiometry and the effects of the substrates.

Annealing

Although the films are superconducting after the air sinter, they have substantially reduced T_c's by both magnetic and transport measurements. Oxygen anneals appear to substantially improve the superconducting properties while not substantially affecting the morphology or stoichiometry. Typical anneals are 15 min. (0.2 µm films) and 30 min (0.7 and 1.5 µm films) at 750 C followed by a furnace cool.

The effects of oxygen annealing may include changes in strain in the films, removal of cation disorder, or changes in the number of oxygen vacancies. Transition temperatures typically increase by 10K. The key points here are that the materials can be either weak (strong magnetic field dependence of J_c) or "strong" (small magnetic field dependence of J_c) linked. The strongly linked films, where the grain boundary Josephson junctions have been suppressed, exhibit significantly enhanced critical currents (Figure 2). In almost all of the materials the quality of the intragrain material is high as evidenced in the magnetization results [16]. The primary factor affecting the superconducting transport is the intergrain properties and like many ceramic materials they are a strong function of the processing of the material.

Materials Properties

Table II below summarizes some of the critical properties for the materials discussed above.

The distinguishing characteristic in our thin films is the concentration of Tl in excess of the phase stoichiometry. Films deficient in Tl are weak linked while the strong linked films to date have had excess Tl. The films were all processed identically as specified above. The magnetization results also support the strongly linked nature of these materials showing that the films are bulk understood. They seem to occur only under circumstances where liquid Tl phases would be present, either by melt processing or through the presence of excess Tl. This is also reflected in the bulk materials where the "melt-processed" ceramics exhibit higher critical currents and a weaker field dependence of J_c than the normal air sintered ceramics. The complex nature of the Tl-O system will make developing reproducible processing protocols difficult.

Table II. Superconducting Properties for Various Materials in the Tl-Ca-Ba-Cu-O system.

Material	Phase	Highest T_c	Best J_c (A/cm^2) (76K)	Weak Links
Sintered Bulk Ceramic	$Tl_2CaBa_2Cu_2O_y$ & $Tl_2Ca_2Ba_2Cu_3O_y$	124K	2000	yes
Melted Bulk Ceramic	$Tl_2CaBa_2Cu_2O_y$ & $Tl_2Ca_2Ba_2Cu_3O_y$	120K	6000	partial
Thick Films	$Tl_2CaBa_2Cu_2O_y$	103K	1200	yes
Thin Films 0.2µm	$Tl_2Ca_2Ba_2Cu_3O_y$	97K	160,000	no
0.7µm	$Tl_2Ca_2Ba_2Cu_3O_y$	111K	240,000	no
0.7µm	$Tl_2CaBa_2Cu_2O_y$	97K	110,000	no
1.5µm	$Tl_2Ca_2Ba_2Cu_3O_y$	105K	_____	yes

While the processing is complicated and difficult to make systematic at the present time, our experiments indicate that the materials are more complex than low temperature superconductors in other ways as well. Figure 3 compares the high-field magnetization to 5T at 5 and 77K for a cut plate of sintered ceramic containing primarily $Tl_2Ca_2Ba_2Cu_3O_y$. The solid triangles represent increasing field strength (following zero field cooling) while the open triangles show decreasing fields. There is no measurable hysteresis above 0.5 T at 77K, while the open loop at 5K suggests a more strongly pinned type II superconductor. Note that the 5K and 77K magnetization scales differ by a factor of 20x. These data are representative of that obtained for thin and thick films as well as single crystals. They indicate that there is very little intragranular pinning at 77K in all of the Tl materials and phases. The critical currents predicted from magnetization data are thus much smaller than the values measured by direct transport. While the agreement is much better at 5K, there is still a substantial difference. For thin films, for example, the magnetization predicts a J_c of 1800 amps/cm^2 at 77K, a factor of 60 below the measured transport critical current.

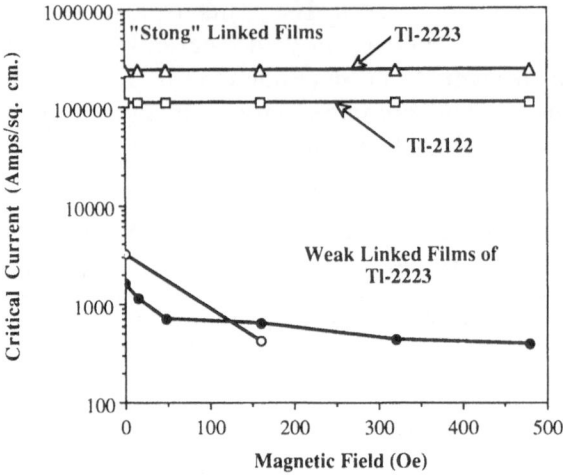

Figure 2. J_c vs in-plane field for a number of thin films of $Tl_2CaBa_2Cu_2O_y$ or $Tl_2Ca_2Ba_2Cu_3O_y$ on $SrTiO_3$ and Y-ZrO_2 showing that the nature of the intergranular links can vary substantially in these materials.

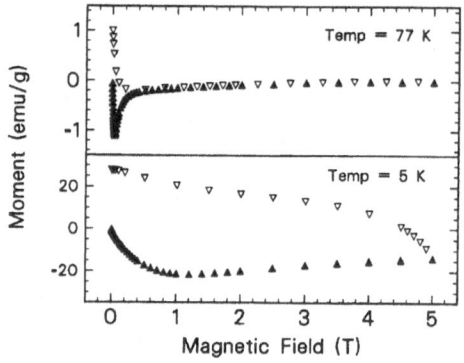

Figure 3. Magnetic moment versus field for a bulk ceramic sample of $Tl_2Ca_2Ba_2Cu_3O_y$. The upper curve is for 77K and the lower for 5K, the lower scale is 20x the upper.

This apparent discrepancy is understandable in the context of the weak pinning observed, since even completely reversible (soft) type II superconductors have non-vanishing critical currents due to the pair-breaking requirement. Thus the critical state model is strictly applicable only to highly hysteretic (hard) superconductors. The observation of substantial critical currents with very little pinning may mean that the introduction of pinning sites through doping or damage may substantially increase J_c [17].

Summary

The properties of all of the Tl materials investigated are quite similar. Their superconducting properties are dominated by intergrain transport. Most of the materials are multiphase due to the complex nature of the Tl-Ca-Ba-Cu-O system and the existence of numerous superconducting phases with similar thermodynamic stabilities. The existence of these multiple superconducting phases may actually be beneficial for producing high quality superconductors from precursors of variable stoichiometry. The bulk intragrain properties of the materials examined are quite good. The grain boundaries however can act as "strong" or weak links. The strength of these links is a strong function of processing, and the liquid phase formation and excess Tl appear to produce strongly linked materials. A comparison of the magnetic and transport properties of these materials indicates that there is very little flux pinning at 77K coexistent with high transport critical current. This suggests a potential lack of applicability of the critical state model to these materials.

The Tl-Ca-Ba-Cu-O system has clearly demonstrated the potential for practical applications. Processing remains a major hurdle in achieving these goals. Two major areas must be addressed. First the optimization of grain boundary properties. This will require a controlled processing of the materials so as to achieve the optimum sintering. To this end more of the Tl-Ca-Ba-Cu-O phase diagram will have to be elucidated and techniques such as rapid thermal annealing developed to minimize Tl loss. Second is the production of pinning centers to maximize the critical currents. This may be accomplished by doping or the introduction of defects in the material. The basic attractions of the Tl-Ca-Ba-Cu-O system remain, but they will only come to fruition with the optimization of processing protocols for the materials.

Acknowledgement

This work at Sandia National Laboratories was supported, in part, by the United States Department of Energy, Office of Basic Energy Sciences, under Contract No. DE-AC04-76DP00789. The technical assistance of T. Castillo and G. Pannell, Jr. is gratefully acknowledged.

References

1. J. G. Bednorz and K. A. Muller, Z. Phys. *B 64*, 189(1986).
2. H. Maeda, Y. Tanaka, M. Fukutomi and T. Asano, Japan J. Appl. Phys. *27*, L209(1987).
3. Z. Z. Sheng, A. M. Hermann, A. El Ali, C. Almasan, J. Estrada, T. Datta and A. Hermann, Nature *332*, 138(1988).

4. M. K. Wu, J. R. Ashburn, C. J. Torng, P. H. Hor, R. L. Meng, L. Gao, Z. J. Huang, Y. Q. Wang and C. W. Chu, Phys. Rev. Lett. *58*, 908(1987).

5. J. F. Kwak, E. L. Venturini, D. S. Ginley, and W. Fu, in Novel Superconductivity, edited by S. A. Wolf and V. Z. Kresin (Plenum, New York, 1987), p. 983.

6. J. F. Kwak, E. L. Venturini, R. J. Baughman, B. Morosin, and D. S. Ginley, Physica C *156*, 103(1988):   and Cryogenics, accepted.

7. D. S. Ginley, B. Morosin, B. J. Baughman, E. L. Venturini, J. E. Schirber and J. F. Kwak, Journal of Crystal Growth, *91*, 456(1988).

8. M. Meada, Y. Tanaka, M. Fukutomi and J. Asamo, Japan J. Appl. Phys., *27*,   2(1988).

9. S. A. Sunshine, T. Siegrist, L. F. Schneemmeyer, D. W. Murphy, R. J. Cava, B. Batlogg, R. B. van Dover, R. M. Fleming, L. H. Glarum, S. Nakahara, R. Farrow, J. J. Krajewski, S. M. Zahurak, J. V, Waszczak, J. H. Marshall, P. Marsh, L. W. Rupp Jr. and W. F. Peck, Phys. Rev. B, *38,* 893(1988).

10. B. Morosin, D. S. Ginley, E. L. Venturini, P. F. Hlava, R. J. Baughman, J. F. Kwak and J. E. Schirber, Physica C-Superconductivity, *152,* 223(1988).

11. D. S. Ginley, J. F. Kwak, R. P. Hellmer, R. J. Baughman, E. L. Venturini and B. Morosin, Appl. Phys. Lett., *53*, 406(1988).

12. D. S. Ginley, J. F. Kwak, R. P. Hellmer, R. J. Baughman, E. L. Venturini, M. A. Mitchell and B. Morosin, Physica C-Superconductivity, *156,* 592(1988).

13. J. H. Kang, R. T. Kampwirth, and K. E. Gray, Phys. Lett. A, *131*, 208(1988).

14. W. Y. Lee, V. Y. Lee, J. Salem, T. C. Huang, R. Savoy, D. C. Bullock and S. S. Parkin, Appl. Phys. Lett. *53*, 329(1988).

15. M. Nakao, R. Yuasa, M. Nemoto, H. Kuwahara, H. Mukaida and A. Mizukami, Jap. J. Appl. Phys., *27*, L849(1988)

16. E. L. Venturini, J. F. Kwak, D. S. Ginley, B. Morosin and R. J. Baughman, Workshop on High Temperature Superconductivity. NIST, Gathersburg, MD, 11-13 Oct. 1988, pg. xxx

17. E. L. Venturini, J. F. Kwak, D. S. Ginley, B. Morosin and R. J. Baughman, Proceedings of Conf. on The Science and Technology of Thin Film Superconductors, R. McConnell ed., Nov. 1988, Colorado Springs, CO pg xxx.

IN-SITU GROWTH OF HIGH T_C THIN FILMS

D. K. Lathrop, S. E. Russek, K. Tanabe[*], and R. A. Buhrman

School of Applied and Engineering Physics
Cornell University
Ithaca, NY

ABSTRACT

A high pressure reactive evaporation process and a high pressure reactive sputtering process have been developed for the growth of high quality thin films of $YBa_2Cu_3O_7$. Both techniques, when used with heated substrates, are effective in the formation of the 123 phase in-situ during the film growth. With reactive evaporation only a cooldown anneal in a higher pressure oxygen ambient is necessary to obtain good superconducting properties. For the reactive sputtering process, the best results are obtained with either a substrate temperature of 720 - 770 C during growth, or lower temperatures during growth and a brief, post growth, rapid thermal anneal. Fully epitaxial growth has been achieved with single crystal MgO substrates. The resultant films, which can be quite smooth and uniform, have been patterned to micron and submicron dimensions and the transport properties of these microstructures have been examined.

INTRODUCTION

Since the discovery of high temperature superconductivity (HTS) in layered copper-oxide based systems[1,2] there has been widespread strenuous efforts to produce thin films of these materials. Essentially all known techniques of thin film synthesis have been applied to this very challenging problem with varying, and improving, degrees of success[3-9]. In this paper we describe recent results on the growth of $YBa_2Cu_3O_7$ thin films that we have obtained with such techniques, high pressure reactive evaporation (HPRE) and high pressure reactive sputtering (HPRS). We have found that both methods when used with carefully heated substrates in the range of 560 to 750 C can result in the in-situ formation during the deposition process of the 123 phase of this material. For the HPRE films a simple "cooldown" anneal in 20 Torr O_2 gives fully oxygenated material with good superconducting properties. HPRS films deposited at above ~700 C show sharp superconducting transitions as deposited, while we have found that a post-deposition rapid thermal oxygen anneal, typically at 850-900 C, for 2 min. is required to produce good superconducting properties on HPRS films deposited with substrate temperatures below 700 C.

For (100) single crystal MgO substrates in the higher temperature range both techniques result in the growth of HTS films with a very high degree of c-axis orientation normal to the plane of the film. For the HPRS process the degree of orientation is particularly high with x-ray pole figures and RBS ion channeling data showing complete epitaxy between film and substrate.

The best HPRE films show sharp resistive transitions with zero resistance between 80-87 K. At 77 K the films on yttria stabilized zirconia (YSZ) have critical current densities J_C > 2 x 10^5 A/cm^2. Films on MgO have lower critical current density at 77 K yet still have J_C values between 2 x 10^6 A/cm^2 and 5 x 10^6 A/cm^2 at 4.2 K, values similar to those obtained with YSZ. The best HPRS films on MgO have a T_C with zero resistance between 75 and 80K. This lower T_C is due to imperfect stoichiometry and large lattice strains in the sputtered films.

The HPRE and HPRS films are comparably smooth and can be readily patterned by conventional photolithography techniques to micrometer and submicrometer dimensions. Ohmic contacts ($\leq$ 3 x 10^{-8} Ω cm^2) can be readily made to the films.

In this paper we review the basic HPRE and HPRS processes and discuss the properties of the resultant films as revealed by various analytical and transport measurements. Emphasis is given to the results of X-ray and transmission electron microscopy studies of film orientation, epitaxy and film-substrate interactions. We also discuss the patterning of the films and describe the transport and "Josephson-like" properties of the resultant microstructures.

THIN FILM GROWTH: HIGH PRESSURE REACTIVE EVAPORATION

The high pressure reactive evaporation process has been described in some detail elsewhere[10]. Briefly, in our current implementation of this technique metallic Y and Ba are evaporated from two electron beam evaporation sources, while metallic Cu is evaporated from a resistively heated thermal source. The three individual deposition rates are controlled by quartz crystal monitors to yield as well as possible the desired 1:2:3 stoichiometry. As discussed below this is the most difficult aspect of the HPRE process.

To obtain the desired 123 phase during evaporation the oxygen pressure is maintained at a constant 0.65 mTorr, which is near the highest pressure at which the electron beam sources will operate reliably.   After the deposition is completed, the chamber is backfilled with oxygen to ~20 Torr while the film cooled. Typical elapsed time in cooling from 650 C to below 200 C is about 20 minutes.

Since in this HPRE process we find that the sticking coefficient of the metallic vapor varies strongly with substrate temperature, to achieve the best reproducibility, the tempera-

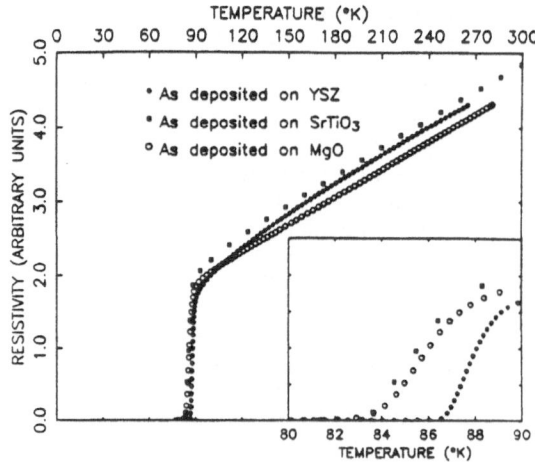

Figure 1. Resistivity vs. temperature for HTS films deposited in-situ on
YSZ, SrTiO$_3$, and MgO substrates.

ture is measured by a chromel-alumel thermocouple embedded in a small gold ball which is bonded to the surface of the substrate. The differences between the temperature measured in this way and the temperature measured on the heater assembly is found to vary widely, depending on the substrate material, the condition of the back surface of the substrate, and how tightly the substrate is clamped to the heater. For MgO substrates with the back surface polished a difference of 10 - 30 C is typical.

Examples of the resistive transitions that we have obtained with the HPRE process are shown in Figure 1. With the use of a cool-down anneal immediately after the deposition good results have been obtained with oriented $SrTiO_3$, YSZ and MgO substrates without a post deposition anneal. The best films produced on YSZ have typical zero resistance T_C's of 83-86 K and room temperature resistivities (ρ_{rt}) of 0.5-3 mΩ-cm. On $SrTiO_3$ a film with T_C = 81 K and ρ_{rt} = 2.5 mΩ-cm is obtained, and on MgO T_C's of 81-83 K are typical.

Generally the HPRE process is used with substrate temperatures in the 625-650 C range. This has yielded our best results, such as those illustrated in Figure 1. We have explored the effects of alternative growth temperatures. Films grown at temperatures as low as 560 C exhibit the 123 phase and superconducting properties, in that a resistive transition is observed, but this transition is broad, not reaching zero resistance until below 30 K. Films grown at substrate temperatures of 580 C are somewhat improved, but not until the substrate temperature during deposition was raised to 600 - 625 C could well oriented films be obtained with sharp resistive transitions.

We find that variation of the substrate temperature for HPRE is a considerable problem since as indicated above we find that the composition of the resultant film changes rapidly if T is varied while the evaporation rates are held constant. On YSZ substrates, temperature gradients of about 20 C can exist across the 1 cm width of the substrate. The Cu/Y ratio can vary from 3.07 to 2.86 over 4 mm, decreasing in the direction of a positive temperature gradient, showing that the sticking coefficient of Cu is a function of the substrate temperature. This observation is consistent with our general observation that for films grown at lower substrate temperatures, the Cu deposition rate must be decreased to produce a film with a stoichiometry close to the desired 3:1 Cu:Y ratio.

THIN FILM GROWTH: HIGH PRESSURE REACTIVE SPUTTERING

Reactive sputtering is a technique that has already seen widespread application and success in HTS thin film research[5,9]. Recently Geerk et al.[11] have demonstrated that a

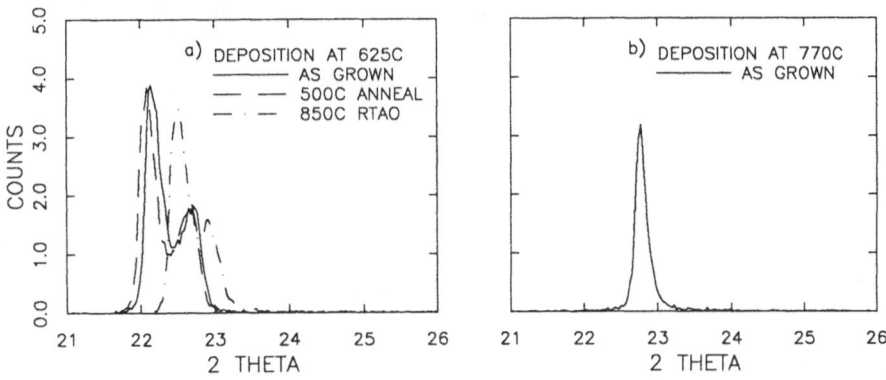

Figure 2. a) X-ray diffraction 2θ scans taken on a mixed orientation HPRS film grown at at 625 C showing the (003) and (100) peaks. After a 2 minute RTOA step the previously expanded c-axis lattice constant contracts from 12.0 Å to 11.80 Å. Also shown is the effect of a longer anneal at 500 C. b) X-ray diffraction 2θ scan of an HPRS film grown at 770 C, only the c-axis is present, and the lattice is not expanded as in a.

comparably high pressure variation of this technique is effective in the in-situ formation of the superconducting 123 phase. We have pursued investigations into the effectiveness of a similar high pressure reactive sputtering technique. Details of this work will be presented elsewhere[12], but the basic process is implemented as follows: A standard 2 inch rf magnetron sputter source is employed with a sintered target of $YBa_2Cu_{4.2}O_x$ composition. During the plasma discharge the system pressure is regulated to a value somewhere in the 250 - 350 mTorr range with a O_2/Ar ambient. Typical discharge parameters that have proven successful are O_2/Ar ratios: 0.1 - 0.23, flow rates: ~ 10 sccm, and rf power density : ~ 5 W/cm^2. This results in a typical deposition rate of 9 nm/min for a target to substrate distance of 45 - 55 mm. During deposition the substrate temperature is typically held at 625 to 775 C.

Films produced on MgO substrates by HPRS are quite uniform in composition over a 1 cm substrate, as measured both by electron microprobe and by resistance measurements on microfabricated test structures across the substrate (see below). Films deposited with the substrate temperature held between 625 and 675 C are generally smooth in appearance but do not have good transitions though they generally are metallic and do exhibit the 123 phase when examined by X-ray diffraction, albeit with a considerably larger than normal c-axis lattice spacing, c = 12.0 Å. In general the c-axis spacing can be readily reduced and the films made to exhibit fairly sharp transitions by the application of a brief rapid thermal oxygen anneal (RTOA) step, which typically involves heating the film to 850 - 900 C for 30 - 120 seconds. The pronounced change in lattice constant that results from such an RTOA process is illustrated in Figure 2a. There the c-axis lattice constant decreased from 12.0 Å to 11.80 Å during a 2 min RTOA at 850 C. Also the resistivity of the annealed film decreased by a factor of 2 - 5 as result of the anneal. We note that the annealed film is still considerably strained since the lattice constant is still greater than the 11.72 Å value measured in the HPRE films and the 11.66 Å value measured in bulk material. Longer anneals such as 1 hr at 500 in O_2 do not relax this lattice strain implying that it is not principally due to O deficiency.

In contrast, films grown with higher substrate temperatures, 725 - 775 C, are matte in appearance, indicating surface roughness. These films do exhibit sharp superconducting transitions as deposited, although the transition temperature is depressed, not reaching zero resistance until 75 - 80 K. They also do not exhibit the expanded lattice constant of the films deposited at lower temperature, instead the x-ray diffraction data shows a c-axis lattice constant of 11.72 Å for the as deposited films (Figure 2b), similar to that measured in the HPRE films.

In the sputtered HTS films we find the deposited film stoichiometry to be a strong function of gas composition, pressure and substrate temperature. Indeed it is by the variation of the gas mixture that we vary the film composition in our attempts to obtain the desired 1:2:3 stoichiometry. The strongest effects are shown in Figure 3. By varying the substrate

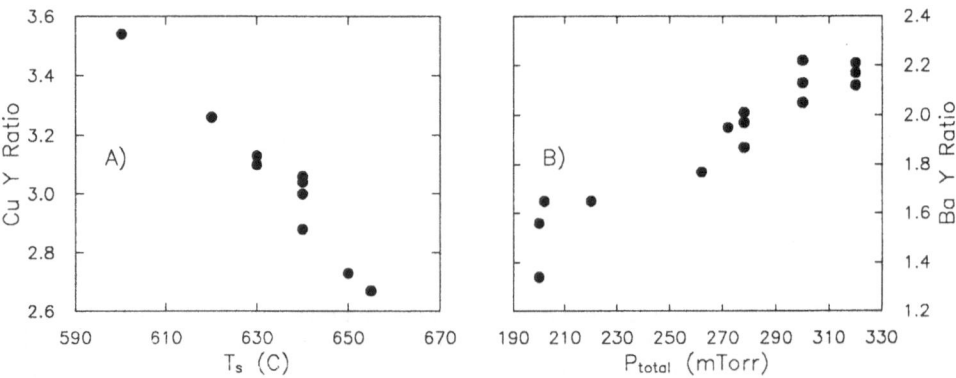

Figure 3. a) Cu Y ratio vs. substrate temperature during growth for HPRS films. b) Ba Y ratio vs. total pressure for HPRS films.

temperature from 600 C up to 660 C, the Cu to Y ratio varies from 3.55 down to 2.65, indicating that, as is also the case in the HPRE films, the Cu sticking coefficient decreases with increasing temperature. In Figure 3b, it can be seen that varying the total pressure from 200 to 300 mTorr has a strong effect on the amount of Ba in the resulting film.

In Figure 4a we show the resistive transition of a 123 film produced by HPRS as deposited and following an RTOA step. This film was deposited at a substrate temperature of 655 C onto a (100) MgO substrate. After the anneal the transition moves to higher temperature, becomes sharper, and begins to exhibit metallic behavior above T_C. In Figure 4b, The RTOA step is seen not to be necessary if the film is deposited at a higher substrate temperature (750 C). But in both cases, these transitions are considerably depressed from the bulk value and, as discussed below, the critical current density of these films, particularly in the region just below T_C, is comparably low. We attribute this to the less than ideal stoichiometry and, perhaps in part, to the still slightly greater than typical c-axis lattice spacing (11.72 - 11.80 Å) of these films, even after the RTOA step. The origin of this expanded lattice constant for the HPRS films is under investigation. It may be due to strain created by the very high pressure ambient during deposition, or by the strong epitaxy (see below) that we observe between the rather mismatched 123 film and MgO substrate (~10% lattice misfit).

THIN FILM ORIENTATION AND EPITAXY

Orientation of both the HPRE and HPRS 123 thin films have been examined with a thin film x-ray diffractometer operating in a 2θ mode, by rocking curve and full pole figure X-ray measurements and by selected area electron diffraction and cross-sectional transmission electron microscopy studies[13]. X-ray 2θ scans reveal that all of the films deposited on YSZ substrates exhibited some degree of mixing between c-axis normal grains and a-axis normal grains, varying from approximately equal amounts of the two growth habits to predominantly c-axis normal growth. Again this variation appears to be strongly correlated with the growth temperature. As T_{sub} is increased to ~ 625 - 650 C the c-axis orientation begins to dominate. These results are confirmed by SAD patterns[10], which also give information about the orientation of the films in the plane of the substrate and whether the film growth is epitaxial to the substrate. For the films grown on YSZ, the areas exhibiting c-axis normal growth are found in most cases not to be well oriented in the plane of the substrate, i. e. the growth is not epitaxial. The regions of a-axis normal growth, however, are registered with the substrate.

In contrast, the 2θ scans of HPRE films grown on MgO substrates were almost entirely composed of the (00k) peaks characteristic of the c-axis normal growth, with an almost, but

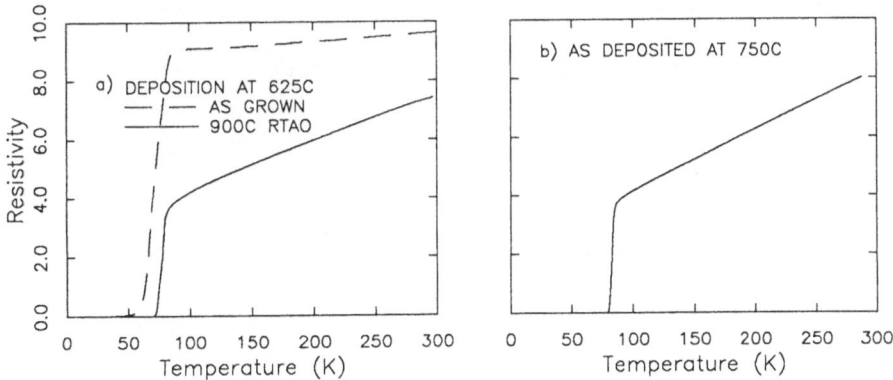

Figure 4. a) Resistive behavior of a HPRS film grown on (100) MgO at 655 C before and after a RTOA treatment. b) Resistive behavior of a HPRS film grown on (100) MgO at 750 C, no anneal was needed to produce this transition.

not quite complete, absence of the (k00) peaks from a-axis normal grains. Furthermore, cross-sectional TEM micrographs of these films showed large areas of clear epitaxial growth with a very clean interface between the substrate and the Cu-O planes of the $YBa_2Cu_3O_7$ lattice. This result is illustrated by the cross sectional TEM micrograph shown in Figure 5.

Even stronger epitaxial growth was observed on MgO substrates with HPRS films. Full pole figure X-ray scans reveal that the superconducting 123 phase and the MgO substrate are completely aligned. No evidence of any other orientation could be detected in this film. This result is supported by preliminary Rutherford backscattering ion channeling experiments, in which the backscattering yield is decreased by a factor of ~1.8 when the irradiating beam is aligned with the MgO substrates crystalline axis. SAD and high resolution cross section TEM studies of these films are now in process and will be reported elsewhere.

THIN FILM UNIFORMITY AND SUBSTRATE REACTION

Auger depth profile and Rutherford backscattering spectroscopy (RBS) have been generally used to investigate the uniformity of the HPRE and HPRS deposited films and the possibility of substrate-film reaction. To the resolution of either technique no evidence for substrate reaction or film non-uniformity have been obtained. However cross-section TEM studies of HPRE films grown on YSZ substrates do reveal the existence of a ~ 5 nm amorphous layer between the substrate and the film. High resolution (~ 1nm) energy dispersive STEM electron microprobe measurements of this interfacial layer reveals the presence of a significant amount of Zr in this amorphous layer and in a portion of crystalline film immediately above the amorphous layer. Since the 123 film has a strong epitaxial relationship to the substrate we suspect that this amorphous layer is formed by out-diffusion from the substrate to the initially crystalline bottom portion of the film, with this out-diffusion occurring gradually during the growth process. As is illustrated in Figure 5 no such interfacial layer is observed for films grown on MgO.

HTS THIN FILM PATTERNING AND PROCESSING

The HPRE and HPRS $YBa_2Cu_3O_7$ films can be readily patterned using conventional

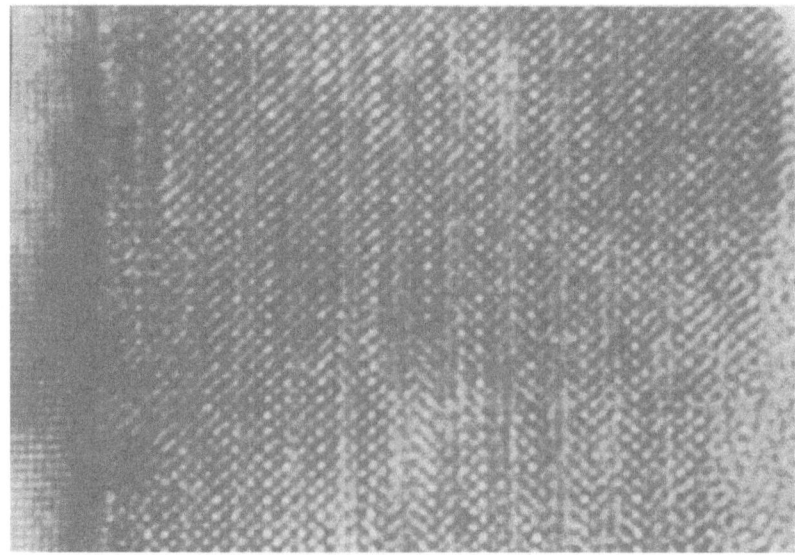

Figure 5. Cross-sectional TEM micrograph of an HPRE film on MgO illustrating the clean film substrate interface.

photolithographic methods and then etched using argon ion milling. Extensive patterns of lines and constrictions with widths ranging from 0.5 µm to 100 µm have fabricated to measure critical current densities, orientation dependence of epitaxial thin film resistance and critical current densities, thin film uniformity and weak link properties. We have also fabricated 5 mm long co-planar transmission lines to study the propagation of ultra fast pulses in high Tc materials[14].

The dense fine grain structure of these films allows for uniform etching and electrical properties down to micron line widths. We have found no degradation in critical current densities going from neighboring 100 µm to 1 µm lines. For HPRE films we find that while critical currents of adjacent lines are very uniform there is usually a large variation of critical current densities across the 12 mm substrates. The best films have critical current densities which vary by a factor of two across the wafer. As noted above uniformity results for the HPRS films deposited on MgO substrates are typically much better. Resistivity variations of less than 20% across a 5 mm sample is typical.

In a process similar to that reported by others[15] low resistance contacts to the $YBa_2Cu_3O_7$ films and test structures are made by depositing Ag after a short argon-oxygen ion mill cleaning of the film's surface. Typically the ion energy is 500 eV. If no further processing is performed the resultant contacts are ohmic but fairly resistant ($\leq 5 \times 10^{-5}$ Ω-cm^2). If the contact is subsequently thermally annealed at 750 C in oxygen for two minutes the contact resistance typically drops to less than 5×10^{-8} Ω-cm^2.

As illustrated by Figure 6 we have found that conventional microfabrication processing which includes solvent cleans, low temperature bakes, and/or thin film depositions in vacuum will not noticeably change the superconducting properties of the $YBa_2Cu_3O_7$ films. The effects of such processing on the surface properties of the films are currently being studied. Rapid thermal anneals in oxygen at temperatures of 700 C to 850 C have slightly improved or slightly degraded the properties of the HPRE films depending on the sample, while RTOA anneals of the HPRS films deposited at T < 700 C is currently required for good results. Much longer anneals at 700 C to 900 C for either class of film cause degradation of the film properties. We have not found extended low temperature (T $\leq$ 550 C) anneals to be efficacious.

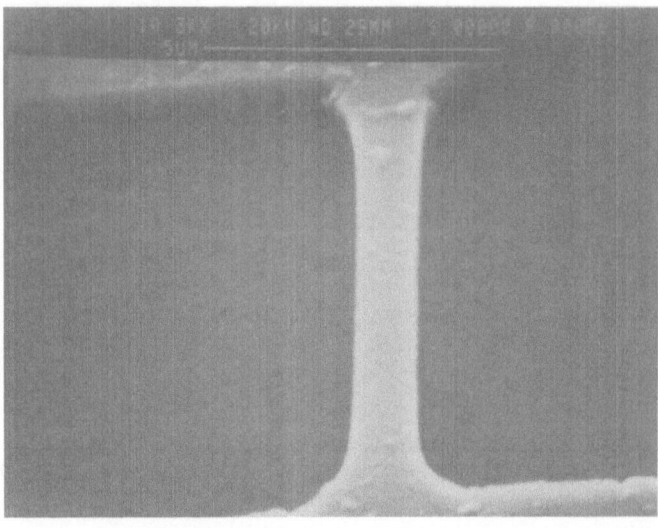

Figure 6. Scanning electron micrograph of a 2 µm constriction patterned into a smooth HTS film grown on MgO by HPRS.

CRITICAL CURRENT AND WEAK LINK BEHAVIOR

Typically, HPRE films on YSZ with a zero resistance Tc of 86 K produce critical currents of 2×10^5 A/cm² at 77 K, with values exceeding 10^6 A/cm² at 4.2 K. On MgO, measurements on both 5 μm and 20 μm width constrictions show that the critical current at 77 K is somewhat lower than that obtained on YSZ, down to 5×10^4 A/cm², but it also rises to reach about 10^6 A/cm² at 4.2 K. This depression of J_C at 77 K is expected due to the lower value of T_C for the films on MgO substrates, 82 K vs. 86 K. In the temperature region near T_C the critical current is clearly limited by weak links between adjacent grains and thus rises slowly below T_C. For the best HPRE films fabricated into 2 μm test structures, J_C values of 5×10^6 A/cm² have been obtained at 4.2 K. J_C vs. T data generally could not be obtained for such films as the microstructure tends to fuse during the first excursion to the normal state when J_C is in this range.

While T_C is further suppressed and the transition is broader for the HPRS films grown on MgO the low temperature critical current density for the better films is still of the order of 5×10^6 A/cm². At higher temperatures, nearer T_C, the reduced critical current densities are presumably caused by weak links between grains due to lack of stoichiometry. We have not observed any clear Josephson effects in the thin film microstructures down to the 1 μm size. Such Josephson effects have been observed in 1μm lines with obvious structural defects as observed by SEM. These defects are caused by electrical stress during testing.

An example of such an effect is shown in Figure 7 where the I-V characteristic of a long 1 μm wide constriction is shown in the presence and absence of 10 GHz microwave radiation. The presence of the Josephson constant voltage steps in the irradiated I-V is clear. Of course the I-V characteristic of the constriction does not exhibit the form typical of a good Josephson weak link, and the effective $I_C R_n$ product of the weak link, ~ 20 μV, is far below the value to be expected from a good high T_C weak link, > 10 mV.

SUMMARY

We have found HPRE to be a successful technique for the in-situ formation of $YBa_2Cu_3O_7$ thin films with good superconductive properties. For best results the technique

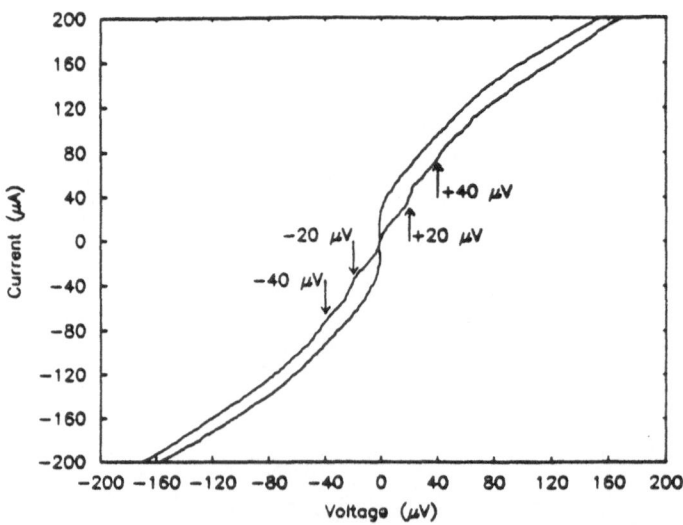

Figure 7. Current-voltage characteristic in the presence and absence of 20 GHz microwave radiation as measured at T = 50 K for a defect induced weak link in a 1 μm wide line. The structure was fabricated in a 0.4 μm thick HTS film grown on MgO by HPRS.

does require precise control of substrate temperature and, of course, film stoichiometry. HPRS has also been demonstrated as being able to provide in-situ formation of the 123 phase although a post-growth rapid thermal anneal step is currently required to obtain the best superconductive properties. When used with YSZ and MgO substrates heated above 625 well oriented film growth is obtained, with (100) MgO substrates providing the best c-axis normal orientation results. HPRS has been found to be particularly effective in yielding complete epitaxy on MgO and in yielding very uniform films. Further improvements in HPRS film stoichiometry should make this a very attractive process.

The HPRE and HPRS thin films are relatively smooth and are readily patterned by photolithography and inert ion beam etching procedures to micrometer and sub-micrometer dimensions. In the less ideal HPRS films ac Josephson weak link effects have been observed in such micro-constrictions in a broad temperature range below T_C.

ACKNOWLEDGEMENTS

We wish to thank J. Silcox and D. H. Shin for providing us with the results of their high resolution STEM electron microprobe studies of the HTS thin films. This research was partially supported by the Office of Naval Research (N00014-85-K-0296), by the Defense Advanced Research Projects Agency (N0014-88-K-0374) and by the National Science Foundation through the use of the National Nanofabrication Facility (ECS-82-00312) and through use of the central facilities of the Cornell Materials Science Center (DMR-85-16616).

REFERENCES

* Permanent address:
NTT Optoelectronics Laboratories
Tokai, Ibaraki 319-11 JAPAN

1. C. W. Chu, P. H. Hor, R. L. Meng, L. Gao, Z. J. Huang, and Y. Q. Wang, Evidence for superconductivity above 40°K in the La-Ba-Cu-O compound system, Phys. Rev. Lett., 58:405 (1987).

2. M. K. Wu, J. R. Ashburn, G. J. Torng, P. H. Hor, R. L. Meng, L. Gao, J. J. Huang, Y. Q. Wang, and C. W. Chu, Superconductivity a 93K in a new mixed phase Y-Ba-Cu-O compound system at ambient pressures, Phys. Rev. Lett., 58:908 (1987).

3. P. Chaudhari, R. H. Koch, R. B. Laibowitz, T. R. McGuire, and R. J. Gambino, Critical-current densities and transport in superconducting $YBa_2Cu_3O_{7-x}$ compounds, Phys. Rev. Lett., 58:2684 (1987).

4. B. Oh, M. Naito, S. Arnason, P. Rosenthal, R. Barton, M. R. Beasley, T. H. Geballe, R. H. Hammond, and A. Kapitulnik, Critical current densities and transport in superconducting $YBa_2Cu_3O_{7-\partial}$ films made by electron beam coevaporation, Appl. Phys. Lett., 51:852 (1987).

5. Y. Enomoto, T. Murakami, M. Suzuki, and K. Moriwaki, Largely Anisotropic Superconducting Critical in Epitaxial Grown $Ba_2YCu_3O_{7-y}$ Thin Film, Jpn. J. Appl. Phys., 26:L1248 (1987).

6. P. K. Mankiewich et. al., Reproducible technique for fabrication of thin films of high transition temperature superconductors, Appl. Phys. Lett., 51:1753 (1987).

7. See also for example the numerous papers on this subject in Applied Physics Letters, Vols 51 and 52, 1988.

8. R. L. Sandstrom et. al., Reliable single-target sputtering process for high-temperature superconducting films and devices, Appl. Phys. Lett., 53:444 (1988).

9. R. T. Kampwirth, J. H. Kang, and K. E. Gray, Superconducting properties of magnetron sputtered high T_C films containing oxide compounds of yttrium, bismuth, or thallium, this volume.

10. D. K. Lathrop, S. E. Russek, and R. A. Buhrman, Production of $YBa_2Cu_3O_{7-y}$ superconducting thin films in-situ by high pressure reactive evaporation and rapid thermal annealing, Appl. Phys. Lett., 51:1554 (1987).

11. J. Geerk et. al., Preparation of Y-Ba-Cu Oxide Superconductor Thin Films by Magnetron Sputtering, presented at the Materials Research Society meeting, Boston, MA, Nov. 30 - Dec. 5, 1987.

12. K. Tanabe, S. E. Russek, and R. A. Buhrman, unpublished.

13. L. A. Tietz, B. C. De Cooman, C. B. Carter, D. K. Lathrop, S. E. Russek, and R. A. Buhrman, Structure of Superconducting Thin Films of $YBa_2Cu_3O_{7-x}$ Grown on $SrTiO_3$ and Cubic Zirconia, Journal of Electron Microscopy Technique, 8:263 (1988).

14. D. K. Dykaar, R. Sobolewski, J. M. Chwalek, J. F. Whitaker, T. Y. Hsiang, G. A. Mourou, D. K. Lathrop, S. E. Russek, and R. A. Buhrman, High-frequency characterization of thin-film Y-Ba-Cu oxide superconducting transmission lines, Appl. Phys. Lett., 52:1444 (1988).

15. J. W. Elkin, A. J. Panson, and B. A. Blankenship, Method for making low resistivity contacts to high T_C superconductors, Appl. Phys. Lett., 51:1753 (1988).

HIGH T_c THIN FILM RESEARCH AT NRL

P. R. Broussard

Naval Research Laboratory
Washington, DC 20375

INTRODUCTION

The potential importance of thin films of high temperature superconductors (HTS) has stimulated numerous approaches to thin film growth. At the Naval Research Lab, film growth of HTS is being studied by several methods: coevaporation[1], flash evaporation[2], chemical vapor deposition[3], along with initial efforts at sputtering. In this paper I would like to summarize some of the work done by the first two efforts.

FILM GROWTH BY COEVAPORATION

Films of the superconductor $Y_1Ba_2Cu_3O_x$ were grown by coevaporation of Y, Cu and BaF_2 in the presence of oxygen, followed by an ex-situ anneal. The advantages of using BaF_2 in place of elemental Ba were pointed out by Mankiewich, et al.[4] and have been substantial in our experience. The details of the film deposition process are given elsewhere.[1] Film thicknesses are typically 5,000 - 10,000 Å after annealing. Our usual annealing procedure was to place the samples into a hot oven at 800 - 900 °C with a flowing gas mixture of O_2 and H_2O (formed by bubbling oxygen through water) for 30 minutes, followed by either a 1 °C/min cool to room temperature in dry oxygen, or a 10 °C/min cool to 550 °C, remaining at 550 °C for 30 minutes, followed by a furnace cool, all in dry oxygen.

Film compositions are determined using elastic backscattering spectroscopy (EBS),[5] which is a variant of RBS but with 6.2 MeV alpha particles instead of the more typical 2 MeV energy. The use of higher energy alpha particles means that some of the lighter elements, such as oxygen and carbon, are no longer Rutherford-like in their backscattering. The higher energy does allow a clearer separation of the signals from the metallic elements, simplifying the composition analysis. The films were also characterized by x-ray diffraction using Cu $K\alpha$ radiation, SEM, optical microscopy, resistivity, and critical current measurements.

We have deposited onto a wide variety of substrates, including (100) SrTiO₃, (100) MgO, (100) cubic zirconia (YSZ), and lithium niobate. Here, I will concentrate on the first three.

Results on Strontium Titanate

Films grown on (100) SrTiO₃ and annealed in the manner described above have very good properties, as shown in our earlier work.[1,6] Figure 1 shows the resistance trace for three samples on SrTiO₃: one on stoichiometry and annealed at 850 °C for 30 minutes (labeled "123"), another with composition Y(20)Ba(30)Cu(50)O$_x$ and annealed as above (labeled Y-rich), and finally one with composition Y(16)Ba(30)Cu(54)O$_x$ annealed at 800 °C for 1 hr and slow cooled. Here the numbers refer to atomic percentages for the metals determined through EBS. The "123" is our best effort to date, with a T$_c$ of 88 - 91 K (complete to onset) and a resistivity ratio (RR) between 300 and 100 K of 2.9. The other films show the effect of compositions deviating from stoichiometry. For the Y-rich film, the onset temperature has remained high, but the complete T$_c$ is now 85 K, with essentially the same RR. For the Cu-rich film, however, the onset T$_c$ is depressed to 82 K, with complete T$_c$ of 80 K, and a larger RR of 3.7. This is consistent with the presence of "248" phase material, identified by Marshall and coworkers.[7,8]

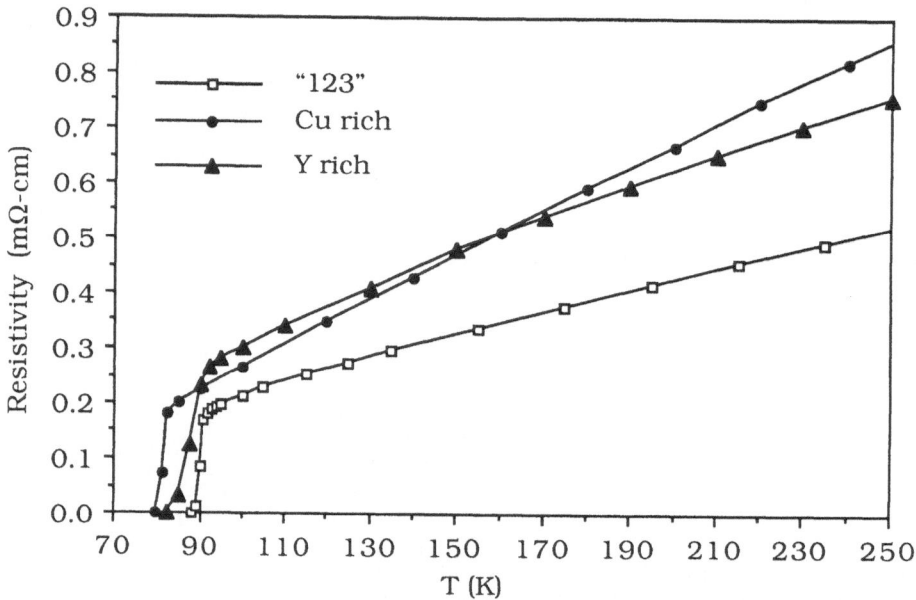

Figure 1. Resistivity vs. temperature for two films grown on SrTiO₃. The film annealed at 800 °C for 1 hour has a composition of Y(16)Ba(30)Cu(54), while the sample annealed at 850 °C for 1/2 hour has a composition of Y(17)Ba(32)Cu(51).

X-ray diffraction for the "123" sample and the Cu rich sample, shown in Fig. 2, show that both samples have the "248" phase along with reflections due to regions of "123" phase material having their a- and c-axes normal to the substrate, along with a third phase. As expected, the Cu-rich sample has a larger fraction of "248" phase, indicated by the ratio of peak intensity between the "248" and "123" reflections. The Y-rich sample (diffraction scan not shown here) only shows peaks from "123" phase material. In Fig. 2a, the on stoichiometric sample also has reflections tentatively identified as Y_2BaCuO_x ("211"). In Fig. 2b, the Cu-rich sample shows evidence for BaF_2, indicating that the wet oxygen anneal was not sufficient to decompose all the deposited BaF_2. This might explain the overall higher resistivity of this sample. Analysis of the reflections normal to the substrate give values for a and c of 3.832 ± 0.004 Å and 11.67 ± 0.02 Å, respectively. The reflections due to the "248" phase can be indexed as (00L) reflections with a lattice constant $c^* = 27.22 \pm 0.02$ Å.

Critical current measurements on these samples have been carried out to examine the dependence on stoichiometry. The details of the measurement have been given elsewhere.[6] Briefly, the samples are patterned after annealing by standard photolithographic techniques and then wet-etched. Here, I only list the results for zero field. For the "123" sample, $J_c(77 \text{ K}) = 3.5 \times 10^3 \text{ A/cm}^2$, and $J_c(4.2 \text{ K}) \approx 10^6 \text{ A/cm}^2$. In contrast, the Y-rich sample had a $J_c(4.2 \text{ K}) \approx 1.8 \times 10^5 \text{ A/cm}^2$. After patterning, this sample's T_c had dropped below 77 K, indicating that off stoichiometric samples are more sensitive to handling. This was also seen in Ref. 6. The Cu-rich sample, with its higher fraction of "248" phase material, has a $J_c(77 \text{ K}) \approx 60 \text{ A/cm}^2$, probably due to its very low T_c (here $t=T/T_c = 0.96$), and $J_c(4.2 \text{ K}) \approx 8 \times 10^5 \text{ A/cm}^2$.

Optical microscopy studies of the films using polarized light allow one to study the degree of grain orientation over a large area of the substrate.[9] Light polarized in the plane of the sample is imaged onto the sample surface and analyzed through a second polarizer. The sample stage can be rotated to vary the angle between the polarization and any preferred direction in the sample plane. Samples with their grains aligned in the plane will show extinction of the light occurring over the entire sample surface at a particular angle. Samples with random orientation in the plane will not show any variation in reflection with sample rotation. Fig. 3a and b show the reflection of a stoichiometric film on $SrTiO_3$ at maximum reflection, and at 45° to the maximum, where the minimum reflection occurs. That the extinction occurs over the entire area (except for isolated spots) shows that the in-plane order extends over the sample surface. This result is consistent with epitaxial growth of the superconductor on the substrate, where the grains can grow matched along either of the (100) axes of the $SrTiO_3$.

SEM studies of the microstructure of on-stoichiometric films, shown in Fig. 3c and d show the presence of a "herring-bone" structure, with rectangular regions of size ≈ 0.5 μm by 3 μm. These regions, which correspond to the a-axis up, keep their relative order over the entire area of the sample, and also show the 4-fold symmetry that has also been seen in x-ray diffraction and optical studies. The deposits on the film surface are probably $BaCuO_2$.

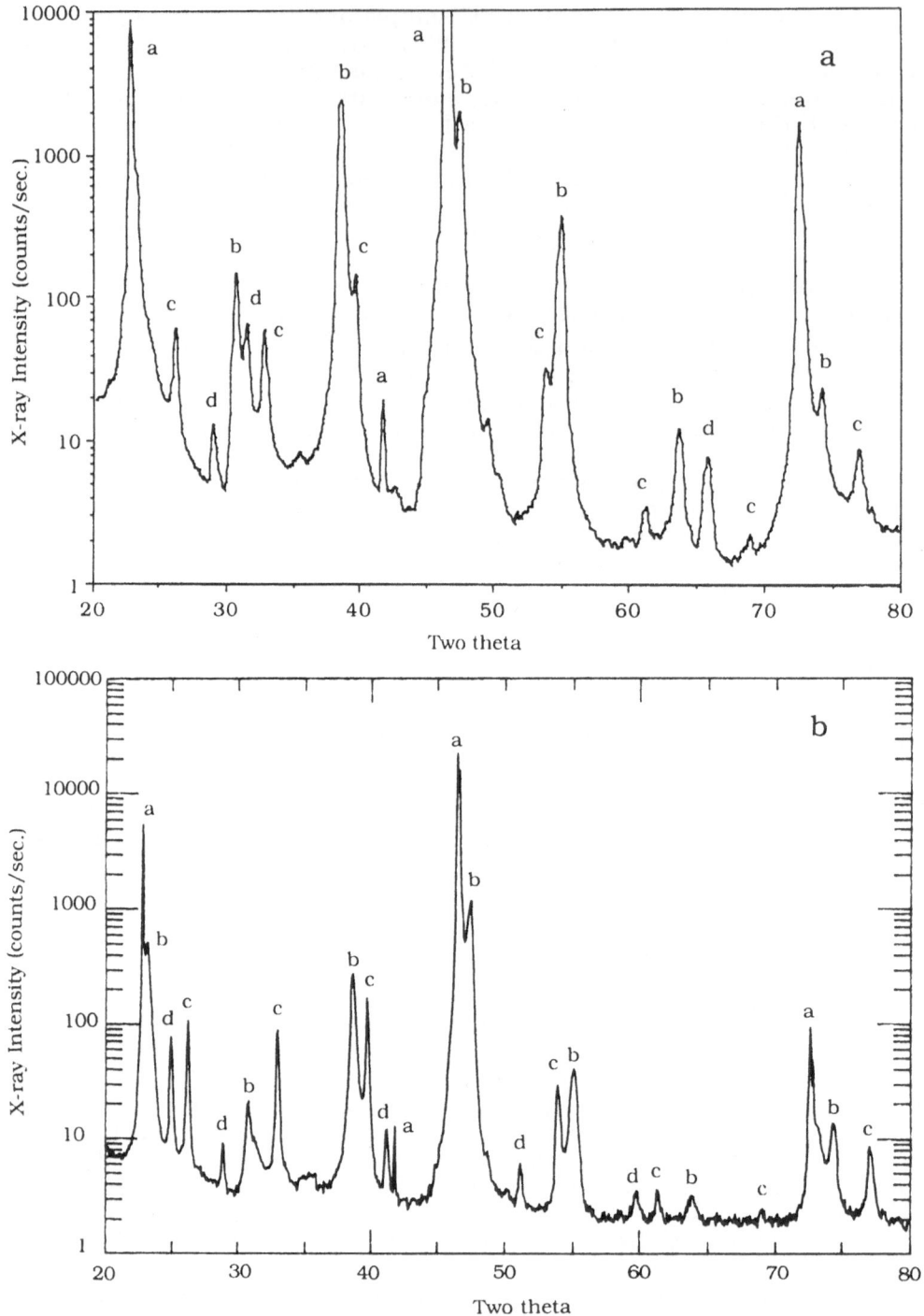

Figure 2. (a) X-ray diffraction scan for a stoichiometric sample on (100) $SrTiO_3$. The labels a, b, c, and d refer to peaks from the substrate, "123" phase material, "248" phase material, and "211" phase material, respectively. (b) Same as (a) for the Cu rich sample in Fig. 1. Here the labels are the same, except d, which refers to peaks arising from BaF_2.

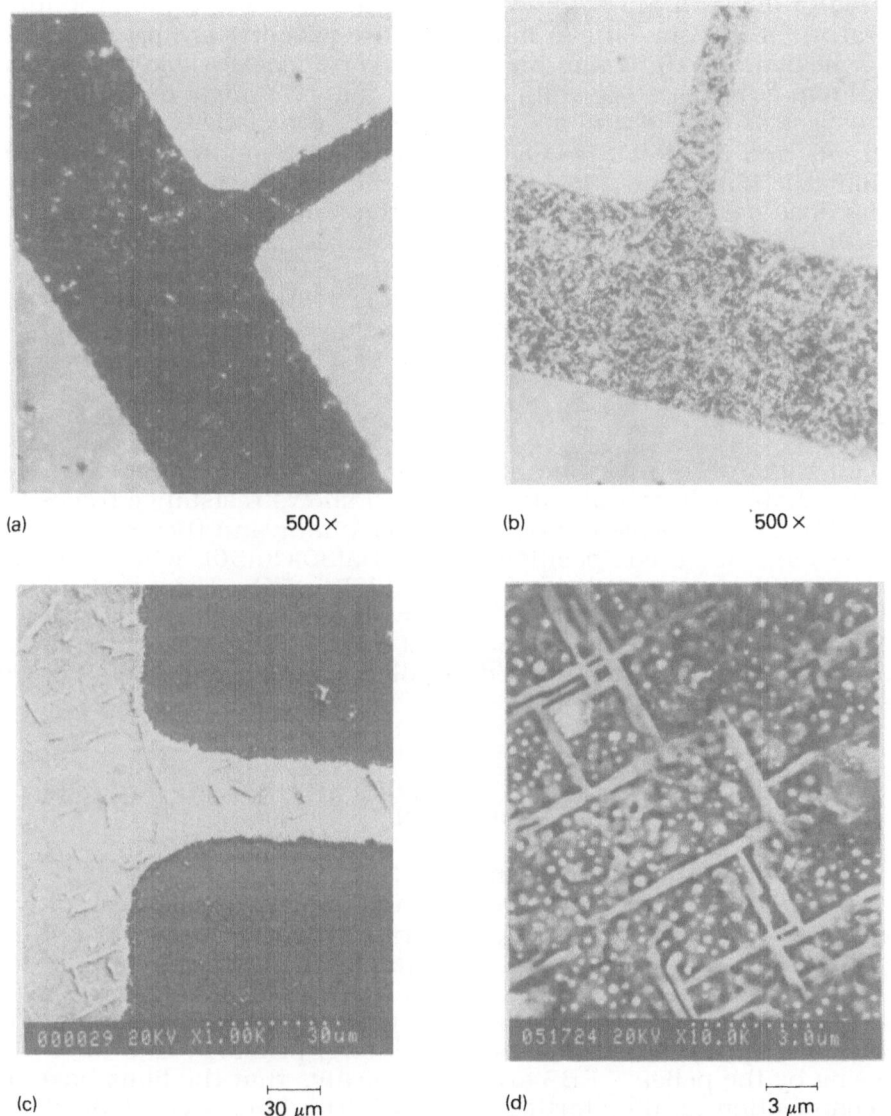

(a)   500×   (b)   500×

(c)   30 μm   (d)   3 μm

Figure 3. (a) and (b) are optical micrographs using polarized light of a stoichiometric film at 2 different angles (45° apart) with respect to the polarization. The near complete extinction across the entire film surface indicates the degree of in-plane order across the film. (c) and (d) are SEM photographs of the same area above at two different magnifications, showing the "herring bone" structure of the film, along with second phase material on the surface. The a-axis up needles are approximately 0.5 μm by 3 μm in size.

Results on MgO

Films deposited on (100) MgO and annealed in the usual manner have higher resistivities and depressed complete transitions as compared to films on SrTiO3.[1] We have found that increasing the initial temperature of the anneal can improve these properties, up to a point. Films annealed at 910 °C and higher have very poor surface morphology, and bad film adhesion. We still find that a 900 °C anneal for 10 minutes in wet oxygen is best, giving a T_c of 80 - 85 K, a resistivity at 100 K of 360 $\mu\Omega$-cm, and a RR of 1.9. Figure 4 shows SEM photographs on two stoichiometric films, one annealed at 900 °C, the other at 850 °C. The first film does not show second phase on the surface, and has a larger grain size than the second, which does shows second phase material on the surface. X-ray diffraction indicates that the films have c-axis growth, but without any in-plane order (fiber texture). There is also evidence for (013) and (026) reflections. Analysis of the (00L) peaks gives a value of c = 11.69 ± 0.01 Å.

Results on YSZ

Our initial work on cubic zirconia has shown it to be much more promising than MgO for BaF_2 use. Figure 5 shows resistance traces for two films on YSZ, annealed at 850 °C for 1/2 hour and then slow cooled. One is Y-rich, with a composition of Y(20)Ba(30)Cu(50), while the other is Cu-rich, with a composition of Y(16)Ba(30)Cu(54). The onset transition temperatures are above 90 K, with our best film having a complete transition by 75 K, with a RR of 1.6. X-ray diffraction indicates that the film is composed of polycrystalline regions along with c-axis textured regions.

FILM GROWTH BY FLASH EVAPORATION

Here films of the bismuth and thallium compounds are being grown utilizing a simple technique. Ceramic pellets of the desired compound, either $Bi_4Sr_3Ca_3Cu_4O_x$ or $Tl_2Ba_2Ca_1Cu_2O_x$, are prepared using standard procedures[10] and cut into pieces of approximately 0.25 gms. These pieces are placed into a Cu electron gun hearth and completely evaporated. The background pressure in the system is 10^{-8} torr, but during deposition it rises to 10^{-5} torr, primarily due to outgassing by the pellets. RBS analysis indicates that the films have the same composition as the starting material. The films were deposited onto (100) oriented MgO substrates kept at 300°C. Typical thicknesses of the deposited films were between 1000 to 3000 Å. The thickness variation is controlled by the number of pellets used in the flash evaporation.

The films are annealed in air at 840°C between 10 minutes and 16 hours in a box furnace and quenched to room temperature. A 10 minute anneal is sufficient to crystallize the film, and produce a T_c of 75 K. A longer anneal, 4 hours, raises the T_c to 78 K, improves the resistance ratio, and produces a fraction of the high temperature phase, with a slight resistance drop between 110 and 115 K. Figure 6 shows the resistance trace for two BiSrCaCuO films, one of 1000 Å thickness and the other of 3000 Å, using different anneals.

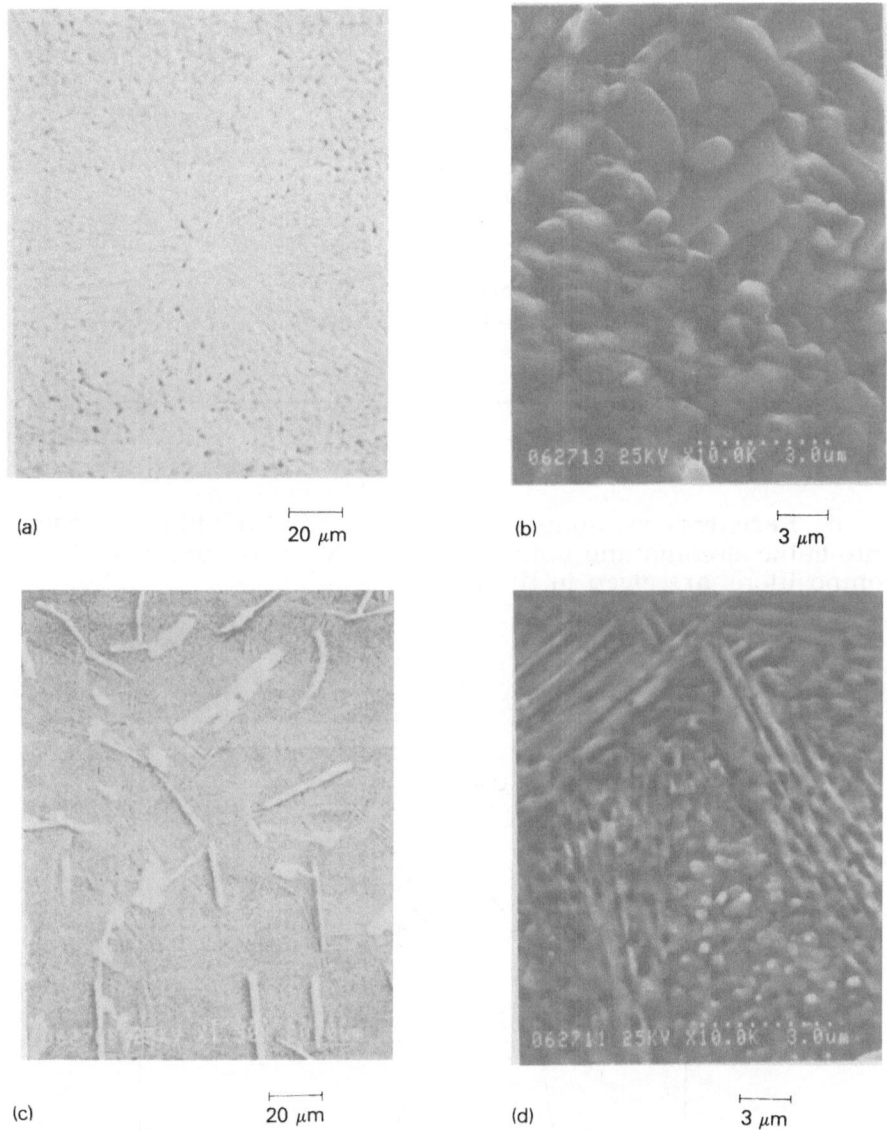

(a)   20 μm

(b)   3 μm

(c)   20 μm

(d)   3 μm

Figure 4. (a) and (b) are SEM photographs at different magnification
for a stoichiometric film deposited on (100) MgO and annealed at
900 °C for 10 minutes in wet oxygen and cooled at 1 °C/min in dry
oxygen. (c) and (d) are SEM photographs of a companion film to
the above annealed at 850 °C for 30 minutes in wet oxygen, and
cooled as above. Notice the large quantity of second phase
($BaCuO_2$) on the top surface.

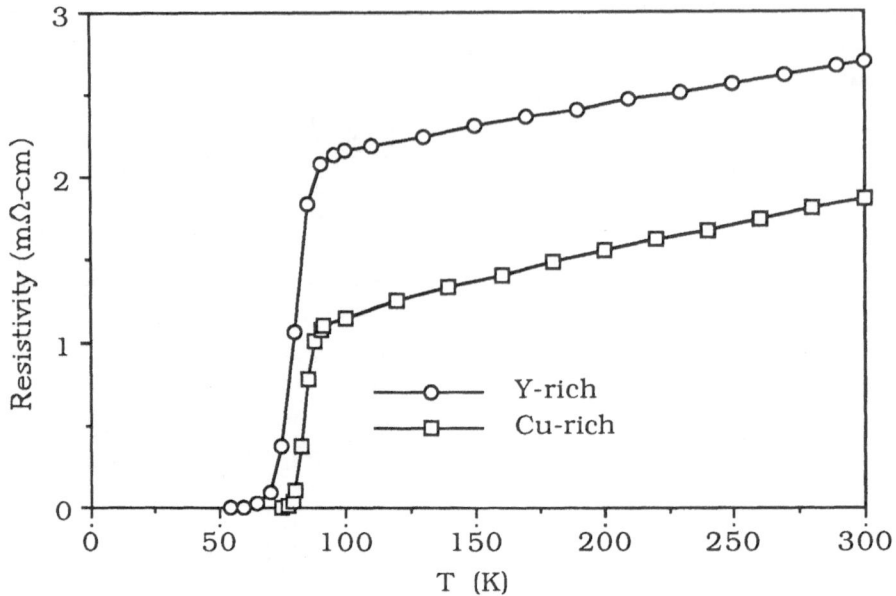

Figure 5. Resistivity vs. temperature for two YBaCuO films codeposited onto cubic zirconia and annealed at 850 °C in wet oxygen. The compositions are given in the text.

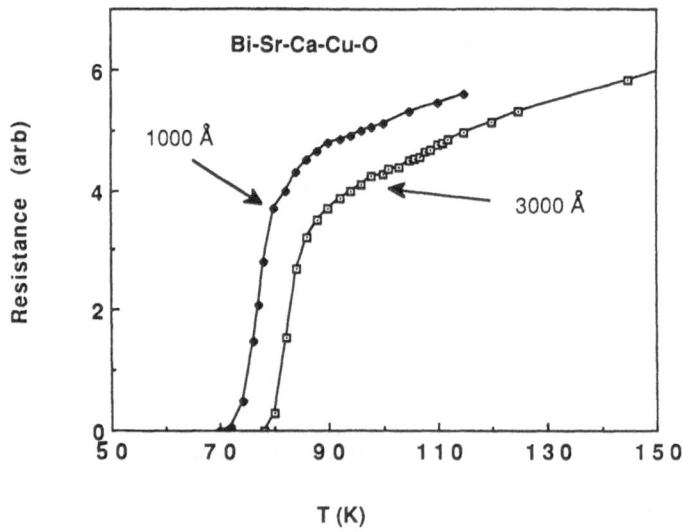

Figure 6. Resistance in arbitrary units plotted vs. temperature for two thin film samples of Bi-Sr-Ca-Cu-O. The 1000 Å thick film was annealed for 10 minutes at 840 °C and the 3000 Å thick film was annealed for 4 hours at 840 °C. Both films were then quenched to room temperature.

X-ray diffraction for these films along the growth direction indicates that the films are textured with the c-axis perpendicular to the substrate. Fig. 7 shows a scan for a TlBaCaCuO film of 2000 Å thickness. The peaks due to the sample can all be indexed to (00L) reflections of $Bi_2Sr_2Ca_1Cu_2O_x$ reported by Zandbergen et al.[11] The value of the c axis lattice parameter was found to be 30.66 Å.

CONCLUSIONS

We have studied the growth of "123" using coevaporation with BaF_2 on a variety of substrates. We find that $SrTiO_3$ is still the substrate of choice for superconducting properties, but cubic zirconia has promise. Our best results on $SrTiO_3$ are T_c = 88-91 K, and a zero field critical current at 4.2 K of $\approx 10^6$ A/cm^2. SEM and optical studies show these films are highly ordered in the plane of growth. Films on magnesium oxide continues to have depressed transition temperatures using this technique, although varying the anneal can improve the film properties. These films show fiber texture, and SEM indicates a surface morphology similar to that of bulk ceramic samples. Using flash evaporation, we have developed a simple technique for depositing a wide range of compositions. The films have transition temperatures of over 70 K for films of 3000 Å, and grow with strong fiber texture about the c axis.

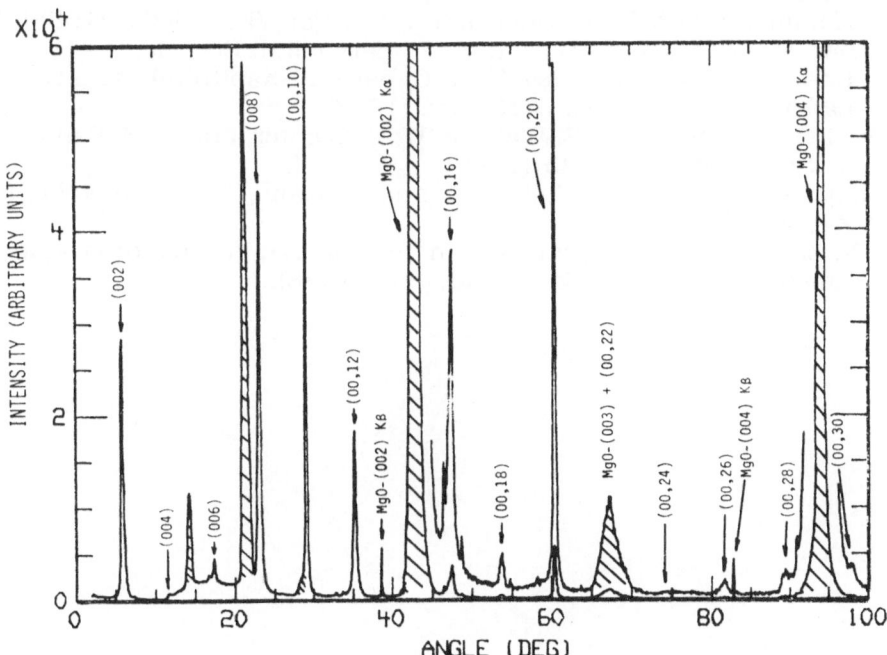

Figure 7. X-ray intensity vs. 2θ for a Tl-Ba-Ca-Cu-O film annealed for 30 sec. at 850 °C and then quenched to room temperature. The reflections from the sample can all be indexed to (00L) reflections, indicating a c-axis textured film. The cross hatched peaks are due to the MgO substrate. Notice that the data for $2\theta > 45°$ is replotted with the ordinate scale factor decreased by 10.

ACKNOWLEDGEMENTS

The work presented here would never have been possible without the efforts of many people at the Naval Research Lab. This group includes J. H. Claassen, M. S. Osofsky, L. H. Allen, C. R. Gossett, W. T. Elam, E. Skelton, H. A. Hoff, B. Bender, and P. Lubitz. This work was funded in part by the Defense Advanced Research Project Agency (DARPA).

REFERENCES

[1] P. R. Broussard, J. H. Claassen, C. R. Gossett, M. S. Osofsky, and W. T. Elam, to be published, *J. Superconduct* **1**, (1988).

[2] M. S. Osofsky, P. Lubitz, M. Z. Harford, A. K. Singh, S. B. Qadri, E. F. Skelton, W. T. Elam, R. J. Soulen, Jr., W. L. Lechter, and S. A. Wolf, *Appl. Phys. Lett.* **53**, 1663 (1988).

[3] A. D. Berry, D. K. Gaskill, R. T. Holm, E. J. Cukauskas, R. Kaplan, and R. L. Henry, *Appl. Phys. Lett.* **52**, 1743 (1988).

[4] P. M. Mankiewich, J. H. Scofield, W. J. Skocpol, R. E. Howard, A. H. Dayem, and E. Good, *Appl. Phys. Lett.* **51**, 1753 (1987).

[5] C. R. Gossett, K. S. Grabowski, and D. Van Vechten, in "Thin Film Superconductors", AIP Conference Proceedings No. 165, p. 443 (New York, NY 1988).

[6] L. H. Allen, P. R. Broussard, J. H. Claassen, and S. A. Wolf, *Appl. Phys. Lett.* **53**, 1338 (1988).

[7] A. F. Marshall, R. W. Barton, K. Char, A. Kapitulnik, B. Oh, R. H. Hammond, and S. S. Laderman, *Phys. Rev. B* **37**, 9353 (1988).

[8] K. Char, Mark Lee, R. W. Barton, A. F. Marshall, I. Bozovic, R. H. Hammond, M. R. Beasley, T. H. Geballe, A. Kapitulnik, and S. S. Laderman, *Phys. Rev. B* **38**, 834 (1988).

[9] H. A. Hoff, A. K. Singh, J. S. Wallace, W. L. Lechter, and C. S. Pande, *J. Superconduct.* **1**, 35 (1988).

[10] H. Maeda, Y. Tanaka, M. Fukutomi, and T. Asano, *Jpn. J. Appl. Phys.* **27**, 2, (1988).

[11] H. W. Zandbergen, P. Groen, G. Van Tendeloo, J. Van Landuyt, and S. Amelinckx, *Sol. St. Comm.* **66**, 397 (1988).

PROPERTIES OF IN-SITU SUPERCONDUCTING $Y_1Ba_2Cu_3O_{7-x}$ FILMS BY MOLECULAR BEAM EPITAXY WITH AN ACTIVATED OXYGEN SOURCE

J. Kwo, M. Hong, D. J. Trevor, R. M. Fleming, A. E. White,
R. C. Farrow, A. R. Kortan, and K. T. Short

AT&T Bell Laboratories
Murray Hill, New Jersey 07974

ABSTRACT

Highly oriented, epitaxial $Y_1Ba_2Cu_3O_{7-x}$ thin films were prepared on MgO(100) by molecular beam epitaxy at a substrate temperature of 550-600°C. The in-situ growth was achieved by incorporating reactive oxygen species produced by a remote microwave plasma in a flow-tube reactor. The epitaxial (001) orientation is demonstrated by X-ray diffraction, and ion channeling. In-situ reflection high energy electron diffraction showed that a layer by layer growth has produced a well ordered, atomically smooth surface in the as-grown tetragonal phase of an oxygen stoichiometry of 6.2-6.3. A 500°C anneal in 1 atm of O_2 converted the oxygen content to 6.7 to 6.8. Typical superconducting transport properties of a $Y_1Ba_2Cu_3O_{7-x}$ film 1000Å thick are ρ(300K)= 325 $\mu\Omega$-cm, ρ(300K)/ρ(100K)= 2.4, T_c(onset) = 92K, and T_c(R=0) = 82K. The transport J_c at 75K is $1\times10^5 A/cm^2$, and increases to $1\times10^6 A/cm^2$ at 70K.

INTRODUCTION

Since the discovery of high temperature superconductivity, rapid progress has been made on the preparation of high T_c superconducting oxide films. For the $Y_1Ba_2Cu_3O_{7-x}$ compound, a variety of techniques have been employed to produce high quality epitaxial films, in which the ability of carrying a high critical current density (J_c) was well demonstrated[1]. The principle process that was common to most film synthesis work reported in 1987 was based on low-temperature (≤400°C) deposition of amorphous oxide films followed by high temperature (≥800°C) O_2 furnace annealing. One notable drawback associated with the high temperature annealing process is that surface segregation occurring during anneal caused alterations of the stoichiometry or the phase near surface region ~2-500Å thick. Additionally, the rough surface morphology on a scale of 0.5-1.0 μm has severely constrained fine-line dimensions in superconducting film devices.

Since the beginning of this year, emphases have been directed toward "in-situ" production of the high-T_c phase at relatively low growth temperature to eliminate completely the high temperature annealing step. The "in-situ" method offers the obvious advantage of fabricating superconducting thin film devices at a temperature compatible with semiconductor processing. It also holds a better prospect of attaining a highly perfect film surface layer exhibiting high T_c superconductivity, which will then be suited for surface-sensitive measurements such as electron tunneling and surface spectroscopy. Furthermore, it promotes the possibility of producing metastable superconducting phases or artificially structured oxides in thin film form.

Several successful attempts of "in-situ growth" have been reported recently. These include laser evaporation[2,3], reactive sputtering[4], and reactive evaporation [5-8]. As is generally agreed, one of the crucial parameters of forming superconducting perovskites by the in-situ process is the abundance of activated oxygen species with a chemical reactivity with metal much greater than molecular oxygen[4]. The commonly known species of reactive oxygen are atomic oxygen, excited oxygen molecule, ozone, and low energy oxygen ions, although there have been no systematic studies to determine which type is most effective for forming the perovskite superconducting phase. These types of activated oxygen are usually present in oxygen plasma with a concentration on a few percent level depending on specific plasma conditions.

We report our recent studies of in-situ low temperature growth of $Y_1Ba_2Cu_3O_{7-x}$ films produced by molecular beam epitaxy (MBE) aided with a reactive oxygen source generated from a microwave discharge in a flow-tube design. The activated oxygen species are predominantly excited molecular oxygen and atomic oxygen. In stead of using $SrTiO_3$ substrates as in previous work[2,3,6,7], MgO was chosen as the epitaxial substrate because the as-produced high T_c films could be more useful for fundamental studies and device applications. X-ray diffraction, and ion channeling demonstrate that a high degree of epitaxial order was achieved between $Y_1Ba_2Cu_3O_{7-x}$ (001) films and MgO(100) substrates in spite of a large lattice mismatch of ~9%. Most importantly, in-situ reflection high energy electron diffraction (RHEED) studies indicated that a layer by layer growth maintained a highly ordered perovskite oxide structure to the upper surface layer. Scanning electron microscopy revealed a smooth surface morphology on a scale finer than 1000Å. The as deposited film forms an oxygen deficient tetragonal cell of an oxygen stoichiometry of 6.2-6.3. After annealing at 500°C for 1 hour, the structure is converted to the orthorhombic symmetry of an oxygen composition of 6.7-6.8. Typical transport properties for nearly stoichiometric films are $\rho(300K)= 325$ μΩ-cm and $\rho(300K)/\rho(100K) = 2.4$. The resistive transition is T_c(onset) = 92K, and $T_c(R=0) = 82K$. The fact that the transport J_c increases rapidly to $1\times10^6 A/cm^2$ at a temperature 15% below T_c suggests that the current carrying capabilities are comparable to state of the art results obtained by high temperature annealing.

EXPERIMENTALS AND OXYGEN PLASMA SOURCE

The samples were prepared in a versatile ultrahigh vacuum molecular beam epitaxy system previously described [10]. Coevaporation of three metal sources were used, i.e. the Y and Cu were from electron-beam heated sources, and the Ba was from an effusion cell oven. No shuttered growth of individual source was employed. The evaporation rate of individual source was monitored and controlled by Inficon Sentinel monitor of which the setting was adjusted according to the calibration by Rutherford backscattering spectroscopy described later. The substrate temperature was calibrated by the optical pyrometer method. Because the substrates were mechanically clamped to the sample block, the actual temperature of the substrate measured by optical pyrometer is usually lower by about 75°C than the reading of thermocouple located in the back of the sample block. The optimal growth temperature extends over a finite range of 550-600°C. In order to achieve a high uniformity of film compositions for both metals and oxygen, the sample block was kept in rotation during growth at a rate of 10 RPM. The overall oxide growth rate was 0.5 Å/sec with total film thickness of ~ 1000Å.

In order to enhance the oxidation of the metal species in the growing film a beam of activated oxygen was directed at the substrate. The activated oxygen species were generated by passing a flow of molecular oxygen through a discharge contained in a flow-tube reactor made of a 10 mm inside diameter untreated quartz tube. The discharge was excited in a McCarral cavity using a microwave power of 120 Watts at 2.45 GHz. Because the cavity is physically located external to the vacuum chamber, this configuration provides a direct access to fine-tuning discharge conditions, thus minimizing the reflected power to as low as 1 watt. The gas flow in the reactor tube was sampled with a 1.6 mm diameter hole located 60 cm from the discharge and directed at the substrate. The distance between the orifice and the center of the substrate is ~2.0 cm.

The flow rate through the small orifice is only 0.2 l/sec. To increase the speed of gas flow, the return path of O_2 flow is connected to a 310 l/sec turbo pump. In the current configuration, about 50% of the total gas flow was through this orifice. The pressure near the discharge region was kept at 400 mTorr, which produces a flux of 2×10^{17} species/cm^2 sec impinging upon the substrate located approximately 2 cm from the orifice. This flux is equivalent to a pressure of 6×10^{-4} Torr, which is two order of magnitude over the pressure of 5×10^{-6} Torr maintained in the growth chamber.

The flow velocity down the tube after the discharge was ~ 280 cm/sec which resulted in a resonance time of the activated species in the flow tube of 0.21 sec. Over this period of time the species undergo deactivation. Two dominant processes are wall recombination and volume recombination. Using literature values[11] for a wall recombination efficiency of 10^{-4}, we estimate the depletion of activated oxygen species to be about ~60%. Volume recombination is dominated by the processes of $O + O_2 + O_2 \rightarrow O_3 + O_2$ and $O + O_3 \rightarrow 2 O_2$. Calculations based on a rate constant[11] of 7×10^7 l^{2}mole^{-2} sec^{-1} of the first reaction give an estimate of the depletion of activated oxygen of only ~1.5%. Typically the production efficiency of atomic oxygen and excited molecular oxygen by the plasma is about 3% and 7%, respectively[12]. Hence the total flux of activated oxygen at the substrate is estimated to be ~ 6×10^{15}/cm^2 sec, which is one order of magnitude over the amount of oxygen required at the current growth rate.

According to isocompositional plots of log $P(O_2)$ vs temperature [13] determined for bulk ceramics of $Y_1Ba_2Cu_3O_{7-x}$, the low partial pressure of oxygen of 10^{-4} Torr during growth resulted in a deficient oxygen stoichiometry of 6.2-6.3 in the as-deposited films. After the growth, the samples were cooled to room temperature in the same oxygen pressure. After removing from the MBE system, they were annealed in 1 atm O_2 flow at 500-550°C for 1 hr, and slowly cooled to room temperature.

SURFACE STRUCTURAL CHARACTERIZATONS

The substrates used in this work are MgO(100), of which the lattice mismatch with the film is about 9%. The surface of the natural cleavage plane (100) is atomically smooth; however, it has many cleavage steps. Both polished and as-cleaved substrates were used, and the results were comparable. Figure 1 shows the in-situ RHEED patterns for MgO(100) surface along azimuthal (a) [100] and (b) [110] axes using 10 KeV electrons of a 1.0° grazing incidence. The diffusive streaks are presumably related to the multi-step surface. After the growth was initiated, the diffraction patters immediately changed to those shown in Figures 1(c), and (d) for a $Y_1Ba_2Cu_3O_{7-x}$ film 150Å thick along [100], and [110], respectively. This is evidenced by the increase of the streaking spacing by about 8%. The in-plane epitaxial relationship is [100] $Y_1Ba_2Cu_3O_{7-x}$ // [100] MgO, and [110] $Y_1Ba_2Cu_3O_{7-x}$ // [110] MgO. The presence of distinct streaks along with Kikuchi arcs suggests that a layer-by-layer growth has produced an atomically smooth, highly ordered surface even for films thinner than 150 Å. Figures 1(e) and (f) are diffraction patterns obtained for an as-grown $Y_1Ba_2Cu_3O_{7-x}$ film of 1000 Å along

[100] and [110], respectively. There is no strong indication of half-integer type reconstructed diffraction pattern. The best determined in-plane lattice parameters are a~b=3.88±0.03Å.

The relative intensities of the diffraction streaks and the background, as a measure of the relative amounts of the $Y_1Ba_2Cu_3O_{7-x}$ phase and impurity components, vary critically with small compositional deviations from the ideal stoichiometry. Over twenty deposition runs, we found that the streaky pattern of $Y_1Ba_2Cu_3O_{7-x}$ preferentially occurred in slightly Cu-rich compositions. The intensity of diffusive background increases with the excessive Cu composition. However, in the case of Y-rich composition, the diffraction pattern of $Y_1Ba_2Cu_3O_{7-x}$ tends to be spotty for thickness less than 100Å, and diminishes rapidly with continuing depositions. The compositional dependence of phase occurrence for $Y_1Ba_2Cu_3O_{7-x}$ appears to be in agreement with thermal-equilibrium phase diagram of the Y-Ba-Cu-O system determined by bulk ceramics synthesis.

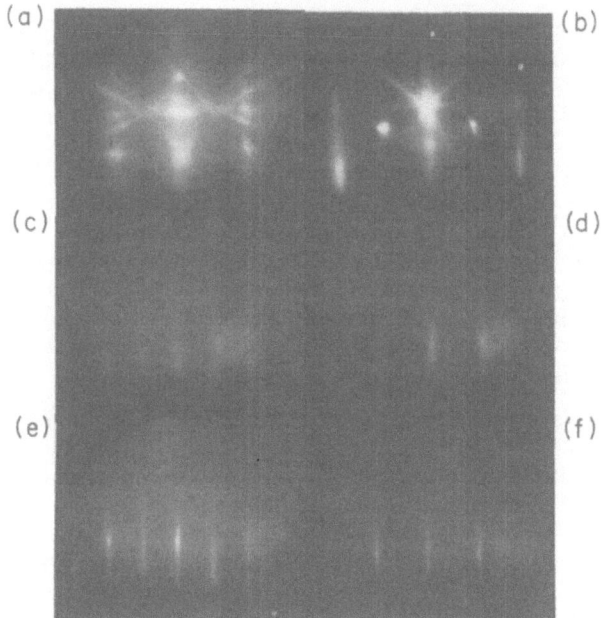

Figure 1   In-situ RHEED patterns along azimuthal directions of (a) [100] MgO, (b) [110] MgO, (c) [100] $Y_1Ba_2Cu_3O_{7-x}$ 150Å thick, (d) [110] $Y_1Ba_2Cu_3O_{7-x}$ 150Å thick, (e) [100] $Y_1Ba_2Cu_3O_{7-x}$ 1000Å thick, and (f) [110] $Y_1Ba_2Cu_3O_{7-x}$ 1000Å thick.

The sample composition was determined from Rutherford backscattering spectrometry (RBS) analysis using a $^4He^+$ ion beam of 2.0-3.0 MeV depending on the film thickness. The spectra of the sample in Fig.1 are plotted in Fig. 2. The composition was measured to be $Y_1Ba_{1.84}Cu_{2.94}$. The ratio of the aligned [100] backscattered yield to the random yield in the Ba region of the spectrum, called χ_{min}, was measured to be 21% at a 2.0 MeV energy of $^4He^+$ ion beam. This value measures the degree of structural alignment along [100] of MgO. Our result compares favorably to the best value of 30% reported for $Y_1Ba_2Cu_3O_{7-x}$ films prepared by high temperature anneal[14]. Moreover, it is also comparable to the value reported for in situ grown films by Meyer et al[15].

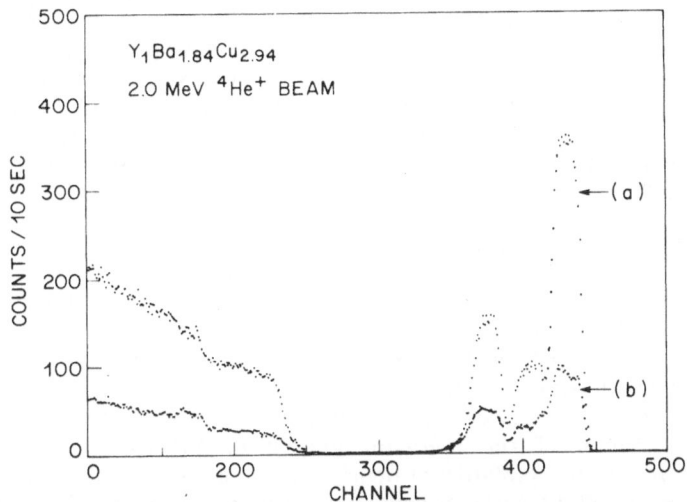

Figure 2   BRS spectra (a) of random yield, and (b) of aligned [100] yield for a sample with a composition of $Y_1Ba_{1.84}Cu_{2.94}O_{7-x}$.

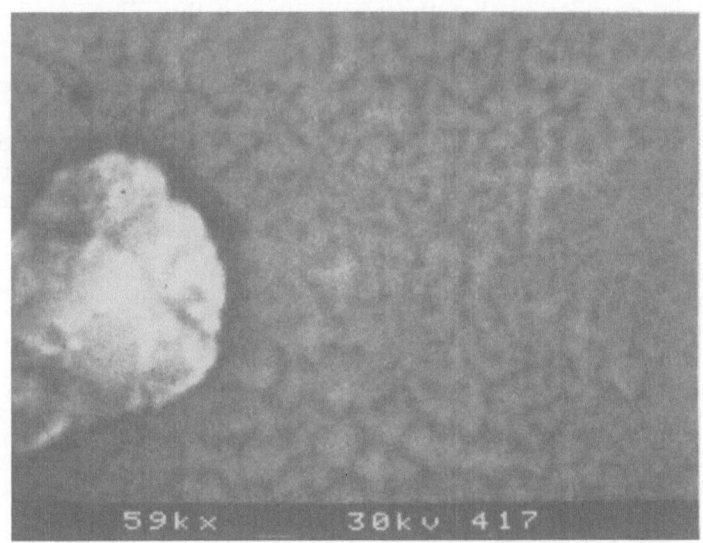

Figure 3   Scanning electron micrograph of the $Y_1Ba_{1.84}Cu2.94O_{7-x}$ sample. At a magnification of 95K, the bar width shown corresponds to 1000 Å. The left region of the photo shows the μm size Cu-Ox precipitate determined by EDAX analysis.

Scanning electron microscopy (SEM) was used to examine the film surface morphology. At a magnification of 95,000, the surface appeared smooth with roughness occurring on a scale finer than 1000Å shown in Fig. 3. This type of morphology is reminiscent of those seen in strain layer epitaxial growth with a large lattice mismatch. Dispersed CuO_x precipitates of submicron size were detected on top of the film surface, with an effective area coverage of ~15%. This partly accounts for the measured χ_{min} being 7 times larger than that measured in single crystals[16].

SUPERCONDUCTING TRANSPORT PROPERTIES

The transport resistivity was measured in a standard 4-point geometry using an AC method using a current of 0.5 μA. The resistivities of as-grown films increase drastically with small deviations of compositions from the ideal ratio. This is because during in situ growth minor compositional deviations cause nucleations of impurity phases that are mostly poor conducting and further lead to drastic deterioration of the epitaxial growth of $Y_1Ba_2Cu_3O_{7-x}$. Typical room temperature resistivity, $\rho(300K)$, of nearly stoichiometric samples are ~ 10-20 mΩ-cm, which is consistent with an oxygen composition of 6.2-6.3 [17]. After annealing at 500°C in 1 atm O_2 for 1 hr, $\rho(300K)$ is reduced to 340 μΩ-cm. The temperature dependence of resistivity shows a metallic behavior with a ratio, $\rho(300K)/\rho(100K)$, of 2.4. Typical resistive superconducting transition is $T_c(onset) = 87K$, and $T_c(R=0)$ = 77K. Additional annealing of the same sample at 550°C for 30 min reduces $\rho(300K)$ slightly to 325μΩ-cm. The superconducting transition temperature, however, is improved to a $T_c(onset) = 92K$, and a $T_c(R=0) = 82K$ as shown in Fig. 4. In one sample of only 150Å thick, a $T_c(R=0)$ of 55K was obtained. This results is remarkable, considering that the composition of that film is very Cu rich of $Y_1Ba_{2.2}Cu_{4.5}O_x$.

The critical current density, J_c was measured by the DC transport method on a mechanically scribed restriction 0.2 mm wide using a criterion of 1 μV/cm. For the film with a R=0 transition at 82K, J_c at 75K is $1\times10^5 A/cm^2$, and at 70K, J_c increases to $1\times10^6 A/cm^2$. The temperature dependence of J_c, as plotted in Fig. 4 inset, is similar to those of state of the art epitaxial films with a T_c of 90K

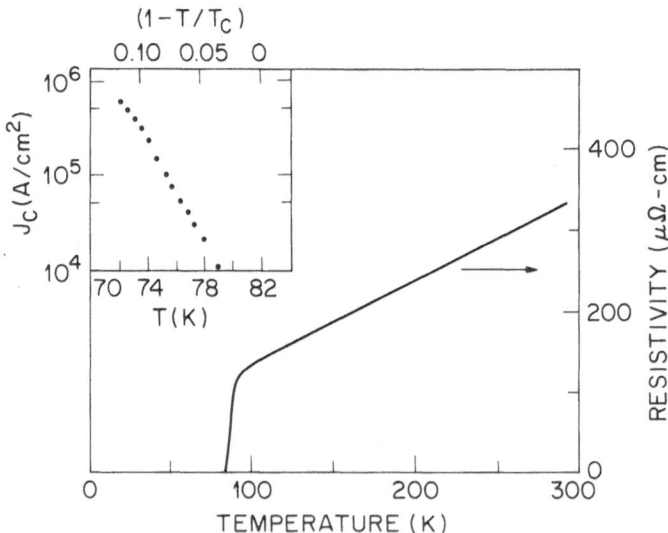

Figure 4   Resistivity vs temperature for a 1000Å $Y_1Ba_2Cu_3O_{7-x}$ film after annealing in O_2 at 550°C for 30 min. The inset plots $\log(J_c)$ vs temperature.

prepared by high temperature anneal[14]. We attribute the slightly low T_c of 82K in our work as opposed to the ideal 92K transition primarily to insufficient oxygenation. However, many studies have shown that the T_c's of films grown on MgO are usually lower than those on SrTiO$_3$ by a few degrees. It appears that other subatrate-related effects are probable as well, including a large mismatch of lattice parameters and thermal expansion coefficients between $Y_1Ba_2Cu_3O_{7-x}$ and MgO.

X-RAY DIFFRACTION STUDIES

X-ray analysis of two $Y_1Ba_2Cu_3O_{7-x}$ films was performed on a four-circle diffractometer with Eulerian geometry. Copper K_α radiation was used with a pyrolytic graphite monohromator and analyzer. The films were mounted with the ϕ-axis of the diffractometer parallel to the film normal, a geometry that allows full access to the reciprocal lattice in a symmetric scattering geometry ($\omega = 0$). Rocking curves at the $Y_1Ba_2Cu_3O_{7-x}$ (005) peak showed a mosaic of about 0.48° as compared to a width of 0.18° measured at the MgO (002) peak. Longitudinal widths of $Y_1Ba_2Cu_3O_{7-x}$ (00ℓ) peaks were about 0.038 Å^{-1} as compared with the resolution width of 0.025 Å^{-1}. This suggest a structural coherence length of the film on the order of 100 Å.

Measurements of the c-axis lattice parameter give a value of 11.72 Å, a value considerably longer than the value of 11.677 Å seen in fully oxygenated ceramics.[17] Based on measurements of the c-axis lattice parameter of ceramics as a function of oxygen content,[17] one would estimate the oxygen content of the film to be near 6.7. A comparison of the superconducting transition temperature of 82 K with transition temperatures in oxygen deficient ceramics[18] suggests an oxygen stoichiometry closer to 6.8.

A characteristic common to all previous films grown on SrTiO$_3$ is the presence of domains with the $Y_1Ba_2Cu_3O_{7-x}$ c-axis parallel to each of the cubic axes of the substrate.[10,14,19] In some cases films could be prepared with the majority of the domains having the c-axis in the plane of the substrate.[10] In the present case of $Y_1Ba_2Cu_3O_{7-x}$ on MgO, no evidence of the c-axis lying in the substrate plane could be observed. Representative scans parallel to <00ℓ> and <h00> are presented in Fig. 5. Fig. 5(a) is a scan along <h00> from (006) to (206). In addition to oriented Bragg peaks, in one sample there is an additional peak at $(\sqrt{2},0,6)$ indicating the presence of domains rotated by 45° about the c-axis. This assumption is confirmed by the presence of peaks at $(\frac{1}{\sqrt{2}}, \frac{1}{\sqrt{2}}, 6)$ and $(\frac{2}{\sqrt{2}}, \frac{2}{\sqrt{2}}, 6)$. The 45°-rotated domains were present in only one of the two films studied and may result from the presence of multi-domains in the MgO substrate used in that film. The diffuse feature in Fig. 5(a) is the tail of the MgO (202) peak. As expected from the cubic substate, domains rotated about the c-axis by 90° are equally populated. This feature is revealed by radial scans through (203) and (023) showing two peaks of equal intensity.

A second, surprising feature of the $Y_1Ba_2Cu_3O_{7-x}$ films on MgO is the presence of a second phase with a c-axis spacing near that of MgO. The presence of this phase is demonstrated in Fig. 5(b) where a scan parallel to <h00> through the MgO (002) and (202) peak is shown. The horizontal axis is labeled in $Y_1Ba_2Cu_3O_{7-x}$ units. (Since the a-axis of MgO is 4.212 Å, the value $\ell = 5.56$ in $Y_1Ba_2Cu_3O_{7-x}$ units corresponds to $\ell = 2$ in MgO units.) Three peaks from MgO can be identified in the scan: (0, 0, 5.56) corresponds to the MgO (002), (.91, 0, 5.56) corresponds to the forbidden MgO peak (102), and (1.83, 0, 5.56) corresponds to the MgO (202). The two peaks marked "X" result from a phase with the c-axis nearly equal to that of MgO an in-plane lattice parameter nearly equal to that of $Y_1Ba_2Cu_3O_{7-x}$. The peak at (1, 0, 5.56) is also seen in the (10ℓ) scan shown in Fig.

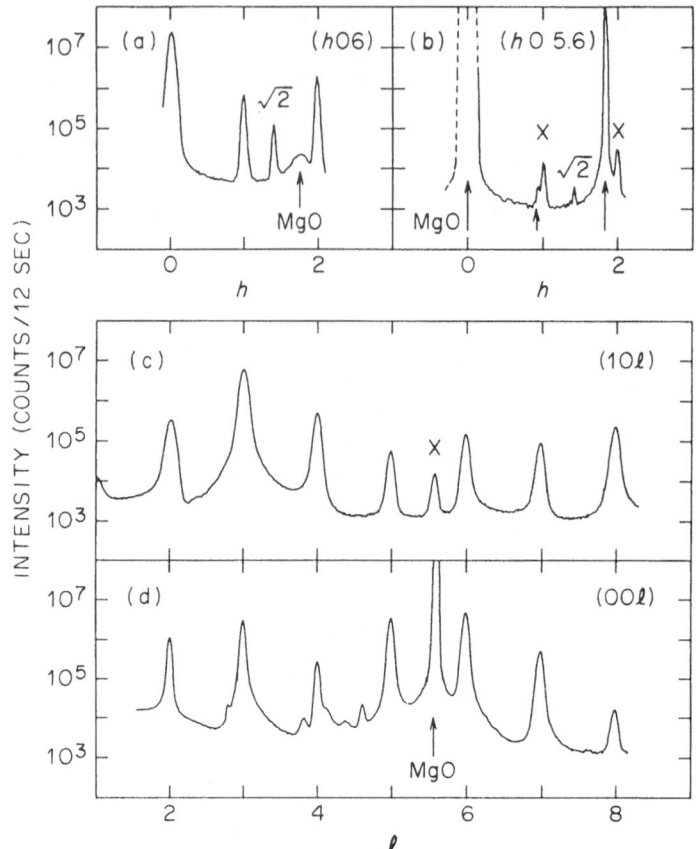

Figure 5   X-ray scans in a $Y_1Ba_2Cu_3O_{7-x}$ film grown on MgO. The peaks marked "×" are from an unidentified, oriented phase with $a_o \approx 3.85$Å and $c_o \approx 4.21$Å. (a) Scan along [h06] showing the absence of a c-axis periodicity in the plane and the presence of a domain rotated 45° about c. The diffuse feature is the tail of the MgO (202) peak. (b) Scan along [h 0 5.56] (equivalent to a [h02] scan in MgO units). (c) Scan along [10ℓ] (d) Scan along [00ℓ].

5(c) and can also be seen in (11ℓ) scans. So far, the phase has not been identified although it appears to be epitaxial with the film as evidenced by the presence of 45°-rotated domains. (Note the peak at ($\sqrt{2}$, 0, 5.56) in Fig. 5(b)). Both MgO and Cu_2O have lattice parameters near 4.2Å, but both materials are cubic and would not produce a reflection with an in-plane lattice of 3.85 Å. An intriguing possibility, however, is the possibility of a phase between the $Y_1Ba_2Cu_3O_{7-x}$ film and the substrate. The presence of an intermediate phase with the same in-plane lattice parameters as $Y_1Ba_2Cu_3O_{7-x}$ might explain the epitaxy in a system with so large a mismatch (~9%). Other defect structures with the same a and b lattice parameters as $Y_1Ba_2Cu_3O_{7-x}$ have been observed. We note that the distance from the Cu-O "chains" to the Cu-O "planes" is about 4.2 Å[20]. Our results would be consistent with the presence of $BaCuO_2$ in an oxygen-deficient perovskite structure. In the bulk, however $BaCuO_2$ exists only as an I-centered cubic cell with lattice parameters near 16 Å, not as a perovskite.

CONCLUSIONS

In conclusion, highly oriented, epitaxial $Y_1Ba_2Cu_3O_{7-x}$ films have been grown on MgO(100) by molecular beam epitaxy aided with a reactive oxygen source generated by a microwave plasma. The relatively low background pressure maintained during growth makes this new technique compatible with standard MBE operation. However, the application can be extended to regular evaporation system as well. The low dielectric constant of the MgO substate implies an useful application of the superconducting films for strip lines and microwave technology. The ability of maintaining a high quality perovskite structure to the upper surface layer represents one significant progress toward meaningful surface spectroscopy studies and tunnel junction fabrications in the future.

ACKNOWLEDGEMENTS

The authors would like to thank helpful discussions with R. A. Gottscho, H. F. Hess, J. A. Mucha, T. Siegrist, and S. Sunshine, and technical assistance from B. A. Davidson, J. J. Yeh, and R. J. Felder.

REFERENCES

1. For a complete review, see D. W. Murphy, D. W. Johnson, Jr., S. Jin, and R. E. Howard, Science, **241**, 922, (1988), and references therein.

2. X. D. Wu, A. Inam, T. Venkatesan, C. C. Chang, E. W. Chase, P. Barboux, J. M. Tarascon, and B. Wilkins, Appl. Phys. Lett. **52**, 754 (1988).

3. S. Witanachchi, H. S. Kwok, X. W. Wang, and D. T. Shaw, Appl. Phys. Lett. **53**, 234 (1988).

4. H. Adachi, K. Hirochi, K. Setsune, M. Kitabatake, and K. Wasa, Appl. Phys. Lett. **51**, 2263 (1987).

5. D. K. Lathrop, S. E. Russek, and R. A. Burhman, Appl. Phys. Lett. **51**, 1554 (1987).

6. T. Terashima, K. Iijima, K. Yamamoto, Y. Bando, and H. Mazaki, Japan J. Appl. Phys. **27**, L91 (1988).

7. R. M. Silver, A. B. Berezin, M. Wendman, and A. L. de Lozanne, Appl. Phys. Lett. **52**, 2174 (1988).

8. R. J. Spah, H. F. Hess, H. L. Stormer, A. E. White, and K. T. Short, Appl. Phys. Lett. **53**, 441 (1988).

9. Similar advantages of using a microwave plasma oxygen source were reported by N. Missert, R. H. Hammond, J. E. Mooij, V. Matijasevic, P Rosenthal, T. H. Geballe, A. Kapitulnik, M. R. Beasley, S. S. Laderman, C. Lu, E. Garwin, and R. Barton, Applied Superconductivity conference, San Franscisco, CA, Aug. 1988, (preprint).

10. J. Kwo, T. C. Hsieh, R. M. Fleming, M. Hong, S. H. Liou, B. A. Davidson, and L. C. Feldman, Phys. Rev. B **36**, 4039 (1987).

11. F. Kaufman, Prog. React. Kinet. **1**, 1 (1961).

12. T. J. Cook, and T. A. Miller, Chem. Phy. Lett. **25**, 396 (1974).

13. P. K. Gallagher, Advances in Ceramic Materials, **2**, 632 (1987).

14. P. M. Mankiewich, W. J. Skocpol, R. E. Howard, A. Dayem, G. J. Fisanick, C. E. Rice, A. E. White, K. T. Short, D. C. Jacobson, J. M. Poate, R. C. Dynes, R. B. Vandover, Proc. Mat. Res. Soc. Symp., **99**, 703(1988).

15. O. Meyer, F. Weschenfelder, J. Geerk, H. C. Li, and G. C. Xiong, Phys. Rev. B **37**, 9757, (1988).

16. N. G. Stoffel, D. A. Morris, W. A. Bonner, and B. J. Wilkens, Phys. Rev. B **37**, 2297, (1988).

17. R. J. Cava, B. Batlogg, C. H. Chen, E. A. Rietman, S. M. Zahurak and D. W. Murphy, Nature **329**, 423 (1987) and Phys. Rev. B **36**, 5719 (1987).

18. H. M. O'Bryan and P. K. Gallagher, Advances in Ceramic Materials **2**, 642 (1987).

19. "Thin Film Research of High T_c Superconductors", M. Hong, J. Kwo, C. H. Chen, R. M. Fleming, S. H. Liou, M. E. Gross, B. A. Davidson, H. S. Chen, S. Nakahara and T. Boone, in *Thin Film Processing and Characterization of High-Temperature Superconductors: Anaheim, CA, 1987*, ed. by J. M. E. Harper, R. J. Colton, and L. C. Feldman, (AIP: New York: 1988), p 12.

20. T. Siegrist, S. Sunshine, D. W. Murphy, R. J. Cava and S. M. Zahurak, Phys. Rev. B **35**, 7137 (1987).

PROPERTIES OF Y-Ba-Cu-O THIN FILMS GROWN IN-SITU AT LOW TEMPERATURES BY CO-EVAPORATION AND PLASMA OXIDATION

A.L. de Lozanne, A.B. Berezin, S. Pan, R.M. Silver, and E. Ogawa

Department of Physics
University of Texas
Austin, TX 78712-1081

ABSTRACT

We report on the synthesis and properties of thin films of Y-Ba-Cu-O grown by co-evaporation and plasma oxidation. A differential pumping scheme is used to obtain high oxygen pressure on the substrate surface, while maintaining a low pressure at the sources and rate monitors. This results in as-deposited films which are superconducting at 73K on silicon, 68K on sapphire, and 84K on strontium titanate substrates. These are the highest transition temperatures reported on bare silicon. We present topographic and spectroscopic data obtained with a low temperature scanning tunneling microscope on films that have never been exposed to atmosphere. The films are also characterized by resistivity, critical currents, x-ray diffraction, energy and wavelength dispersive spectroscopy (EDS and WDS), inductively coupled plasma spectroscopy (ICP), SEM and TEM.

DEPOSITION CHAMBER

Our deposition chamber, described briefly elsewhere[1,2], is a commercial unit[3] that has been modified to allow large oxygen pressures on the substrate surface. This is accomplished by the addition of a sub-chamber which houses the substrates, as shown in figure 1. The sub-chamber is pumped by a 500 l/s turbopump that is filled with fomblin oil in order to handle pure oxygen. The main chamber is pumped by a 2700 l/s diffusion pump without special provision for oxygen handling. A hole, 9 cm in diameter, connects the two chambers. This differential pumping scheme allows us to increase the oxygen pressure to 3 mTorr in the substrate sub-chamber, while keeping the main chamber below $7x10^{-5}$ Torr.

The sources and rate monitors are contained in the main chamber, which has a base pressure of $5x10^{-7}$ Torr. Pure yttrium and copper are evaporated from two electron beam sources, while pure barium is evaporated from a resistively heated boat. Three independent quartz crystal rate monitors and feedback electronics control the stoichiometry of the growing film. The relatively low oxygen pressure achieved in the main chamber during deposition results in longer filament lifetime for the e-beam guns (no filaments have burned out in more than a year of operation). The rate monitors also operate more reliably since there is less scattering (which reduces cross-talk) and less oxidation of the metals deposited on them.

Oxygen is fed through a quartz nozzle directed at the substrates. An oxygen plasma is created in the nozzle by RF excitation with a frequency near 14 MHz. Originally[1] we used a coil wrapped around the nozzle to excite the plasma, an idea that was quickly duplicated by others[4]. More recently we found it easier to use parallel plates for this purpose. No special provisions were taken to prevent recombination inside the quartz tube[5], since in our case the plasma is excited only 2 cm from the nozzle exit, and the substrates are 2 to 3 cm beyond that. In fact, a strong plasma glow extends all the way to the nozzle exit. We believe that this is the best way to ensure that all excited oxygen species reach the growing surface, although at this time it is not known which species are responsible for producing in-situ growth of the 123 phase. Our method is more difficult to implement compared to the creation of a plasma outside the chamber[5], because the RF power is transmitted through a vacuum feedthrough and vacuum coaxial cable, with concomitant impedance mismatches. Once the technical problems are solved, however, our method gives a very reliable and strong oxygen plasma near the substrates.

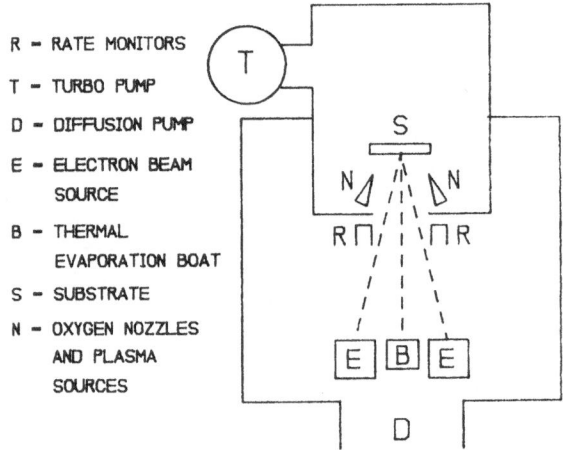

R — RATE MONITORS

T — TURBO PUMP

D — DIFFUSION PUMP

E — ELECTRON BEAM
    SOURCE

B — THERMAL
    EVAPORATION BOAT

S — SUBSTRATE

N — OXYGEN NOZZLES
    AND PLASMA
    SOURCES

Fig. 1. Schematic of the differentially pumped evaporator. Only two of the three rate monitors are shown for clarity.

The substrates are attached to a copper holder which is attached to a massive (about 2 Kg) cylindrical copper block. This block is heated by five vertical quartz halogen lamps, 500 W each (the lamps are not run at full power). The lamps are housed in a water cooled shroud in a way such that there is no light reaching the substrates. The massive block has a one-inch hole, 5 inches deep, which we use to measure temperature with a pyrometer. The pyrometer is completely shielded from all light sources and has been calibrated with two thermocouples mounted in place of a substrate. The thermocouples are not used on a typical deposition run because we rotate the substrates and we load them together with the massive block through a load lock.

We typically grow our films at a rate of 0.6 nm/s, up to a thickness of 300 nm. The substrates are heated to a fixed temperature between 540C and 580C, and rotated at 6 r.p.m. While the oxygen pressure in the sub-chamber is 3 mTorr, as measured with a capacitance manometer, the actual pressure on the substrate surface may be up to ten times higher due to the fact that the oxygen is directed at the substrates by the nozzle. After deposition the substrate temperature is lowered to 400C and the substrates are left rotating in the plasma for 20 min. Following this the substrates are pulled into the load lock and are allowed to cool to room temperature in 20 psi of pure oxygen for about two hours. There is no annealing of any of the films discussed here.

LOW TEMPERATURE SCANNING TUNNELING MICROSCOPE

The scanning tunneling microscope (STM) used for these studies operates in ultra-high vacuum at low temperatures. The pressure is typically in the low 10^{-10} Torr regime in the room temperature section, while the STM itself is below 10^{-11} Torr. This instrument has a number of unique capabilities:
  * Temperature range: 10K to 400K
  * Topographic imaging and I-V characteristics
  * In-situ Auger analysis, capable of scanning
  * In-situ LEED
  * In-situ ion milling, annealing and thermal evaporation
  * Capability to cleave samples in-situ at low temperature
  * Load-lock exchange of samples and tips
The last feature has been important in measuring high T_c materials since it has allowed us to measure a large number of combinations of tips and samples in a short time[6-8]. A transfer vessel with manipulators has been implemented in order to transfer samples from the synthesis chamber to the STM chamber without exposure to atmosphere. The films are always in pure oxygen or in vacuum. The time that the films spend in vacuum at room temperature is minimized to avoid loss of oxygen, and is typically less than 3 minutes. The motivation for this in-situ transfer is to reduce surface contamination in order to be able to image the surface and have confidence that the spectroscopic features observed are not due to adsorbed contaminants.

CHARACTERIZATION

Spectroscopic measurements in the STM were obtained with aluminum tips, while electrochemically-sharpened tungsten tips were used for topographic images. Aluminum tips are used for spectroscopy because we have found that hard tips produce smearing of the features due to the high current densities achieved (order of 10^6 A/cm^2)[6-8]. The soft aluminum tip conforms to the surface as it gently touches it, thus increasing the tunneling area and reducing the current density, and its native oxide is an excellent tunneling barrier.

Thus far only two of our good thin film samples has been successfully transferred to the STM without exposure to atmosphere. These samples were deposited on SrTiO$_3$(110) at a temperature of 570C; the resistive transition showed an onset at 90K and zero resistance at 84K. Spectroscopic measurements obtained with an aluminum tip are shown in figures 2 and 3, which show a gap and behavior reminiscent of SIN tunneling. The value of the gap seen in figure 2, $\Delta \sim 40$ mV, is similar to what we often obtain for this material, yielding a ratio $2\Delta/kT_c \sim 11$. The sub-gap conductance is not as low as we have observed in the best data obtained with cleaved sintered samples[7] or thin films that were exposed to

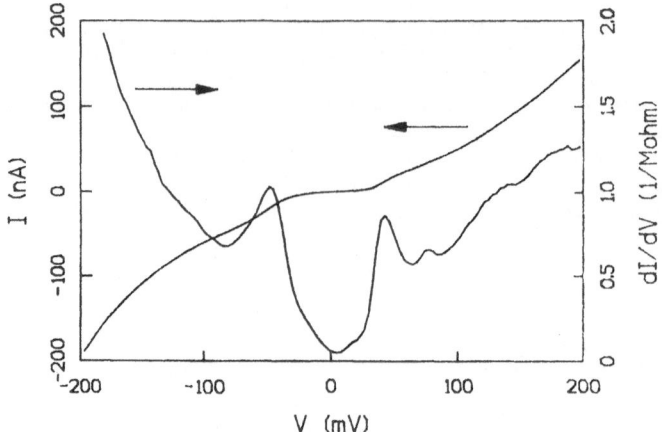

Fig. 2. I-V and differential conductance (dI/dV) versus voltage for a thin
film sample of $YBa_2Cu_3O_{7-y}$ on $SrTiO_3$ (110) and an aluminum tip .
The sample was transferred from the deposition chamber to the STM
chamber without exposure to atmosphere.

atmosphere. We believe, however, that as we obtain more data on in-situ
transferred samples more high quality curves will be obtained. An example
of a less ideal tunneling characteristic is shown in figure 3, where
multiple peaks appear at V=-129, -39, 36, and 149 mV. Such peaks may be
due to single particle charging effects, as observed recently by Barner
and Ruggiero[9] with sandwich junctions, and by van Bentum et al.[10] with a
point contact. For particle charging the peaks in dI/dV should be at
voltages of e/C, ±3e/C, ±5e/C,...., where C is the capacitance of the
small particle and e is the electronic charge. This roughly agrees with

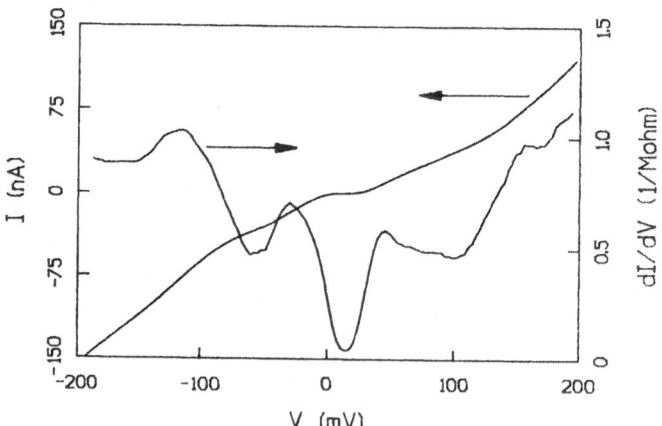

Fig. 3. Same sample and tip as figure 4, on a different surface location

114

the peaks in figure 3, with $C = 4\times10^{-18}F$. This is not observed in figure 2, however. Unfortunately, it is not always possible to increase the voltage up to three times the voltage of the first peak because this often "burns" the junction.

The issue of particle charging can be best addressed by looking at the surface topography with the STM. While one can not image exactly the same area used in spectroscopic measurements, due to the need to change from aluminum to tungsten tips, fairly general conclusions can be reached by imaging large areas. Figure 4 shows two topographic images, 12 nm on a side, of one of the in-situ transferred samples. These were chosen out of over 300 images of this sample because they show features that might indicate particles on the surface. The features seen in figure 4 are smooth mounds with diameters between 1.5 and 3 nm. These mounds are not,

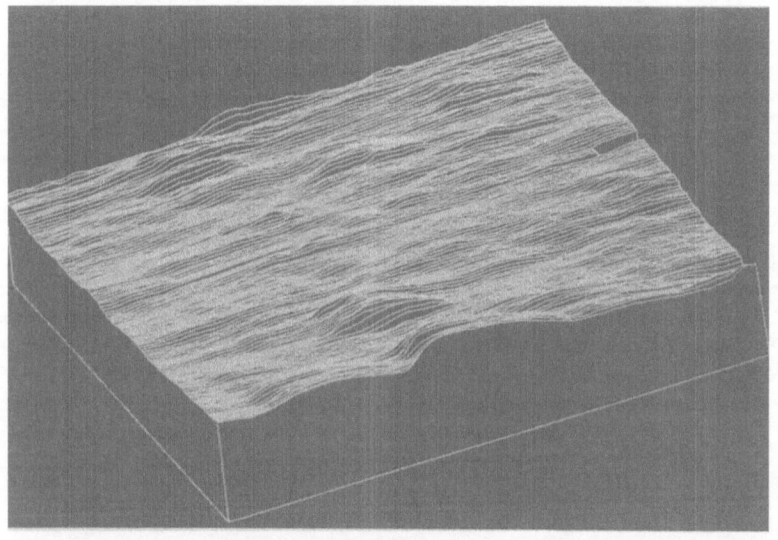

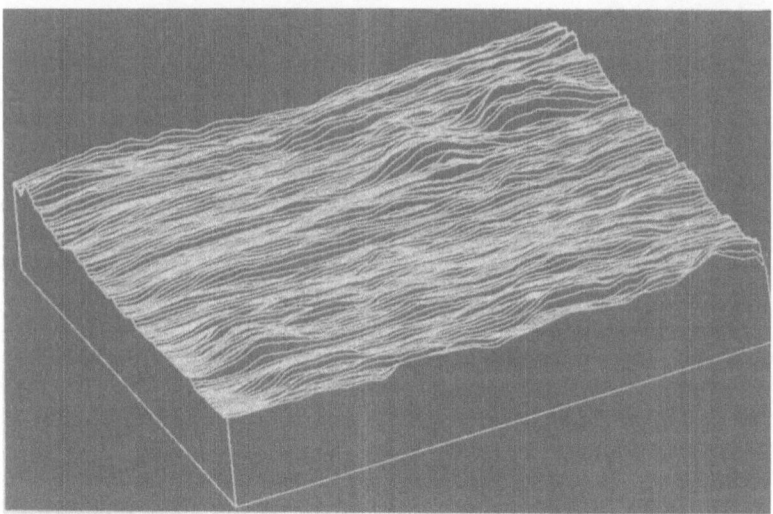

Fig. 4. Surface topography of a sample similar to that in figure 2. Two neighboring regions are shown, each region is 12 nm x 12 nm.

however, what one would expect a particle to look like on the surface. In order to have charging effects it is necessary to have an insulator between the particle and the surface, so one would expect a more discontinuous boundary around the particle. We must also emphasize that the great majority of the images have much smoother topographies. Based on this we conclude that particle charging is not a viable explanation for the large gaps seen on these materials. The only case where particles may play a role is in multiple peak features, like figure 3. It remains to be shown that these large gaps are related to superconductivity, although it is unlikely that such a large feature in the electronic spectrum does not have an effect on the superconducting properties.

Surface morphology on a larger scale was studied by SEM, as shown in figure 5 for a sample similar to the ones discussed above. The films are smooth, with a light surface roughness of about 100nm lateral size. Films on silicon show similar lateral features but a substantially smoother surface. Part of the surface roughness of the films on $SrTiO_3$ is due to our repolishing of these substrates. TEM selected area diffraction patterns and bright field images[1] show that films on $SrTiO_3(110)$ grow epitaxially. X-ray diffraction also shows excellent orientation, as shown in figure 6. Films on silicon substrates, on the other hand, grow with a mixture of a and c axis perpendicular to the surface[1].

Resistivity versus temperature measurements were made using a standard four point probe and a DC power supply with a current usually around 10 microamps. Gold contacts were used for their low contact resistance. Fig. 7 shows a resistivity versus temperature plot for a film on bare silicon with an onset at 90K, a completed superconducting transition at 73K, and metallic behavior above T_c. The silicon orientation did not play an important role in our work since we did not

Fig. 5. SEM image of a sample similar to that in fig. 2. The bar is a one micron marker.

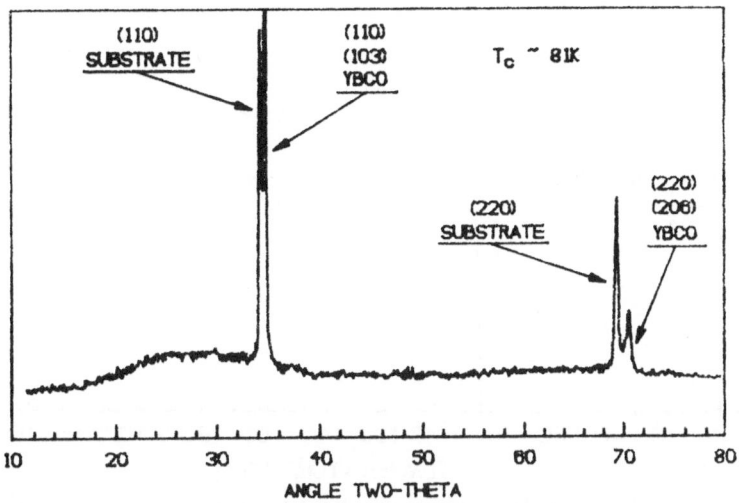

Fig. 6. X-ray data (Cu K_α radiation) for a thin film sample of $YBa_2Cu_3O_{7-y}$ on SrTiO3(110).

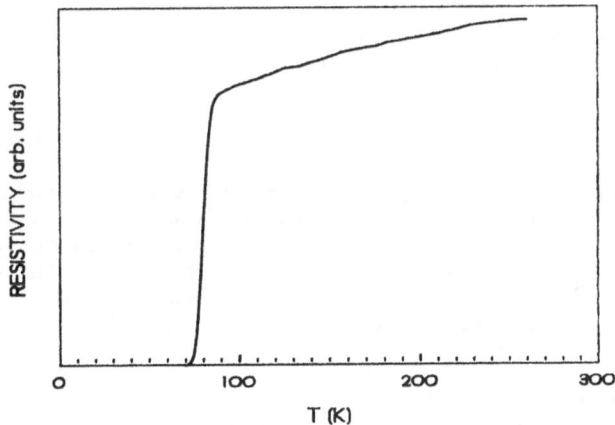

Fig. 7. Resistance vs. temperature for a thin film sample of $YBa_2Cu_3O_{7-y}$ on bare silicon.

remove the natural oxide layer from the silicon for most runs. Further, when the natural oxide layer was removed similar results were obtained; this is probably due to the rapid oxide growth which occurs when silicon is submersed in an oxygen plasma. Films on Al_2O_3 showed very similar critical temperatures to those on silicon, although the resistivity versus temperature curves had different slopes above the critical temperature[1,2]. Shown in Fig. 8 is a curve for a film on $SrTiO_3(110)$. These samples typically have a completed superconducting transition around 84K, while films on $SrTiO_3(100)$ usually had a completed superconducting transition at 70K, similar to Al_2O_3 results[1].

The films were patterned to make critical current measurements. We used standard positive photolithographic techniques and a 1.5 percent solution of phosphoric acid to etch the films; this had no adverse effects on their transition temperatures. Our pattern is a 15 micron-wide

117

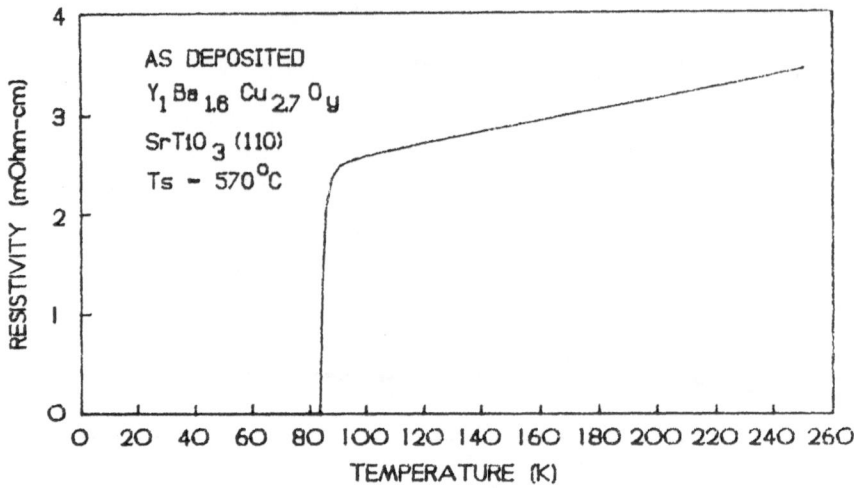

Fig. 8. Resistance vs. temperature for a thin film sample of $YBa_2Cu_3O_{7-y}$ on SrTiO3(110).

strip, 100 microns long, with two voltage pads and two current pads[1]. The critical currents for films on $SrTiO_3$(110) were 1×10^6 A/cm^2 at 4.2K and 1×10^4 A/cm^2 at 74K. Measurements for films on silicon typically yield 1×10^4 A/cm^2 at 4.2K. The temperature dependence of the critical current of a film on $SrTiO_3$(110) was above 1×10^5 A/cm^2 and fairly linear below 60K, while it fell rapidly above 70K[1].

CONCLUSION

Our differentially pumped evaporator with plasma-excited oxygen produces films of $YBa_2Cu_3O_{7-y}$ in-situ at low deposition temperatures, without the need for annealing. The transition temperatures on bare silicon and sapphire are the highest ever reported. Films on $SrTiO_3$(110) are very highly oriented, perhaps single crystal. These films promise to be important for applications and fundamental studies.

ACKNOWLEDGEMENTS

We gratefully acknowledge the donation by Motorola Inc. of the evaporator used in this experiment. We are also indebted to S. Sutton for assistance with EDS and WDS, and L. Brashear and L. Deavers for technical assistance. This work is supported by a Presidential Young Investigator Award (A.L. de Lozanne) with matching contributions from Texas Instruments, Bell Communications Research, Kodak and the Microelectronics and Computer Technology Corporation (MCC). R.M. Silver is supported by the Air Force Office of Scientific Research (87-0228).

REFERENCES

1. R.M. Silver, A.B. Berezin, M. Wendman, and A.L. de Lozanne, As-deposited Superconducting Y-Ba-Cu-O Thin Films on Si, Al_2O_3, and $SrTiO_3$ Substrates, Appl. Phys. Lett., 52:2174, (1988).
2. R.M. Silver, A.B. Berezin, E. Ogawa, and A.L. de Lozanne, Properties of In-situ Superconducting Thin Films of Y-Ba-Cu-O on Si, Al_2O_3,

and $SrTiO_3$ Substrates, Proceedings, Applied Superconductivity (1989). To appear.

3. The evaporator is a surplus Balzers BAK 550 donated by the Austin division of Motorola, Inc.

4. R.J. Spah, H.F. Hess, H.L. Stormer, A.E. White, and K.T. Short, Parameters for in situ Growth of High T_c Superconducting Thin Films using an Oxygen Plasma Source, Appl. Phys. Lett., 53:441 (1988).

5. N. Missert, R. Hammond, J.E. Mooij, V. Matijasevic, P. Rosenthal, T. H. Geballe, A. Kapitulnik, M. R. Beasley, S.S. Laderman, C. Lu, E. Garwin, and R. Barton , In Situ Growth of Superconducting YBaCuO Using Reactive Electron-Beam Coevaporation Proceedings, Applied Superconductivity Conf., Aug. 1984, San Francisco, CA. IEEE Trans. Magn. MAG-25, (1989). To appear.

6. S. Pan, K. W. Ng, A. L. de Lozanne, J. M. Tarascon, and L.H. Greene, Measurements of the superconducting gap of La-Sr-Cu-O with a scanning tunneling microscope, Phys. Rev. B 35:7220 (1987).

7. K. W. Ng, S. Pan, and A. L. de Lozanne, Tunneling spectroscopy of High Tc oxides with a scanning tunneling microscope, Proc. 18-th Int. Conf. on Low Temp. Phys., Aug 22, 1987, Kyoto, Japan. Jap. J. Appl. Phys, 26 (supplement 26-3):993 (1987).

8. A.L. de Lozanne, K.W. Ng, S. Pan, R.M. Silver and A. Berezin, Tunneling Spectroscopy of High Temperature Superconductors, Proc. Third Int. Conf. on STM, Oxford, 4-8 July, 1988. To appear in J. of Microscopy, Nov. 1988.

9. J.B. Barner and S.T. Ruggiero, Observation of the incremental charging of Ag particles by single electrons, Phys. Rev. Lett. 59:807 (1987).

10. P.J.M. van Bentum, R.T.M. Smokers, and H. van Kempen, Incremental charging of single small particles, Phys. Rev. Lett. 60:2543 (1988).

$YBa_2Cu_3O_7$ THIN FILMS PREPARED BY EVAPORATION FROM YTTRIUM, COPPER AND BaF_2

I. D. Raistrick, F. H. Garzon, J. G. Beery, D. K. Wilde,
K. N. Springer, R. J. Sherman, R. A. Lemons and A. D. Rollett

Los Alamos National Laboratory
Los Alamos NM 87545

INTRODUCTION

The recent discovery of high-temperature superconductivity in mixed-valence copper oxides has stimulated great interest in the development of thin films suitable for a variety of electronic applications. These applications range from simple electronic interconnects and transmission lines, through junction devices, such as IR detectors and SQUIDS, to complex Josephson structures for memory and logic applications. Although each application imposes different materials constraints, most have a number of requirements in common. Foremost among these is the ability to make thin, superconducting films with highly controlled properties. Although the relationships between, for example, microstructure and electronic properties are not well understood at present, it is clear that fabrication and processing variables must be specially optimized for this new class of materials.

Most attention has been devoted to the 90 K superconductor $YBa_2Cu_3O_7$, and high quality films have been prepared by a variety of techniques. Generally, the most successful methods have been physical deposition processes, including evaporation, sputtering and laser ablation.

In this paper we focus attention on the preparation of thin films of this material by co-evaporation of Y, BaF_2 and Cu, and the subsequent synthetic reaction of the as-deposited films with water vapor and oxygen to produce superconducting phases [1]. The use of BaF_2 instead of barium metal reduces the barium reactivity of the as-deposited pre-annealed films with the atmosphere, facilitating the handling and processing of these materials. We have studied the effect of the reaction conditions on the occurence of particular phases, the interactions of the superconductor with the substrate and the film texture.

EXPERIMENTAL

$YBa_2Cu_3O_7$ thin films were prepared by electron-beam evaporation from copper and yttrium metal targets and thermal evaporation of barium fluoride onto room-temperature substrates of $SrTiO_3$ and sapphire. The deposition was performed at a base pressure of 10^{-7} torr at a rate of 40 Å/sec. Film thicknesses of 1000 - 15000 Å have been produced. The highest quality films have been between 2500Å and 1 μm in thickness. The evaporation process produced almost amorphous films. Subsequent reaction with oxygen and water vapor was carried out at various temperatures between 750°C and 1000°C for 30 min or one hour, followed by 3 hr anneals in dry oxygen at 750°C and 400°C. The sample was then furnace-cooled to room temperature. Compositional data were collected by Rutherford backscattering (RBS) using 8.8 MeV α particles, [2] and energy dispersive analysis of x-rays (EDAX) produced by 30 keV electron illumination. The EDAX analysis was calibrated against bulk standards of

$YBa_2Cu_3O_7$, Y_2BaCuO_5 and $YBa_3Cu_2O_{6.5}$ using Y $K\alpha$, Ba $L\beta_2$ and Cu $K\alpha$ emission lines, with background subtraction. Excellent agreement (less than 2% variation) between RBS and EDAX chemical analysis was found for samples deposited on sapphire. RBS analysis of yttrium and copper on films deposited on $SrTiO_3$ suffers from interference by the strontium and titanium scattering edges. Analysis of the barium peak shape, however, provides useful information about the barium depth concentration profile. The microstructure of the films was studied using scanning electron microscopy, and the phases present were identified using a conventional x-ray powder diffractometer (Siemens D500). Qualitative information about the texture of the films was obtained by comparing the relative intensities of (00l) and (h00) reflections of $YBa_2Cu_3O_7$.

More detailed texture analysis was carried out on some films using a pole figure analysis procedure. Intensities were measured in full circles out to a tilt of 85°. A correction curve for defocussing was generated by measuring the fall-off in intensity with tilt for a sample of the same material that was known not to have any preferred orientation.

The resistances of the unpatterned films were measured using a conventional 4-probe conductivity arrangement, and critical currents were measured as a function of magnetic field using a Quantum Designs SQUID magnetometer. The films were patterned into a bridge structure and metallized using a photolithographic process described elsewhere [3]. Transport critical currents, I-V behavior and optical response were measured on these structures.

CONSTITUENT PHASES

A comparison of the composition of the films, as determined by EDAX and RBS, with the actual phases present, as determined by x-ray powder diffraction, indicated that only at high temperatures (>850°C) was the equilibrium distribution of phases present, as expected from the ternary phases diagram published by Roth et al. [4].

At temperatures of 800°C and below, unreacted BaF_2 was seen in the films deposited on $SrTiO_3$. Between 700°C and 750°C, the predominant superconducting phase is $Y_2Ba_4Cu_8O_{20-x}$, first reported by Marshall et al. [5]. This phase forms even in barium-rich films, suggesting that the reaction involving BaF_2 is rate-controlling at these lower temperatures. This leads to an effective composition that is barium-deficient, which favors nucleation of the 2-4-8 stoichiometry. This hypothesis is supported by the observation that films deposited on sapphire also often show the presence of $Y_2Ba_4Cu_8O_{20-x}$. Here, however, the 'barium deficiency' is produced by reaction of the barium with Al_2O_3 to produce $BaAl_2O_4$, which is easily seen in the x-ray diffraction patterns. CuO is also produced by this reaction and is seen as approximately spherical grains on the film surfaces. The reactivity of $YBa_2Cu_3O_7$ with Al_2O_3 and other substrate materials is predictable on the basis of its known thermodynamic properties [6]. RBS analysis of the films prepared on $SrTiO_3$ indicates barium diffusion into the substrate, the extent of which increases with increasing temperature. This is expected, since the phase $Ba_{1-x}Sr_xTiO_3$ shows almost complete solid-solution behavior [7], and it also known that $YBa_2Cu_3O_7$ shows extensive solubility for strontium [8]. Films annealed at the highest temperatures (>950°C) show a significantly reduced T_c which may well be due to strontium diffusion into $YBa_2Cu_3O_7$. We have also detected strontium in thinner films using EDAX.

At higher temperatures (>800°C) $Y_2Ba_4Cu_8O_{20-x}$ gives way to $YBa_2Cu_3O_7$ as the predominant superconducting phase. This may be because of an inherent instability of $Y_2Ba_4Cu_8O_{20-x}$ at higher tempertures, (which would also account for the difficulty of preparing this material as a bulk phase) or a consequence of the complete dissociation of BaF_2 which allows the relatively more barium-rich material $YBa_2Cu_3O_7$ to form. If the films are barium-rich overall, $BaCuO_2$ and Y_2BaCuO_5 can be detected as impurities, in good agreement with the equilibrium phase diagram [4]. Impurity phases often appear to be rejected to the surface of the films.

In addition, we have found that an additional phase is sometimes formed in films that lie approximately between $YBa_2Cu_3O_7$ and Y_2BaCuO_5 in composition. Although structural and compositional characterization of this phase is as yet incomplete, it appears to become superconducting at about the same temperature as $YBa_2Cu_3O_7$.

MORPHOLOGY

After reaction at elevated temperatures in H_2O/O_2 mixtures, the films exhibit a needle-and-plate morphology that depends on the annealing temperature and duration, and on substrate orientation. On <100> $SrTiO_3$ at lower temperatures (750 - 850°C) highly developed needles are arranged in mutually perpendicular directions. The pole figure analysis, discussed in more detail below, showed that the long axes of the grains were well aligned with the equivalent <010> and <001> substrate directions. On misoriented or rough samples the relative orientations of the needles becomes more random.

As the temperature is raised to 900°C and 950 °C, grain growth is evident, and the crystallite structure becomes denser and more plate-like. At higher temperatures (e.g. >1000°C), the films are no longer conducting, and SEM shows a lack of connectivity of the grains, perhaps due to partial melting.

The SEM photographs and the x-ray diffraction patterns also suggest a change in preferred orientation of the grains, with an increase in the proportion of c-normal texture as the annealing temperature is increased. This was confirmed by an analysis of the (001), (002) and (012)/(102) pole figures of two samples that were annealed at 800 and 900°C. The (001) and (002) reflections do not overlap with any other peaks from either $YBa_2Cu_3O_7$ or from the substrate. They indicate a strong alignment of the c axis perpendicular to the plane of the film. The degree of alignment is higher in the 900°C material than in 800°C material. This conclusion is based on the intensity of these reflections being sixty times random in the former material, and only 35 times random in the latter. The 900°C material showed an intensity maximum only in the center of the figure, whereas satellite peaks also appeared in the 800°C material, indicating that there are some crystals whose orientation is not perpendicular to the film plane. In order to obtain direct information about the a-axis perpendicular component, we also examined the (012)/(102) pole figure. This is a reflection that does not overlap with any $SrTiO_3$ peak and which can distinguish a-axis from c-axis oriented material. Taking the ratios of the peak heights, there is 5 times as much c-axis as a-axis material in the 900°C material. The 800°C material, however, has more a-axis than c-axis oriented grains: the ratio of peak intensities is 2:1. In summary, a change in 100°C in the annealing temperture causes an order of magnitude change in the degree of alignment of the crystals.

The tendency toward c-normal texture of the $YBa_2Cu_3O_7$ films on <100> $SrTiO_3$ at high temperatures may be understood in terms of the variation with temperature of the lattice parameters of the substrate [9] and the $YBa_2Cu_3O_7$ thin films. In Fig. 1 the percentage lattice mismatch is plotted versus temperature for the c-axis and the a-axis. The anisotropic axial thermal expansion data for $YBa_2Cu_3O_7$ used in this plot are from high-temperature x-ray diffraction studies performed by Gallagher *et al.* [10]; high-temperature neutron diffraction studies performed by Jorgensen *et al.* [11] show similar trends. At low temperatures a morphology with the c-axis parallel to the substrate has a lower mismatch while at higher temperatures an a-axis parallel texture is favored, consistent with the experimental data.

The morphology and texture of films deposited on <111> and <110> $SrTiO_3$ has also been investigated. The films deposited on these substrates and annealed for one hour at 850°C exhibit plate-and-needle morphology similar to the films deposited on <100> strontium titanate. In the case of the other orientations, however, the preferred textures were different. In both cases the texture closely followed the orientation of the substrate. Strong <113> texturing was evident from the x-ray diffraction pattern of the material deposited on <111> $SrTiO_3$, and a hexagonal or trigonal pattern is seen in the SEM photographs. Similarly, on the <110> substrate, the <110> $YBa_2Cu_3O_7$ orientation is the predominant alignment.

ELECTRONIC PROPERTIES

After annealing at 850°C, the films prepared as described above, on <100> single crystal

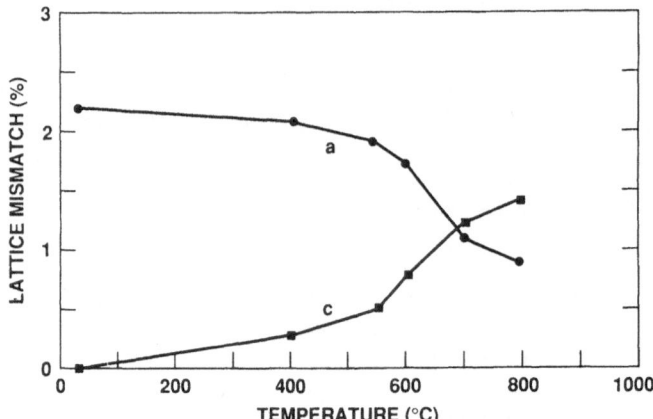

Figure 1. Percentage $YBa_2Cu_3O_7$ a- and c-axis mismatch to $SrTiO_3$ plotted as a function of temperature. At lower temperatures the $YBa_2Cu_3O_7$ c axis is a closer match to the $SrTiO_3$ lattice parameter, whereas, at higher temperatures the converse is true.

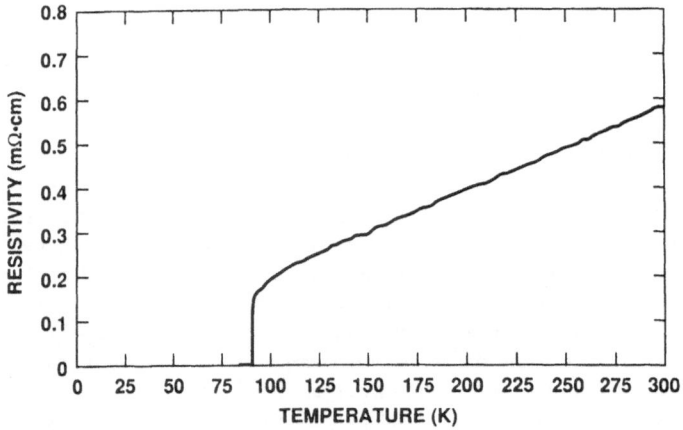

Figure 2. Four-probe resistivity measurement on unpatterned film annealed at 850°C.

SrTiO$_3$ surfaces are fully superconducting at temperatures of about 90 K, as indicated by a four-probe dc resistivity measurement (Fig. 2). Normal state resistivities at 100 K are about 0.2 mΩ cm and at 298 K are 0.6 mΩ cm. Films prepared on unoriented SrTiO$_3$ typically show transition temperatures a few degrees lower.

Superconducting films can be prepared over a fairly wide range of as-deposited compositions but, as mentioned above, superconductivity is not always principally due to the presence of YBa$_2$Cu$_3$O$_7$. Films prepared on sapphire substrates may also be superconducting, although transition temperatures are typically much reduced and transition widths are much greater. When subject to more rigorous tests of quality, such as RF conductivity or critical current measurements, films prepared on Al$_2$O$_3$ are invariably much inferior to those prepared on SrTiO$_3$.

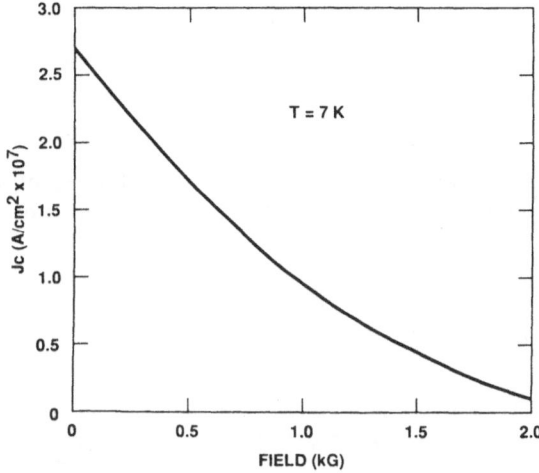

Figure 3. Magnetic field dependence of J$_c$, determined by magnetic hysteresis measurement.

Critical current densities have been measured on as-prepared films at low temperatures using a SQUID magnetometer. At 7 K, a film that showed a prodominantly c-axis perpendicular orientation with respect to the substrate (c $\perp$/ c $\parallel$ $\approx$ 5/1) had a J$_c$ of 2.7$\times$10^7 A/cm^2. The magnetic field dependence of J$_c$ for this film, however, was pronounced (Fig. 3).

The critical currents were also measured *after* patterning into suitable bridge structures, and subsequent metallization. The critical currents for several small bridges are shown in Fig. 4, as a function of temperature. The two bridges (100$\times$100 μm and 30$\times$30 μm) with the higher critical currents were on the same device, which had an average composition very close to the 1-2-3 stoichiometry. The other data is a result from a 30$\times$30μm bridge on a different device with a less satisfactory stoichiometry. On the better devices, critical currents exceed 10^5 A/cm^2 above 77K. Films with more c axis perpendicular to the substrate surface have higher critical current densities, but the absolute values appear to depend significantly on the composition of the films. Films annealed at 950°C and above, perhaps due to Sr/Ba interdiffusion, have significantly worse normal-state and superconducting properties than those prepared at 850°C or 900 °C.

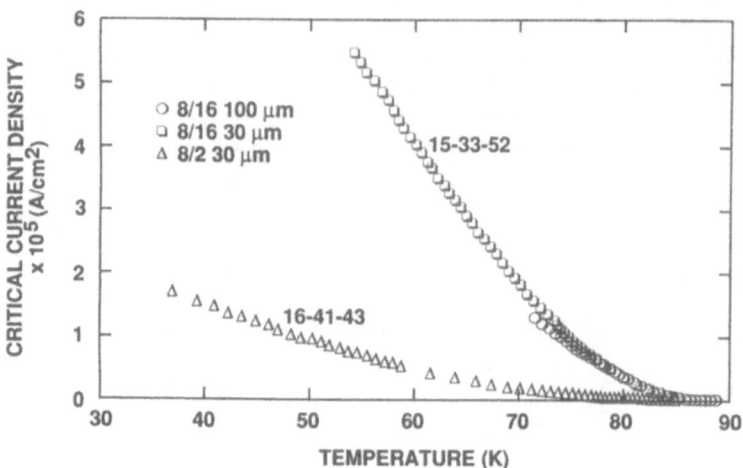

Figure 4. Critical current measurement on several small bridges of $YBa_2Cu_3O_7$ made by patterning. The two curves on the right were two bridges on the same device.

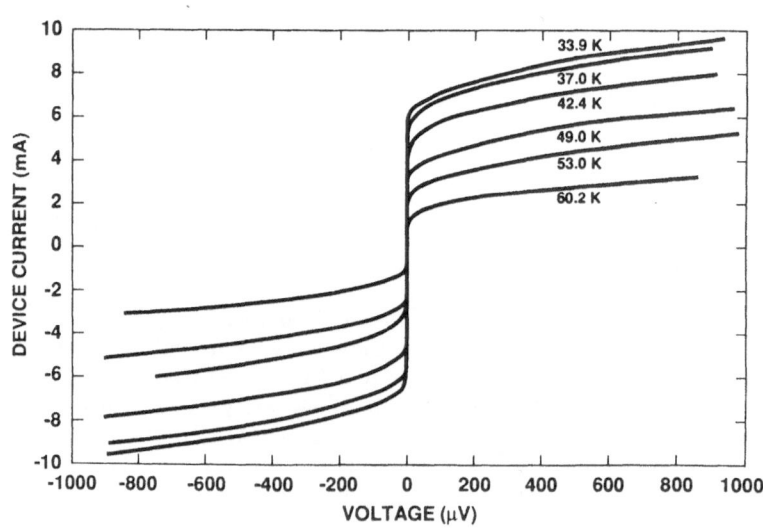

Figure 5. Current – voltage characteristics of a typical bridge at a number of temperatures.

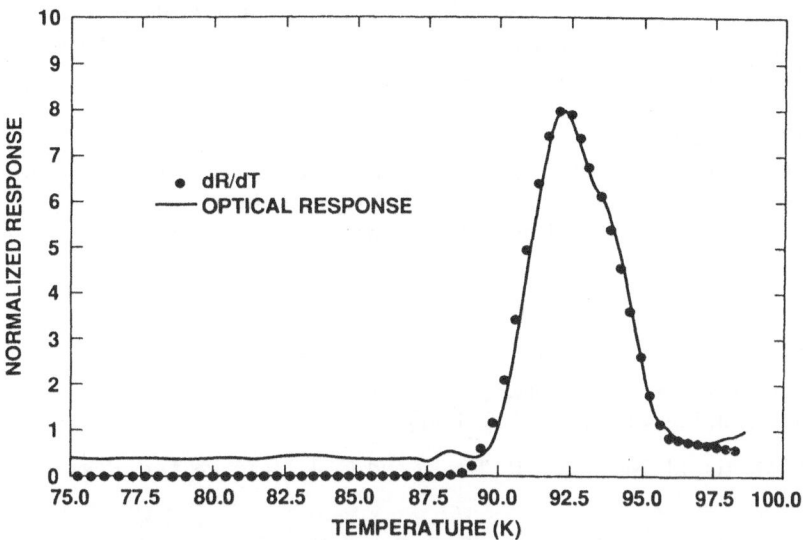

Figure 6. Optical response of device to 632 nm light (solid curve). The superimposed points correspond to the computed derivative of the resistance with respect to temperature.

We have fitted the (transport) J_c – T curve, close to J_c to the expression:

$$J_c = J_o \{ 1 - \frac{T}{T_c} \}^\alpha$$

The fitting procedure gives a value for the exponent, α, very close to 2.0. This is in contrast to a higher exponent value recently found by Allen et al. [12]. It is, however, in agreement with a theoretical prediction of Deutscher & Müller [13], in which weak link character was associated with internal twin boundaries. It should be noted that other possible scenarios also predict an exponent of 0.5, for example an array of SNS junctions may also exhibit this behavior [14].

The typical current-voltage characteristic of these bridges is shown for several different temperatures in Fig. 5. The shape of these curves is similar to that expected from an array of weakly coupled Josephson junctions, in which the coupling energy is less than the thermal energy [15]. Alternatively, or in addition, a distribution of weak link properties, such that a distribution of J_c's is present, may also be expected to show this type of behavior.

Lastly, we have attempted to characterize the response of the films to incident light. Earlier work on granular (low temperature) superconductors [16] and on $BaPb_x Bi_{1-x} O_3$ [17] has shown that optical response may be obtained by a pair-breaking mechanism. We have characterized a number of patterned films, of different quality and type, over the temperature range from 35 to 100 K, with chopped, incident light at $\lambda = 632$ nm. The devices were current biased at a constant level and the temperature varied through T_c. Changes in voltage were measured using phase-sensitive detection. A typical response curve is shown in Fig. 6. On the same curve we have overlayed the derivative of the resistance with respect to temperature. As may be seen there is a direct correspondence between the two, which strongly suggests a bolometric response. The response is a maximum at the inflection point of the superconducting transition. So far we have been unable to detect any contribution from other mechanisms in these films. This result is largely in agreement with other work [18] although a non-bolometric response has been suggested by Leung et al. [19] and Osterman et al. [20]

on 'poor' quality films (i.e. films with low T_c's and J_c's that might be expected to have particularly weakly coupled junctions). We are currently investigating methods to modify the weak link structure of the films.

Acknowledgements

We gratefully acknowledge the assistance of Carl Necker and Raul Bolmaro in aquiring the x-ray pole figure data, and of Joe Thompson for performing the magenetic susceptibilitiy measurements. We would also like to thank Dipen Sinha for many suggestions and helpful discussions.

References

[1] P.M. Mankiewich, J. H. Scofield, W. J. Skocpol, R. E. Howard, A. H. Dayem and E. Good, Appl. Phys. Lett. **51**, 1753 (1987).

[2] J. A. Martin, M. Nastasi, J. R. Tesmer and C. J. Maggiore App. Phys. Lett. **52** 2177 (1988).

[3] C. B. Mombouquette, J. G. Beery, D. R. Brown, R. A. Lemons and I. D. Raistrick. *These proceedings.*

[4] R. S. Roth, J. R. Dennis and K. L. Davis, *Phase Diagrams for Ceramists*, American Ceramic Society, Columbus, Ohio (1987).

[5] A. F. Marshall, R. W. Barton, K. Char, A. Kapitulnik, B. Oh, R. H. Hammond, and S. S. Laderman, Phys.Rev. B **37**, 9353 (1988).

[6] F. H. Garzon, J. G. Beery, D. M. Brown, R. J. Sherman and I. D. Raistrick, submitted to Appl. Phys. Lett.

[7] J. A. Basmajian and R. C. DeVries, J. Am. Cer. Soc., **40**, 374 (1957).

[8] J. M. Tarascon, L. H. Greene, B. G. Bagley, W. R. McKinnon P. Barboux and G. W. Hull in *Novel Superconductivity* eds. S. A. Wolf and V. Z. Kresin. Plenum, New York (1987).

[9] K. Wasa, M. Kitabatake, H. Adachi, K. Setsune and K. Hirochi in *Thin Film Processing and Characterization of High Temperature Superconductors* eds. J. M. E. Harper, R. J. Colton and L. C. Feldman, American Inst. Pysics, NY 1988.

[10] P.K. Gallagher, H. M. O'Bryan, S. A. Sunshine and D. W. Murphy, Mat. Res. Bull. **22**, 995 (1987).

[11] J. D. Jorgensen, M. A. Beno, D. G. Hinks, L. Soderholm, K. J. Volin, R. L. Hitterman, J. D. Grace, I. K. Schuller, C. U. Serge, K. Zhang and M. S. Kleefisch, Phys. Rev. B **36**, 3608 (1987).

[12] L H. Allen, P. R. Broussard, J. H. Claassen and S. A. Wolf Appl. Phys. Lett. **53** 1338 (1988).

[13] G. Deutscher and K. A. Müller, Phys. Rev. Lett. **59** 1745 (1987).

[14] P. G. de Gennes, Rev. Mod. Phys. **36** 225 (1964).

[15] C. M. Falco, W. H. Parker, S. E. Trullinger and P. K. Hansma, Phys. Rev. B **10** 1867 (1974).

[16] M. Leung, U. Strom, J. C. Culbertson, J. H. Claassen, S. A. Wolf and R. W. Simon, Appl. Phys. Lett. **50** 1691 (1987).

[17] Y. Enomoto and T. Murakami, J. Appl. Phys. **59** 3807 (1986).

[18] M. G. Forester, M. Gottlieb, J. R. Gavaler and A. I. Braginski Appl. Phys. Lett. **53** 1332 (1988).

[19] M. Leung, P. R. Broussard, J. H. Claassen, M. Osofsky, S. A. Wolf and U. Strom, Appl. Phys. Lett. **51** 2046 (1987).

[20] D. P. Osterman, R. Drake, R. Pratt, E. K. Track, M. Radparvar and S. M. Faris, To be published.

A SIMPLE TECHNIQUE TO PREPARE HIGH QUALITY SUPERCONDUCTING Bi—Sr—Ca—Cu—OXIDE THIN FILMS

H.—U.Habermeier,W.Sommer, and G.Mertens

Max—Planck—Institut für Festkörperforschung
Heisenbergstr. 1,D—7ooo Stuttgart, FRG

INTRODUCTION

The discovery of high transition temperature superconductors as bulk ceramics [1,2] and subsequently the development of techniques to grow these materials as thin films [3] has opened up different areas of possible applications at working temperatures above 77 K.The study of thin films of compounds with high T_c is important for both, fundamental physics as well as applications. One first large scale application of the new high T_c materials is expected to be in thin film devices like SQUIDs ,wiring interconnections in conventional devices and some new device types interfacing semiconductor and superconductor systems. Consequently, the development of techniques facilitating a large scale production of films with reproducible physical properties is of major importance.The rare earth free Bi—Sr—Ca—Cu—Oxide (BSCCO) superconductor system is a good candidate for economic large scale applications due to its ease of thin film preparation and chemical stability. The discovery of the Bi—Sr—Ca—Cu—Oxide system with several superconducting phases [4—6] generated much fundamental as well as practical interest. It was readily seen that at least two superconducting phases of this compound exist with critical temperatures of 85 K, and 11o K, respectively.Structural analysis of these compounds shows that the occurrence of CuO planes in an unit cell is associated with the high critical temperature, the number of CuO planes is proposed to be correlated with the critical temperature[7]. Associated with the number of CuO planes in the unit cell is the c—axis lattice constant of the unit cell , two CuO planes in the unit cell result in a c—axis of 3.07 nm,and three planes result in a c—axis of 3.7 nm. Basic research on this complicated system should best be performed on single crystal bulk material or highly oriented single phase thin films in order to investigate the anisotropy of the electrical properties and the correlation of structure and electron—phonon coupling.Additionally, the role of point like defects such as cation disorder, interstitials,and vacancies with respect to the electrical properties (T_2, B_{c2}) is of vital importance.From the point of view of practical interest previous work on Y—Ba—Cu—Oxide thin films has demonstrated large critical currents in highly oriented or epitaxially grown thin films, which is a prerequisite for any promising implementation of these films in future electronic devices[8].

131

Numerous groups have succeeded in producing thin films of Bi–Sr–Ca–Cu–Oxide on $SrTiO_3$ and MgO single crystals. The deposition techniques include sputtering[9], electron beam deposition[10], pulsed laser evaporation[11] as well as evaporation techniques using $e^- -$ beam heated sources simultaneously with thermally heated sources[12]. Previously, we have reported the possibility of sequential evaporation of the metals to form thin films of BSCCO with critical temperatures exceeding 7o K[13].

In this paper we apply the method of sequential deposition combined with a solid state reaction method for the oxidation to form single phase high T_c BSCCO thin films. Our films are characterized by x–ray diffractometry as well as electrical measurements. As a first step to establish a microfabrication technology using superconducting BSCCO thin films as layers to be patterned we investigate conventional lithographic methods in combination with chemical wet etching to prepare micropatterns.

EXPERIMENTAL PROCEDURES

To deposit the proper amount of the metallic constituents for a stoichiometric $Bi_2Sr_2Ca_1Cu_2O_{8+\delta}$ (2212) compound we use a conventional high vacuum evaporation system (Balzers BA 510) with a base pressure of 10^{-4} Pa. The system is equipped with two resistively heated evaporation sources powered sequentially. Sophisticated equipment for composition control or a complicated set of different evaporation sources is not required. To obtain stoichiometric composition of the metallic constituents the proper amount of Ca and Cu is filled into an alumina crucible and the corresponding amount of Bi and Sr is filled into a molybdemum crucible. For a stoichiometric $Bi_2Sr_2Ca_1Cu_2O_{8+\delta}$ thin film of thickness 2 μm we use 208.9 mg Bi, 87.6 mg Sr, 2o mg Ca and 63.5 mg Cu and evaporate first the Ca/Cu mixture completely followed by a complete evaporation of the Bi/Sr. Reactions of the metals with the crucible material are not observed. As substrates we use either one side polished (100) – oriented $SrTiO_3$ single crystals or (100) oriented MgO single crystals; the deposition temperature is 260° C. After deposition the vacuum chamber is bled with oxygen and the substrate is cooled down to room temperature within one hour. Direct after deposition the films are electrically conducting with a resistivity in the mΩcm range. To form the high T_c superconducting phase a post deposition annealing procedure is necessary. This step turned out to be the most critical to obtain single phase high T_c material. We use conventional furnace annealing with dry oxygen flowing through a heated quartz tube. The best results are obtained if a two step annealing is employed. Fig 1 shows a typical temperature time profile of the annealing procedure with T_1 and T_2 as variables in an optimization process; the nesting times at T_1 and T_2 were 3o minutes, respectively. Attempts to prepare high quality BSCCO thin films using rapid thermal annealing (RTA) failed, the superconducting phase could be formed, however a complete transition to superconductivity is not achieved so far (c.f. Fig. 2). This is in contrast to the post deposition annealing of Y–Ba–Cu–O thin films where high quality thin films can be prepared successfully by the RTA technique[14,15].

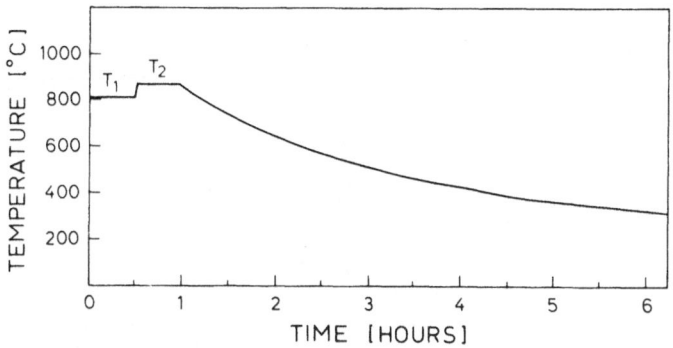

Fig. 1. Temperature –time profile for thin film furnace annealing.

The electrical contacts were made by vacuum deposition of 2oo nm Au thin film pads of 1 mm diameter, current and voltage leads are attatched by silver epoxy.Measurements of the transition temperature are done in a Helium cryostat with standard four point probe technique, a calibrated Pt resistance thermometer is used for temperature measurements.

Patterning of the film is performed in the usual way by spinning Shipley AZ 135o J photoresist onto the film, prebake at $90^{o}C$, expose through a photomask, develop, postbake at $9o^{o}$ C and etch the film in diluted HCl.

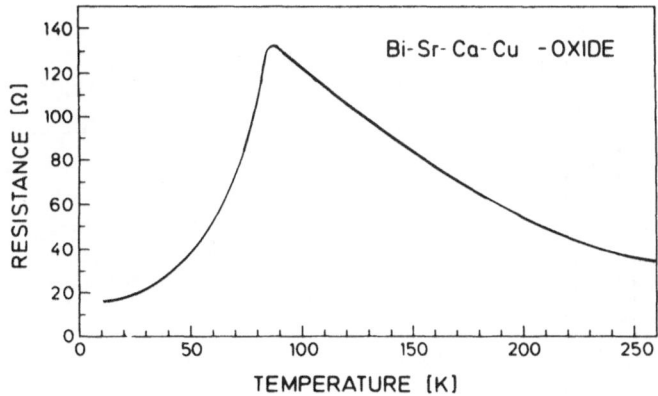

Fig.2. Resistance versus temperature curve of an RTA
annealed Bi–Sr–Ca–Cu–Oxide thin film (5 min at $820^{o}C$)
(broad and incomplete transition to superconductivity).

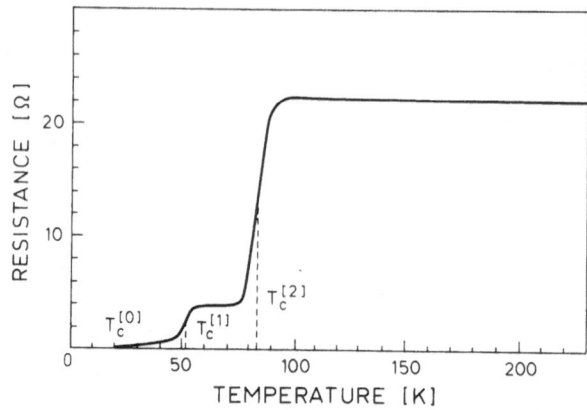

Fig.3 . Temperature dependence of the electrical resistance of a $Bi_2Sr_2Ca_1Cu_2O_{8+\delta}$ thin film oxidized 3o min at 820° C and 30 min at 870^oC.

EXPERIMENTAL RESULTS

Fig. 3 shows the temperature dependence of the electrical resistance of a 2212 specimen having different superconducting phases. The different transitions are characterized by the temperature for zero resistance $T_c^{[0]}$ and the temperatures of the inflection points $T_c^{[1]}$ and $T_c^{[2]}$. The shape of the

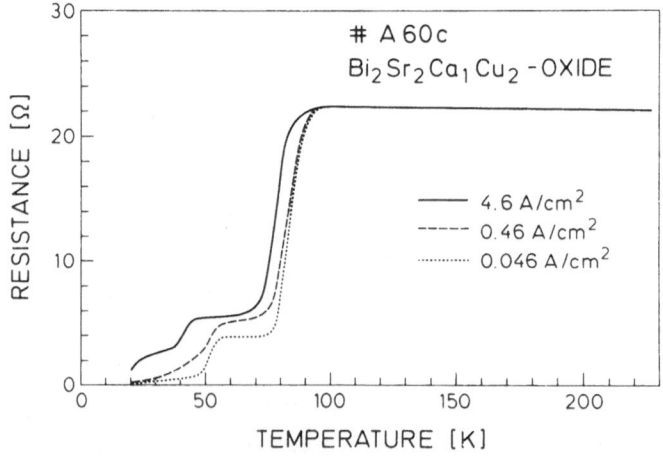

Fig.4. Transition curves of a 2212 multiphase thin film measured with different current densities.

transition curve of such specimens with different superconducting phases depend on the measurement current density as shown in Fig. 4. The shift of the inflection points to lower temperatures with increasing current density and the incomplete transition to superconductivity for higher current densities indicate a very weak coupling between the grains. X–ray diffractometry of such specimens indicate the occurrence of tetragonal phases with c– axis lattice constants of 2.46 nm as well as 3.07 nm in comparable intensities. Varying the annealing temperatures T_1 and T_2 systematically we obtain specimens covering the range of single phase 2212 material with high T_c and metallic characteristic to specimens with a negative temperature coefficient of resistance and no resistance anomalies, including specimens with multistep transitions. The data for the annealing conditions and the resulting electrical properties are listed in Tab. I for specimens with a nominal composition 2212. Films with one transition above 80K and zero resistivity above 75 K can be reproducibly prepared if the first annealing step is performed at 810° C and the second at 860° C– 870° C. In Fig. 5 the temperature dependence of the electrical resistivity is given for two different specimens, prepared in two different runs and exposed to two separate annealing procedures with identical temperature/time profiles, indicating the same characteristic for the transition to superconductivity. The x–ray diffraction pattern of the film # A 65 a is given in Fig.6 . The material is single phase $Bi_2Sr_2Ca_1Cu_2O_{8+\delta}$ showing some texture in the direction of the c–axis.

Table I. Data for different annealing conditions and resulting electrical parameters of the films (2212– composition)

specimen #	T_1 [^{o}C]	T_2 [^{o}C]	$T_c^{[0]}$ [K]	$T_c^{[1]}$ [K]	$T_c^{[2]}$ [K]
A 59 c	800	840	semiconducting, no resistance anomaly		
	800	870	–	40	81
A 63 b	810	860	77	–	82
A 63a,65a	810	870	83	–	90
A 44a	810	880	–	–	82
A 46b	820	860	–	–	73
A 60a	820	870	4.2	62	88
A 60d	830	865	20	–	83
A 59 d	830	870	–	–	78
A 62 a	840	860	semiconducting, anomaly at 85 K		

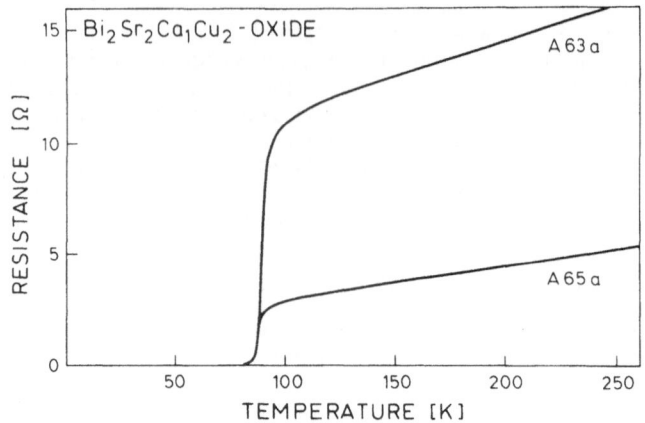

Fig. 5. Temperature dependence of the electrical resistance of
two different Bi—Sr—Ca—Cu—O thin films oxidized 3o min at
$810^{\circ}C$ and 30 min at $865^{\circ}C$.

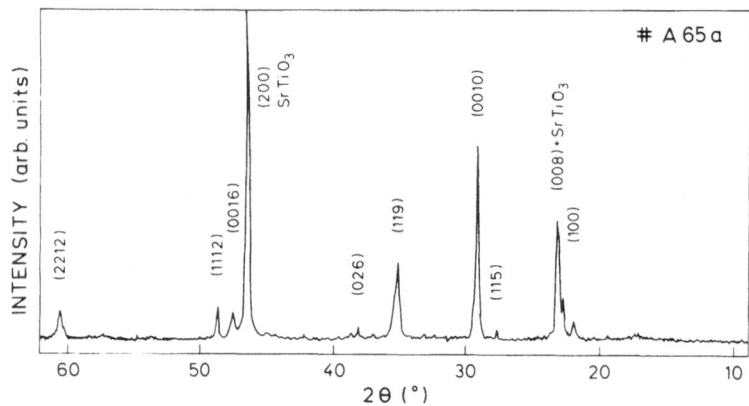

Fig.6. X—ray diffraction pattern of the film # A 65 a
(see Fig.5).

Careful analysis by an least square fit of the x—ray diffraction data of
specimens with one transition to superconductivity shows a correlation of
the length of the c—axis as well as the a— axis and the transition
temperature. In Fig. 7a the midpoint of the transition to superconductivity
is plotted versus the length of the c— axis, Fig. 7b shows the result for the
a— axis. There is a maximum for the transition temperature for a c— axis of
($3.06 \pm .005$) nm; lattice expansion or quenching results in a decreased T_c.

Fig. 8 demonstrates an example for patterning of a Bi_2—Sr_2—Ca_1—Cu_2—$O_{8+\delta}$
thin film by conventional lithography. Before and after patterning the film
has transition curves as given in Fig. 5.

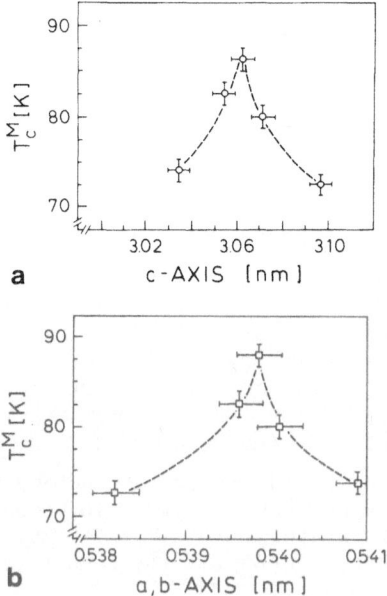

a

b

Fig 7.   Midpoint of the transition to superconductivity as a
function of the dimensions of the unit cell.
Fig 7a: T_c vs. c—axis,
Fig. 7b: T_c vs. a—axis

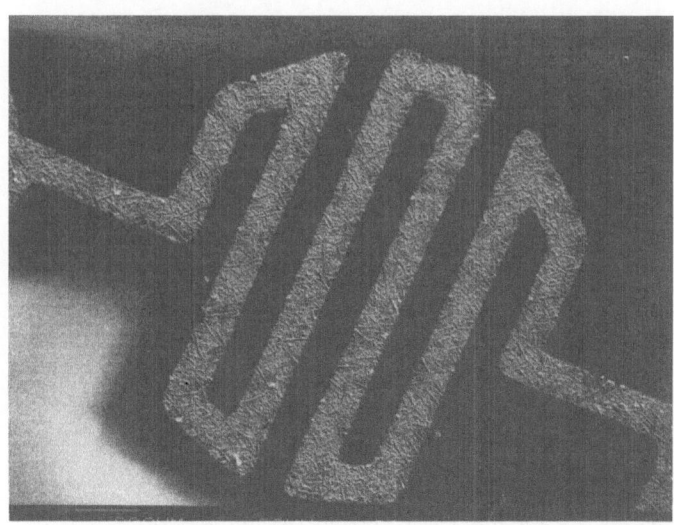

Fig. 8.   Pattern of a Bi—Sr—Ca—Cu—O thin film prepared by
conventional lithography.

DISCUSSION

The experiments demonstrate the possibility to produce single phase high T_c thin films of the BSCCO system by sequential deposition of the metallic constituents combined with a subsequent furnace annealing. The annealing procedure turned out to be the critical step in forming single phase material. Experiments designed to prepare fully superconducting films with only one annealing step at $860^{\circ}C$–870° C failed, the films showed inhomogeneous decomposition. Attempts to isolate the superconducting phase with a c–axis of 3.7 nm by changing the composition to a more copper rich compound (2223 or 1113) failed so far. Fig. 9 gives an example of the transition curve of a film with the composition 1113 , annealed identical to specimen A 65a. The result is a multiphase material showing in the x–ray diffraction pattern reflections due to the 2.46 nm c–axis as well as to the 3.07 nm c–axis. This result indicates that the details of the solid state reaction of the components with oxygen play the key role in the process forming the different phases with different numbers of Cu–0 planes in the unit cell. During the solid state reaction the chemically most stable oxides will be formed in the first annealing step around $800^{\circ}C$, during the second step the correct lattice type will be formed. To prepare BSCCO– thin films with zero resistance temperatures above 100 K the solid state reactions of the metals with oxygen must be known and the temperature /time profiles of the annealing must be designed carefully.

The analysis of the x– ray data and the dependence of the transition temperature with lattice parameters indicate the sensitivity of the film properties with respect to small changes of the lattice parameters and thus to even weak disorder of the perfect lattice. Disorder can arise from deviations of the oxygen stoichiometry as well as from cation antisite disorder , interstitials or vacancies. A detailed study of disorder in epitaxially grown thin films of BSCCO is required to optimize the electronic properties of these promising compounds.

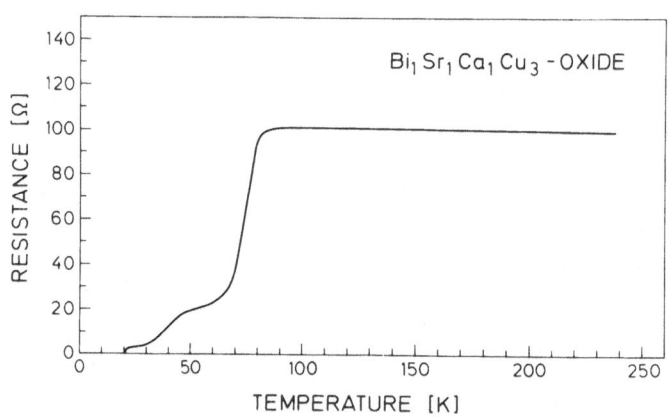

Fig. 9 Temperature dependence of the electrical resistance of a $Bi_1Sr_1Ca_1Cu_3O$ – thin film oxidized 3o min at $810^{\circ}C$ and 30min at 865° C.

REFERENCES

1. J.G. Bednorz and K.A. Müller, Possible Hiigh T_c Superconductivity in the Ba—La—Cu—O System, Z. Phys. B 64: 189 (1986)

2. M.K.Wu,J.R.Ashburn,C.J.Torng,P.H.Hor,R.L.Meng,L.Gao,Z.J. Huang,Y.Q.Wang,and C.W.Chu, Superconductivity at 93 K in a New Mixed Phase Y—Ba—Cu—O Compound System at Ambient Pressure,Phys. Rev. Lett. 58: 908,(1987)

3. R.B.Laibowitz,R.H.Koch,P.Chaudhari,and R.J.Gambino, Thin superconducting oxide films,Phys. Rev.B: 8821, (1987)

4. H.Maeda,Y. Tanaka,M.Fukutomi,and T. Asano, A New High T_c Oxide Superconductor Without Rare Earth Elemenmt,Jpn. J.Appl. Phys. 27: L 209,(1988)

5. M.A. Subramanian,C.C.Torardi, J.C.Calabrese ,J.Gopalakrishnan , K.J.Morrissey,T.W. Askew,R.B. Flippen,U. Chowdry,A.W. Sleight, A New High—Temperature Superconductor: $Bi_2Sr_{3-x}Ca_xCu_2O_{8+y}$, Nature: 1015(1988)

6. C.W.Chu,J.Bechthold,L.Gao,P.H.Hor,R.L.Meng,Y.Q.Wang,and Y.T.Xue, Superconductivity up to 114 K in the Bi—Al—Ca—Sr—Cu—O Compound System Without Rare Earth Elements,Phys. Rev.Lett. 60,941 (1988)

7. R.M.Hazen,L.W.Finger,R.J.Angel,C.T.Prewitt,N.L.Ross, C.G.Hadidiacos, P.J.Heamy,D.R.Veblen,Z.Z.Shen,A.El Ali, and A. Hermann. 100K Superconducting Phases in the Tl—Ca—Ba Cu—O System, Phys.Rev.Lett. 60: 1657 (1988)

8. P.Chaudhari,R.H.Koch,R.B.Laibowitz,T.R. McGuire,and R.J.Gambino, Critical—Current Measurements in Epitaxial Films of $YBa_2Cu_3O_{7-x}$ Compound , Phys. Rev.Lett. 58: 2684 (1987).

9. H.Koinuma,M. Kawasaki,S.Nagata,K.Takeuchi, and K. Fueki,Preparation of High T_c Bi—Ca—Sr—Cu—O Superconducting Thin Films by AC Sputtering,Jpn. J. Appl. Phys. 27: L 376 (1986)

10. T.Yoshitake,T.Satoh,Y.Kubo,and H. Igarashi,Preparation of Thin Films by Coevaporation and Phase Identification in Bi—Sr—Ca—Cu—O System, Jpn.J. Appl. Phys. 27: L 1089 (1988)

11. D.K. Fork,J.B. Boyce,F.A.Ponce,R.J.Johnson,G.B.Anderson,G.A.N. Connell,C.B.Eom,and T.Geballe, Preparation of oriented Bi—Ca—Sr—Cu—O Thin Films Using Pulsed Laser Deposition,Appl. Phys. Lett 53: 337 (1988).

12. V.Pascrin.M.A. Rashti, and D.E. Brodie, Vacuum Deposition of Multilayer Bi—Ca—Sr—Cu—O Superconducting Thin Films,Appl.Phys.Lett.53:624(1988)

13. H.—U.Habermeier,W.Sommer,and G. Mertens, Preparation of Superconducting Bi—Sr—Ca—Cu—Oxide Thin Films by Thermal Evaporation.J. Appl. Phys. in the press

14. D.K. Lathrop,S.E. Russek, and R.A. Buhrmann,Production of $YBa_2Cu_3O_{7-y}$ Superconducting Thin Films in situ by High Pressure Reactive Evaporation and Rapid Thermal Annealing,Appl. Phys. Lett. 51: 1554 (1988)

15. H.-U. Habermeier,S. Kalt,G. Wagner, and G. Mertens, Optimization of Rapid Thermal Processing to Prepare Superconducting Y-Ba-Cu-O Thin Films, Proceedings of the Symposium on Trends and New Applications of Thin Films, Regensburg, February 1989

PREPARATION OF THIN FILMS YBaCu(F)O WITH HIGH T_c AND J_c

X.K. Wang, S.J. Lee, K.C. Sheng, Y.H. Shen,
S.N. Song, R.P.H. Chang and J.B. Ketterson

Materials Research Center
Northwestern University, Evanston, Il. 60208

ABSTRACT

Thin films of YBaCu(F)O were deposited on (100) $SrTiO_3$ substrates by a three e-gun, multisubstrate, computer-controlled evaporator. Several substrate materials, including MgO, ZrO, and $SrTiO_3$, were studied. Annealed films deposited on $SrTiO_3$ substrates were found to be epitaxial and had the best superconducting properties. Transport measurements on the best films show a sharp resistive transition at a temperature $T_c(R=0)$ of 90 K and have critical current densities of 2.9×10^6 A/cm^2 at 4.2 K and 5×10^4 A/cm^2 at 77 K. X-ray diffraction studies reveal that the resulting structure has the a-axis perpendicular to the substrate. SEM micrographs show a morphology consisting of an array of orthogonal, interconnecting bars with well developed junctions. The preparation process, the effects of the substrate, and the correlation between the microstructure and the superconducting properties are discussed.

A required step in pursuing fundamental studies and practical applications involving high quality thin films is to develop reproducible procedures to prepare them. Although a great effort has been made to prepare superconducting thin films of the YBaCuO family,[1-6] the consistent growth of films with high T_c and J_c still presents problems. This is primarily due to difficulty in controlling the exact composition during deposition and an interaction of the films with the substrate during annealing. Off-stoichiometry and substrate/film interdiffusion may cause the formation of other phases. These factors correlate with a broadening of the resistive transition and a degradation of the critical current density. Crystallinity of the film appears to be crucial in making good films: superior electrical properties require oriented films with strong contacts between the individual crystallites. Crystallinity is affected by the preparation conditions including such factors as: the precise composition, the substrate material, whether barium fluoride is used as a source material, the substrate temperature, the deposition rate, and the annealing temperature vs. time history.[7,8] Another problem with films prepared from metallic Y, Ba, and Cu is their sensitivity to environmental conditions, which leads to a degradation of T_c or a complete loss of superconductivity with time, possibly involving a conversion of some of the barium to barium oxide and the subsequent reactivity of this compound with water and carbon dioxide. Rapid degradation of a sample under ordinary conditions makes it useless for serial measurements or practical applications and a number of workers have addressed this problem.[9,10]

Thin films prepared with BaF_2, rather than metallic Ba, have a greatly reduced sensitivity to moisture.[9] In our experiments, for instance, a typical film was submerged in water for 64 hours before annealing and then exposed to the atmosphere for 120 hours after annealing without any degradation of its superconducting properties.

We report here the preparation of thin film YBaCu(F)O using an in-house constructed three e-gun, multisubstrate, computer-controlled evaporator. Rather than codepositing simultaneously from all three e-guns, as other workers have, our films are prepared by sequentially depositing layers of the three constituents. Basically we prepared an artificial superlattice, however interdiffusion of the constituents results in a homogeneous structure after annealing. Each layer is kept thin (on the order of 100 A) and their relative thickness is adjusted to achieve the correct stoichiometry.

The films were deposited using Y, BaF_2, and Cu as source materials in an atmosphere of 5×10^{-5} Torr of O_2. The substrates were $SrTiO_3$; they were mounted on a substrate wheel which was driven by a computer-controlled stepping motor. The wheel was radiantly heated by a tungsten wire of 450°C. Any of the (up to 20) substrates could be positioned over any of the three e-guns. A second computer-controlled stepping motor drove a shutter wheel which allowed the flux from any one of the three e-guns to reach the substrate directly above it. The flux of all three e-guns was monitored by individual quartz crystal sensors which controlled their respective fluxes via feedback to the e-gun power supply. In addition, when the accumulated thickness of the sensor associated with the e-gun depositing a given layer of the multilayer structure reached a preset thickness, the computer was activated to advance the substrate to the next e-gun (and the monitor was reset to zero thickness). A complete superlattice was deposited on a given substrate before commencing deposition on the next substrate. All deposition rates were less than 3 A/sec. Since the thickness of the individual layers deposited can be controlled very accurately, this technique permits the concentration of the three constituents to be adjusted very precisely.

The as-deposited films are smooth, shiny, insulating and disordered. They were subsequently annealed in a furnace constructed as follows: The heater is a strip of thin Pt foil wound on a 16 mm diameter quartz tube which was positioned coaxially with a second longer quartz tube which was 36mm in diameter; rubber stoppers, through which the thermocouple and the input and output electrical and gas lines passed, sealed the ends of the longer tube. Samples were annealed in a two-step procedure. The first part was carried out in a flowing atmosphere of O_2 saturated with H_2O at 860°C for 1/2 hour. In the second part the H_2O component was removed and the temperature was kept at 860°C for another 1/2 hour and then reduced linearly to room temperature at 2°/min. The water vapor was found to be an essential component in achieving good films.

The films were characterized by conventional four probe resistivity measurements, x-ray diffraction, scanning electron microscopy (equipped with EDAX), and transmission electron microscopy.

X-ray diffraction analysis was performed using Cu-K$_\alpha$ radiation. The diffraction pattern of a film deposited on SrTiO$_3$ is shown in Fig. 1. Only the (n00) reflections of the YBaCu(F)O crystal structure are clearly observed, i.e. the film is a highly-oriented with the a-axis perpendicular to the substrate. We also observed a small peak (visible in Fig. 1) indexing with the (220) plane of BaF$_2$. This indicates that some of the fluorine, present as BaF$_2$, is left unreacted.

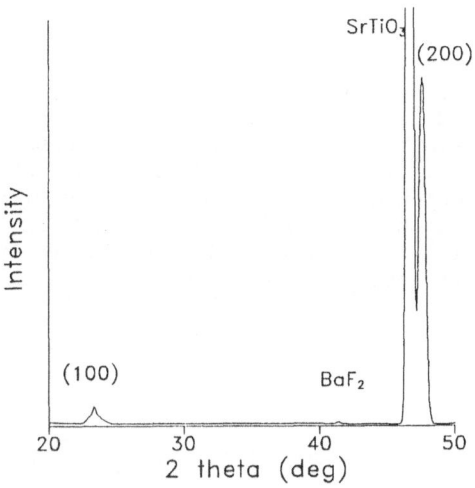

Fig. 1. X-ray diffraction pattern of an annealed film on SrTiO$_3$(100). Only the (n00) reflections of the sample are observable. A small peak indexes with the (220) plane of BaF$_2$.

Fig. 2 is a scanning electron micrographs of the microstructures of thin films of YBaCu(F)O deposited on MgO, ZrO, and SrTiO$_3$ substrates. All films were deposited under the same conditions (identical compositions, layer thicknesses, substrate temperature and O$_2$ partial pressure) and were annealed in the same run. Fig. 2(a) shows the resulting morphology of an annealed film deposited on a (100) MgO

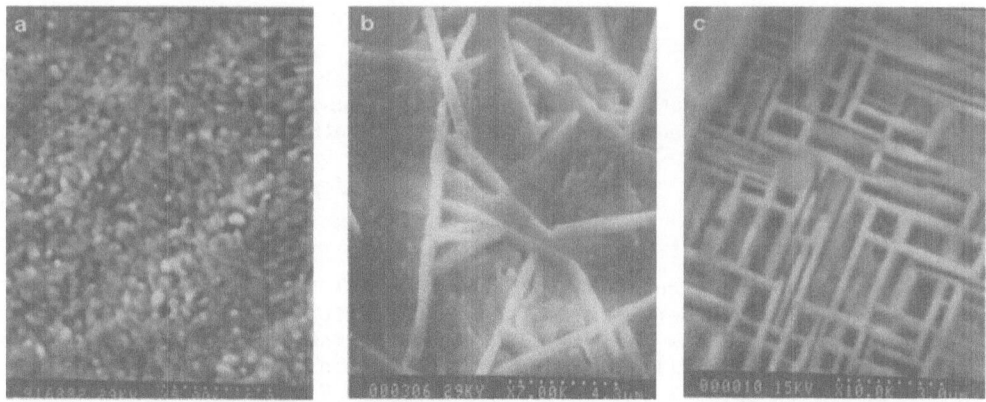

Fig. 2. SEM photographs of superconducting films of YBaCu(F)O on: (a) MgO, (b) ZrO, (c) SrTiO$_3$.

substrate; we observe a disordered array of the grains, 0.6x1 μm in size. Fig. 2(b) shows the morphology of a film deposited on a (100) ZrO substrate showing a disordered distribution of needlike grains, 0.5x4 μm in size. In contrast to the previous two substrate materials, Fig. 2(c) shows the novel morphology of a film deposited on a (100) SrTiO$_3$ substrate; it consists of an array of orthogonal, interconnecting bars with well developed junctions. Direct atomic imaging (not shown) using transmission electron microscopy reveals an atomically-abrupt junction with the axis of the bars growing parallel to the crystallographic b axis.[11]

The temperature dependence of the resistance was measured by the conventional d.c. four probe method. Fig. 3 shows the transition curves for the films on MgO, ZrO, and SrTiO$_3$; as with the SEM samples, these films all experienced identical deposition and annealing conditions. They all exhibit a similar linear resistivity versus temperature above the onset temperature. The film on SrTiO$_3$ shows a sharp resistive transition as a function of temperature with $T_c(R=0)$ at 90 K. In contrast, the best T_c obtained using the MgO and ZrO substrates was 70 K and 76 K respectively. The corresponding resistive transitions were broader and the onset temperatures lower.

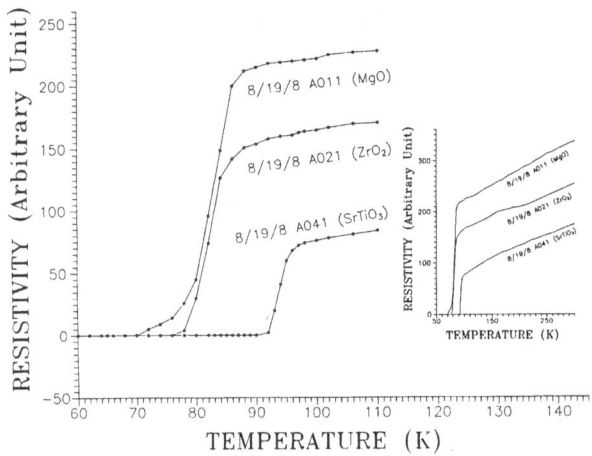

Fig. 3. The temperature dependence of the resistance of identically prepared and annealed films of YBaCu(F)O on: MgO, ZrO, and SrTiO$_3$.

The procedure described by Chaudhari et al.[6] was used to measure the temperature dependent critical current density of the film deposited on SrTiO$_3$. The film was positioned so that the magnetic field was perpendicular to the film surface, and the half of the magnetic hysteresis loop up to 2 Tesla was traced at 4.2 K with a SHE VTS-50 susceptometer. After the magnetic field was reduced to zero, the temperature dependence of the residual magnetization was measured up to the transition temperature. The critical current density was calculated according to Bean's expression, $J_c = 30M/r$, where M is the residual magnetization in emu/cm^3, r is an effective film radius in centimeters, and J_c is in A/cm^2. The calculated results are shown in Fig. 4. The critical current density was 2.9×10^6 A/cm^2 at 4.2K and zero magnetic field, and the sample still maintained a J_c of 5×10^4 A/cm^2 at 77K. Noting that an activation process could occur in the flux lattice,[12] and that the zero resistance transition temperature for this particular sample was only 85.5K, the above values of the critical current density represent a lower limit relative to our best (90 K) film (which was, unfortunately, destroyed attempting a direct measurement of J_c).

In summary, we have been able to reproducibly fabricate high T_c and J_c, oriented, superconducting thin films of YBaCu(F)O by multilayer

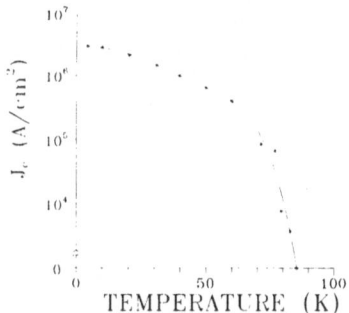

Fig. 4.   The temperature dependent critical current density at zero magnetic field for a thin film deposited on $SrTiO_3$ with $T_c(R=0) = 85.5K$.

e-beam deposition using BaF_2, rather than metallic barium, as a source material. The resulting morphology of the annealed films associated with three different substrate materials, MgO, ZrO, and $SrTiO_3$, have been examined. A correlation between the novel crystal microstructure and the excellent superconducting properties for films deposited on $SrTiO_3$ has been noted. On the basis of our results, crystallinity and connectivity of the films appear to be the crucial factors in achieving high performance superconducting thin films.

ACKNOWLEDGEMENTS

This work was supported by the Northwestern University Materials Research Center under grant DMR-85-20280 and by the Office of Naval Research under grant N00014-88-K-0106.

REFERENCES

1.   B. Oh, M.Naito, S. Arnason, P. Rosenthal, R. Barton, M.R. Beasley, T.H. Geballe, R.H. Hammond, and A. Kapitulnik, Appl. Phys. Lett. **51**, 852 (1987).
2.   B-Y. Tsaur, M.S. Dilorio, and A.J. Strauss, Appl. Phys. Lett. **51**, 858, (1987).

3. S.J. Lee, E.D. Rippert, B.Y. Jin, S.N. Song, S.J. Hwu,
    K. Poeppelmeier, and J.B. Ketterson, Appl. Phys. Lett. **51**, 1194 (1987).
4. X.K. Wang, K.C. Sheng, S.J. Lee, Y.H. Shen, S.N. Song, D. X. Li,
    R.P.H. Chang, and J.B. Ketterson, submitted to Appl. Phys. Lett.
5. A.M. DeSantolo, M.L. Mandich, S. Sunshine, B.A. Davidson,
    R.M. Fleming, P. Marsh, and T.Y. Kometani, Appl. Phys. Lett. **52**, 1995
    (1988).
6. P. Chaudhari, R.H. Koch, R.B. Lacbowitz, T.R. McGuire, and
    R.J. Gambino, Phys. Lett. Lett. **58**, 2684 (1987).
7. J.G. Huang, X.P. Jiang, J.S. Zhang, Y.Z. Wang, H.Q. Hao,
    M. Giang, Y.L. Ge, G.W. Qiao, and Z.Q. Hu, Superconductor
    Science and Technology, **1**, 110 (1988).
8. Y. Hakuraku, F. Sumiyoshi, and T. Ogushi, Appl. Phys. Lett. **52**, 1582
    (1988).
9. P.M. Makiewich, J.H. Scofield, W.J. Skocpol, R.E. Howard,
    A.H. Dayem, and E. Good, Appl. Phys. Lett. **51**, 1753 (1987).
10. Chin-An Chang, Appl. Phys. Lett. **53**, 1113 (1988).
11. X.K. Wang. D.X. Li, Y.H. Shen, S.J. Lee, K.C. Sheng,
    R.P.H. Chang, and J.B. Ketterson, to be published.
12. S.N. Song, Q. Robinson, S.J. Hwu, D.L. Johnson, K.R. Poeppelmeier, and
    J.B. Ketterson, Appl. Phys. Let. **51** 1376 (1987).

SUPERCONDUCTING PHASE CONTROL FOR RARE-EARTH-FREE

HIGH-Tc SUPERCONDUCTING THIN FILMS

Kiyotaka Wasa, Hideaki Adachi, Yo Ichikawa, Kumiko Hirochi, and Kentaro Setsune

Materials Science Laboratory, Central Research Laboratories
Matsushita Electric Ind. Co., Ltd.
Moriguchi, Osaka 570, Japan

INTRODUCTION

Much attention has been paid to the rare-earth-free high-Tc oxide superconductors of the Bi-Sr-Ca-Cu-O and Tℓ-Ba-Ca-Cu-O. Recent studies suggest that these rare-earth-free superconductors show a layered structure comprizing Bi-O and Cu-O_2 layer and the critical temperature Tc varies with the numbers of the Cu-O_2 layer. Typical rare-earth-free high-Tc oxide superconductors are listed in Tab. 1 [1-11]. The $Bi_2Sr_2CaCu_2O_x$ comprising two layers of the Cu-O_2 exhibits Tc ≃ 80 K (2-2-1-2 structure) and the $Bi_2Sr_2Ca_2Cu_3O_y$ comprizing three layers of the Cu-O_2 exhibits Tc ≃ 110 K (2-2-2-3 structure). Further increase of the Cu-O_2 layers is expected to increase the Tc.

The thin films of the Bi and/or Tℓ system were prepared by a conventional deposition process including electron beam deposition, sputtering, and chemical vapour deposition [12], [13], [14]. However, there exists a limitation in a fine control of the Cu-O_2 layer. In this paper first we consider the basic thin film processing for controlling the numbers of the Cu-O_2 layers in the conventional deposition process. Secondly we describe a layer-by-layer deposition for a fine control of the Cu-O_2 layer.

BASIC THIN FILM PROCESSING

Thin film processing for the rare-earth high-Tc superconductors is classified into three processes: (1) deposition at a low substrate temperature followed by a postannealing at around 900°C; (2) deposition at a crystallizing temperature 600 - 800°C, followed by the postannealing; (3) deposition at the crystallizing temperature under oxydizing atmosphere [15].

The thin film process (1) and/or (2) are used for the deposition of the rare-earth-free high-Tc superconductors. Sputtering and electron beam deposition were widely used for the deposition of the thin films. However, the resultant films often showed mixed phases comprising 2-2-1-2 and 2-2-2-3 structure.

Table 1. Rare-earth-free high-Tc superconductors

Bi-system : $Bi_2O_2 \cdot 2SrO \cdot (n-1)Ca \cdot nCuO_2$

		Tc (K)	Institute	Date	Ref.
$Bi_2Sr_2CuO_6$	(2 2 0 1)	7~22	Caen Univ. (France) Aoyamagakuin Univ. (Japan)	1987.5	[1],[2]
$Bi_2Sr_2CaCu_2O_8$	(2 2 1 2)	80	National Res. Institute for Metals (Japan)	1988.1	[3]
$Bi_2Sr_2Ca_2Cu_3O_{10}$	(2 2 2 3)	110	National Res. Institute for Metals (Japan)	1988.3	[3]
$Bi_2Sr_2Ca_3Cu_4O_{12}$	(2 2 3 4)	~90	Matsushita Elec. (Japan)	1988.9	[4]

Tℓ-system : $Tℓ_2O_2 \cdot 2BaO \cdot (n-1)Ca \cdot nCuO_2$

		Tc (K)	Institute	Date	Ref.
$Tℓ_2Ba_2CuO_6$	(2 2 0 1)	20~90	Institute for Molecular Sci. (Japan) Arkansas Univ. (U.S.A.)	1987.12	[5],[6]
$Tℓ_2Ba_2CaCu_2O_8$	(2 2 1 2)	105	Arkansas Univ. (U.S.A.)	1988.2	[7]
$Tℓ_2Ba_2Ca_2Cu_3O_{10}$	(2 2 2 3)	125	Arkansas Univ. IBM (U.S.A.)	1988.3	[7],[8]

: $TℓO \cdot 2BaO \cdot (n-1)Ca \cdot nCuO_2$

		Tc (K)	Institute	Date	Ref.
$TℓBa_2CaCu_2O_7$	(1 2 1 2)	70~80	IBM (U.S.A.)	1988.5	[9]
$TℓBa_2Ca_2Cu_3O_9$	(1 2 2 3)	110~116	IBM (U.S.A.)	1988.3	[10]
$TℓBa_2Ca_3Cu_4O_{11}$	(1 2 3 4)	120	ETL (Japan)	1988.5	[11]
$TℓBa_2Ca_4Cu_5O_{13}$	(1 2 4 5)	<120	ETL (Japan)	1988.5	[11]

PHASE CONTROL IN A CONVENTIONAL PROCESS

1) Bi-Sr-Ca-Cu-O thin films

Thin films of the Bi system are prepared by a conventional rf-planar magnetron sputtering. Typical sputtering conditions are shown in Tab. 2. The target is complex oxides of Bi-Sr-Ca-Cu-O. The composition is around 1-1-1-2 ratio of Bi-Sr-Ca-Cu. The process (1) and/or (2) are used for the deposition. Single crystals of (100) MgO are used as the substrates. The superconducting properties are improved by the postannealing at 850 - 900°C in 5 hr in O_2 [16]. These sputtered films show a mica-like structure as shown in Fig. 1.

The experiments suggested that the superconducting properties were strongly affected by the substrate temperature during the deposition. Figure 2 shows typical X-ray diffraction patterns with resistivity-temperature characteristics for the Bi-Sr-Ca-Cu-O thin films of around 0.4 μm thick deposited at various substrate temperature. It is seen that the films deposited at 200°C exhibit $Bi_2Sr_2CaCu_2O_x$ structure with the lattice constant $C \simeq 30$ Å which corresponds to the low Tc phase. The films show the zero resistance temperature of $\simeq 70$ K [Fig. 2 (a)].

When the substrate temperature is raised up during the deposition the high Tc phase with Tc $\simeq$ 110 K, the $Bi_2Sr_2Ca_2Cu_3O_x$ structure with the lattice constant $C \simeq 36$ Å, is superposed on the X-ray diffraction pattern [Fig. 2 (b)]. At the substrate temperature of around 800°C a single high-Tc phase is observed. The films show the zero resistance temperature of $\simeq 104$ K [Fig. 2 (c)].

These experiments suggest that the superconducting phases of the Bi-Sr-Ca-Cu-O thin films are controlled by the substrate temperature during the deposition.

Table 2. Sputtering conditions

Target	Bi:Sr:Ca:Cu:=1-1.7:1:1-1.7:2	100 mm in diameter
Sputtering gas	Ar/O_2=1-1.5	
Gas pressure	0.5 Pa	
rf input power	150 W	
Substrate temperature	200 - 800°C	
Growth rate	80 Å/min	

2) Tℓ-Ba-Ca-Cu-O thin films

Similar to the Bi-Sr-Ca-Cu-O system, thin films of Tℓ-Ba-Ca-Cu-O system are prepared by the rf-magnetron sputtering on the MgO substrate. Typical sputtering conditions are shown in Tab. 3. However, their chemical composition is quite unstable during the deposition and the postannealing process due to the high vapour pressure of Tℓ. Thin films of the Tℓ system are deposited without intentional heating of substrates (∿ 200°C) and annealed at 890 - 900°C in Tℓ vapour [17].

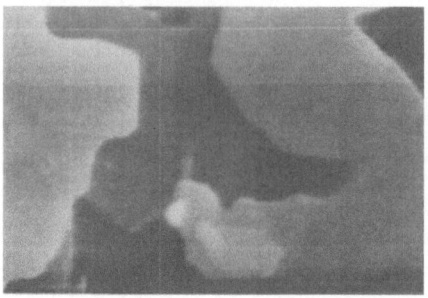

1μm

Fig. 1. SEM image of sputtered Bi-Sr-Ca-Cu-O thin films.

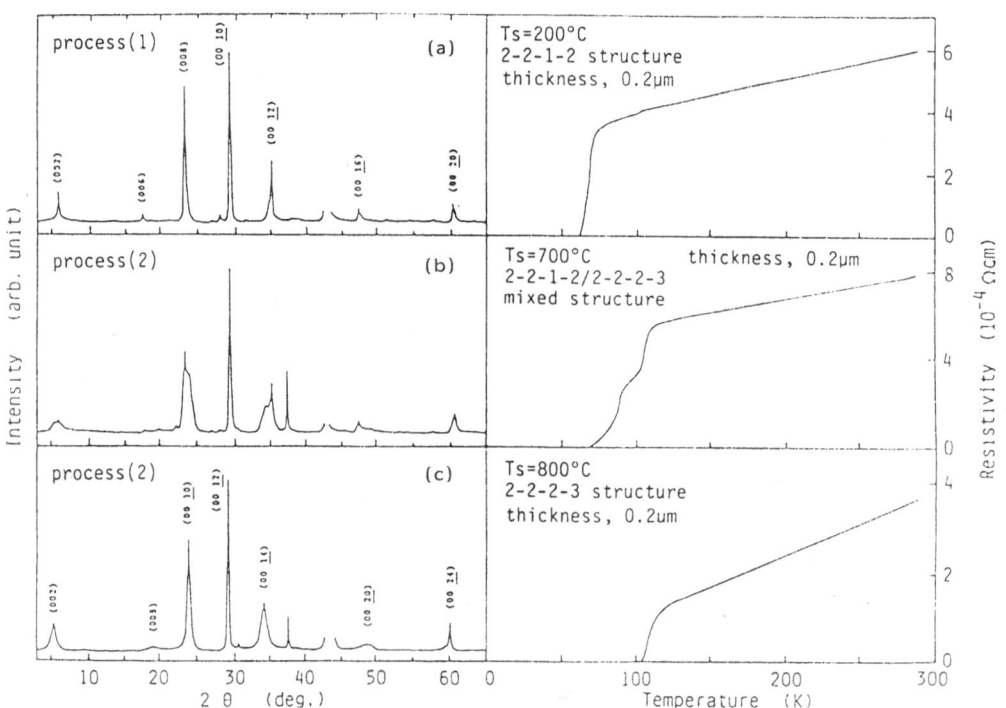

Fig. 2. X-ray diffraction patterns with resistivity-temperature
characteristics for the sputtered Bi-Sr-Ca-Cu-O thin films.

Figure 3 shows typical X-ray diffraction patterns with resistivity-temperature characteristics for the Tℓ-Ba-Ca-Cu-O thin films annealed at different conditions.

The 0.4 μm thick film exhibits the low temperature phase, $Tℓ_2Ba_2CaCu_2O_x$ structure, with the lattice constant $C \simeq 29$ Å after slight annealing at 900°C 1 min [Fig. 3 (a)]. The 2 μm thick films heavily annealed at 900°C 13 min show the high temperature phase, $Tℓ_2Ba_2Ca_2Cu_3O_x$ structure, with the lattice constant $C \simeq 36$ Å [Fig. 3 (b)]. In the specific annealing condition, without Tℓ vapour, the other superconducting phase $TℓBa_2Ca_3Cu_4O_x$ structure with the lattice constant $C \simeq 19$ Å is also obtained [Fig. 3 (c)]. These sputtered films show rough surface morphology as shown in Fig. 4.

Table 3. Sputtering conditions

Target	Tℓ:Ba:Ca:Cu:=2:1-2:2:3	100 mm in diameter
Sputtering gas	$Ar/O_2=1$	
Gas pressure	0.5 Pa	
rf input power	100 W	
Substrate temperature	200°C	
Growth rate	70 Å/min	

PHASE CONTROL BY LAYER-BY-LAYER DEPOSITION

The microstructure of these rare-earth-free superconducting thin films comprises the different superconducting phase. TEM image suggests the sputtered Bi-Sr-Ca-Cu-O thin films comprize 2-2-1-2, 2-2-2-3, and 2-2-3-4 structure as shown in Fig. 5. The presence of the mixed phase is also confirmed by the spreading skirt observed in the X-ray diffraction pattern at the low angle peak around $2\theta \simeq 4°$.

It is reasonably considered that the presence of the mixed phases results from the specific growth process of the present rare-earth-free superconducting thin films: The rare-earth-free superconducting thin films may be molten during the annealing process. The superconducting phase will be formed during the cooling cycle. The crystallinity during the deposition affects the superconducting phase of the postannealed films. The multi-phase is frequently observed for the films annealed from amorphous films which are prepared by the process (1). For the crystallized films which are prepared by the process (2) the appearance of the multi-phase will be suppressed during the annealing.

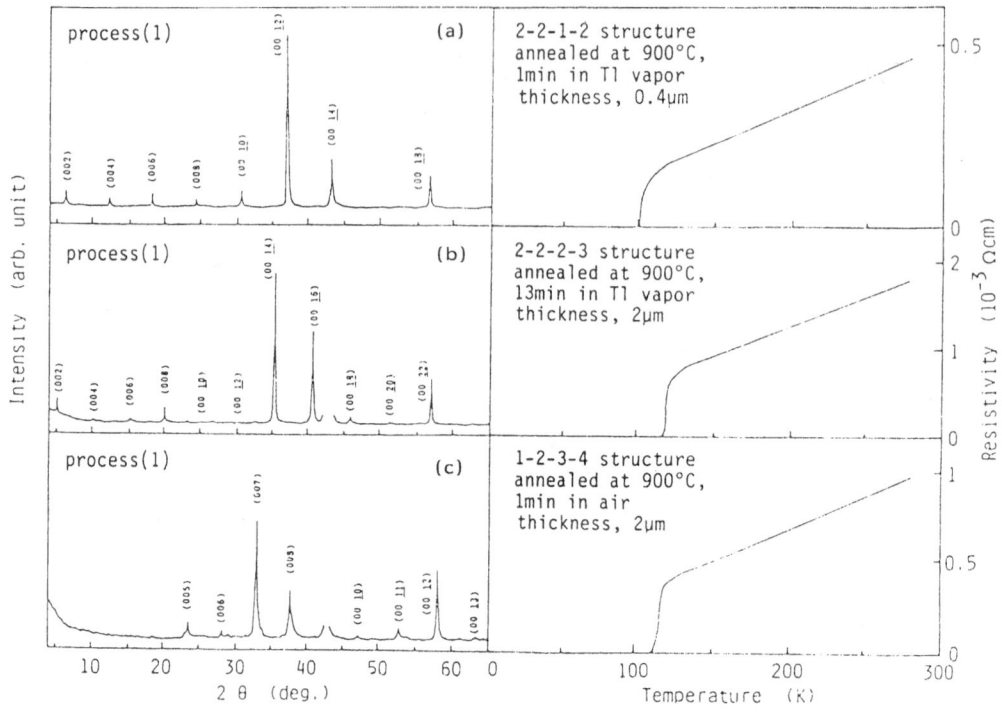

Fig. 3. X-ray diffraction patterns with resistivity-temperature
characteristics for the sputtered Tℓ-Ba-Ca-Cu-O thin films.

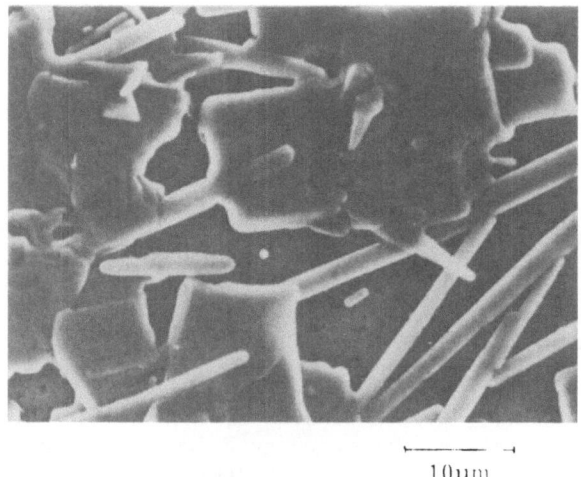

10µm

Fig. 4. SEM image of sputtered Tℓ-Ba-Ca-Cu-O thin films.

XPS measurements for the crystallized Bi-Sr-Ca-Cu-O films suggest that the annealing process modifies the crystal structure near the Cu-O$_2$ layer, increase of density of Cu^{3+}. The Bi-O layered structure is stable during the annealing [19]. This implies that the single superconducting phase will be synthesized when the Bi-O basic structure is crystallized and the stoichiometric composition is kept for the unit cell of the Bi-Sr-Ca-Cu-O.

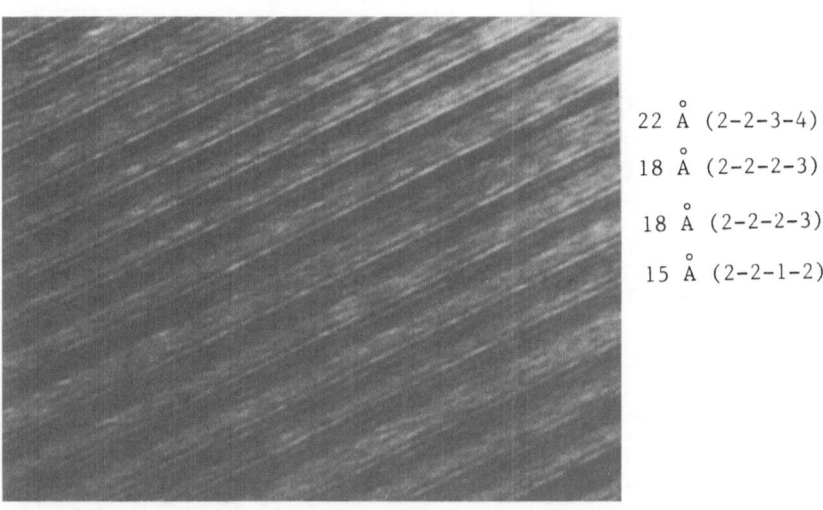

22 $\overset{\circ}{A}$ (2-2-3-4)

18 $\overset{\circ}{A}$ (2-2-2-3)

18 $\overset{\circ}{A}$ (2-2-2-3)

15 $\overset{\circ}{A}$ (2-2-1-2)

Fig. 5. TEM image of sputtered Bi-Sr-Ca-Cu-O thin films.

These considerations have been confirmed by the experiments: The layer by layer deposition was conducted by the multi-target sputtering system shown in Fig. 6. Typical sputtering conditions are shown in Tab. 4. The deposition rate is selected so as to pile up the Bi-O, Sr-O, Cu-O$_2$, and Ca layer in an atomic scale range. The substrate temperature was kept around crystallizing temperature of 650°C. Figure 7 shows the typical results for the layer by layer deposition. It is noted that the phase control is achieved simply by the amounts of Cu-Ca-O during the layer by layer deposition. The experiments show that the Tc does not increase monotonously with the numbers of the Cu-O layer. In the Bi bi-layer system the Tc shows maximum, 110 K, at three layers of Cu-O, Bi$_2$Sr$_2$Ca$_2$Cu$_3$O$_x$. At the four layers of Cu-O, Bi$_2$Sr$_2$Ca$_3$Cu$_4$O$_x$, Tc becomes 90 K [4].

CONCLUSIONS

The present layer by layer deposition is one of the most promising process for a fine control of the superconducting phase of the rare-earth-free superconductors. Man-made high-Tc superconductors will be synthesized by the layer-by-layer deposition.

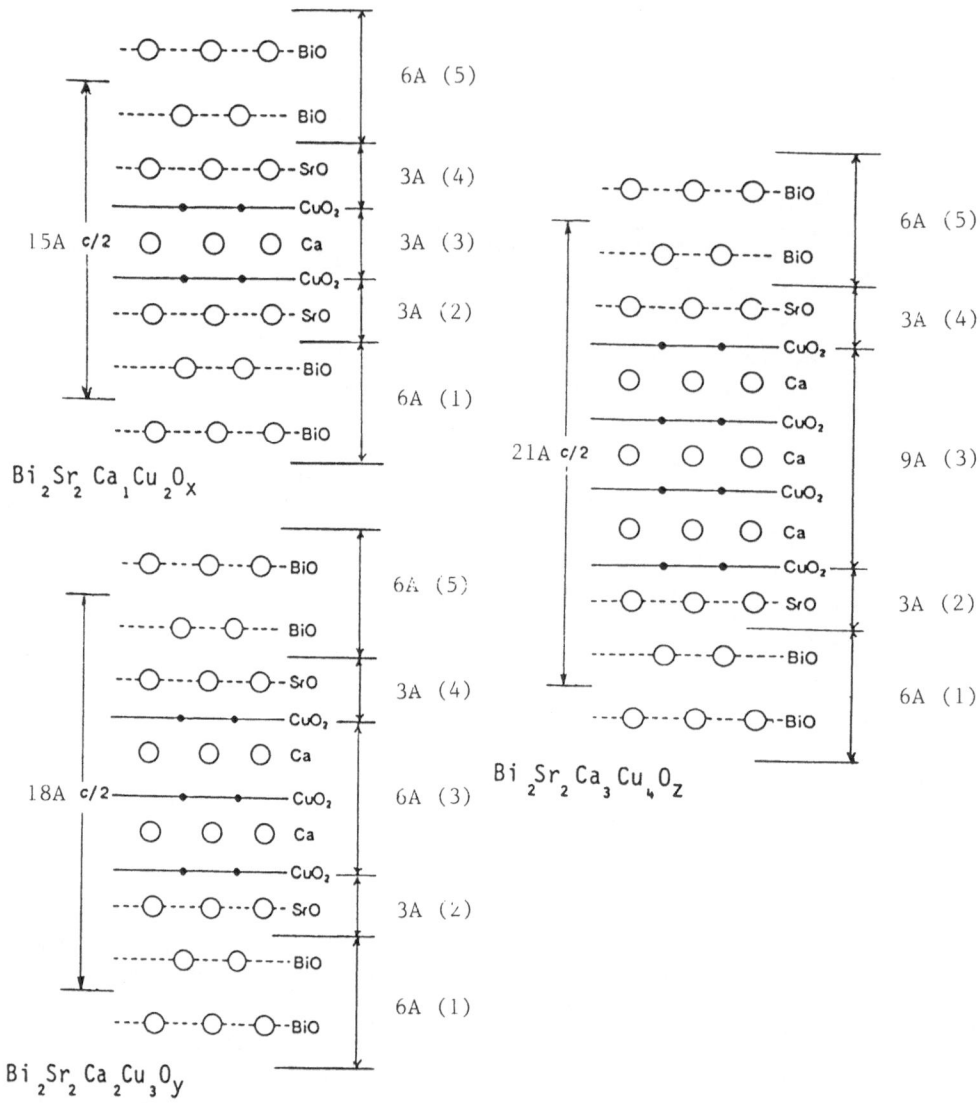

Fig. 6. Layer-by-layer deposition by a multi-target sputtering: alternative deposition in the order (1) → (2) → (3) → (4) → (5).

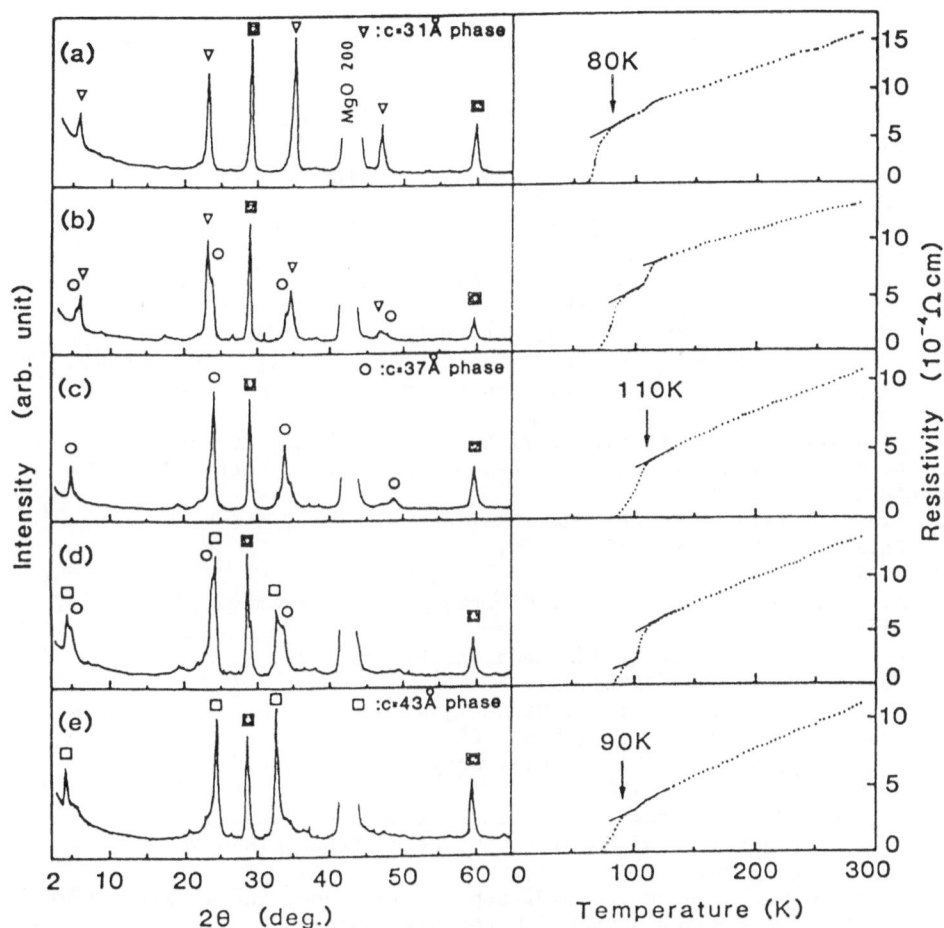

Fig. 7. X-ray diffraction patterns with the resistivity-temperature characteristics for the phase-controlled Bi-Sr-Ca-Cu-O thin films. (a) $Bi_2Sr_2CaCu_2O_x$, (b) $Bi_2Sr_2CaCu_2O_x/Bi_2Sr_2Ca_2Cu_3O_y$, (c) $Bi_2Sr_2Ca_2Cu_3O_y$, (d) $Bi_2Sr_2Ca_2Cu_3O_y/Bi_2Sr_2Ca_3Cu_4O_z$ and (e) $Bi_2Sr_2Ca_3Cu_4O_z$.

ACKNOWLEDGEMENTS

The authors thank S. Kohiki for his XPS analyses. They also thank S. Hayakawa and T. Nitta for their continuous encouragements.

REFERENCES

[1]   C. Michel, M. Hervieu, M.M. Borel, A. Grandin, F. Deslandes, J. Provost, and B. Raveau, Z. Phys. B, 68 (1987) 421.
[2]   J. Akimitsu, A. Yamazaki, H. Sawa, H. Fujiki, Jpn. J. Appl. Phys. 26 (1987) L2080.
[3]   H. Maeda, Y. Tanaka, M. Fukutomi, and T. Asano, Jpn. J. Appl. Phys. 27 (1988) L209.
[4]   H. Adachi, S. Kohiki, K. Setsune, T. Mitsuyu, and K. Wasa, Jpn. J. Appl. Phys. 27 (1988) (in press).
[5]   S. Kondoh, Y. Ando, M. Onoda, M. Sato, and J. Akimitsu, Solid State Commun. 65 (1988) 1329.
[6]   Z.Z. Sheng, A.M. Hermann, A. El Ali, C. Almasan, J. Estrada, T. Datta, and R.J. Matson, Phys. Rev. Lett. 60 (1988) 937.
[7]   Z.Z. Sheng and A.M. Hermann, Nature 332, 138 (1988).
[8]   S.S.P. Parkin, V.Y. Lee, E.M. Engler, A.I. Nazzal, T.C. Huang, G. Gorman, R. Savoy, and R. Beyers, Phys. Rev. Lett. 60 (1988) 2539.
[9]   R. Beyers, S.S.P. Parkin, V.Y. Lee, A.I. Nazzal, R. Savoy, G. Gorman, T.C. Huang, and S. LaPlaca, Appl. Phys. Lett. 53 (1988) 432.
[10] S.S.P. Parkin, V.Y. Lee, A.I. Nazzal, R. Savoy, and R. Beyers, Phys. Rev. Lett. 61 (1988) 750.
[11] H. Ihara, R. Sugise, M. Hirabayashi, N. Terada, M. Jo, K. Hayashi, A. Negishi, M. Tokumoto, Y. Kimura, and T. Shimomura, Nature 334 (1988) 510.
[12] H. Adachi, K. Wasa, Y. Ichikawa, K. Hirochi, and K. Setsune, J. Cryst. Growth, 91 (1988) 352.
[13] Y. Ichikawa, H. Adachi, K. Hirochi, K. Setsune, S. Hatta, and K. Wasa, Phys. Rev. B., 38 (1988) 765.
[14] J.H. Kang, R.T. Kampwirth, K.E. Gray, S. Marsh, and E.A. Huff, Physics Lett., 128 (1988) 102.
[15] K. Wasa, M. Kitabatake, H. Adachi, K. Setsune, and K. Hirochi, American Institute of Physics Conference Processings No.165, New York 1988, p.38.
[16] Y. Ichikawa, H. Adachi, K. Hirochi, K. Setsune, and K. Wasa: Proc. MRS Int. Meeting on Advanced Materials, June 1988, Tokyo (in press).
[17] Y. Ichikawa, H. Adachi, K. Setsune, S. Hatta, K. Hirochi, and K. Wasa, Appl. Phys. Lett., 53 (1988) 919.
[18] J. Zhou, Y. Ichikawa, H. Adachi, T. Mitsuyu, and K. Wasa, Jpn. J. Appl. Phys., (to be submitted).
[19] S. Kohiki, K. Hirochi, H. Adachi, K. Setsune, and K. Wasa, Phys. Rev. B, 38 (1988) (in press).

AS-DEPOSITED SUPERCONDUCTING Y-BA-CU-O THIN FILMS ON Si, SiO_2, GaAs AND Ni/Cu SUBSTRATES BY HIGH PRESSURE DC SPUTTERING PROCESS

R.J. Lin and P.T. Wu

Materials Research Laboratories
Industrial Technology Research Institute
Chutung, Hsinchu 31015, Taiwan, R.O.C.

ABSTRACT

The superconducting Y-Ba-Cu-O thin films on Si, SiO_2, GaAs and Ni/Cu substrates have been reproducibly prepared by high pressure DC sputtering process without further annealing treatment. The superconductivity of the films is Tc(onset)=95K and Tc(zero)=70K. The targets were Y-Ba-Cu-O compounds made by solid state reaction. The substrate temperature is lower than 450°C. X-ray analysis indicates that the 0.6~2 μm films consist of predomintly the orthorhombic superconducting phase of YBa_2Cu_3Ox with traces of CuO. The effect of processing parameters on the superconductivity of the films will be discussed.

INTRODUCTION

Generally, high Tc superconducting films are prepared by annealing at 900°C in O_2 atmosphere after deposition. This high temperature promotes the interdiffusion between semiconductor substrates and film. The superconducting films can be prepared on them by using a buffer layer[1] between film and substrate. For the sake of compatibility with present semiconductor processing, a low temperature procedure to limit interdiffusion is more desirable.

It is well known[2] that the high temperature stable compounds can be synthesized at low temperature by plasma enhanced processes. The sputtering process is one of them. Meanwhile, we know that the ionization degree of gas[3] and the preferential ionization factor[4] of O_2 in Ar increase with increasing total pressure during sputtering. These favor the formation of superconducting film at low temperature. Therefore, we try to deposit Y-Ba-Cu-O films by high pressure DC sputtering process. The as-deposited superconducting Y-Ba-Cu-O films on SiO_2[5] and Si[6,7] have been successfully prepared. In this paper, the effect of important parameters, gas pressure and target composition, on superconductivity and composition of the films, the interface between film and substrate, and magnetization of the films will be discussed.

EXPERIMENTS

The Y-Ba-Cu-O films were prepared by the high pressure DC planar diode sputtering process from a sintered Y-Ba-Cu-O compound target. The target(diameter 4.5cm; thickness 0.4cm) was made by a solid state reaction of Y_2O_3, $BaCO_3$ and CuO in the stoichiometric ratio of Y:Ba:Cu=1:2:0.8, 1:2:1.5, 1:2:3. The base pressure of the vacuum system prior to deposition was 1×10^{-3} torr. During deposition, the sputtering atmosphere was a mixture of Ar and O_2 gases whose ratio of volume flow rate was 1. The deposition pressure was 1~2.5 torr. The target-to-substrate separation was 2 cm. The substrates were (100)Si wafer, amorphous quartz, (100)GaAs wafer, and amorphous P-containing Ni film on Cu foil(thickness 1mm). Ni films were grown by electroless plating method[8]. The substrate was heated to 380°C by IR quartz heater. The stable temperature was 400~430°C. The reported temperature was measured at back side of film-growing face of substrate with K type thermocouple. The naked measuring head was kept close touch with substrate surface. The sputtering voltage and current were 240~260 volt and 0.7~0.8 amper, respectively. The thickness of the films was 1~2μm and the deposition rate was 2~4A°/sec. After deposition, the gas ionization system and rotary vacuum pump were turned off. Immediately, the chamber was backfilled with oxygen to above 600 torr and the film was cooled to below 100 C in this oxygen environment. The typical cooling time from deposition temperature to below 100 C was about 40 min..

The structure of the films was characterized by X-ray diffractometer (Philips PW 1700) with monochromated CuKα radiation(40KV; 30mA). The composition of the films was examined by energy dispersive spectrometer (EDAX 9100/70). The depth profile of film composition was analyzed by Auger spectrometer(V.G. Scientific microlab III). The thickness of the films was determined by the surface profilometer. The resistance of the films was measured by AC four point method(Linear Research LR400) using Ag paste contacts. The magnetization of the films was measured by a SQUID(Quantumn Design) magnetometer.

RESULTS AND DISCUSSION

The effect of sputtering gas pressure on the resistance-temperature curves of the films is shown in Fig.1. The superconductivity of the films is strongly affected by the sputtering gas pressure. There is an optimum pressure value in the range of 1 to 2.5 torr. This result can be attributed to the dependence of compositions and phases of the films on the gas pressure, which is shown in Table 1 and Fig.2. The Cu content and atomic ratio between Ba and Y, Ba/Y, increase with increasing sputtering gas pressure. This is consistent with the X-ray diffraction patterns of the films. The nonsuperconducting impurity phases change from $Y_2Ba_1Cu_1O_5$ and CuO to $BaCuO_2$ and CuO with increasing the sputtering gas pressure. The semiconducting behavior and long tail in the resistance-temperature curves of the films are caused by these impurity phases.

The effect of the target composition on the resistance-temperature curves of the films is shown in Fig.3. The films prepared from the targets with Ba/Y=2 have a resistance drop near 90K. The temperature of zero resistance can be greatly improved by adjusting the Cu content in the target. Fig.3 shows that Tc(zero)=62K and Tc(zero)=70K have been obtained by using YBa_2Cu_3Ox and $YBa_2Cu_{1.5}Ox$ target, respectively. This results is consistent with the dependence of compositions and phases of the films on the target composition, as shown in Table 2 and Fig.4. The Cu content in the films decreases with decreasing the Cu content of the target. The X-ray diffraction patterns show that the nonsuperconducting impurity phases are CuO and $Y_2Ba_1Cu_1O_5$.

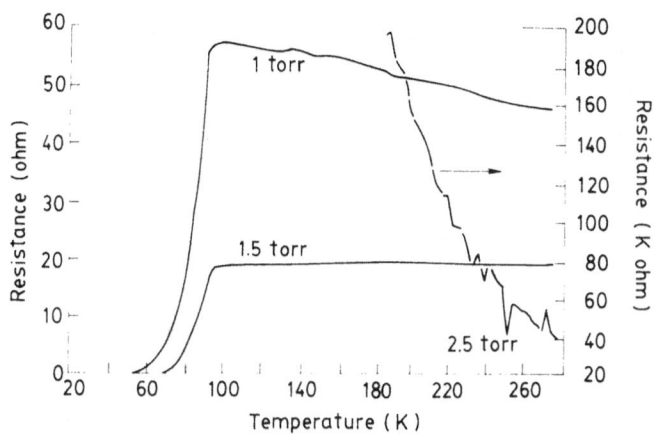

Fig.1 The effect of the sputtering gas pressure on the
resistance-temperature curve of the Y-Ba-Cu-O films.

Table 1 The effect of the sputtering gas pressure on the film
composition (target composition YBa$_2$Cu$_3$Ox).

No	gas pressure (torr)	Film Composition					
		ratio			atomic % (Y+Ba+Cu=100%)		
		Y	: Ba	: Cu	Y	Ba	Cu
1	1	1	1.1	3.4	18.2	20	61.8
2	1.5	1	1.9	8.8	8.5	16.2	72.5
3	2.5	1	5.5	26.2	3.1	17.3	82.6

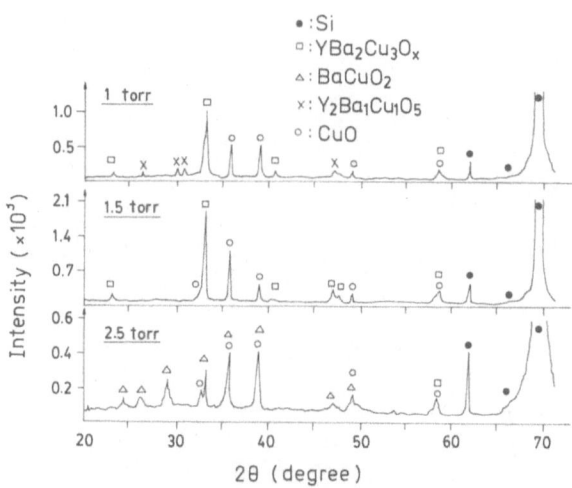

Fig.2 The effect of the sputtering gas pressure on the
diffraction patterns of the Y-Ba-Cu-O films.

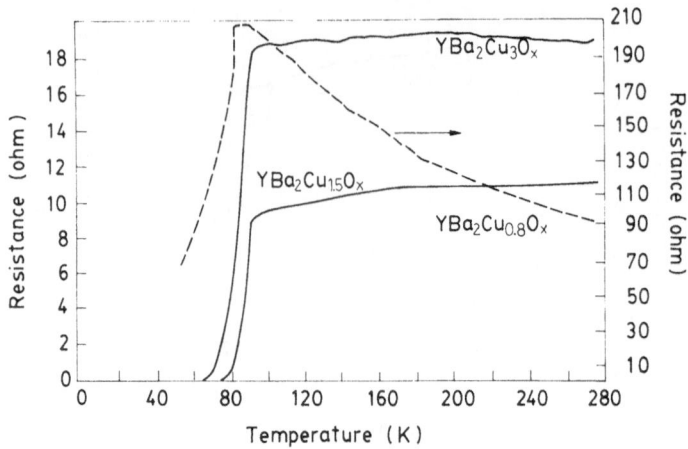

Fig.3 The effect of the target composition on the
resistance-temperature curve of the Y-Ba-Cu-O films.

Table 2 The effect of the target composition on the film
composition (gas pressure 1.5 torr).

No	Target	Film Composition					
		ratio			atomic % (Y+Ba+Cu=100%)		
		Y :	Ba :	Cu	Y	Ba	Cu
1	YBa$_2$Cu$_{0.8}$ Ox	1	1.6	5.2	12.8	20.5	66.7
2	YBa$_2$Cu$_{1.5}$ Ox	1	1.6	6.2	11.4	18.2	70.4
3	YBa$_2$Cu$_{3.0}$ Ox	1	1.9	8.8	8.5	16.2	72.5

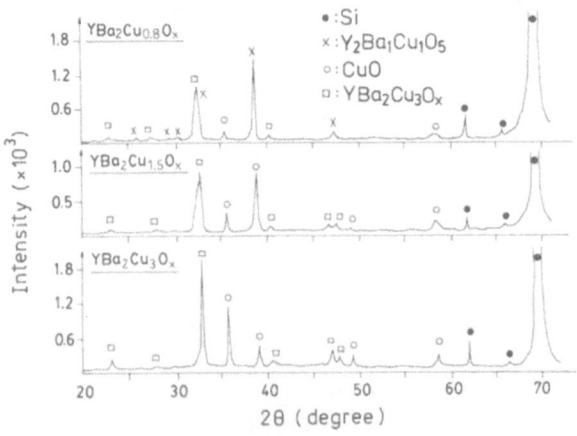

Fig.4 The effect of the target compositions on the diffraction
patterns of the Y-Ba-Cu-O films.

The above-mentioned results show that the Cu content in the films is much higher than that of the target. This agrees with the composition variation on the target surface before and after sputtered, shown in Table 3. The Cu and Ba content on the target surface greatly reduces, but the Y content increases largely after the target is sputtered. It is not easy to explain the relationship between film composition and sputtering parameters without in situ monitoring the glow-discharge phenomena. We conjecture that possible factors are the followings: the differences of vapor pressure[9], sputtering yield[10], atomic or ionic radius[11] and ionization energy[11] among Y, Ba and Cu components; strong plasma environment and bombardment of substrate surface by high energy particles[12]; high temperature of target surface caused by high glow-discharge current density.

Table 3 Variation of target composition ($YBa_2Cu_{1.5}Ox$) after sputtered (gas pressure 1.5 torr).

| No | | Composition of Target Surface | | | | | |
| | | ratio | | | atomic % (Y+Ba+Cu=100%) | | |
		Y	: Ba	: Cu	Y	Ba	Cu
1	near edge on surface	1	1.2	0.2	41.6	50	8.4
2	center on surface	1	1.3	0.3	38.5	50	11.5
3	inside the target (unaffected area)	1	2.1	1.4	22.2	46.7	31.1

Fig.5 indicates that the reproducibility of the process to prepare the as-deposited superconducting films on Si substrate by a single compound target. It shows that the films of three runs almost have same superconducting behaviors. This is consistent with their X-ray diffraction patterns, shown in Fig.6. This is very interesting because the previous results show the composition of target surface dramatically change before and after sputtered. The possible reason to obtain good reproducibility is that the composition of the target surface may rapidly attain the steady state during sputtering. This can be verified by the analysis of AES depth profile of Y-Ba-Cu-O film on Si substrate, shown in Fig.7. As seen in the figure, the concentration of Y, Ba, Cu and O are constant throughout the entire film. In addition, the interdiffusion between Y-Ba-Cu-O film and Si substrate is very limited from estimate of sputtering etch time. This is consistent with the result of a cross-section TEM micrograph of a thin film on Si, shown in Fig.8. The little diffusion between film components and substrate occurs, so the Y-Ba-Cu-O films can be successfully grown on Si substrate.

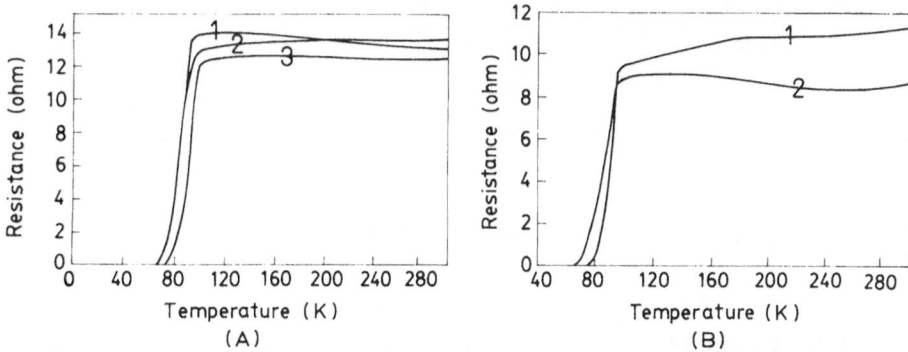

Fig.5 Reproducibility of process to prepare as-deposited films on Si; (A) YBa_2Cu_3Ox target (B) $YBa_2Cu_{1.5}Ox$ target.

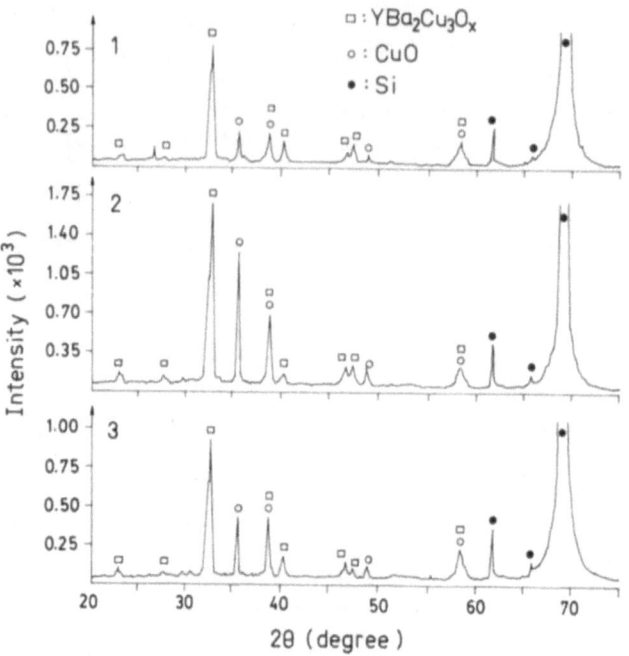

Fig.6 X-ray diffraction patterns of reproducible films, shown in Fig.5(A).

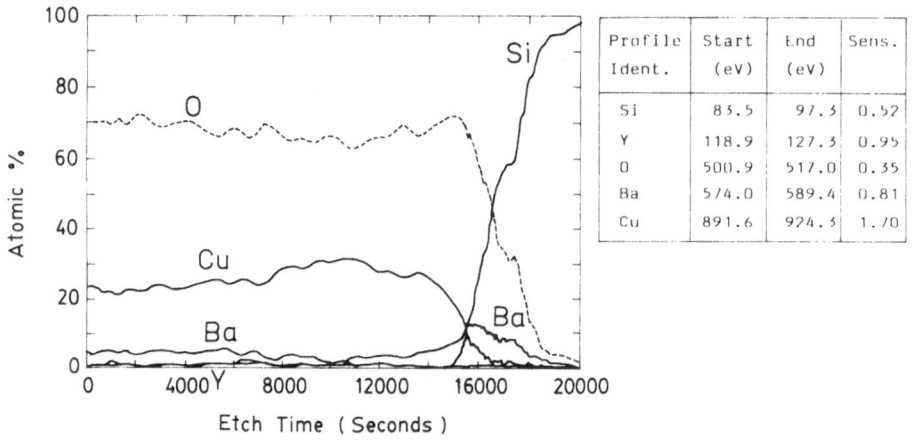

Profile Ident.	Start (eV)	End (eV)	Sens.
Si	83.5	97.3	0.52
Y	118.9	127.3	0.95
O	500.9	517.0	0.35
Ba	574.0	589.4	0.81
Cu	891.6	924.3	1.70

Fig.7 AES profile of a as-deposited film on Si.

Except Si substrate, the superconducting Y-Ba-Cu-O films also have been successfully deposited on GaAs, quartz and Ni/Cu substrates. Their typical resistance-temperature curves are shown in Fig.9. They all have marked sharp resistance drop near 90K. It means that their superconducting phase is same.

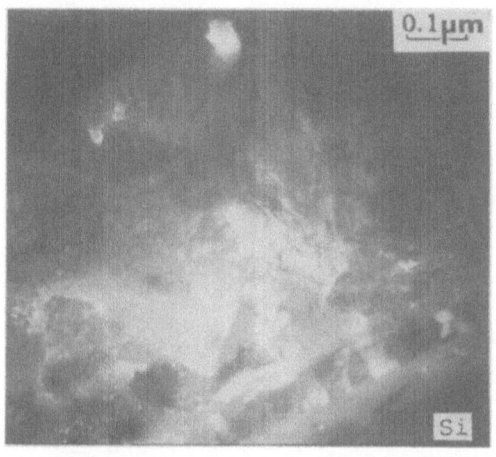

Fig.8 A cross-section TEM micrograph of a Y-Ba-Cu-O
      film on Si substrate.

The semiconducting behavior in the normal state and long tail are caused by
the deviation of film composition from YBa_2Cu_3Ox and generation of nonsuperconducting impurity phase. The magnetization of the film on GaAs substrate
is shown in Fig.10. There are strong diamagnetic shield and Meissner effect.
This confirms the superconductivity of as-deposited Y-Ba-Cu-O films.

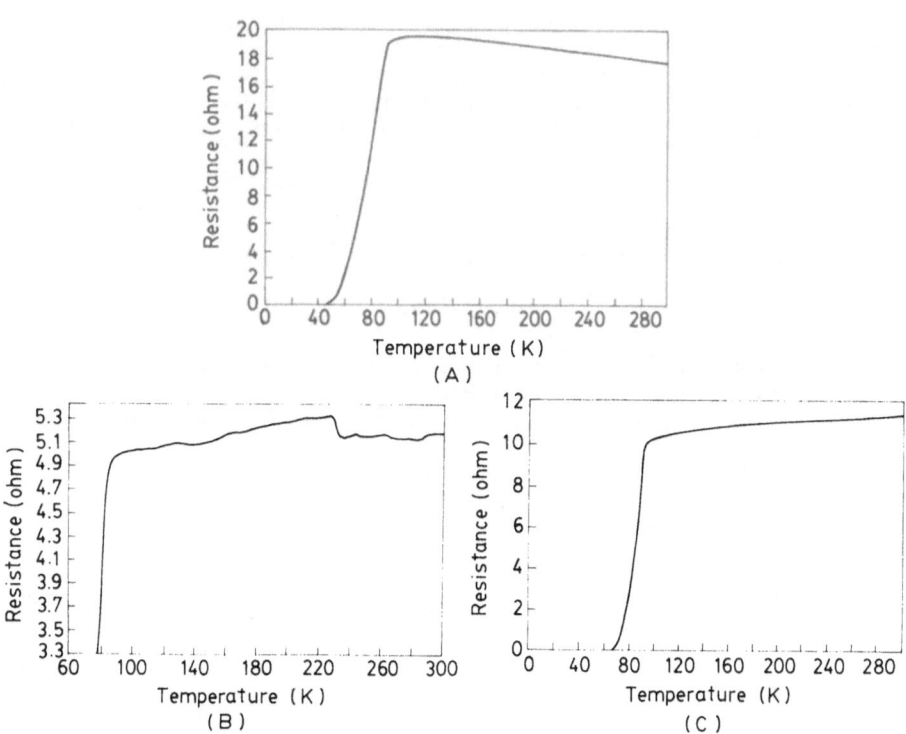

Fig.9 Typical resistance-temperature curves of the as-deposited
      Y-Ba-Cu-O film on (A) GaAs substrate (B) Ni/Cu (C) SiO_2.

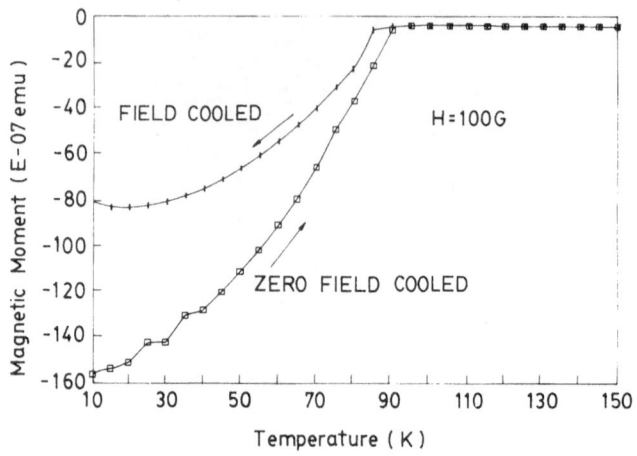

Fig.10 Magnetization curve of Y-Ba-Cu-O film on GaAs substrate.

In summary, as-deposited superconducting Y-Ba-Cu-O films have been reproducibly prepared on Si, SiO$_2$, GaAs and Ni/Cu substrates by high pressure Dc sputtering system. The superconductivity, composition and phases of the films are strongly affected by sputtering gas pressure and target composition. The interdiffusion between film and substrate is limited. The improvement of film quality and film characterization are in progress.

REFERENCES

1. A. Mogro-campero and L.G. Turner, Appl. Phys. Lett. 52:1185(1988).
2. V. Valvoda etal., Thin Solid Films 156:53(1984).
3. Brian Chapman, "Glow Discharge Processes", John Wiley & Sons, p.78(1980).
4. M. Hecq, etal., Thin Solid Films 115:L45(1984).
5. R.J. Lin, Y.C. Chen, J.H. Kung and P.T. Wu, Mat. Res. Soc. Symp. Proc. 99:319(1987).
6. R.J. Lin, J.H. Kung and P.T. Wu, Physica C 153-155:796(1988).
7. R.J. Lin, J.H. Kung and P.T. Wu, Proc. MRS Int. Meet. on Advanced Materials, Tokyo, MRS(1988).
8. R.J. Lin and John Lin, Proc. MRS Int. Meet. On Advanced Materials, Tokyo, MRS(1988).
9. CRC Handbook of Chemistry and Physics, edited by R.C. Weast, 58th edition (1978).
10. John A. Thornton, in Deposition Technologies for Films and Coating, edited by R.F. Bunshah, Noyk, Park Ridge(1982).
11. W.D. Kingery, et al., in "Introduction to Ceramics", John Wiley & Sons (1976).
12. Norio Terada, et al., Jpn. J. Appl. Phys., 27:L639(1988).

SUPERCONDUCTING PROPERTIES OF MAGNETRON SPUTTERED Bi-Sr-Ca-Cu-O AND Tl-Ba-Ca-Cu-O THIN FILMS

R.T. Kampwirth, J.H. Kang, and K.E. Gray

Argonne National Laboratory
9700 S. Cass Ave.
Argonne, IL 60439

ABSTRACT

Thin films of high temperature superconducting Bi-Sr-Ca-Cu-O and Tl-Ba-Ca-Cu-O have been made using multiple source magnetron sputtering. We will discuss preparation and annealing and how they affect the superconducting properties. The Bi-based films form in at least two compounds, a lower T_c phase with a 2212 composition and $T_{c0} \approx 80$ K and a higher T_c phase with a nominal composition of 2223 and a $T_c \approx 105$ K. The higher T_c phase is very difficult to form so that a complete superconducting path exists, requiring long annealing times at temperatures somewhat below the formation temperature. Although superconducting properties are less sensitive to stoichiometry than $YBa_2Cu_3O_{7-x}$, annealing temperatures must be held in a narrow range around 865°C to get the best results. Films prefer growing on (100) MgO in a highly textured polycrystalline form, with the c-axis perpendicular to the plane of the film. The Tl-based films form at least three compounds (2212, 1223, 2223) with T_c's ranging from ≈ 100 K to 114 K and grow best on (100) and polycrystalline $Y-ZrO_2$ substrates. Annealing conditions are particularily difficult because of the volatility of thallium and require samples to be annealed in sealed capsules with a controlled atmosphere of thallium and oxygen. Upper critical field measurements show a large anisotropy (≈ 15 for Bi-Sr-Ca-Cu-O and ≥ 70 for Tl-Ba-Ca-Cu-O) in the parallel to perpendicular field slope ratio as might be expected from a highly textured anisotropic structure.

INTRODUCTION

With the discovery of high T_c superconductors (HTS) , particularily $YBa_2Cu_3O_{7-x}$ (YBCO), it soon became evident that films of this material made by a variety of techniques including coevaporation[1], ion-beam sputtering[2], magnetron sputtering[3] and laser evaporation[4] could have critical current densities $J_c \approx 10^6$ A/cm^2 at H=0T in liquid nitrogen, suggesting possible applications in, e.g., superconducting interconnects and SQUIDS. However the requirements to hold stoichiometry in YBCO films to within $\approx 1\%$ of the correct 123 value and to grow films on single crystal $SrTiO_3$ in order to achieve the highest J_c's, could limit potential applications.

Recently superconductivity was found in bulk samples of two different phases of Bi-Sr-Ca-Cu-O compounds (BSCCO) one with T_c of ≈ 80K and the other of ≈ 110K, and methods similar to those used to prepare YBCO thin films were employed to make thin films of BSCCO compounds[5-7]. Even though there have been numerous reports about the successful preparation of thin films of BSCCO compounds, it has been known that it is very difficult to prepare thin films of the higher T_c phase. Compared to the strict requirements on stoichiometry needed to get good quality YBCO films, good quality lower T_c phase BSCCO thin films have been reported in a rather broad range of metallic compositions.

Most recently, Hazen, et.al[8] showed that replacing Bismuth and Strontium with Thallium and Barium raised T_{c0} in bulk samples to about 125K. Since then Kang, et.al[9] and Ginley, et.al[10] reported films of $Tl_2Ba_2Ca_1Cu_2O_x$ (T2212) with $T_{c0} \approx 100$K, and W.Y. Lee, et.al[11] reported films of $Tl_2Ba_2Ca_2Cu_3O_x$ (T2223) with $T_{c0} \approx 120$K. Toxicity and volatility of Tl are pointed out to be major obstacles in making good quality thin films of Tl-Ba-Ca-Cu-O compounds (TBCCO).

In this paper we report the conditions necessary in preparing BSCCO and TBCCO thin films using magnetron sputtering, and their superconducting properties.

SAMPLE PREPARATION AND CHARACTERIZATION

High T_c superconducting thin films were prepared by using a three-gun dc magnetron sputtering system equipped with a turbomolecular pump, which provided a typical base pressure of low 10^{-8} torr. The three dc magnetron sputtering guns are aimed at a common point about 6 inches above the sources, which provides compositional uniformity to ± 1 % over a 2 cm^2 substrate area. Targets of Bi, Cu and a 1:1 SrCa mixture for BSCCO thin films, and Tl, Cu and a 1:1 BaCa or a 2:3 BaCa mixture for TBCCO thin films were simultaneously sputtered in a 20 mTorr argon atmosphere with an oxygen partial pressure of ≈ 0.1 mTorr being introduced directly adjacent to the substrates. A load-lock mechanism permitted changing samples without breaking vacuum in the growth chamber, which was crucial to maintaining reproducible conditions in successive runs, by providing less contaminated target surfaces. A quartz crystal monitor was placed next to the substrates to determine the sputtering rates of each source prior to starting a deposition. Substrates were mostly kept between 50 and 200°C during the deposition. Outgassing substrates at 400-500°C prior to deposition was determined important to obtaining good film adherence. Ex-situ post-annealing treatment was done in a flowing oxygen atmosphere. Film thicknesses ranged from 300-700 nm.

BSCCO thin films were deposited onto (100) oriented single-crystal MgO substrates. These films showed $T_{c0} \approx 80$K and ICP-AES average composition to be $Bi_{1.4}Sr_1Ca_{0.96}Cu_{1.53}O_x$. The typical schedule for post-annealing was 1 hr at 770 °C followed by 5 min. at 865 °C and 1 hr at 835 °C in flowing oxygen. The film quality was most sensitive to the highest temperature, which occurred during the 5 min. step. Other batches of films where the variations of average metallic compositions are more than 50% show T_{c0} values distributed between 69 K and 80 K, indicating the relaxed stoichiometry requirements in making superconducting BSCCO thin films. The use of air instead of oxygen resulted in the degradation of the surface, although T_c was unchanged.

TBCCO thin films were deposited onto (100) oriented single crystal or polycrystalline ZrO_2-9%Y_2O_3 substrates. In order to avoid the loss of the highly volatile Tl during the annealing process, the films were made with excess Tl and sealed in a closed Au crucible, then placed in a flowing oxygen tube furnace and annealed at 850-890 °C for about 5-30 minutes.

Alternatively, a small quantity of bulk TBCCO powder was added to the crucible containing the sample and annealed. The results were similar for both methods. A 1:1 BaCa mixture sputtering target was used to make T2212 thin films and a 2:3 BaCa mixture sputtering target was used to make T1223 and T2223 thin films. Higher annealing temperatures and longer annealing times were required to make T2223 thin films. Scanning electron micrographs of TBCCO films appear to indicate a smooth dense surface. There is an interconnected backbone structure, which looks topologically similar to that found in epitaxial $YBa_2Cu_3O_{7-y}$ films on $SrTiO_3$ substrates[12], except for the lack of epitaxial orientation for our Tl-Ba-Ca-Cu-O films. The composition of TBCCO thin films measured by using energy dispersive x-rays agrees reasonably well with each structure.

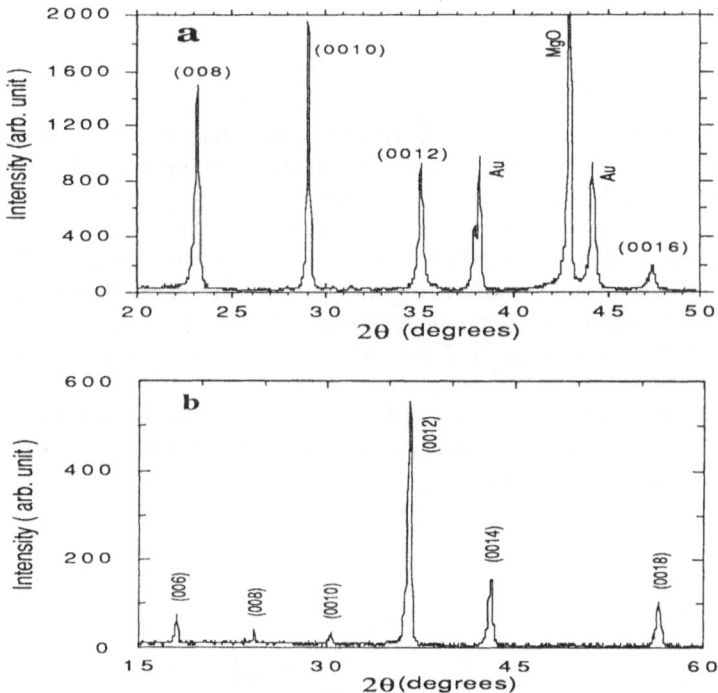

Fig. 1. X-ray diffraction patterns for BSCCO film(a) and TBCCO film (b). In (a) there are only multiples of the (002) line reported[2] for the 84K superconducting phase, expect for peaks from the substrate and Au contact pads. Also in (b) only the multiples of the (002) line are shown, indicating a strong c-axis orientation perpendicular to the plane of the film.

X-ray diffraction of our BSCCO and TBCCO thin films indicates a high degree of orientation with the c-axis perpendicular to the substrate, as shown in Fig. 1 by the θ-2θ scan normal to the film surface. Rocking curves made on our best samples gave a full width half maximum of 0.7°. In fact, it has not been possible to produce BSCCO and TBCCO films with other than a c-axis orientation, which is in stark contrast to YBCO films which are difficult to grow in the preferred c-axis orientation.

EXPERIMENTAL TECHNIQUE

The resistance was measured vs. temperature on high T_c thin films using a standard four-probe method and a current density of 1-10 A/cm². Electrical contacts were made with silver paint. The upper critical fields $B_{c2}(T)$ were measured resistively in dc magnetic fields up to 12 T in a gas-flow helium cryostat. The films were in contact with exchange gas and the temperature was controlled by feedback from a capacitance thermometer to a heater, both of which were directly attached to the sample block. Sample temperatures were measured using a magnetoresistance corrected carbon glass thermometer also attached to the sample block. Data was taken by sweeping the temperature at a fixed magnetic field. Films were mounted with the plane of a film either perpendicular or parallel to the direction of the magnetic field. In both orientations, the current flow was in the plane of the film and perpendicular to the field direction.

RESULTS

Bi-Sr-Ca-Cu-O Films

Annealing. In BSCCO thin films, the post-annealing conditions for high-quality films were particularly stringent. For example, it was found that annealing at 865°C gave a higher T_{c0} than annealing at 855 or 875°C. X-ray diffraction indicated impurity peaks for the samples annealed at 855 °C and 875 °C, which were absent in the sample annealed at 865 °C. Similar trends were observed in other batches of films in which the T_{c0} values were distributed between 69 K and 80 K. It is sometimes possible to improve T_{c0} by long annealing times at a temperature slightly lower than the optimum temperature to form the superconducting phase. For example, Fig.2 shows resistivity vs. T for a BSCCO film annealed using the times and temperatures previously discussed to obtain maximum T_{c0} (curve I) and the improvement obtained by annealing for 24 hrs. in air at 825°C (curve II). It should be noted that the 110K phase of BSCCO has begun forming, suggesting that this phase can grow out of the lower T_c phase if annealing times are increased.

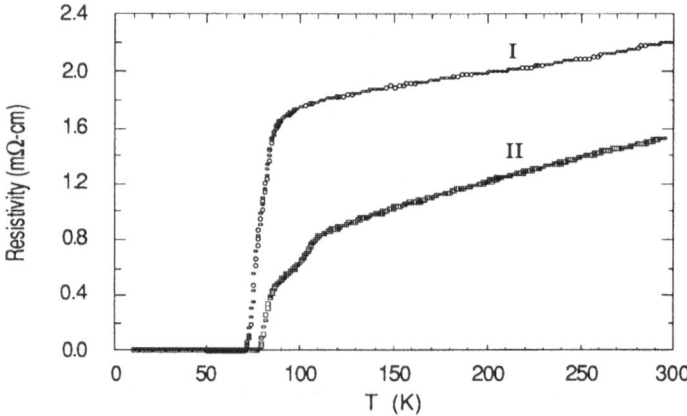

Fig.2. Resistivity vs. temperature for the lower T_c phase of BSCCO (I) using the recommended annealing program described in the text and, (II) the same film after reannealing at 825°C in air for 24 hrs. Note the 110K phase has begun forming after reannealing.

<u>Upper Critical Field</u>. Fig.3 shows the resistive transition as a function of H for each orientation. The determination of B_{c2} from the data shown in Fig.3 is not straightforward due to the resistance 'tails' developing at low temperatures and higher values of B. Since the curves are neither sharp, nor similarly shaped, the value of B_{c2} will depend on the criterion used. The onsets of superconductivity are extremely insensitive to field, while the zero-resistance points are extremely sensitive. The origin of these tails is not completely understood and suggestions include inhomogeneities and magnetic flux creep[13]. Since neither of these are intrinsic properties of the pure superconducting material, but rather depend on defects within a given sample, it is argued here that measurements at resistances above the tail are more appropriate for studying a materials intrinsic anisotropy.

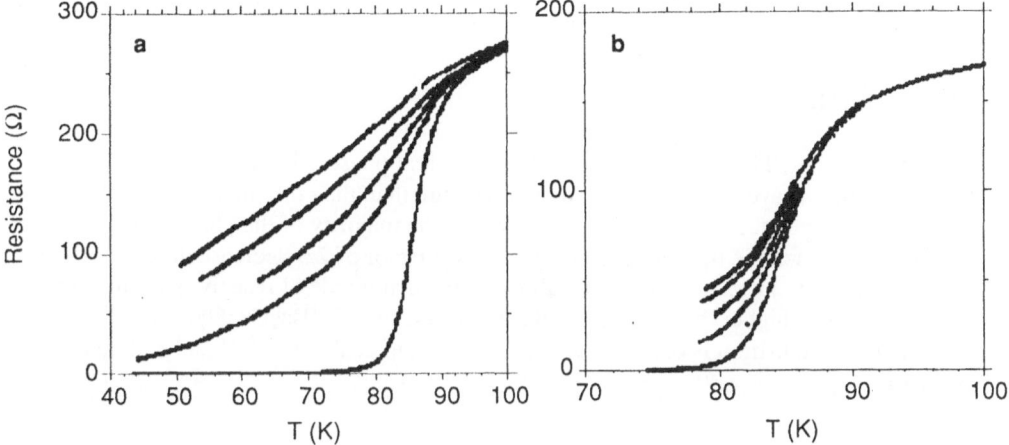

Fig.3. Resistive transitions as a function of applied field (0,2,4,8, and 12 T) for BSCCO film. (a) Field perpendicular to the film plane and parallel to the c-axis. (b) Field parallel to the film surface and the a-b planes.

As a result of these considerations, data for the midpoints of the resistive transition (50% of the normal state resistance just above T_c) are presented in Fig.4, and show a large temperature dependence for fields parallel to the a-b planes, i.e., parallel to the film surface and perpendicular to the c-axis. In Fig. 4, we find $B_{c2\parallel}' = -(dB_{c2}/dT)_{T \approx T_c} = 8.5$ T/K, and $B_{c2\perp}' = 0.56$ T/K giving an anisotropy of ≈ 15 which is much larger than in YBCO films. Critical fields are used to calculate the coherence length, ξ, from the Ginzburg-Landau relation: $B_{c2} = \Phi_0/2\pi\xi^2$, where $\Phi_0 = 2.07 \times 10^{-15}$ Tm^2 is the flux quantum. In the a-b plane of the HTS, the superconducting properties are reasonably isotropic and are given by a single $B_{c2\perp}$ value, so that $\xi_{ab} = (\Phi_0/2\pi B_{c2\perp})^{1/2}$. In parallel fields, the coherence length perpendicular to the a-b plane is given by[14] $\xi_c = (\Phi_0/2\pi B_{c2\parallel})^{1/2} (B_{c2\perp}/B_{c2\parallel})^{1/2}$. Using these expressions, our data result in $\xi_c = (0.18$ nm$)/(1-t)^{1/2}$ and $\xi_{ab} = (2.7$ nm$)/(1-t)^{1/2}$, where $t = T/T_c$. Note that the anisotropic effects of the thin film geometry are expected to be negligible because their thicknesses are much larger than ξ_c, except within ≈ 0.001 K of T_c.

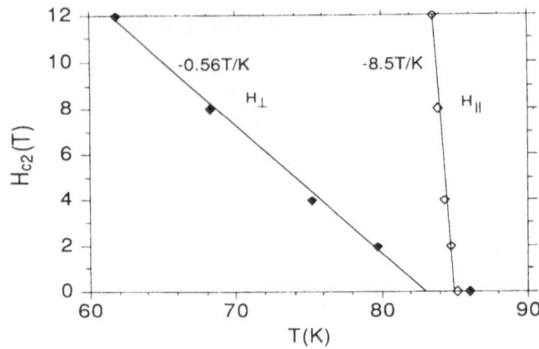

Fig.4. Upper critical fields of BSCCO film obtained by taking the midpoint of the resistive transition for the perpendicular and parallel field direction. Note that the anisotropy of ≈15 is considerably larger than that found in YBCO films.

Tl-Ba-Ca-Cu-O films

We prepared TBCCO thin films in three different phases, T2212, T1223 and T2223. T2223 phase films have a higher transition temperature than the other phases and Fig. 5 shows resistivity vs. temperature for one of the films where the majority of the phase is T2223. The film shows good metallic behavior, as evidenced by a factor of 2.5 decrease in resistivity from room temperature to just above the onset, although the magnitude of resistivity is almost an order of magnitude higher than the best bulk, single crystals of $YBa_2Cu_3O_{7-y}$. A single superconducting transition is observed in Fig. 5 beginning at about 125 K and ending at T_{c0}=114 K.

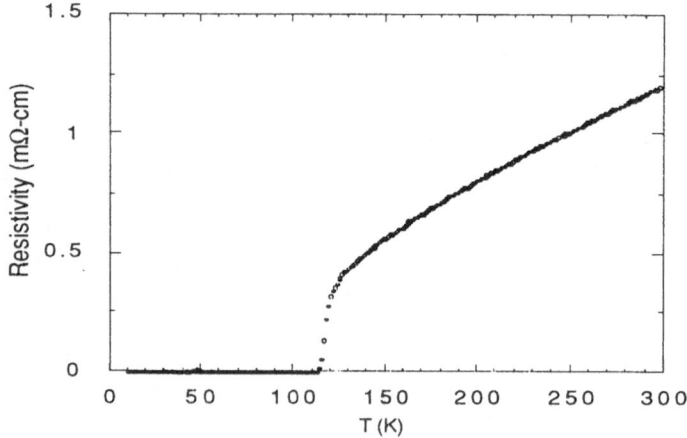

Fig.5. Resistivity vs. temperature for a 2223 TBCCO film. A single superconducting transition is observed which begins at about 125 K and goes to zero resistivity at 114 K. Note the excellent metallic behavior in the normal state.

In Fig. 6, the upper critical fields $B_{c2}(T)$ (taken at 50% of resistive transition) of the TBCCO thin film for each orientation are shown. From Fig. 6 both the anisotropy factor and the slope of the parallel critical field are much higher than previously reported for YBCO HTS films[15] or bulk single crystals[16]. We find $B_{c2\parallel}' \geq 70$ T/K, and $B_{c2\perp}' = 1$ T/K giving an anisotropy of at least 70. By the same analysis used above for BSCCO thin films, our data result in $\xi_c = (0.03$ nm$) /(1-t)^{1/2}$ and $\xi_{ab} = (2.0$ nm$) /(1-t)^{1/2}$.

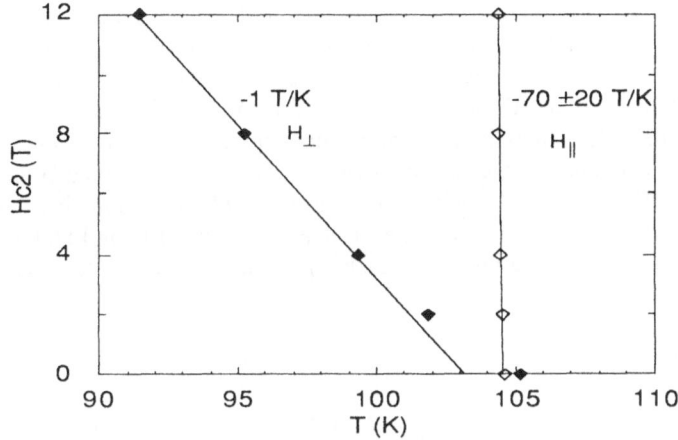

Fig. 6. Upper critical field of 2212 phase TBCCO film obtained by taking the midpoint of the resistive transition. The anisotropy of this film and other TBCCO films of ≈70 is unusually large.

SUMMARY AND DISCUSSIONS

We have prepared thin films of BSCCO and TBCCO on (100) MgO and (100) oriented or polycrystalline ZrO_2-9%Y_2O_3. The 80K phase of BSCCO films were grown on MgO substrates, with a predominant c-axis orientation perpendicular to the film surface. The requirements on stoichiometry to get good quality BSCCO thin films are less stringent than YBCO films, however the annealing temperature must be held within a narrow range around 865°C. TBCCO films of three difference phases, T2212, T1223, and T2223, were grown on ZrO_2 substrates. TBCCO films show much better metallic behavior than other films, suggesting less grain boundary problems in this compound. The upper critical magnetic field measurements of TBCCO and BSCCO thin films show a very high anisotropy in the critical field slope (≥70 for TBCCO and ~15 for BSCCO), as might be expected from highly oriented materials. For single crystals of $YBa_2Cu_3O_x$ the most common values[16] indicate $B_{c2\parallel}' \sim 3$-3.8 T/K, and $B_{c2\perp}' \sim 0.54$-0.9 T/K giving an anisotropy of ~3.3-7. For oriented thin films[16] of $YBa_2Cu_3O_x$, there is a greater variation with $B_{c2\parallel}' \sim 3$-6.8 T/K, and $B_{c2\perp}' \sim 0.9$-2.5 T/K giving an anisotropy of ~1.8-4.4. Recent measurements[17] on a single crystal of BSCCO gave a $B_{c2\parallel}'$ of 45 T/K and an anisotropy of about 60. These comparisons show that our results in TBCCO and BSCCO oriented films are quite remarkable when compared with other films and comparable to the best single crystal results.

Such large B_{c2} anisotropies in our films might also result from anisotropic thermally activated flux creep[18], with a more isotropic intrinsic B_{c2}. Although our transitions cannot be convincingly fit to the Tinkham model[18] due to the long extended tails at low temperature (see

Fig.3), we cannot rule out that this tail is an inhomogeneity feature and that the Tinkham model does apply. For reduced resistances above 50%, the individual curves can be fit, but B_{c2} is proportional to $(1-t)^{1.2}$ instead of the predicted[18] 1.5 power. Although samples without such a tail would be necessary to unambiguously decide on this issue, note that most results on TBCCO and BSCCO films, unlike YBCO, exhibit such tails.

These large anisotropies might be explained by using a two-dimensional (2D) superconductor picture. In a 2D superconductor found for a superconductor thickness $d \ll \xi$, $B_{c2\parallel}$ follows the $(1-t)^{1/2}$ dependence with a formula $B_{c2\parallel} = \sqrt{3}\, \Phi_0/\pi d \xi_{ab}$[19], where ξ_{ab} comes from the previous 3D analysis of $B_{c2\perp}$. An example of this is the similar B_{c2} behavior observed[20] in very thin multilayers of superconducting NbN (5-30 nm thick) and insulating AlN (2 nm thick). The $B_{c2\perp}$ agreed with the 3D analysis, but for parallel fields, $B_{c2\parallel}$ followed the $(1-t)^{1/2}$ dependence of a (2D) superconductor. For a multilayer system, the validity of this 2D model depends on the strength of the interlayer coupling (between Cu-O planes in the HTS case). If the Cu-O conducting planes are viewed as 2D structures, then using $d{\sim}0.5$ nm, one finds $B_{c2\parallel} \sim (1000\ \text{T})\,(1-t)^{1/2}$. Considering the accuracy of the carbon glass thermometer used in this measurements, the present data suggests that the sample used in Fig. 6 could be modeled as a 2Dsuperconductor. Further work at much higher fields may help to clarify this point.

ACKNOWLEDGEMENTS

The authors wish to thank David Day for developing the data acquisition system, E.D. Huff for the compositional analysis, Richard Lee for EDX measurements, and D.M. Mckay and S.J. Stein for invaluable technical assistance. This work is supported by the U.S. Department of Energy, BES-Materials Sciences, under contract #W-31-109-ENG-38.

REFERENCES

1. P. Chaudhari, R.H. Koch, R.B. Laibowitz, T.R. McGuire, and R.J. Gambino, Phys. Rev. Lett. 58:2684 (1987).
2. P. Madakson, J.J. Cuomo, D.S. Yee, R.A. Roy and G. Scilla, J. Appl. Phys. 63:2046 (1988).
3. M. Hong, S.H. Liou, J. Kwo, and B.A. Davidson, Appl. Phys. Lett. 51:694 (1987).
4. X.D. Wu, D. Dijkkamp, S.B. Ogale, A. Inam, E.W.Chase, P.F. Micchi, C.C. Chang, J.M. Tarason, and T. Venkatesan, Appl. Phys. Lett. 51:861 (1987).
5. J.H. Kang, R.T. Kampwirth and K.E. Gray, Appl. Phys. Lett. 52:2080 (1988).
6. Y. Ichikawa, H. Adachi, K. Hirochi, K. Setsune, S. Hatta and K. Wasa, Phys. Rev.B 38:765 (1988).
7. C.E. Rice, A.F.J. Levi, R.M. Fleming, P. Marsh, K.W. Baldwin, M. Anzlower, A.E. White, K.T. Short, S. Nakahara and H.L. Stormer, Appl. Phys. Lett. 52:1828 (1988).
8. R.M. Hazen, L.W. Finger, R.J Angel, C.T. Prewitt, N.L. Ross, C.G. Hadidiacos, P.J. Heaney, D.R. Veblen, Z.Z. Sheng, A. El Ali and A.M. Herman, Phys. Rev. Lett. 60:1657 (1988).
9. J.H. Kang, R.T. Kampwirth, K.E. Gray, Phys. Lett.A 131:208 (1988).
10. D.S. Ginley, J.F. Kwak, R.P. Hellmer, R.J. Baughman, E.L. Venturini and B. Morosin, Appl. Phys. Lett. 53:406 (1988).
11. W.Y.Lee, W.Y.Lee, J. Salem, T.C. Huang, R. Savoy, D.C. Bullock, and S.S.P. Parkin, Appl. Phys. Lett. 53:329 (1988).
12. S.H. Liou, M. Hong, B.A. Davidson, R.C. Farrow, J. Kwo, T.C. Hsieh, R.M. Fleming, H.S. Chen, L.C. Feldman, A.R. Kortan and R.J. Felder, in Proc. of Amer.

Vac. Soc. Topical Conf. on "Thin Film Processing and Characterization of High Temperature Superconductors", Amer. Institute of Phys., New York, (1988).

13. Y. Yeshurun and A.P. Malozemoff, Phys. Rev. Lett. 60:2202 (1988).

14. A. Umezawa, G.W. Crabtree, J.Z. Liu, in Proc. Intl. Conf. on High-Temperature Superconductors and Materials and Mechanisms of Superconductivity; and A. Umezawa, G.W. Crabtree, J.Z. Liu, T.J. Moran, S.K. Malik, L.H. Nunez, W.L. Kwok and C.H. Sowers, private communication.

15. P. Chaudhari, R.T. Collins, P. Freitas, R.J. Gambino, J.R. Kirtley, R.H. Koch, R.B. Laibowitz, F.K. LeGoues, T.R. McGuire, T. Penney, Z. Schlesinger, A.P. Segmueller, S. Foner and E.J. McNiff, Jr., Phys. Rev.B 36:8903 (1987).

16. J.S. Moodera, R. Meservey, J.E. Tkaczyk, C.X. Hao, G.A. Gibson and P.M. Tedrow, Phys. Rev.B 37:619 (1988).

17. T.T.M. Palstra, B. Batlogg, L.F. Schneemeyer, R.B. van Dover and J.V. Waszczak, Phys. Rev. B38:5102 (1988).

18. M. Tinkham, Phys. Rev. Lett. 61:1758 (1988).

19. M. Tinkham, "Introduction to Superconductivity," McGraw-Hill, New York (1975).

20. J.M. Murduck, D.W. Capone II, I.K. Schuller, S. Foner and J.B. Ketterson, Appl. Phys. Lett. 52:504 (1988).

FILM FABRICATION OF ARTIFICIAL (BiO)/(SrCaCuO) LAYERED STRUCTURE

J. Fujita, a)T. Tatsumi, T. Yoshitake, and H. Igarashi

Fundamental Research Laboratories
a)Microelectronics Research Laboratories
NEC Corporation
4-1-1 Miyazaki, Miyamae-ku, Kawasaki, 213, Japan

ABSTRACT

Artificial (BiO)/(SrCaCuO) layered structures epitaxially grown on (100) MgO substrate by using oxygen-gas-reactive dual ion beam sputtering were studied. Films with artificial periods of 12, 15 and 18Å (which are identical to bulk 24, 30 and 36Å phases) were selectively formed by adjusting the thickness of SrCaCuO layer sandwiched by BiO bi-planes. Epitaxial films designed to have high Tc phase period (36Å) with total thickness of about 300Å showed a onset Tc of 110K and a zero resistivity temperature of 45K without post annealing. The layer by layer growth by the shuttering technique was found to be very effective to form smooth and fine superconductive thin films.

1. INTRODUCTION

Epitaxially grown films of the recent high Tc oxide superconductors [1,2,3] are quite important not only for application to electrical devices but also for studying superconducting properties of the oxides and thus contributing to the basic understanding related to superconducting mechanism; i.e. anisotropy of energy gap[4], critical magnetic field[5], etc. In particular, the Bi based oxide systems have quite unique layered structure where BiO bi-planes sandwiches SrCaCuO perovskite type layers. The 80K and 110K superconductivity were observed for 2212 structure (30Å phase) and 2223 structure (36Å phase), respectively, and the origin of the high temperature superconductivity is considered to relate to the layered structure consisting of BiO bi-planes and SrCaCuO perovskite layers. In Bi based oxide system, however, neither perfect epitaxial film nor high quality bulk single crystal which seems to be sufficient for the physical measurements have ever successfully obtained. Our goal of this research is to achieve completely controlled artificial structure of the oxide superconductors. In this study, we used oxygen-gas-reactive[6,7,8] ion beam sputtering[9,10]. Ion beam sputtering seems to provide the best controllability for flux rate because the beam is quite stable and it is easy to adjust the beam power, furthermore it has an advantage that the substrate is completely separated from the plasma so that we could avoid the influence of ionized particle to film growth.

2. EXPERIMENTAL TECHNIQUE

Film fabricating conditions were listed in table 1. This sputtering system shown in Fig. 1 is provided with two sputtering sources, two shutters controlled by microcomputer connected with film thickness monitors and an electron gun for in situ reflected high energy electron diffraction (RHEED) observation. Two Kaufman type ion sources having hollow cathode (Ion Tech) were used for sputtering and the space inside the vacuum chamber was neutralized by a hollow cathode neutralizer. Oxygen gas was directed to the substrate from the special gas nozzle which encircles the substrate in order to achieve the uniformity of gas pressure. The substrate temperature and the oxygen gas pressure at the substrate were fixed to typical values of 605°C and 6×10^{-3} torr respectively and the targets of Bi_2O_3 and $Sr_2Ca_2Cu_3O_x$ were used. Film growth rate was about 0.18Å for Bi oxide and 0.10Å for SrCaCuO. Since this growth rate is quite sensitive to oxygen gas pressure, feedbacked power control of ion beam was required in order to keep the deposition rate constant. Since the deposition rate monitor (INFICON XTC) has a resolution of 1Å, further precise thickness control was needed to achieve the artificial layered film growth by shuttering. So the computer always calculated the mean flux rate every 4 minutes in order to get the accuracy of about 0.1Å and the film thickness was controlled by the shutter opening time which is calculated from the mean flux rate and the sticking rate during film growth process.

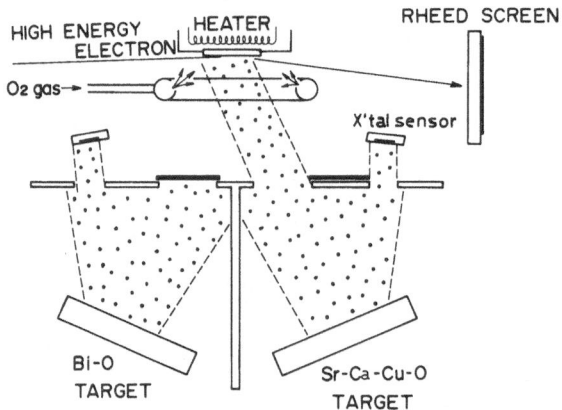

Fig. 1 Schematic illustration of dual ion beam sputtering system.

Table 1 Film Growth Condition

substrate	(100) MgO single crystal
substrate temp.	$600 - 610$°C
oxygen pressure	6×10^{-3} torr
chamber background	2×10^{-8} torr
target 1	Bi_2O_3
target 2	$Sr_2Ca_2Cu_3O_x$
ion beam	Ar^+
ion beam current	$20 - 50$ mA
ion beam voltage	600 V
film growth rate 1	0.1-0.2 Å/sec for BiO
film growth rate 2	0.05-0.15 Å/sec for SrCaCuO
film composition	SrCaCuO $= 2.2 : 2.0 : 2.7$

Fig. 2 shows the film growth sequence. We additionally put 4 minutes intermission between each deposition. Because the X-ray diffraction patterns suggested the 4 minutes intermission drastically improved the crystal quality of the layered structure. In this study, all film structure was fixed to first 12Å of SrCaCuO buffer layer and additional 16 periods of (BiO)/(SrCaCuO) layers which is identical to 8 unit cells of bulk superconductor. After film growth, films were quickly cooled down to below 150°C within 60 minutes at the same oxygen pressure for deposition.

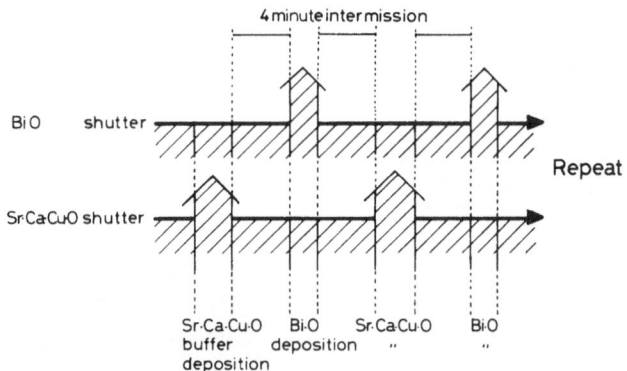

Fig. 2 Film growth sequence.

3. BUFFER LAYER EFFECT

The choice of buffer layer was important to get high quality epitaxial films on MgO substrate. Fig. 3 shows the in situ RHEED patterns of MgO substrate, SrCaCuO buffer layer with 6, 12, 18, and 30Å thickness and Fig. 4 shows the in situ RHEED patterns of MgO substrate, BiO buffer with 12 and 30Å thickness on (100)MgO substrate. Both SrCaCuO and BiO well grow epitaxially on MgO, but the surface morphology and the orientation were different. The SrCaCuO first tightly grows on MgO with the same orientation for MgO ($<100>$SrCaCuO//$<100>$MgO) and the same in-plane lattice spacing of 4.2Å in Fig. 3b. Then the lattice spacing was gradually expanded to 4.7Å as shown in Fig. 3c with increasing the buffer thickness to 12Å. The broadening of RHEED spots were also observed at the same time. The structure of buffer was considered to be fcc-like judging from RHEED pattern, though the SrCaCuO layer in the bulk superconducting phase was bcc perovskite structure. The RHEED pattern changed to spotty one for the SrCaCuO buffer over 18Å thickness. Some phase separation seems to start to appear resulting in a surface roughness. In contrast, BiO buffer perfectly grew on MgO with the orientation of BiO $<100>$ axis parallel to MgO $<110>$ axis with the in plane lattice spacing of 5.8A ($4.2 \times \sqrt{2}$) and this BiO buffer seems to form rock salt type fcc structure. The BiO lattice spacing was not varied even when the buffer thickness was increased further.

The final surface morphology of layered structure grown on each buffers strongly depended on the thickness of SrCaCuO buffer. Fig. 5 shows the schematic structure of film and its RHEED pattern. The 6Å epitaxial buffer tightly grown on MgO produced two type of epitaxial orientation films a-axis film parallel to $<100>$ MgO axis and a-axis parallel to $<110>$ MgO axis accompanied with second phase shown in Fig. 5a and we observed the same film growth pattern on BiO buffer. The best epitaxial film with uniform orientation in which the a- and b-axis of film parallel to $<100>$ axis of MgO was grown on 12Å buffer as shown in Fig. 5b. The surface roughness of buffer remained unchanged throughout the deposition if once the surface roughness was produced as shown in Fig. 5c. Therefore, controlling of buffer layer thickness is of crucially important to obtain clear epitaxial films and the 12Å SrCaCuO buffer seems to the best for the epitaxial growth on (100) MgO.

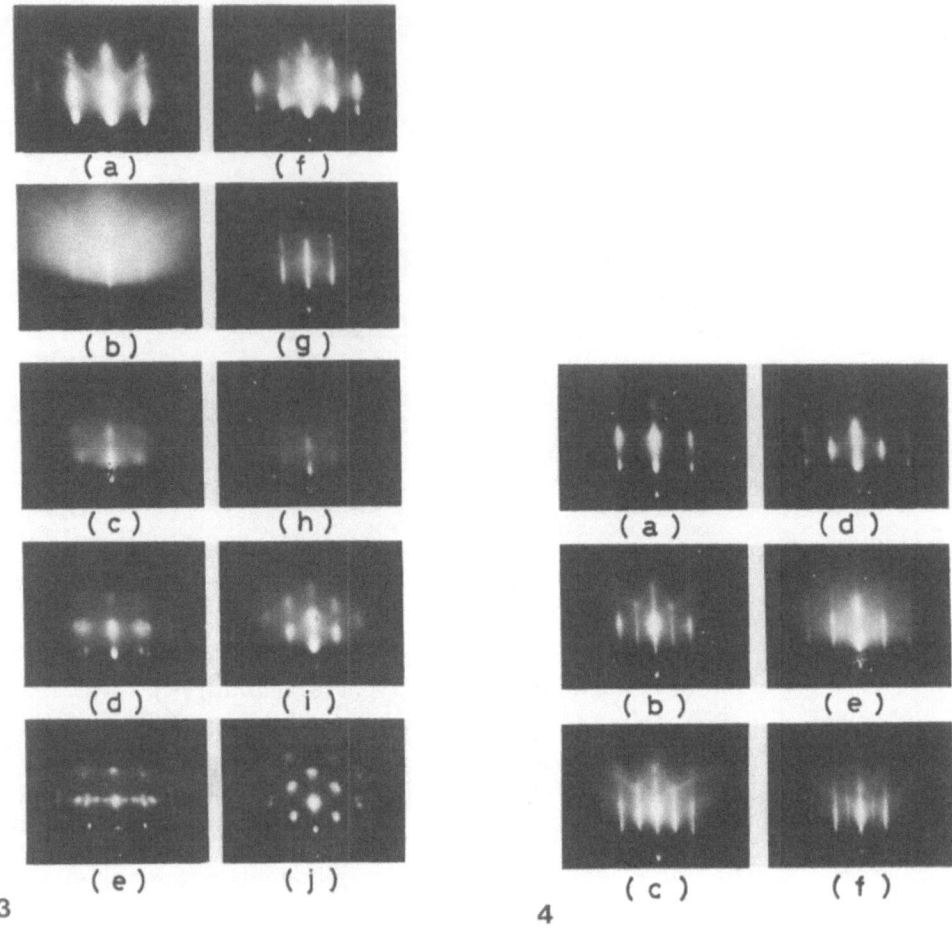

Fig. 3   Sequential RHEED pattern of SrCaCuO buffer layer on MgO. (a) (b) (c) (d) (e) show the patterns of <100> MgO azimuth and (f) (g) (h) (i) (j) show the <110> MgO azimuth.

(a) and (f) are the RHEED patterns of (100) MgO substrate.
(b) and (g) are the RHEED patterns of 6Å SrCaCuO buffer on substrate.
(c) and (h) are the RHEED patterns of 12Å SrCaCuO buffer on substrate.
(d) and (i) are the RHEED patterns of 18Å SrCaCuO buffer on substrate.
(e) and (j) are the RHEED patterns of 30Å SrCaCuO buffer on substrate.

Fig. 4   Sequential RHEED pattern of BiO buffer layer on MgO. (a) (b) (c) show the patterns of <100> MgO azimuth and (c) (d) (e) show the <110> MgO azimuth.

(a) and (d) are the RHEED patterns of (100) MgO substrate.
(b) and (e) are the RHEED patterns of 12Å BiO buffer on substrate.
(c) and (f) are the RHEED patterns of 30Å BiO buffer on substrate.

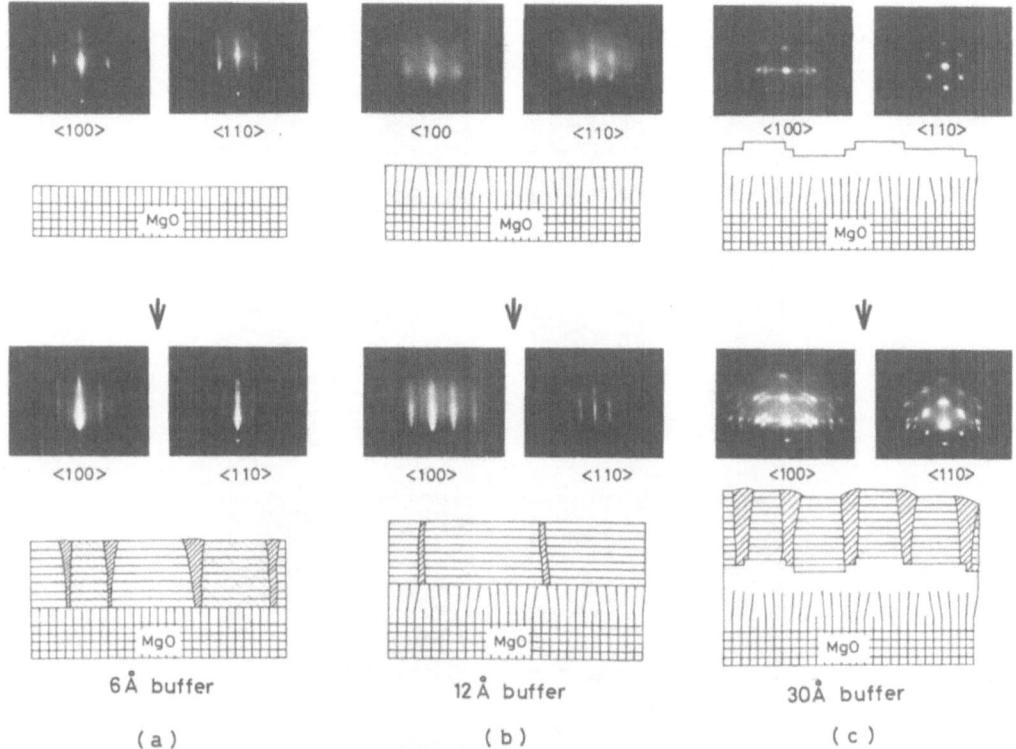

Fig. 5   Schematic illustration of film growth structure and its RHEED pattern
on (a) 6Å, (b) 12Å, and (c) 30Å of SrCaCuO buffer layers
(a)   Two types orientation were observed; a-axis of film parallel to <100>
        MgO axis and a-axis parallel to <110> MgO axis.
(b)   Film shows the unique orientation of a- and b-axes parallel to <100>
        MgO axis, which indicated the twining of crystal.
(c)   The surface roughness of buffer remains to the end of deposition.

4. ARTIFICIAL (BiO)/(SrCaCuO) LAYERED STRUCTURE

Artificial (BiO)/(SrCaCuO) layered structures were successfully performed
on (100)MgO surface with buffer. Fig. 6 shows the sequential change of RHEED
patterns along <100> and <110> MgO azimuths during film growth of[(Bi-O)
5.5Å]/[(Sr-Ca-Cu-O) 10Å]. Fig. 6(a) and (e) are the patterns of the substrate with
<100> azimuth and <110> azimuth, respectively. Fig. 6(b) and (f) are the
patterns after growing the 12Å of SrCaCuO buffer layer. Fig. 6(c) and (g) show
the pattern after the growth of 8 unit layers (corresponding to 4 unit cells in bulk
phase) on buffer. Fig. 6(d) and (h) show the pattern after 16 unit layers. The
center lines and the strong streaks on the both sides are indexed as (000), ($\bar{2}$00),
and (200). The weak lines between strong streaks are indexed as (100) which is
commonly observed along a-axis in electron diffraction of bulk single crystal.
The clear but weak superspots due to the incommensurate modulation which
exist along b-axis in a bulk single crystal also appeared along MgO <100>
azimuth.   Hence the film contains the two equivalent orientation of
incommensurate structure and actually the same RHEED pattern was observed
for the 90 degree rotation around the surface normal axis. It is quite noticeable
that the streak pattern with superspots appears clearly after only 2 unit layers
stacking on buffer and all of epitaxial films in these experiments show the streak
RHEED patterns with the superspots.

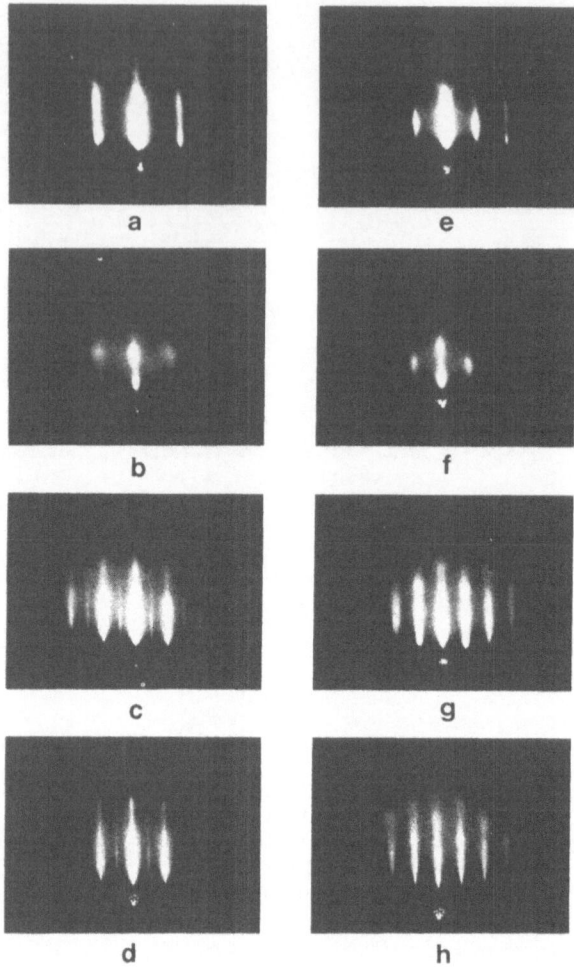

Fig. 6　Sequence of RHEED pattern of perfect epitaxial film.
(a) (b) (c) (d) show the patterns of <100> MgO azimuth and (e) (f) (g) (h) show
the <110> MgO azimuth.

　　　Fig. 7 shows the Scanning Electron Microscopy (SEM) images of a typical
film surface and its RHEED pattern from <100> MgO azimuth. Fig. 7(a)
shows the film surface of low Tc phase. The cubic like block is not second phase
but some dust on surface, which was included for focusing. Fig. 7(b) shows the
example of appearance of second phases on film surface. In this case, the spacing
between second phase spots in RHEED indicates 2.4Å and X-ray diffraction
pattern also showed the second phase with 2.405Å spacing (2 theta = 37.36
degree), which is identical to CaO.
　　　In the present study, we successfully performed the epitaxial film growth
with the artificial periodicity according to the design. Fig. 8 shows the X-ray
diffraction patterns of the film with periodicity of (a) 39Å, (b) 36.8Å, (c) 34Å, (d)
30.1Å, (e) 27Å, and (f) 24Å which are selectively grown by controlling the
SrCaCuO layer thickness and these result are summarized in Table.2.
　　　Some of diffraction patterns show the intensity oscillation at low angle
diffraction caused from extremely thin and smooth film. The peak to peak angle
of low angle oscillation indicated the film thickness about 300Å which is
identical to the designed value of the film thickness. we can also see the
intensity oscillation beside the higher angle diffraction peaks caused by Laue
function oscillation for finite film thickness in Fig. 8(d), (e) and (f).

(Hereafter the periodicity shows the doubled value calculated from diffraction pattern according to common manner in bulk studies.)

a   **b**

Fig. 7 Typical SEM Image of Film.

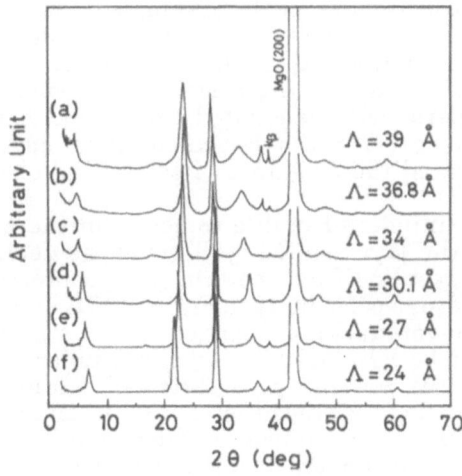

Fig. 8 Typical X-ray diffraction patterns.

Table 2 Results of Artificial Film Growth

Sample	Design		Structure			
No	<BiO/SrCaCuO> (Å) (Å)	Periodicity $\Lambda \times 2$(Å)	Phase	Second phase		Tc (k) onset-end
(g)	<5.5/15>	44	HH	+ + +		- - -
(a)	<5.5/13>	39	?	+ +		80 - 21
(b)	*<5.5/13>	36.8	H	+ +		110 - 45
(c)	<5.5/10>	34	?	- - -		90 - 22
(d)	<5.5/8.5>	30.1	L	- - -		70 - 25
(e)	*<5.5/8.5>	27	?	+		55 - 9
(f)	**<5.5/6.5>	24	I	- - -		- - -

symbol"+" means the quantity of second phase.
symbol"HH","H","L","I" mean the 44Å phase, high Tc phase(36Å), low Tc phase(30Å) and insulator phase(24Å),respectively.
*   6Å buffer was used
** $Po_2 = 9 \times 10^{-3}$torr

5. DISCUSSION

(Structure of artificial layered film)

All diffraction patterns in Fig. 8 seem to be those of single phase epitaxial films. However the diffraction profiles are classified to two groups; (1) the films having the same periodicities for bulk compound and (2) the films having the intermediate periodicities. The diffraction peaks of group(1) are perfectly indexed by [00n] as shown in Fig. 8 (b) 36.8Å, (d) 30.1Å, and (f) 24Å, in contrast the diffraction peaks of group (2) having the intermediate periodicities of (a) 39Å, (b) 34Å, and (c) 27Å as shown in Fig. 8, which periodicity we did not commonly observe them in bulk samples.

For example, the diffraction peaks of sample (c) with 34Å periodicity could be indexed by neither the periodicity of high Tc phase (36Å) nor the periodicity of low Tc phase (30Å), furthermore this diffraction profile could not be indexed by self periodicity of 34Å. But the intermediate periodicity seems to be surely correct because the periodicity calculated from peak to peak of higher angle diffraction shows good agreement to that calculated from first peak. The shift of peak positions indicated the existance of some kind of interface irregular in layered structure.

We think such unindexed profile comes from the microscopic structural phase mixing along the layer stacking direction as depicted in Fig. 9, which stacking is probably resulted from the film growth process closing to layer by layer growth. The average periodicity in Fig. 9 is 17Å and the structure contains phase mixing along layer stacking direction. Such structural phase mixing is sometimes observed in both Bi and Tl system bulk superconductors[11]. It seems natural to think that the one unit perovskite cell is a minimum order in layer growing process and in fact that the difference between high Tc phase and low Tc phase in bulk Bi system superconductor is the one unit copper perovskite layer. Therefore, when the total amount of SrCaCuO was not enough to cover the surfaces as a single phase, there should appear some terrace on surface. Such terrace structure also possibly occurs as a result of substrate roughness. Once this type terrace comes into surface, phase mixing along stacking direction remains unchanged to the end of deposition. So It is also well understood that the intermission between shuttering greatly helps the migration of source material on terrace.

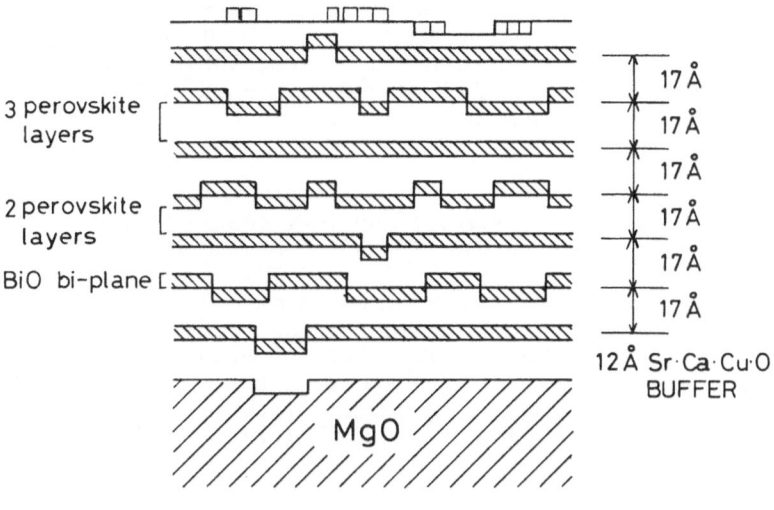

Fig. 9 Film Growth Model

(Relating aspects of film growth)

The layer by layer film growth seems to apparently achieved. However, it should be noted that the film growth process strongly dominated by natural law; chemical composition, processing temperature, and oxidation, etc. Table. 3 shows the composition effect for co-sputtering with BiO and SrCaCuO target.

Table 3

| Sample | Flux Rate | | | Structure | |
No	BiO ($Å/s$)	SrCaCuO ($Å/s$)	Phase	Second Phase	Correspondence to Artificial Structure
(h)	0.042	0.104	weak H(36Å)	many	(a) (39Å)
(i)	0.058	0.108	L(30Å)	many	(b) (34Å)

Composition ratio was adjusted by controlling the sputtering ion beam current. The 4 minutes intermission was also put between each 2 minutes deposition as performed in shuttering process and the total film thickness was fixed to 300Å. The composition of sample(h) is identical to sample (a) in Table. 2, which (a) forms the 39Å periodicity, and the composition of sample (i) is identical to sample (c) having 34Å periodicity. The changes of chemical composition apparently affect to the phase forming; low BiO composition leads to high Tc phase (36Å) ,in contrast high BiO composition leads to low Tc phase (30Å). Both film contained many second phases and did not show clear streak RHEED pattern. Hence the chemical composition obviously affect the phase formation. But the intermediate periodicity was never formed in natural way, which supports the film growth is based on the layer by layer epitaxy in shuttering process.

183

Another example of the natural law in film growth was seen in BiO bi-layer forming. While the attempt was done to make mono- and tri-BiO layers artificial structure, non of them showed the evidence forming the new structures and only shows the series of BiO Bi-layer structure, which fact means the film growth is strongly dominated by natural law.

6. SUMMARY

We studied the buffer layer effect and demonstrated the artificial (BiO)/(SrCaCuO) layered structures epitaxially grown on (100) MgO substrate by using oxygen-gas-reactive dual ion beam sputtering. Films with artificial periods of 24 to 39Å were selectively formed by controlling the SrCaCuO layer thickness. The epitaxial film having high Tc phase with about 300Å thickness show the Tc onset of 110K, and the Tc end of 45K without post annealing. The shuttering is very effective to form the smooth and fine superconductive epitaxial thin films. The layer by layer film growth may be achieved, although this film growth is strongly dominated by the natural law of structure forming of superconducting phase.

7. ACKNOWLEDGMENT

The authors would like thank T. Satoh for composition analysis, M. Sugimoto for electrical measurements, and Y. Kubo for helpful discussion. We are also grateful for the support from M. Yonezawa during the course of this research.

REFERENCE

1. J.G. Bednorz and Müller, Z. Phys. B64, 189(1986).
2. M.K. Wu, J.R. Ashburn, C.J. Torng, P.H. Hor, R.L. Meng, L. Gao, Z.J. Haung, Y.Q. Wang, and C.W. Chu, Phys. Rev. Lett. 58, 908(1987).
3. H. Maeda, Y. Tanaka, M. Fukutomi, and T. Asano, Jpn. Appl. Phys. Lett. 27, L209(1988)
4. J.S. Tsai, I. Takeuchi, J. Fujita, T. Yohitake, S. Miura, S. Tanaka, Y. Bando, K. Iijima, and K. Yamamoto. Proceedings of the international Conference on "High Temperature Superconductors and Materials and Mechanisms of Superconductivity" Part 2, 1385. February 28 - March 3, 1988, Interlaken, Swizerland.
5. T. Yoshitake, T. Satoh, Y. Kubo, T. Manako, H. Igarahsi, Jpn. J. Appl. Phys. Lett. 27, L1094(1988)
6. D.K. Lathrop, S.E. Russeck, and R.A. Buhruman, Appl. Phys. Lett. 51, 1554(1987)
7. T. Terashima, K. Iijima, K. Yamamoto, Y. Bando, and M. Mazaki, Jpn.J.Appl.Phys.Lett.27,L91(1988).
8. N. Missert, R. Hammond, J.E. Mooij, V. Matijasevic, P.Rosenthal,T.H.Geballe A.Kapitulnik, M.R.Beasley, S.S.Laderman, C. Lu, E. Garwin, R. Barton. preprint
9. J. Fujita, Y. Yoshitake, A. Kamijyo, T. Satoh, and H. Igarashi, J. Appl. Phys. 64,1292(1988).
10. J. Fujita, Y. Yoshitake, A. Kamijyo, T. Satoh, and H. Igarashi Conference Proceedings on Ion Beam Modification of Material, June 4, 1988, Tokyo, Japan. submitted for publication.
11. S. Iijima, T. Ichihashi, Y. Shimakawa, T. Manako, and Y. Kubo, Jpn. J. Appl. Phys. Lett. 27, L837(1988).

MAGNETRON SPUTTERING OF THIN FILMS OF Tl-Ca-Ba-Cu-O

P. Arendt, W. Bongianni, N. Elliott, and R. Muenchausen

Materials Science and Technology Division
Los Alamos National Laboratory
Los Alamos, NM, 87545

ABSTRACT

Superconducting thin films of the Tl-Ca-Ba-Cu-O system have been prepared by magnetron sputtering from a single compound target and post deposition annealing in an overpressure of Tl. The films are deposited on <100> MgO and <100> yttria stabilized zirconia (YSZ). The anneals were done at atmospheric pressure and the anneal protocols used result in both the 2122 and 2223 phases. The best films are grown on the YSZ substrates and have zero resistance at 107°K. RF surface resistance was measured as a function of the oxygen partial pressure during the post deposition anneals. The lowest surface resistance measurements were obtained for the 100% oxygen partial pressure anneals.

INTRODUCTION

Thin films of the Tl-Ca-Ba-Cu-O system have been synthesized by a variety of physical vapor deposition methods.[1-5] Thin films with high, nearly field independent, critical currents[6] and transition temperatures up to 120°K have been demonstrated. These films have potential application in many areas including RF linear accelerating cavities. To this end, we have been investigating the fabrication of thin films of these materials and have been measuring their materials and superconducting properties by X-ray fluorescence (XRF), backscattering spectrometry, X-ray diffraction (XRD), four point probe and microwave stripline absorption.

In this report, we discuss the microwave stripline measurement technique used to determine the superconducting films' RF surface resistance. A simple, single gun, magnetron sputter system used to fabricate thin films of the Tl-Ca-Ba-Cu-O system is described. A target-substrate geometry which results in acceptable film stoichiometries is detailed. A method for generating a steady, long term overpressure of Tl during the post deposition anneal is discussed. Finally, we report on variations in the material and superconducting properties of the films based on the anneal protocols employed.

EXPERIMENTAL SET-UP

The surface resistance measurement is based on a microwave bandstop resonator technique used to measure epitaxial yttrium iron garnet in a microstrip geometry[7]. More recently this technique has been used for characterization of high temperature superconductors[8].

In practice, the film/substrate is cut into a rectangular geometry with a length to width aspect ratio of between 3:1 to 10:1. The film is placed down on a 50 ohm microstrip substrate with the length parallel to the microstrip line. The film forms a quarter wave resonator which magnetically couples to the microstrip line, with coupling strength determined by the film to microstrip separation.

When an RF signal generator is swept across an appropriate frequency band, a number of absorption lines will appear which represent microwave absorption by the film corresponding to the quarter wave resonances in the sample. The width of the resonance and depth of the signal will yield the unloaded Q (Q_u) of the sample. Replacing the sample with a copper reference of the same dimensions allows the measurement of the Q_u of the copper at the same frequency. The ratio of Q_u's yields the surface resistance of the sample relative to copper.

A small, single magnetron sputter gun system was built and installed in a HEPA filtered ventilation hood. In order to minimize the size of the system within the hood, the pumping is done with a single, three stage mechanical pump. The system base pressure is in the 10^{-5} torr range. Post deposition annealing is done in a crucible furnace which is also located in the HEPA filtered hood. This containment of the deposition and annealing system is used to minimize the risk of thallium contamination.

The sputter target is a 5 cm diameter, arc melted alloy of $Tl_2Ca_2Ba_2Cu_3$. The target is conductive so that a DC power source may be employed to sputter the material. Initially, the substrates were not rotated. The capability to rotate the substrates was added in order to improve the spatial uniformity of the film stoichiometries. Table 1 describes the deposition parameters employed to deposit the films.

Table 1. Process parameters for single gun sputter deposition system

Target	$Tl_2Ca_2Ba_2Cu_3$
Input power (W/cm^2)	10
Target-substrate distance (cm)	4
Sputter gas/pressure	Ar/14×10^{-3} torr
Deposition rate (nm/min)	120
Film thickness (nm)	500-1500

The low input power to the target is needed to avoid overheating of the alloy target. The sputter pressure used results in film compositions which are closest to the desired 2223 phase. The film stoichiometries are determined by backscattering spectrometry and the phases are determined by XRD.

When a quantitative compositional analysis of the films as a function of deposition parameters is required, the films are deposited on silicon so that the substrate will not interfere with the backscattering spectrum. The film stoichiometries are measured by backscattering spectrometry using 6 MeV helium ions. This energy is needed in order to separate the back-scattering spectra of the high Z elements (Ba and Tl) in films that are approximately 500 nm thick. The scattering cross sections for copper, barium and thallium are Rutherford at this energy[9]. However, the calcium scattering cross section is non-Rutherford for 6 MeV helium ions. We have measured the relative change in scattering cross section for Ca at 6 MeV and thus are able to quantitatively measure the calcium in the films.

XRF is also used for a rapid semiquantitative measurement of the film compositions. XRF is most useful when a quick assessment of film composition (e.g. thallium content as a function of anneal protocol) is required. The thallium $L_{\alpha 1}$, the barium $L_{\alpha 1}$, the copper $K_{\alpha 1}$ and the calcium $K_{\alpha 1}$ are the peaks that are used in these XRF measurements.

RESULTS

Initially, the substrates were fixed during the deposition process. Using the deposition parameters of table 1, typical stoichiometries of the as deposited films are shown in table 2. The substrate radial positions are centered directly below the target center.

Table 2. Film stoichiometry vs position
(Substrates fixed)

Radial position (cm)	Stoichiometry Tl:Ca:Ba:Cu
0	2.34:2.31:2.38:3.0
1	2.24:2.03:1.75:3.0
2	2.25:1.55:1.70:3.0

Relative to the copper, the calcium and barium content of the films decreases through the 2:3 ratios from the center to the edge of the substrate holder. The thallium content is relatively constant and remains greater than the target ratio of 2:3. The thallium content of the as deposited films is less important because, as shown below, it may be readily varied during the annealing of the film.

The post deposition anneal protocol used for these films is similar to that described by Ginley[6], where a 15 minute air anneal at 840-870°C is followed by a rapid quenching of the films. A subsequent oxygen anneal at 750°C for 1-3 hours is then followed by a furnace cool.

Initially, to maintain a thallium overpressure during the anneals, the films were annealed face down against bulk ceramic pellets of the desired 2223 phase. The film-pellet was wrapped in gold foil and placed in an alumina crucible with a close fitting lid. The crucible was then placed in the furnace. XRF of the films from successive anneals showed that the thallium content steadily decreased indicating a loss in the thallium content in the bulk pellet material. In some cases the thallium content of the films decreased by 50% from one run to the next.

An alternate method for maintaining the thallium partial pressure during the film annealing was done by adding bulk 2223 material and Tl_2O_3 powder to alumina crucibles with close fitting lids. The films are positioned in the crucibles on a wire screen which is fitted midway in the crucibles. A suitable screen material, which is not rapidly eroded by the thallium vapor, was found to be gold coated copper. The crucibles are conditioned by placing them in the furnace at the maximum anneal temperature for one hour. The crucibles and bulk material used are porous and provide a large surface area upon which the Tl can condense and reevaporate. Small amounts of thallium do escape through the crucible-lid seal but the rate is sufficiently slow so that many successive anneals may be done before thallium needs to be added to the system. One further advantage of this annealing system is that it is amenable to annealing large area films, which will be required for some of our projected RF cavity surface resistance measurements.

Using the anneal protocol described above, the best thin films have transitions near 120°K and are fully superconducting by 107°K. Figure 1 illustrates a resistance vs temperature transition for one of our films prepared on a ö YSZ substrate. XRD shows the films to consist primarily of the 2122 phase after the first air anneal. The lower temperature oxygen anneal then results in the formation of increasing amounts of the 2223 phase. The lower temperature oxygen anneal also results in an increase of the temperature at which the films become superconducting. For poorer quality films this can be as much as 20°K. Scanning electron micrographs of the annealed films indicate a surface much smoother than our 123 films[10]. The surfaces of the films indicate a polycrystalline microstructure consisting of submicron sized grains. Microwave stripline absorption surface resistance measurements on films annealed with the air-oxygen protocol gave a best value of 36 mohms at 6GHz, at 77°K. The microwave stripline values measured were quite scattered (varying by an order of magnitude) for films made during the same deposition so attempts were first made to obtain more consistent compositions in the films.

Substrate rotation was done in order to deposit films that were closer to the desired 2223 composition, as well as to obtain more consis-

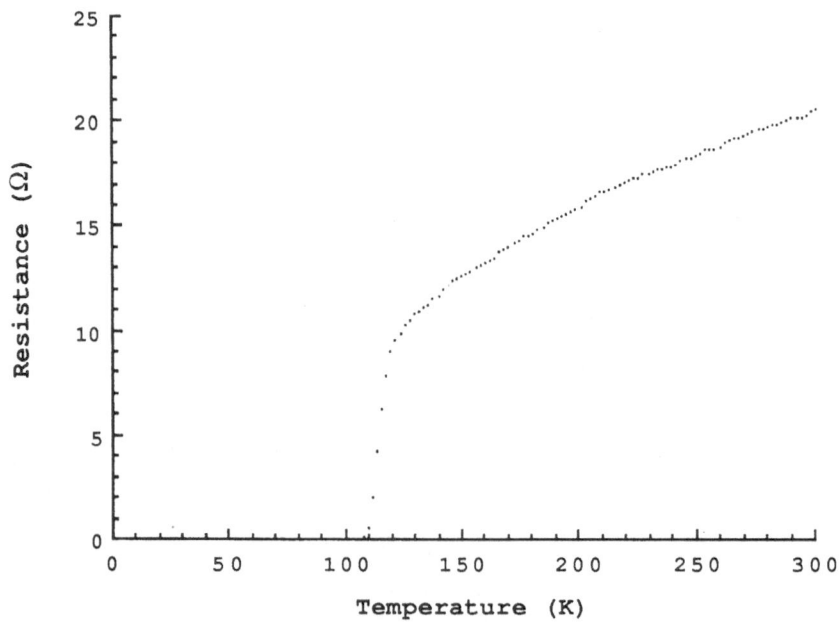

Figure 1. Resistivity vs temperature for a $Tl_2Ca_2Ba_2Cu_3$ and a $Tl_2Ca_1Ba_2Cu_2$ film deposited on <100> YSZ. The film resistance vanished, within the resolution of the apparatus, at 107°K.

tency in the film stoichiometry The center of rotation was offset from the target center by 2.5 cm. The film compositions resulting from this target-substrate geometry change are listed in table 3.

Table 3. Film stoichiometry vs position
(Substrates rotated)

Radial position (cm)	Stoichiometry Tl:Ca:Ba:Cu
0	2.37:1.93:2.09:3.0
1	2.49:1.69:2.06:3.0
2	2.51:1.78:1.99:3.0

It is seen that the barium is now within 5% of the desired 2:3 ratio and the calcium is near the desired 2:3 ratio at the center and no less than 15% below it for films radially outward from the center.

Investigations of film quality as a function of oxygen partial pressures during the post deposition anneal have been done for the Bi-Sr-Ca-Cu-O system[11], with low (5%) partial pressures of oxygen resulting in the highest quality films. To determine if the oxygen partial pressures during the anneal were a factor in film quality we under took a similar study for our Tl based films. A 100 nm thick film was deposited on a 2.5x2.5 cm square substrate (rotating) of <100> YSZ. The film was then sectioned into rectangular strips with a diamond wire saw. The films were then annealed in mixtures of Ar/O_2. The annealing cycle was a ramp from room temperature to 865°C in approximately 20 min, a soak at 865° for 15 min and a 4 hour furnace cool back to room temperature. Before the anneal cycle began, the furnace was purged with the desired Ar/O_2 mixture for thirty minutes. Microwave stripline RF resistance measurements were then done on the films with the results (4.1-4.3GHz) listed in table 4.

Table 4. Film RF surface resistance vs oxygen partial pressure

Oxygen partial pressure (%)	RF surface resistance (mΩ)
5	40
10	27
14	24
100	23

It is seen that the RF surface resistance decreases rather quickly with increasing oxygen partial pressure. This trend in film quality is quite surprising when compared with the results of Face[11] for the Bi based films where 100% O_2 anneals resulted in semiconducting films.

XRD of this series of films was also done. The results are shown in Figure 2 for the 2θ range of 25 to 40 degrees. It is seen that, as indicated by the constant intensity of the (109) reflection, the volume sampled from each film remained fairly constant.

Texturing is produced in the films as the O_2 content increases from 5 to 15%. At 100% O_2 this effect is reversed, This is shown by the relative intensities of the 200 and 211 reflections to the 109 reflection.

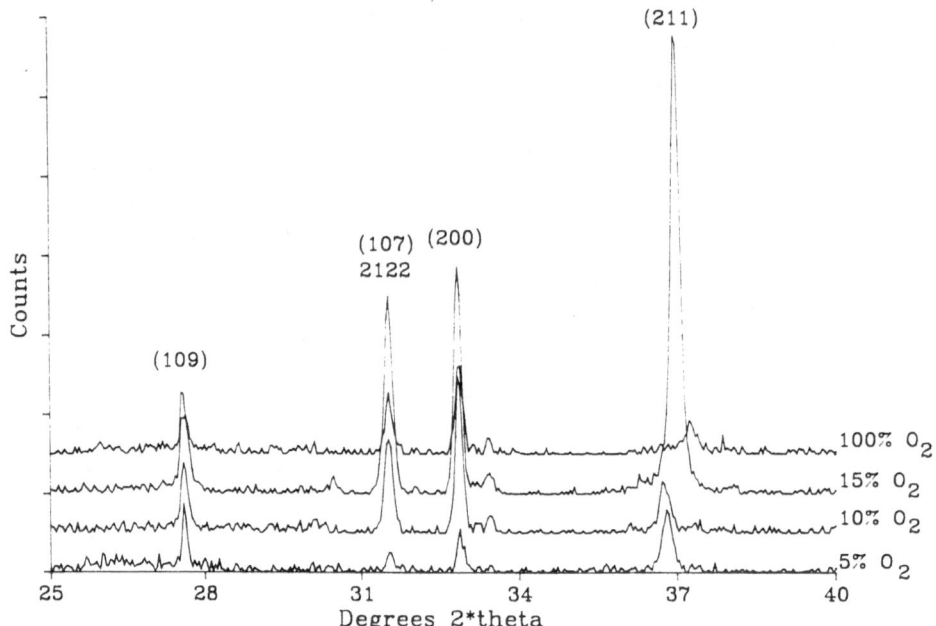

Figure 2. X-ray diffraction pattern showing the variations in line intensities as a function of oxygen partial pressure during the anneal

The amount of 2122 phase is found to remain fairly constant from examination of all reflections from that phase. There is a similar texturing produced in the 107 plane of 2122 that is also reversed at 100% O_2.

In addition to texturing, the 211 reflection indicates a change in d-spacing from 2.443 to 2.445 as the O_2 partial pressure increases from 5 to 100%. Similar effects have been seen in 123 compounds[12]. This is presumed to be caused by increasing oxygen content in the films.

SUMMARY

In summary, superconducting thin films of Tl-Ca-Ba-Cu-O have been prepared by DC magnetron sputtering from a compound target of $Tl_2Ca_2Ba_2Cu_3$. The film stoichiometry improves if the substrates are offset from the target center and rotated during deposition. A reliable method of producing repeatable thallium overpressures during the post deposition anneals is obtained by using ceramic crucibles which are conditioned with bulk material and thallium oxide powder. Our best films become superconducting at 107°K. RF surface resistance decreases with increasing oxygen partial pressure during the post deposition anneal. XRD of the films show them to be a mixture of the 2223 and the 2122 phases. XRD of the films also shows increased texturing with increasing oxygen partial pressure but this trend is reversed if the oxygen partial pressure is 100%.

REFERENCES

1. J. H. Kang, R. T. Kampwirth, and K. E. Gray, Phys. Lett. A, vol. 131, pp. 208-210 (1988).
2. D. S. Ginley, J. F. Kwak, R. P. Hellmer, R. J. Baughman, E. L. Venturini, and B. Morosin, Appl. Phys. Lett., vol. 53, pp. 406-408 (1988).
3. W. Y. Lee, V. Y. Lee, J. Salem, T. C. Huang, R. Savoy, D. C. Bullock, and S. S. P. Parkin, Appl. Phys. Lett., vol. 53, pp. 329-331 (1988).
4. I. Shih and C. X. Qiu, Appl. Phys. Lett., vol. 53, pp. 523-525 (1988).

5. Y. Ichikawa, H. Adachi, K. Setsune, S. Hatta, K. Hirochi, and K. Wasa, Appl. Phys. Lett., vol. 53 pp. 919-921 (1988).

6. D. S. Ginley, J. F. Kwak, R. P. Hellmer, R. J. Baughman, E. L. Venturini, M. A. Mitchell and B. Morosin, Submitted to Physica C.

7. J. L. Archer, W. L. Bongianni, and J. H. Collins, J. Appl. Phys., Vol.41, No. 3, pp.1359-1360, (1970).

8. A. Fathy, D. Kalokitis, and E. Belohoubek, Microwave Jour., Vol. 1, No. 10, pp. 75-94, Oct. (1988).

9. J. A. Martin, M. Nastasi, J. R. Tesmer, and C. J. Maggiore, Submitted to Appl. Phys. Lett.

10. P. N. Arendt, N. E. Elliott, R. E. Muenchausen, M. Nastasi, and T. Archuleta, submitted to "1988 AVS Topical Conference on High T_c Superconductors".

11. D. W. Face, J. T. Kucera, J. Crain, M. M. Matthiesen, D. Steel, G. Somer, J. Lewis, J. M. Graybeal, T. P. Orlando, and D. A. Rudman, to appear in IEEE Transactions on Magnetics, MAG-25 (1989).

12. P. K. Gallagher, et al. Mat. Res. Bull., vol. 22, pp. 995-1006 (1987).

INTERACTION OF HIGH TEMPERATURE SUPERCONDUCTORS WITH CONDUCTIVE OXIDES*

A.C. Greenwald, E.A. Johnson, J.S. Wollam, A.J. Gale,[+] and
N.K. Jaggi[++]

Spire Corporation, Bedford, MA 01730
[+] Ion Optics, Stoneham, MA 02180
[++] Northeastern University, Boston, MA 02115

Transparent conductive oxides, SnO_2 and TiO_2, were tested as contacts
to $YBa_2Cu_3O_7$ and $BiSrCaCu_2O_4$. After sintering the superconducting material, tin diffused into the bismuth material preferentially replacing
bismuth and calcium. Tin diffused into the 1-2-3 material but did not
show preferential replacement. Titanium diffused into the 1-2-3 compound
preferentially replacing copper. Contact resistance could not be measured as the films were not superconducting after sintering. Titanium
dioxide did act as a partial diffusion barrier on top of the bismuth compound when the film was not sintered. The results imply that surface
chemistry dominates the performance of transparent oxide contacts and
diffusion barrier layers for the different perovskite superconductors.

INTRODUCTION

The objective of this work was to determine how to fabricate low resistance contacts to the new superconducting materials. Loading the materials with silver and coating the desired surface areas with silver is one
known technique[1]. This research was performed to find oxide materials
which would be stable at high temperature, not likely to react with the
1-2-3 compound, have low resistance and block the diffusion of oxygen into
and out of the superconductor. Transparent conductive contacts can also
be used in optical sensor designs using thin films as detector elements.

SAMPLE PREPARATION

Conductive oxide films were deposited by reactive vacuum deposition
techniques on sapphire and quartz, or on top of the superconductive material. Quartz slides were used for measurements of the films properties.
Titanium oxide was deposited by evaporation of pure titanium in an atmosphere of oxygen at 10^{-5} torr to a total film thickness of 200nm. Tin
oxide was deposited by reactive sputtering of pure tin, in an atmosphere
of 10^{-4} torr argon and oxygen in equal proportions, to a total film

---

*This research was supported by NSF Small Business Innovative Research
Program grant ISI 8760790.

thickness of 1200nm. The films were ion implanted with fluorine at 35 keV to a total dose of 10^{15} ions/cm^2 and heated at 150°C for annealing. The resistivity of the TiO$_2$ film was 80 ohm-cm while the resistivity of the tin oxide was 0.01 ohm-cm. The total conductivity of the tin oxide film could have been increased if the entire film thickness were doped.

Superconducting films were deposited by ion beam sputtering in a mixed atmosphere consisting of equal parts oxygen and argon, at 10^{-5} torr. Two different types of ion beam sources were used. One was a high energy beam of 15 keV Ar ions at a current of 1 mA focused to a spot 1 cm diameter on a small target (2.5 cm diameter pellet). The ion beam was incident upon the target at an angle of 45° and the substrates were placed 5 cm from the target, parallel to the target plane and centered over the ion beam focus. The second ion source was an 85 mm diameter collimated beam with 30 mA at 1 keV ion energy sputtering a 100 mm diameter target with 1-2-3 stoichiometry. The larger beam was normally incident upon the target with the samples arranged at varying angles to the target surface (Figure 1).

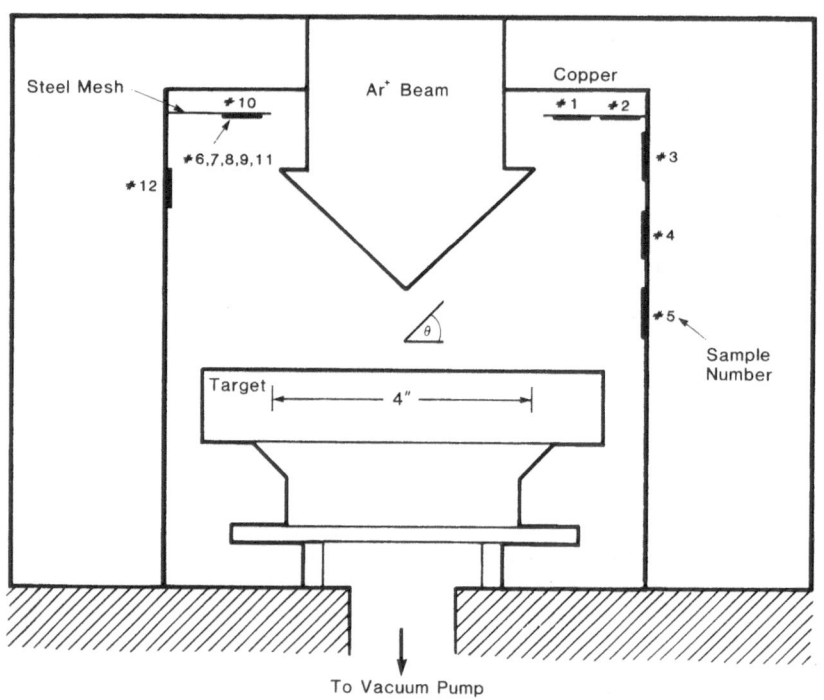

FIGURE 1.   EXPERIMENTAL GEOMETRY FOR LOW ENERGY ION BEAM SPUTTER DEPOSITION OF YBa$_2$Cu$_3$O$_7$.

Deposited film thickness for the small high energy beam followed a cosine distribution. The thickness was proportional to the area of the target observed at the point of measurement. There were no significant changes in composition as a function of the angle from the target. The deposited films showed the start of the superconducting transition at 70 K but did not go to zero resistance above 4 K. The result was attributed to contamination. The films were of the correct stoichiometry but contained 1.2% aluminum after annealing, attributed to diffusion from the sapphire substrate.

Deposited film thickness and composition varied as a function of angle from the normal for the second (larger) ion beam (Table 1). As the angle from the normal increased, the amount of argon retained in the film increased, copper concentration decreased, and the yttrium concentration increased. The concentration of barium did not seem to change. The inclusion of argon in the film is a thermal effect, but the change in composition for the other elements is not clear. The substrates were not deliberately heated during deposition. However, samples mounted over the target were heated by radiation from the top collimator plate (Figure 1) which intercepted much of the ion beams energy. These warmer samples had lower argon content.

TABLE 1.   COMPOSITION AS A FUNCTION OF ANGLE (theta) FROM TARGET
IN FIGURE 2.

Angle (degrees)	Y	Ba	Cu	Ar
1	33.8	37.4	29.8	11.7
16	37.2	37.3	25.6	4.2
38	36.3	33.6	30.0	4.9
43	34.3	35.1	30.7	3.6
47	31.7	34.9	33.5	2.9
53	31.7	34.5	33.8	2.1
64	27.3	34.5	38.1	2.1

Atomic concentrations of Y, Ba, Cu normalized to 100%, but the concentration of Ar is actual value.

The low energy ion beam was collimated to a 5 cm diameter and used to sputter a $BiSrCaCu_2O_6$ oxide target placed at 45° to the beam axis with the substrate parallel to the target. The film had a surplus of copper because the ion beam was slightly off center and impinged on a copper support for the target.

CONDUCTIVE OXIDE CONTACTS

Tin oxide and titanium oxide were studied as contact materials both on top of superconducting films and as a substrate coating for high temperature superconducting (HTSC) films. In all cases, superconducting properties were destroyed after the combination of HTSC and conductive oxide films were sintered. Typical time temperature profiles were 15 minutes at 850°C followed by a slow cool.

Because interdiffusion of components during sintering was suspected, all films were analyzed for composition versus depth by secondary ion mass spectroscopy (SIMS)[2]. Key results are shown in Figure 2, 3, 4, and 5. Without good standards for comparison, absolute numbers could not be placed upon the relative concentrations of the constituents. Also, data displayed in Figures 2 through 5 have a depth scale fixed assuming that the depth changes linearly with sputtering time. This is not true when sputtering through an interface of different materials, but the endpoint is correct.

Figures 2 and 3 show the concentration of Y, Ba, Cu, Ti, and O for a sample from low energy ion beam sputtering at an angle of 60° (Table 1) as coated with 50 nm of TiO_2 and the same sample after sintering. There is a sharp interface before heating, and after heating interdiffusion of

copper (into the surface film) and titanium (into the superconductive
material) has occurred. A data analysis interference exists in that some
isotope combinations of titanium and oxygen have the same gross mass as
copper. Thus, in Figure 2, copper follows the titanium curve near the
surface. In Figure 3, the separation of the curves after heating and the
different shape imply that nearly all of the copper diffused toward the
surface. The mechanism for diffusion (such as grain boundary diffusion
or intergrain diffusion) is not known. The effect of a departure from
perfect stoichiometry is not known. Clearly the materials are not stable
at 850°C in contact with each other.

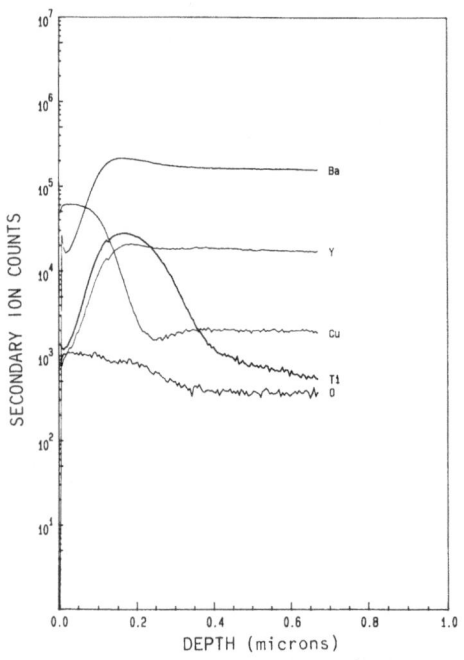

FIGURE 2. COMPOSITION VERSUS
DEPTH OF TiO$_2$ ON
Y-Ba-Cu-O MATERIAL
BEFORE SINTERING.

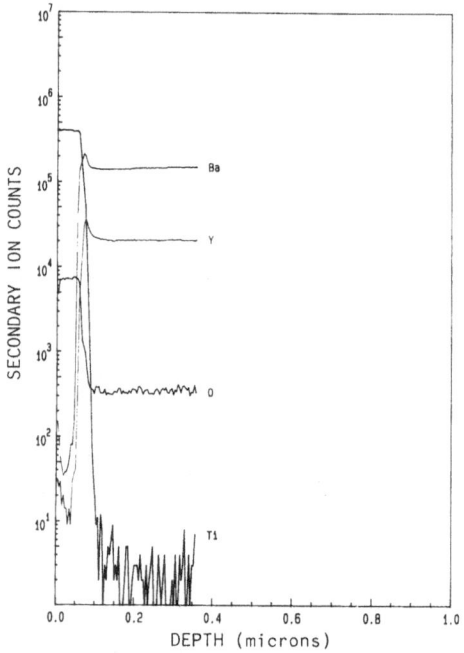

FIGURE 3. COMPOSITION VERSUS
DEPTH OF THE SAMPLE
IN FIGURE 2 AFTER
SINTERING.

Figure 4 shows the effect of heating a similar Y-Ba-Cu-O film on a
layer of tin oxide. This time there does not appear to be a differentia-
tion of diffusion of one component of the superconductor, i.e. Y, Ba and
Cu appear to have moved together. A recent report implies that 1-2-3
material may be stable on indium-tin-oxide[3]. The data reported herein
indicates that there is interdiffusion.

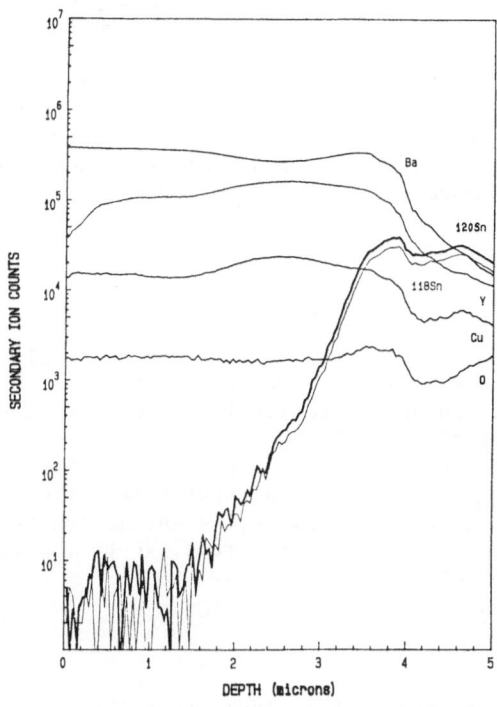

FIGURE 4. COMPOSITION VERSUS DEPTH OF 1-2-3 FILM ON TOP OF SnO_2 LAYER
AFTER SINTERING.

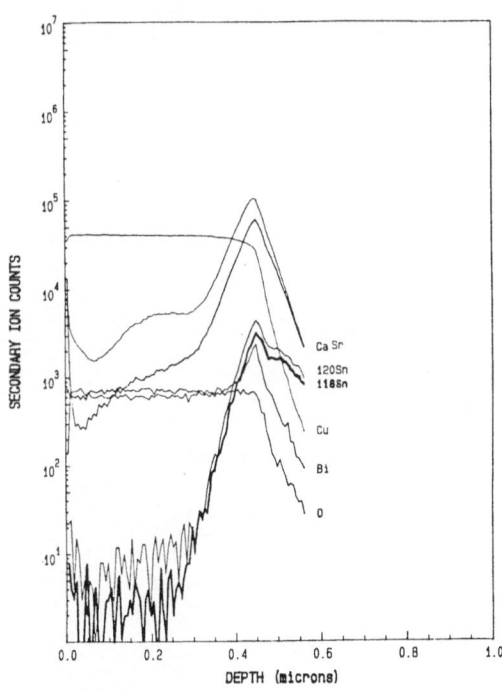

FIGURE 5. COMPOSITION VERSUS DEPTH OF COPPER RICH Bi-Ca-Sr-Cu-O ON TOP
OF SnO_2 AFTER SINTERING.

Figure 5 shows the effect of heating on a copper rich bismuth compound on top of tin oxide. The exact shape of the curve is distorted at the boundary between the bismuth and the tin oxide film. This is referred to as a matrix effect in the analysis of the data. A significant change in composition may increase or decrease the sputtering rate so that the signal may increase or decrease in a manner not related to the actual relative composition. The graph clearly shows that Ca and Sr have diffused relatively more than bismuth or copper.

CONCLUSIONS

The data presented here shows that titanium reacts with the 1-2-3 compound while tin, though diffusing into the material, does not show preferential reactions. Tin does react with the barium compounds, showing preferential diffusion. Electrical contact resistance could not be measured as the diffusion of the compounds prevented the superconductive transition. Experiments on ion beam sputtering show that high energy ions tended to show less differential sputtering effects from a target with atoms of very different mass. Differential ion beam sputtering effects, creating an excess of copper relative to the target, were also reported by Lorentz and Sexton[4].

REFERENCES

1. Y. Tzeng, A. Holt, and R. Ely, Appl. Phys. Lett. 52, 155(1988).
2. Charles Evans and Assoc., CA
3. Nippon Sheet Glass announcement in Superconductor Week, 31 October(1988).
4. R.D. Lorentz and J.H. Sax, Appl. Phys. Lett. 53, 1654(1988).

PREPARATION AND CHARACTERIZATIONS OF Bi-Pb-Sr-Ca-Cu-O SUPERCONDUCTING THIN FILMS

G.C. Xiong, H. Gu, T.X. Lin, S.Z. Wang, and D.L. Yin

Department of Physics
Peking University
Beijing 100871, P.R. China

ABSTRACT

Textured superconducting thin films of Bi-Pb-Sr-Ca-Cu-O (BPSCCO) have been prepared by DC magnetron sputtering using a sintered target. The films were deposited on $SrTiO_3$ and MgO substrates at room temperature. After deposition the films were annealed to 830-880°C in various partial pressures of oxygen. Our best films have zero-resistance temperatures of 77 K. Some samples had high Tc steps above 100 K. The x-ray diffraction (XRD) data show that the annealed superconducting films have the tetragonal phase (a=b=3.8Å, c=30.7Å) and the c-axis was perpendicular to the substrate surface. Transmission electron microscope (TEM) observations suggest that the textured superconducting thin films have stacking layer structure. Twist boundary, stacking faults and other defects were revealed directly by the TEM images.

INTRODUCTION

Since the discovery of superconductivity in the Bi-Sr-Ca-Cu-O (BSCCO) system[1], superconducting thin films in this system have been prepared by a variety of techniques, including coevaporation[2], multilayer deposition[3], magnetron sputtering[4,5] pulsed laser evaporation[6]. Recently, Pb-doped BSCCO bulk samples with Tc(end)=110 K was reported by several groups[7,8]. The high Tc phase was determined by X-ray diffraction techniques and TEM [8]. The superconducting phases are tetragonal with a=b=3.8Å, c=30.7Å and c=37.1Å. The properties of superconductivity are very sensitive to the heat treatment parameters. The superconducting phase are formed in a narrow temperature range, just below the melting point. For achieving a high zero-resistance temperature sample a long time annealing procedure, over several days, is necessary.

The high Tc superconducting thin films have promising future for applications. For an application of superconducting thin films to the electronic devices, not only information on superconductivity, but also knowledges of the microstructure are required. We tried the fabrication of the BPSCCO thin films using DC magnetron sputtering and observed their microstructures by scanning electron microscope (SEM) and TEM.

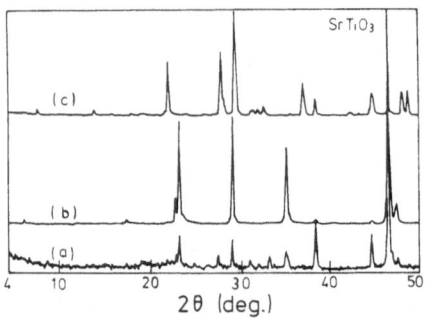

Fig. 1. X-ray diffraction patterns for annealed at (a) 850°C,
(b) 860°C and (c) 870°C.

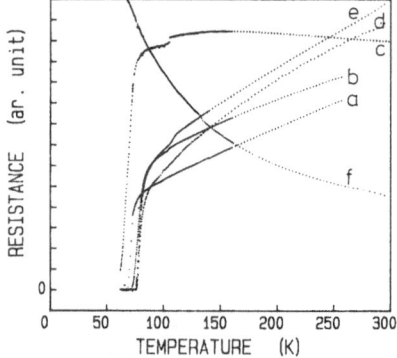

Fig. 2. Temperature dependences of the resistances with
following annealing conditions: (a) 850°C for 10 min
(b) 860°C for 10 min, (c) 855°C for 20 min, (d) 855°C
for 5 hr, (e) 860°C for 14 hr, (f) 870°C for 10 min.

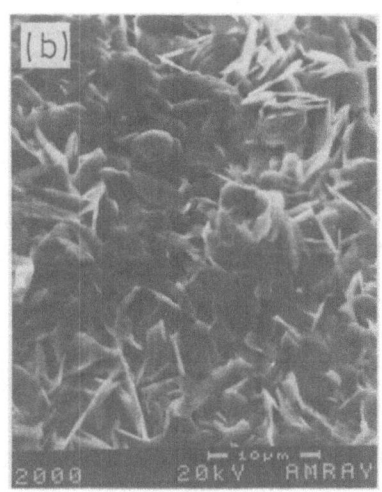
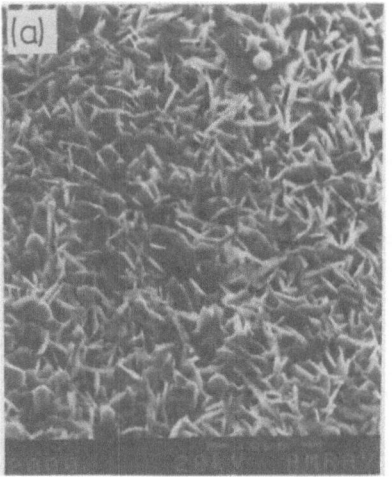

Fig. 3. Scanning electron microscope showing the surface
topography of various superconducting thin films :
(a) annealing at 855°C for 5 hr, (b) annealing at
860°C for 20 hr.

EXPERIMENTAL

A planar target 4cm in diameter is used in this DC magnetron sputtering system. The powder of the BPSCCO was prepared by gel-method. The target preparation procedure was first preheating the powder at 810°C in air and then pressing and sintering at 850°C for 14 hr. The nominal composition of the target is (Bi,Pb):Sr:Ca:Cu=2:2:2:3 . The films were deposited on SrTiO₃ and MgO substrates. The substrates were held at ambient temperature during the deposition. Pure argon gas flowed over the target which gave an operating pressure of $3*10^{-1}$ torr. A typical sputtering condition was 150 V and 200 mA with the sputtering rate of 2Å/sec. All of the as-deposited films were insulating and had smooth and shiny surfaces.

The films were annealed in a furnace tube with mixtures of oxygen and argon. Annealing was carried out in the range of 830°C to 880°C for 10 min to several hours. After annealing, a XRD study was performed using a Rigaku diffractometer. The superconducting properties were examined through DC conductivity measurements by the standard four-probe method. Using a plastic protection technique large scale transparent TEM specimens were obtained. SEM and TEM observations were performed on a KYKY 1000-B + TN 5500 SEM and JEOL JEM 200-CX TEM.

RESULTS AND DISCUSSIONS

Superconducting thin films were obtained by annealing the as-deposited films at temperature range from 850°C-860°C. The XRD data show that the superconducting phase is tetragonal with a=b=3.8Å and c=30.7Å. The XRD patterns of films, corresponding for different annealing temperature, are shown in Fig. 1. Annealing at 850°C for 10 min the tetragonal superconducting phase emerged. The XRD pattern for a sample annealed at 850°C is shown in Fig. 1 (a). Fig. 1 (b) is a representative XRD pattern for films deposited on single crystal (100) TiSrO₃ substrates and annealed at 860°C. The pattern shows that the dominant peaks are (00L) lines. This suggests that as increasing the annealing temperature to 860°C the superconducting films are highly textured with c-axis perpendicular to the substrate surface. When annealing temperature higher than 880°C, the post-annealed films became semiconducting or insulating. Fig. 1 (c) is the XRD pattern of such a semiconducting film which was annealed at 870°C for 10 min in 5% oxygen.

For different annealing conditions the respective resistance versus temperature results are shown in Fig. 2. At suitable conditions a ten-minute annealing is enough for forming the superconducting phase. Fig. 2 (a) and (b) are the temperature dependences of the resistance for samples which were annealed at 850°C or 860°C in 20% oxygen for 10 minutes, respectively. Fig. 2 (c) shows a high Tc drop at 104 K . This sample was annealed at 855°C for 20 minutes. Increasing annealing time, the zero-resistance temperatures increase. Fig. 2 (d) and (e) are the transition curves with Tc(end)=77 K. The samples were annealed in pure oxygen at 855°C for 5 hr and at 860°C for 20 hr, respectively. The sample annealed at 870°C for 10 minutes showed a semiconducting behaviour. Fig. 2 (f) is the temperature dependences of the resistance. The corresponding XRD is Fig. 1 (c).

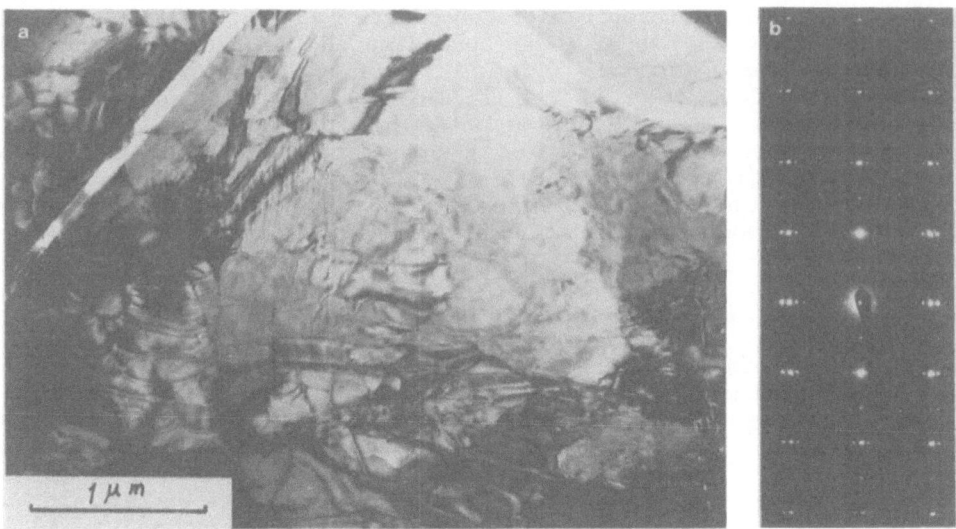

Fig. 4. (a) Transmission electron microscope for a Bi-Pb-Sr-
Ca-Cu-O superconducting thin film, (b) the corresponding
electron diffraction pattern.

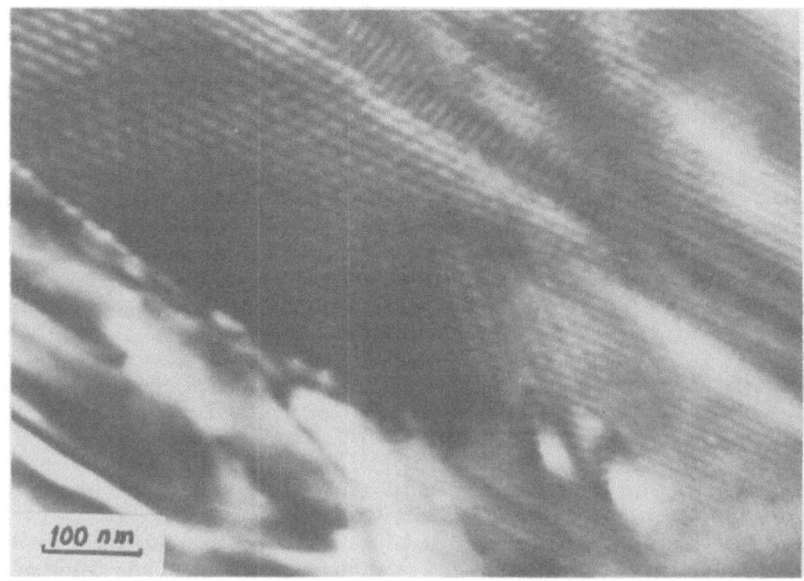

Fig. 5. A moire pattern of the transmission electron microscope.

After annealing the smooth and shiny as-deposited films became gray black. Fig. 3 show the SEM images of two superconducting films, annealed at (a) 850°C and (b) 860°C respectively. The corresponding XRD patterns are Fig. 1 (a) and (b). Comparing the two SEM images it seems that as increasing the annealing temperature the average particle diameter gets larger.

Some typical microstructural features of the films are shown by the TEM observation in Fig. 4 (a). The electron beam was perpendicular to the film surface. Most of grains are sheet form and stacked with their (001) basal planes nearly parallel to the film. This fact can be recognized from the electron diffraction pattern shown in Fig. 4 (b). It agrees with the observed strong c-axis texture in the XRD observations. So the film can be expected as a Josephson-tunneling coupled system. Different kinds of lattice defects such as glide dislocations and stacking faults were seen. Fig. 5 shows the moire patterns due to a small angle (001) twist grain boundary, which is very popular in our films.

In summary, Bi-Pb-Sr-Ca-Cu-O thin films were prepared by the DC sputtering. After annealing at 860°C in pure oxygen the films had Tc(end)=77 K. Some samples had resistance drop above 100 K. The superconducting films had c-axis texture and microstructure in the form of stacked sheets. The glide dislocations on basal planes, stacking faults and twist boundary are some of the most frequently observed defects in the Bi-Pb-Sr-Ca-Cu-O superconducting thin films.

REFERENCES

1. H. Maeda, Y. Tanaka, M. Fukatomi and T. Asano, Jpn. J. Appl. Phys. Lett., 27, 209, (1988).

2. C.E. Rice, A.F.J. Levt, R.M. Fleming, P. Marsh, K.W. Baldwin, M. Anzlowar, A.E. White, K.T. Short, S. Nakahara, and H.L. Stormer, Appl. Phys. Lett., 52, 1828, (1988)

3. B.B. Jie, S.L. Wang, Z.L. Bao, F.R. Wang, C.Y. Li, S.Z. Wang and D.L. Yin, IEEE Transaction on Magnetics, MAG-25, (1988).

4. S.I. Shah, G.A. Jones and M.A. Subramanian, Appl. Phys. Lett., 53, 429, (1988).

5. Brian. T. Sullivan, N.R. Osborne, W.N. Hardy, J.F. Carolan, B.X. Yang, P.J. Michael and R.R. Parsons, Appl. Phys. Lett., 52, 1992, (1988).

6. C. Richard Guarnieri, R.A. Roy, K.L. Saenger, S.A. Shivashankar, D.S. Yee and J.J. Cuuomo, Appl. Phys. Lett., 53, 532, (1988).

7. R.M. Hazen and C.W. Chu et al., Phys. Rev. Lett., 60, 1174, (1988).

8. X. Zhu, S.Q. Feng, J. Zhang, and Z.Z. Gan, IEEE Transactions on Magnetics MAG-25, (1988).

Bi-Sr-Ca-Cu-O COMPOSITIONAL MODULATION AND MULTILAYERED FILM SYNTHESIS BY PLASMA CONTROLLED SPUTTERING

Hideomi Koinuma, Hirotoshi Nagata[*], Akihiro Takano, Masashi Kawasaki[**], and Mamoru Yoshimoto

The Research Laboratory of Engineering Materials
Tokyo Institute of Technology
Nagatsuta 4259, Midori-ku, Yokohama, Kanagawa, Japan

Compositionally modulated Bi-Sr-Ca-Cu-O films were prepared by a method in which the target area to be sputtered can be controlled by varying the distribution of magnetic field over the target. When a binary target comprised of an inner $Ca_{1.0}Cu_{1.5}O_x$ and an outer $Bi_{1.6}Sr_{1.0}O_y$ disks was used, the deposited film composition could be varied successively from Bi-Sr rich to Ca-Cu rich by controlling the magnetic coil currents. Based on this result, films accumulating predominant (Bi-Sr-O) and (Ca-Cu-O) bilayers were deposited continuously on MgO(100) substrates by periodical change of the coil currents. The film designed to have 100 repetitions of $10Å$ thick Bi-Sr rich and $10Å$ thick Ca-Cu rich layers has an actual periodicity of about $19Å$ by the analysis with low angle X-ray scattering. The films deposited at 200°C were almost insulating in the as-grown state, but they turned to be superconducting by the annealing at temperatures of 880-900°C in air. The annealed films have strong c-axis orientation normal to the substrate, resistivities of a few $m\Omega cm$ at room temperature, and zero resistivity temperatures higher than 70K.

INTRODUCTION

Ever since the superconducting Bi-Sr-Ca-Cu-O (BSCCO) system was first discovered by Maeda,[1] extensive efforts have been made to isolate the high T_C(110K) phase of this system. We can find several reports describing of the success in the growth of almost pure high T_C phase which is generally noted to have a composition of $Bi_2Sr_2Ca_2Cu_3O_y$. For instance, bulk BSCCO with a zero resistivity temperature ($T_{C,0}$) higher than 100K was prepared

---

[*] Central Research Laboratory, Sumitomo Cement Co. Ltd.
[**] Department of Industrial Chemistry, Faculty of Engineering, University of Tokyo

by PbO addition to $Bi_{2-x}Sr_2Ca_2Cu_3O_y$ and sintering it under reduced oxygen partial pressure.[2] Thick films with $T_{C,0} \gtrsim 100K$ were also prepared by the spray-pyrolysis of aqueous solutions containing Bi,Sr,Ca,Cu ions in an atomic ratio of 2:2:2:3 and a small amount of Pb ions.[3] In spite of these general notion for high T_C $Bi_2Sr_2Ca_2Cu_3O_y$ phase, it still remains a matter of controversy what the real composition and structure of high T_C phase are. Another characteristic feature of BSCCO is the apparent correlation between T_C and the number of CuO_x planes in a layered perovskite slab sandwiched by two Bi_2O_2 layers; $T_{C,0}$ increases in the order of $0 \sim 20K$ for $Bi_2Sr_2CuO_x$, $<80K$ for $Bi_2Sr_2Ca_1Cu_2O_y$ and $<110K$ for $Bi_2Sr_2Ca_2Cu_3O_z$. Thus, the development of methods which enable easy control of composition and layer structure in multi-component oxides should be an attractive research subject from the viewpoint not only of searching for new higher T_C materials but also of elucidating structure-property relationship in new type of ceramics. As an approach to this end, we employed the plasma controlled sputtering method. This method developed originally by Hata[4] depends on the principle that the spacial distribution of plasma generated in a dc or rf sputtering system can be controlled effectively by using another magnetic coil in addition to a conventional magnetron coil. The use of a target with different compositions along the radius makes it possible to deposit compositionally modulated films by a simple current control.

Here, we report the compositional modulation and multi-layered film synthesis of Bi-Sr-Ca-Cu-O system by the plasma controlled sputtering method. Annealing effect on the deposited layered films is also presented in relation to the realization of superconductivity of the films.

EXPERIMENTAL

Figure 1 shows the plasma controlled sputtering system used in this work. The binary target was composed of an inner $CaCu_{1.5}O_x$ disk ($\phi40x^t3.5$ mm) fixed on a central part of an outer $Bi_{1.6}SrO_y$ disk ($\phi70x^t3.5$ mm). These oxide disks were made from $CaCO_3$, CuO, Bi_2O_3 and $SrCO_3$ by sintering properly mixed powders. The binary target was sputtered with plasma generated by applying a 100W rf-power under 5mTorr of Ar/O_2 (8/2) mixed gas atmosphere. The ring-shaped emissing plasma is formed above the target and its diameter is continuously modulated by controlling the currents in two magnetic coils placed under (coil B_M) and around (coil B_C) the target. As the plasma is focused, the mainly sputtered area moves towards the center of the binary target to vary the composition of the deposited film continuously. When the sputtered area is varied periodically between two extremes of outer and inner parts of the binary target, the deposited film can have accumulated (Bi,Sr) rich and (Ca,Cu) rich layers. In this study, multilayered films were prepared on MgO(100) substrates. The substrate was not heated intentionally, but its temperature reached at $200 \sim 300°C$ during the sputtering. The distance between the target and the substrate was 40mm.

Deposited films were annealed in air at temperatures in the range of 700~900°C to make them superconducting. Film structure was determined before and after the annealings by an X-ray diffractometer with monochromated CuKα radiation using an MAC Science model MXP-3. Film resistivities were measured by the conventional dc four probe method with evaporated gold terminals on the specimens. Film composition was determined by inductively coupled plasma emission analysis (ICP) using a Seiko model SPS-1200. Film thickness was measured by a Taylor Hobson's Talystep to calculate the deposition rate. The surface morphology of some of the films was observed by a scanning electron microscope (SEM) using a JEOL model JSM-T100.

RESULTS AND DISCUSSION

Film Preparation

The deposited film composition was successively changed by the simple control of a coil current in the apparatus shown in Fig.1. Figure 2 shows the variation of film composition depending on the current I_C in the compressive magnet coil B_C. The current I_M of the magnetron coil B_M was fixed at 1A. As the I_C value changed from 1.0 to 4.5A, the plasma was focused toward the center and the predominantly sputtered area shifted from the outer to inner part of the binary target to modulate the film

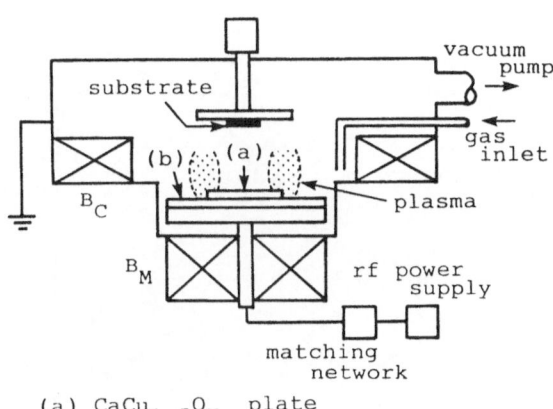

(a) $CaCu_{1.5}O_X$ plate
(b) $Bi_{1.6}SrO_Y$ plate

Fig. 1. Schematic diagram of the plasma controlled sputtering apparatus. Ring-shaped plasma is generated on the binary target consisted of (a) an inner $CaCu_{1.5}O_X$ and (b) an outer $Bi_{1.6}SrO_Y$ plates. B_M and B_C are magnetic coils to control a magnetic field distribution over the target.

composition from (Bi,Sr) rich to (Ca,Cu) rich. Multilayered films were prepared by alternatingly switching the coil current between the conditions (a) I_C=1A and (b) I_C=4A. The film composition and deposition rate were $Bi_{2.6}Sr_{2.0}Ca_{1.6}Cu_{0.7}O_x$ and 1.95Å/s for condition (a) and $Bi_{1.0}Sr_{2.0}Ca_{5.0}Cu_{3.3}O_y$ and 0.76Å/s for condition (b). The multilayered film was designed to have a periodicity of 20Å (10+10Å) and total thickness of 2000Å by repeating the current switching 199 times manually. The total thickness was measured to distribute in the range of 2100~2250Å from place to place. The composition of the multilayered film was determined to be $Bi_{2.2}Sr_{2.0}Ca_{2.2}Cu_{3.4}O_x$, being in good agreement with the calculated average composition of $Bi_{2.1}Sr_{2.0}Ca_{2.3}Cu_{2.7}O_x$. The periodicity of 20Å was designed in view of the c-axis length of 18Å between two Bi_2O_2 layers in the (2223) structure.

The low-angle X-ray scattering of the as-deposited multilayered film is shown in Fig.3. The first order scattering peak appeared at d=19.9Å to reinforce that the actual periodicity agreed well with the designed value. However, higher order peaks were weak and broad, probably because of a fluctuation at the layer interfaces. In a multilayered film designed to have a 200Å(50Å+150Å) periodicity, a clear low-angle X-ray scattering pattern was obtained as shown in Fig.4[5]. The actual periodicity for this

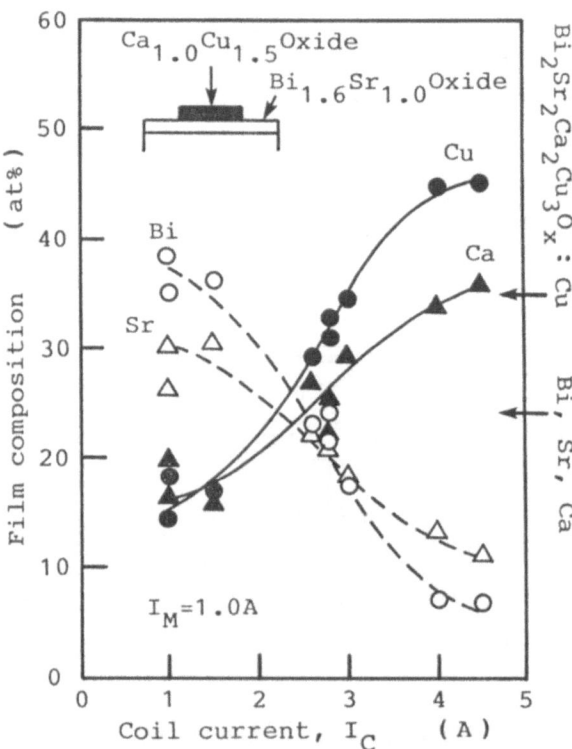

Fig. 2. Variation of film composition deposited under controlled magnetic field. Magnetic field is altered by coil current I_C with I_M fixed to 1A.

film was calculated to be 236Å by applying the Modified Bragg's equation to the data in Fig.4. The broad X-ray diffraction pattern for 10Å/10Å layered film is considered to come from (1) mixing and/or knock-on effect of sputtered particles during the deposition and (2) compositional mixing on the target during the sputtering. Sharper interface structures are expected to be formed by decreasing the sputtering energy and/or by the optimization of current condition and target dimension.

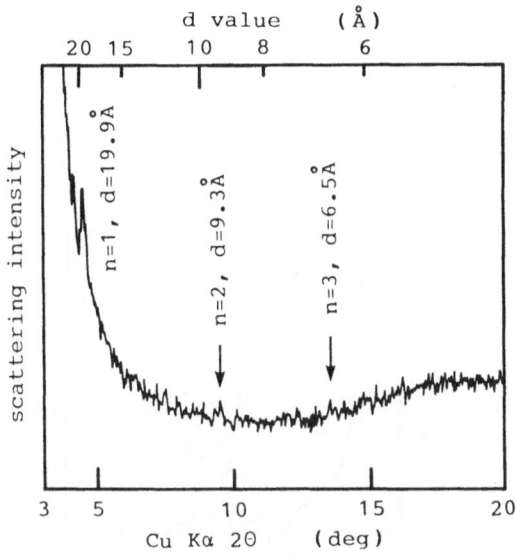

Fig. 3. Low-angle X-ray scattering pattern of the as-deposited film designed to have a periodicity of 20Å.

Annealing Effect and Superconductivity

The as-deposited films have dark brown color and resistivities higher than 600Ω cm at room temperature. To gain superconductivity in the films, annealing in air was carried out under several temperature conditions.

When the multilayered films have a periodicity close to the actual spacing along the c-axis of (2223) crystal structure, annealing is expected to provide such effects on the crystal growth in the films as the preferential growth of the (2223) phase and the decrease of growth temperature of superconducting phases than the usually employed 850~890 °C[6,7].

Figure 5 shows the resistivities of the 10Å/10Å multilayered and homogeneous composition films measured at room temperature after annealing at 880°C for various periods. The homogeneous film was prepared by fixing T_C at 2.8A so that it had almost the same composition as the average composition of multilayered film (cf Fig.2). The resistivity of the homogeneous film dropped very quickly to be less than 10mΩcm in 30min. For the multilayered film, annealing time longer than 2h was necessary to decrease its resistivity in a range of mΩcm. This result may be related to the nucleation in a crystal growth. The optimal composition for superconducting $Bi_2Sr_2Ca_1Cu_2O_x$ or $Bi_2Sr_2Ca_2Cu_3O_y$ should be satisfied at every place in the homogeneous film, while it should be only around the layer interface. Therefore, the sites of nucleation for crystal growth may be limited at the interface region.

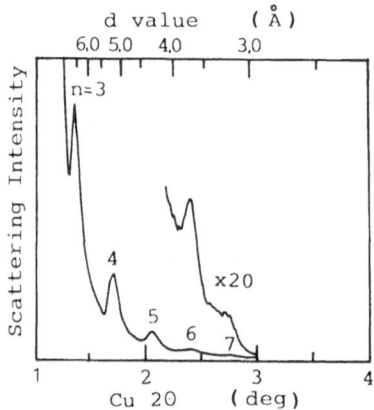

Fig. 4. Low-angle X-ray scattering pattern of the as-deposited film designed to have a periodicity of 200Å. n indicates the reflection order. The actual periodicity was calculated to be 236Å using Modified Bragg's equation.

Zero-resistivity temperature ($T_{C,0}$) of the annealed multilayered film at 880°C for 2h was 41K as shown in Fig.6. The films annealed for shorter than 2h had semiconductor-like resistivity profiles, although highly c-axis oriented $Bi_2Sr_2CaCu_2O_y$ (2212) phase was detected in XRD. An SEM observation revealed that the annealed multilayered film had lower degree of grain growth than the higher $T_{C,0}$ superconducting film prepared from the homogeneous film.

Figure 6 shows the temperature-resistivity relationships for the multilayered films annealed for 30min at temperatures higher than 880°C. $T_{C,0}$ was obtained at 71K for the films annealed at 885 and 890°C. The

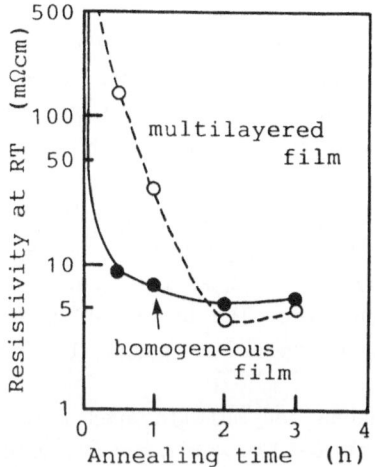

Fig. 5. Annealing time dependence of the résistivity at room
temperature for the multilayered and the homogeneous films
annealed at 880°C in air.

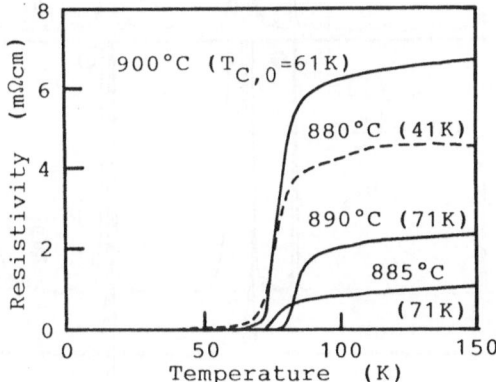

Fig. 6. Temperature dependence of the resistivity for the
multilayered films annealed at 885°C, 890°C, and 900°C for
30min in air. The dashed line is for the film annealed at
880°C for 2h. Values in ( ) indicate zero-resistivity
temperatures.

both annealed films had mainly the (2212) phase predominantly and c-axis orientation normal to the substrate surface. Annealing at temperatures higher than 890°C deteriorated the film superconductivities because of the film evaporation and a chemical interaction with MgO substrate[8].

Possibility of low-temperature annealing was also investigated for the multilayered films. Annealing at 700°C maintained a smooth surface of the as-grown film and produced a small amount of $Bi_2Sr_2CuO_z$ (2201) phase, but it did not work to grow the high Tc (2212) or (2223) phase. Figure 7 shows a crystal growth in the homogeneous and the multilayered films by annealing at 750°C for up to 300h. At this annealing temperature, the formation of (2201) phase was detected by XRD as a peak at $2\theta=7.2°$. In the multilayered film, the growth of additional XRD peaks were observed by the annealing. The broad XRD peak at $2\theta=6.2°$ shifted to $2\theta=5.8°$, which corresponded to (2212) phase, by further annealing.

Both of the homogeneous and multilayered films annealed at 750°C for 300h showed semiconductor-like resistivity profiles and their room temperature resistivities were about ten times higher than those of films annealed at temperatures higher than 880°C. Clear resistivity drop at about 90K observed in the multilayered film was considered to be

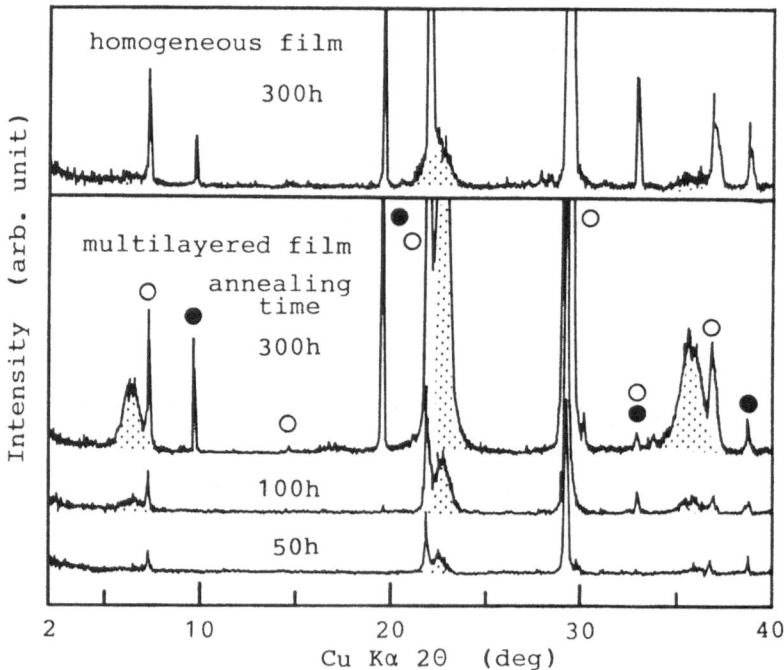

Fig. 7. XRD patterns of the homogeneous and the multilayered films annealed at 750°C for up to 300h in air. Peaks (O) are for a $Bi_2Sr_2CuO_z$ phase and peaks (●) are for a Bi-Ca-O phase. Dotted peaks are for a phase having c-axis longer than that of $Bi_2Sr_2CuO_z$.

correlated with an unknown phase shown as broad peaks in Fig.7. Superconductivity was improved by annealing at 800°C for 20h to show a metallic resistivity profile and zero-resistivity at 61K. On the basis of these results, the formation of higher T_C phases at temperatures below 800 °C is expected by the optimization of the initial periodicity and composition in the deposited films, as well as of the annealing conditions.

SUMMARY

Multilayered films were prepared on MgO(100) substrates using the plasma controlled sputtering to have periodicities of about 20Å and 200Å. As-grown insulating films with a 20Å lattice were annealed at temperatures ranging from 700°C to 890°C to form (2212) phase and show zero-resistivity at temperatures as high as 71K. Low-temperature annealing of the multilayered films formed a phase which could not be detected in the homogeneous composition films.

REFERENCES

1. H.Maeda, Y.Tanaka, M.Fukutomi, and T.Asano, "A New High-Tc Oxide Superconductor without a Rare Earth Element," Jpn. J. Appl. Phys., **27**: L209 (1988).

2. U.Endo, S.Koyama, and T.Kawai, "Preparation of the High-Tc Phase of Bi-Sr-Ca-Cu-O Superconductor,"Jpn. J. Appl. Phys.,**27**: L1476 (1988).

3. H.Nobumasa, K.Shimizu, Y.Kitano, and T.Kawai, "Formation of a 100K Superconducting Bi(Pb)-Sr-Ca-Cu-O Film by a Spray Pyrolysis," Jpn. J. Appl. Phys., **27**: L1669 (1988).

4. T.Hata, Y.Kamide, S.Nakagawa, and K.Hattori, "Heat-resistance of hydrogenated amorphous silicon films prepared by compressed magnetic field-magnetron sputtering with He and Ar gases," J. Appl. Phys., **59**: 3604 (1986).

5. H.Koinuma, H.Nagata, A.Takano, M.Kawasaki, and M.Yoshimoto, "Preparation of Compositionally Modulated Bi-Sr-Ca-Cu-O Multilayered Films by Plasma Controlled Sputtering," to be published in Jpn. J. Appl. Phys., **27** (1988).

6. T.Kato, T.Doi, T.Kumagai, and S.Matsuda, "Superconducting Bi-Sr-Ca-Cu-O Thin Films by Sputtering," Jpn. J. Appl. Phys., **27**: L1097 (1988).

7. Brian T.Sullivan, N.R.Osborne, W.N.Hardy, J.F.Carolan, B.X.Yang, P.J.Michael, and R.R.Parsons, "Bi-Sr-Ca-Cu-oxide superconducting thin films deposited by dc magnetron sputtering," Appl. Phys. Lett., **52**: 1992 (1988).

8. H.Nagata, A.Takano, M.Kawasaki, M.Yoshimoto, and H.Koinuma, "Preparation of Bi-Sr-Ca-Cu-O Thin Films by Sputtering under Variable Magnetic Field," submitted to Communications of J. Amer. Cer. Soc..

HIGH T_c Y-Ba-Cu-O FILMS PREPARED BY MULTILAYER, REACTIVE SPUTTERING

FROM SEPARATE Y, Cu, and $Ba_{0.5}Cu_{0.5}$ TARGETS

S.J. Lee, K.C. Sheng, Y.H. Shen, E.D. Rippert,
X.K. Wang, R.P. VanDuyne, R.P.H. Chang and
J.B. Ketterson

Materials Research Center
Northwestern University, Evanston, IL 60208

ABSTRACT

In order to thoroughly investigate the phase diagram in films, and to perform destructive tests (e.g. various annealing procedures), a set (nearly ninety) of films having a composition $Y_xBa_2Cu_yO_z$ with $0.6 < x < 1.9$ and $2.5 < y < 4.5$ (centered around the $Y_1Ba_2Cu_3O_{7-\delta}$ compound in the ternary phase diagram) were prepared on MgO. Multilayer deposition of the constituents at ambient temperature was performed by reactive dc-magnetron sputtering from separate Y, Cu, and $Ba_{0.5}Cu_{0.5}$ targets in an atmosphere of 6×10^{-5} Torr O_2 and 1.0×10^{-2} Torr Ar. Two kinds of post-deposition heat treatment were applied: rapid thermal annealing between 950 - 1000°C and slow annealing between 840 - 860°C. Superconducting films with T_c (R=0) between 80 K and 90 K were routinely produced by the rapid thermal annealing process, if the stoichiometry lay in the range of $0.9 < x < 1.4$ and $3.2 < y < 3.5$. The interrelation between stoichiometry, heat treatment, T_c (R=0), chemical reactions with the substrate, and the presence of phases other than the known 1-2-3 superconductor was explored; Raman spectroscopy was employed to study the latter two properties. A strong correlation was found between the Raman spectra and T_c (R=0), which suggests the potential utility of Raman scattering as an in-situ probe to characterize the growth of Y-Ba-Cu-O films. We also found that the rapid thermal annealing process is a very efficient way to reduce chemical reactions of the film with its substrate.

Since the discovery of the high T_c oxide superconductors,[1] research on thin films of these materials has progressed at an unprecedented pace. Various deposition techniques have been employed or developed to make high quality films on various substrates. However many film processing conditions and parameters still are not well understood. Therefore, systematic studies are necessary in order to produce films with reproducibly high quality. Two of the most important parameters are the film stoichiometry and the heat treatment. According to the ternary phase diagram of Y_2O_3, CuO, and BaO in the bulk Y-Ba-Cu-O system[2,3,4] $YBa_2Cu_3O_{7-\delta}$ is surrounded with other phases: CuO, $BaCuO_2$, Y_2BaCuO_5, and $YBa_3Cu_2O_x$ which are either insulating or semiconducting.

Two important problems encountered in achieving high quality films of this compound are the need to achieve strict stoichiometry (so as to avoid second phases) and the need to reduce interdiffusion between the substrate and the deposited film. Small variations in composition can result in the formation of undesired phases, particularly at the grain boundaries, which may lead to a long tail in the resistive transition below a sharp onset. A reduced T_c onset can be caused by chemical reaction between the 1-2-3 phase and the substrate. The heat treatment is also an important factor in obtaining the correct phase, with the correct oxygen stoichiometry, and to minimize the interdiffusion between the film and the substrate.

In this paper, we report the successful fabrication of an Y-Ba-Cu-O superconducting film with a $T_c(R=0)$ of 90 K prepared by ambient temperature sputter-deposition on an MgO substrate followed by a rapid thermal annealing; to our knowledge, films deposited under ambient conditions on MgO have never been reported to attain $T_c(R=0)$ higher than 80 K, regardless of the post-deposition heat treatment. In addition, the effects of stoichiometry and various heat treatments on T_c and the evolution of the Raman spectrum under annealing will be presented.

Films were deposited by dc-magnetron sputtering from separate metal targets: Y, Cu, and $Ba_{0.5}Cu_{0.5}$. Previous results using a stoichiometric $YBa_2Cu_3O_{7-\delta}$ compound target were reported elsewhere.[5,6] Multilayer deposition on MgO, by alternatively passing a rotating substrate platform over the three targets, was carried out at ambient temperature in an atmosphere consisting of a mixture of 6×10^{-5} Torr of O_2 and 1.0×10^{-2} Torr of Ar. Reactive sputtering was chosen to incorporate oxygen during deposition. The introduction of oxygen decreases the deposition rate from all targets and significantly alters the ratio between Ba and Cu from the $Ba_{0.5}Cu_{0.5}$ target. The oxygen partial pressure was fixed at 6×10^{-5} Torr for most depositions. Films having different metallic stoichiometry were prepared with Y : Ba : Cu = 0.6 - 1.9 : 2: 2.5 - 4.5, where the compositions were determined by electron microprobe analysis. Typically, films had a thickness of 1.8 - 2.5 μm with each sublayer not exceeding 300 Å. As-deposited films were amorphous and highly resistive.

Two kinds of heat treatment were used: a "long-annealing" at 840°C for 1 hour and a "rapid thermal annealing" at 1000°C followed by a long period at much lower temperature (shown in Fig. 1). All annealed films

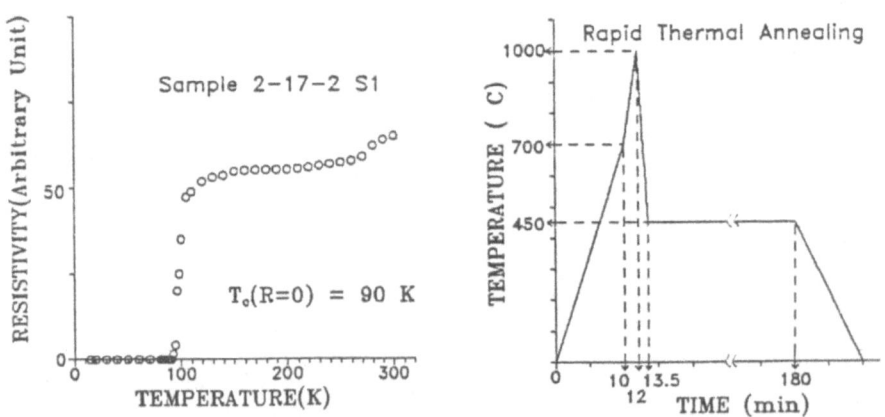

Figure 1. Resistive Transition Curve of Sample #2172S1 and Its Annealing Profile.

Table 1. Films Annealed at 840°C for 1 hour

Sample Number	$T_c(R=0)$	Composition Y : Ba : Cu		
#2171A1	77	1.4 :	2 :	3.5
#206C1	74	1.1 :	2 :	2.9
#2171P1	74	1.3 :	2 :	3.8
#2171N1	73	1.3 :	2 :	3.2
#210K1	73	1.0 :	2 :	3.4
#201B1	70	1.0 :	2 :	3.1
#2171Q1	69	1.0 :	2 :	3.1
#207G2	64	1.2 :	2 :	3.1
#216B2	62	1.2 :	2 :	3.1
#207A1	55	1.0 :	2 :	3.1

Table 2. Films Annealed at 1000°C Rapidly

Sample Number	$T_c(R=0)$	Composition Y : Ba : Cu		
#2172S1	90	1.4 :	2 :	3.2
#210J1	89	1.2 :	2 :	3.5
#210P1	88	0.9 :	2 :	3.2
#211B11	84	1.4 :	2 :	3.5
#2172T1	84	1.4 :	2 :	3.4
#2171M1	82	1.4 :	2 :	3.2
#211K2	76	0.9 :	2 :	3.0
#2172J1	76	1.5 :	2 :	3.7
#2171L1	73	1.7 :	2 :	3.5
#210B1	73	1.4 :	2 :	3.4

have a morphology consisting of needles and platelets (SEM). The
resistive transition of the annealed samples was measured with the dc
four probe method. Films having higher $T_c(R=0)$'s are listed with their
metallic compositions in Tables 1 and 2. From these results, several
features are observed.

1. Superconducting film with $T_c(R=0)$ = 90 K can be made on
    MgO by rapid thermal annealing of a film deposited at ambient
    temperature.
2. High T_c (60 K) films can be made if the composition lies in
    the range Y : Ba : Cu = 0.9 - 1.7 : 2 : 2.8 - 3.8. By
    using the rapid annealing process (Table 2) films with a wide
    range of composition can have T_c over 70 K. This may
    suggest the possible presence of (meta-sable) superconducting
    phases other than 1-2-3 phase;[7,8,9] a TEM study to search
    for such phases is under way.
3. The rapid thermal annealing is more efficient than the more
    conventional long annealing.

One reason that the rapid annealing gives the best results could be
that it minimizes the chemical reactions with the MgO substrate.
Evidence supporting this conjecture comes from the Raman spectra. Raman
scattering is reported to be very sensitive to the presence of impurity
phases in the Y-Ba-Bu-O system.[10] All Raman spectra were taken at room

temperature. The Raman line from the superconducting phase, at about 500 cm^{-1}, is very weak compared to those from other phases. Figs. 2 and 3 show the Raman spectra from a high T_c (90 K) film, annealed rapidly at 1000°C, and a low T_c (55 K) film, annealed at 840°C for 1 hour. Fig. 4 shows the Raman spectra of a film before and after annealing; the annealed film had a T_c(R=0) = 90 K.

Sample 2-17-2S1

Interface: Air/123

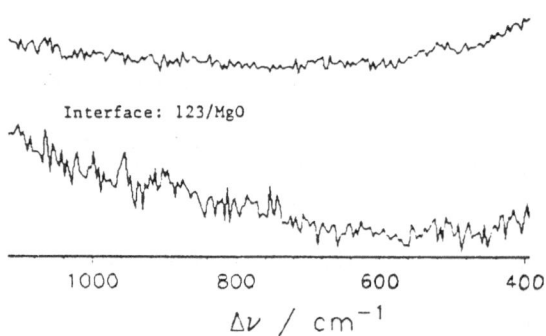

Interface: 123/MgO

$\Delta\nu$ / cm^{-1}

Figure 2. Raman spectra of a film having a T_c(R=0) of 90 K.

Sample 2-7A1

Interface: Air/123

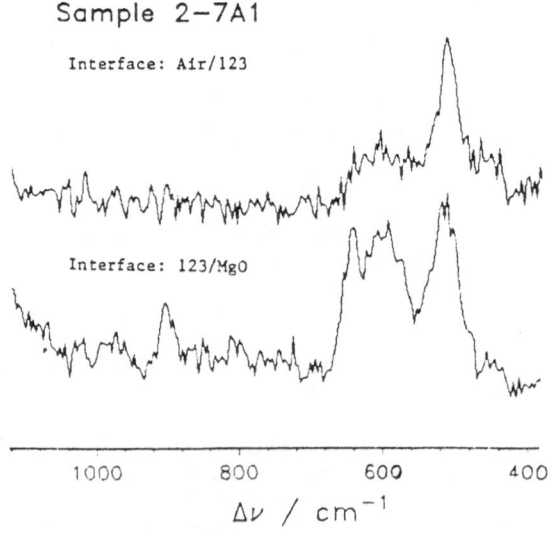

Interface: 123/MgO

$\Delta\nu$ / cm^{-1}

Figure 3. Raman spectra of a film having a T_c(R=0) of 55 K.

Several conclusions can be drawn:

1. Films with a high T_c generally have a simple Raman spectrum in relation to those with a low T_c.
2. The spectrum of the free surface (air/film) and the interface (MgO/film) can differ significantly. The interface spectrum displays more spurious lines in low T_c films.
3. High T_c films prepared by rapid thermal annealing show fewer spurious lines, which may imply less chemical reactivity between a film and its substrate.
4. Unannealed (amorphous) films show a single broad line.
5. A strong correlation between the Raman spectra and $T_c(R=0)$ suggests the potential utility of Raman scattering as an in-situ probe to characterize the growth and annealing of Y-Ba-Cu-O films during deposition and heat treatment.

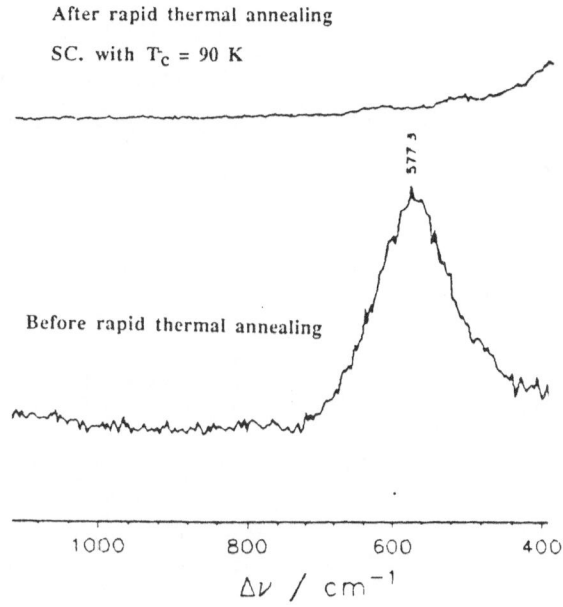

Figure 4. Raman spectra of the film with $T_c(R=0) = 90$ K before and after the rapid thermal annealing.

ACKNOWLEDGEMENTS

The work was supported by the Northwestern University Materials Research Center under NSF grant DMR-85-20280, and by the Office of the Naval Research under grant N00014-88-K-0106.

We would like to thank B.Y. Jin and S.N. Song who contributed to the early phases of this work.

REFERENCES

1. G.D. Bednorz and K.A. Muller, Z. Phys., **B64**, 189 (1986).
2. G. Wang, S.J. Hwu, S.N. Song, J.B. Ketterson, L.D. Marks, K.R. Poeppelmeier, and T.O. Mason. Adv. Cer. Mat., **2**, 313 (1987).

3. B.Y. Jin, S.J. Lee, S.N. Song, S.J. Hwu, K.R. Poeppelmeier, and
    J.B. Ketterson, Adv. Cer. Mat., **2**, 436 (1987).
4. S.J. Lee, E.D. Rippert, B.Y. Jin, S.N. Song, S.J. Hwu,
    K.R. Poeppelmeier, and J.B. Ketterson, Appl. Phy. Lett., **51**, 1194
    (1987).
5. K. Char, M. Lee, R.W. Barton, A. F. Marshall, I. Bozovic,
    R.H. Hammond, M.R. Beasley, T.H. Gaballe, and A. Kapitulnik,
    Phys. Rev. **B38**, 834 (1988).
6. A.F. Marshall, R.W. Barton, K. Char, A. Kapitulnik, B. Oh, and
    R.H. Hammond, Phys. Rev., **B37**, 9353 (1988).
7. D.D. Berkley, D.H. Kim, B.R. Johnson, A.M. Goldman,
    M.R. Mecartney, K. Beauchamp, and J. Maps, Appl. Phy. Lett.,
    53, 708 (1988).
8. A. Mascarenhas, S. Geller, and L.C. Xu, Appl. Phy. Lett.,
    52, 242 (1988).

SUPERCONDUCTING Tl-Ca-Ba-Cu-O THIN FILMS

R.J. Lin and P.T. Wu

Materials Research Laboratories
Industrial Technology Research Institute
Chutung, Hsinchu 31015, Taiwan, R.O.C.

ABSTRACT

The superconducting Tl-Ca-Ba-Cu-O thin films (0.6~2μm) on (001)MgO and YSZ single crystal substrates have been prepared by diffusion and interaction between Tl_2O_3 vapor and sputtered Ca-Ba-Cu-O films in the flowing O_2 furnace. The best superconductivity of the films is Tc(onset) =120K and Tc(zero)=78K. The superconducting phase is highly oriented $Tl_1Ca_1Ba_2Cu_2Ox$.

INTRODUCTION

Since the discovery of a new class of Tl-Ca-Ba-Cu-O superconductors with remarkably high Tc values above 100K[1], their superconducting films have been prepared by sputtering[2,3,4], sequential e-beam evaporation[5,6] and sequential thermal evaporation[7]. The most drawback about these processes is the contamination of toxic Tl-containing compounds in the deposition chamber. It may cause great harm for researchers due to absorbing the toxic dust through the skin or inhaled. Simultaneously, thallic oxide has high vapor pressure at temperature above 700°C[8] and low melting point (717°C)[9]. Therefore, we try to develop the safer process to prepare the superconducting Tl-Ca-Ba-Cu-O films. It contains Tl_2O_3 deposition and liquid phase sintering by diffusion between Tl_2O_3 and Ca-Ba-Cu-O films. The detailed process and results will be described.

EXPERIMENTS

The Tl-Ca-Ba-Cu-O films were prepared by interaction and diffusion between Tl_2O_3 vapor and sputtered Ca-Ba-Cu-O films in annealing furnace with flowing O_2. The Ca-Ba-Cu-O films were made by the high pressure DC sputtering process[10,11,12] from sintered compound targets. The targets (diameter 4.5cm; thickness 0.3cm) were made by a solid state reaction of $CaCO_3$, $BaCO_3$ and CuO at 900°C for 10~15hr in air. The stoichiometric ratios of Ca:Ba:Cu were 1:1:1.5, 1:0.75:1.5, 1:0.75:1 and 1:1.5:2, respectively. The base pressure of the vacuum system prior to deposition was 1×10^{-3} torr. During deposition, the sputtering atmosphere was the mixture of Ar and O_2 gases whose ratio of volume flow rate was 1. The deposition pressure was 1.5 torr. The target-to-substrate separation was 2cm. The substrates

(10x3x1mm) were (001)MgO and YSZ single crystals. The substrate temperature was 350°C. The sputtering voltage and current were 220 volt and 0.6 amp, respectively. The thickness of the Ca-Ba-Cu-O films was 0.6~2μm and deposition rate was about 100A°/min.

The schematic structure of the system forming Tl-Ca-Ba-Cu-O films was shown in Fig.1. The Tl_2O_3 powder(0.3~0.6g) was set on the alumina boat which located at the highest temperature(middle) region in the furnace. The Ca-Ba-Cu-O films were annealed by two ways. First, the film was set at the A region near the edge of the furnace. It took 1 hr to heat the furnace to attain the controlled temperature 925°C in flowing O_2 atmosphere. Followingly, the film was rapidly drawn into the B region, shown in Fig.1 and hold for 15~60 min.. Then, the film was cooled in the furnace with following O_2 by turning off the furnace power. It took 3 hours to cool down 200°C. The film was drawn out at below 100°C. Second, the film was located near the Tl_2O_3 source. The Tl_2O_3 source-to-film separation was 4 cm. It took 25 min. to heat the furnace to attain the controlled temperature 820~840°C. The film was hold for 15~60 min.. Then, the above-mentioned procedures were followed.

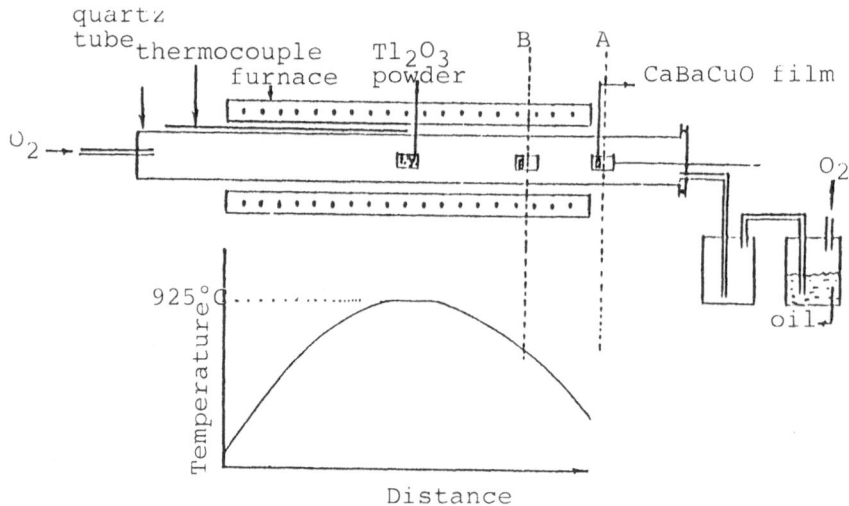

Fig.1 Schematic structure of system of forming Tl-Ca-Ba-Cu-O films by diffusion method.

The structure of the films was characterized by X-ray diffractometer (Philips PW 1700) with monochromated CuKα radiation(40KV; 30mA). The composition of the films was examined by the EDAX 9100/70 energy dispersive spectrometer. Quantitative analysis was performed using standardless techniques with ZAF correction factors[13]. The thickness of the film was determined by the surface profilometer. The resistance of the films was measured by AC four point method using Ag paste contacts.

RESULTS & DISCUSSION

The main phase of the sintered Ca-Ba-Cu-O targets is $BaCuO_2$, shown in Fig.2(A). The as-deposited Ca-Ba-Cu-O films are grey crystalline insulators. Their typical X-ray diffraction patterns are shown in Fig.2(B). The patterns of unknown Ca-Ba-Cu-O phase is similiar to those of $BaCuO_2$. These results suggest that there is large solubility for Ca element in the $BaCuO_2$ crystal.

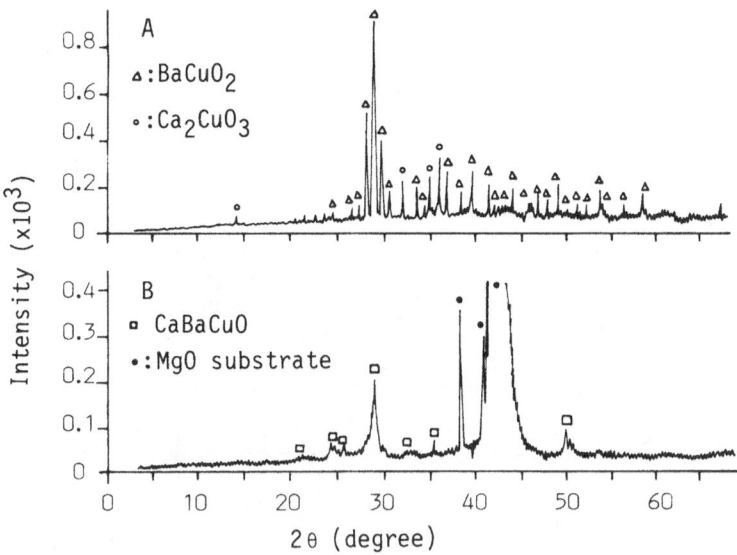

Fig.2 X-ray diffraction patterns of sintered $Ca_1Ba_{0.75}Cu_{1.5}Ox$ target(A) and as-deposited film(B).

The effect of annealing way on the superconductivity of the Tl-Ca-Ba-Cu-O films is shown in Fig.3. The A and B films were annealed by the first and second way, indicated in the section of experiments, respectively. It shows that the film prepared by the second way has higher transition temperature Tc(onset). This is attributed to the different superconducting phase in the films, shown in Fig.4. The superconducting phase of A and B film is $Tl_1Ca_1Ba_2Cu_2Ox$[14] and $Tl_2Ca_1Ba_2Cu_2Ox$, respectively. In addition, the C film, shown in Fig.3 and 4, was prepared by the first way but quenched from annealing temperature. It also shows superconducting behavior. The superconducting phase is $Tl_1Ca_1Ba_2Cu_2Ox$. In the following, only Tl-Ca-Ba-Cu-O films prepared by the first way are discussed.

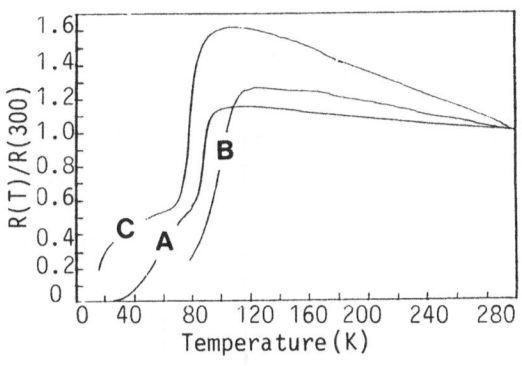

Fig.3 Effect of annealing way on the superconductivity of Tl-Ca-Ba-Cu-O films. Treating condition: (A) first way (B) second way, (C) first way, quenched from annealing temperature; at 805± 5°C, annealing time 15 min., Tl_2O_3 powder 0.4g.

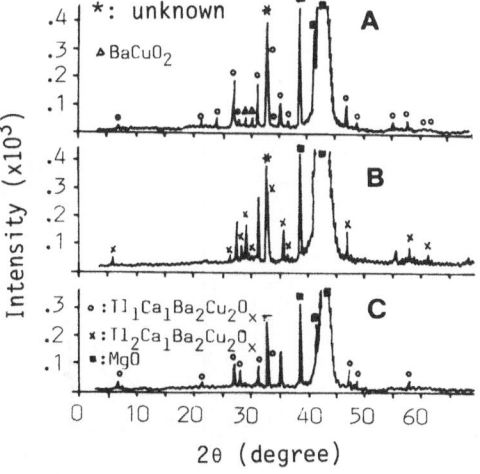

Fig.4 Effect of annealing way on the X-ray diffraction patterns of Tl-Ca-Ba-Cu-O films. Treating condition is shown in Fig.3.

The effect of Ca-Ba-Cu-O target composition on the composition of Tl-Ca-Ba-Cu-O films is shown in Table 1. The film composition shifts from that of targets becoming Ba and Cu-rich. Simultaneously, the Tl content in the Tl-Ca-Ba-Cu-O films, which were formed by treating Ca-Ba-Cu-O films in Tl_2O_3 vapor and O_2 atmosphere, increases with increasing the Ba content in the Ca-Ba-Cu-O films. We conjecture that this is due to the difference of reactivity between Tl and Ca, Ba or Cu.

Table 1 Effect of composition of Ca-Ba-Cu-O target on composition of Tl-Ca-Ba-Cu-O films; treating condition: substration temperature $805 \pm 5°C$, 0.4g Tl_2O_3 powder and treatment time 15 min..

No	target composition ratio			film composition ratio			
	Ca :	Ba :	Cu	Tl :	Ca :	Ba :	Cu
A	1	1	1.5	1.2	1	1.3	2.2
B	1	0.75	1.5	0.6	1	0.9	2.4
C	1	0.75	1	0.8	1	1.2	1.6
D	1	1.5	2	1.6	1	1.9	4.5

In addition, the superconductivity of the Tl-Ca-Ba-Cu-O films is also affected by the composition of Ca-Ba-Cu-O films, shown in Fig.5. The temperature of sharp resistance drop is strongly affected by the Ca content in the Tl-Ca-Ba-Cu-O films. The more the Ca content is, the higher the temperature of sharp drop is. The semiconducting behavior in the normal state and long tail in the resistance vs temperature curves are due to the bad treatment conditions. The films contain large amount of semiconducting or insulating phases. They are $BaCuO_2$, CuO and unknown phase($2\theta=32.8$), marked *, shown in Fig.6. The * phase is identified as an insulator because it exists in the typical insulating samples with small amount of Tl in the films, shown in Fig.7. The X-ray diffraction patterns of superconducting phase of these films are consistent with those of reported $Tl_1Ca_1Ba_2Cu_2Ox$ phase with a single Tl-O layer[4,14]. At the same time, the extraporated temperature of zero resistance of these films agrees with the reported temperature 65~90K[14,16,17].

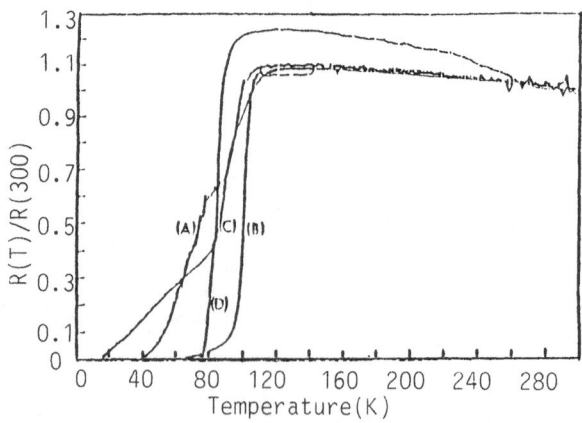

Fig.5 Effect of target composition on the superconductivity of Tl-Ca-Ba-Cu-O films. Treating condition is shown in Table 1.

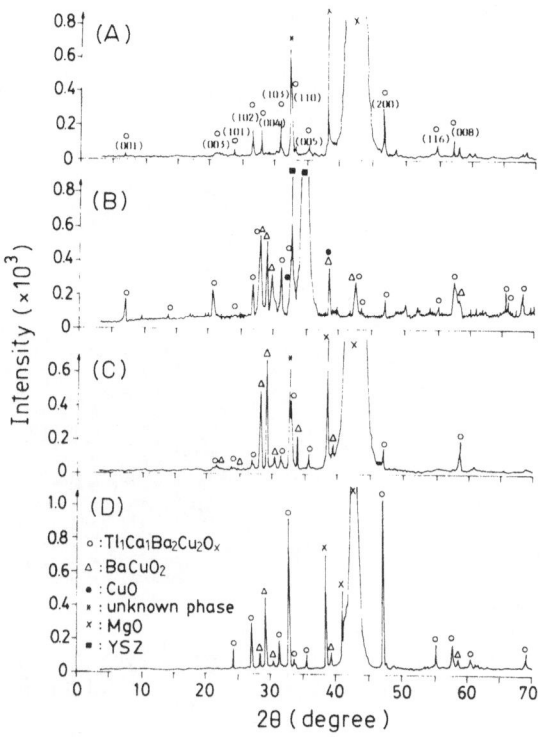

Fig.6 Effect of target composition on the X-ray
diffraction patterns of Tl-Ca-Ba-Cu-O films.
Treating condition is shown in Table 1.

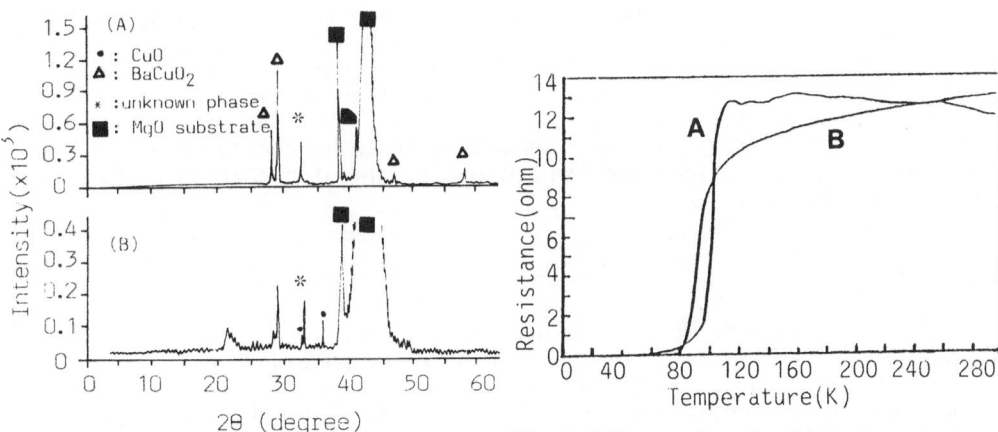

Fig.7 X-ray diffraction patterns
of typical insulating Tl-Ca-
Ba-Cu-O films;
(A) $Tl_{0.1}Ca_1Ba_{0.7}Cu_{1.9}Ox$ film
(B) $Tl_{0.1}Ca_1Ba_{0.9}Cu_{2.7}Ox$ film.

Fig.8 Effect of annealing temperature
on the superconductivity of Tl-
Ca-Ba-Cu-O films. (A) 805±5°C
(B) 815±5°C; at Tl_2O_3 0.4g,
annealing time 15 min.,
$Ca_1Ba_{0.75}Cu_{1.5}Ox$ target.

The effect of annealing temperature on the superconductivity of Tl-Ca-Ba-Cu-O films is shown in Fig.8. It indicates that the superconductivity of Tl-Ca-Ba-Cu-O films can be improved by increasing the treatment temperature. By far, the best superconductivity of the film is Tc(onset)=120K and Tc(zero)=78K, shown in Fig.8(B). In addition, its X-ray diffraction patterns, shown in Fig.9, indicate that it is highly oriented $Tl_1Ca_1Ba_2Cu_2Ox$ superconductor.

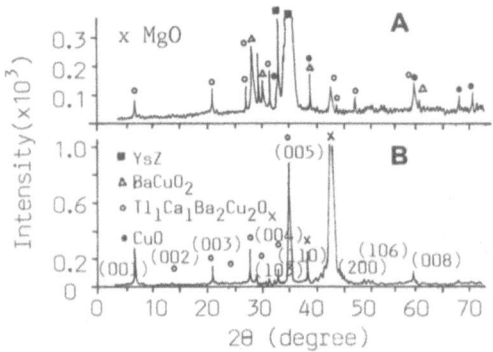

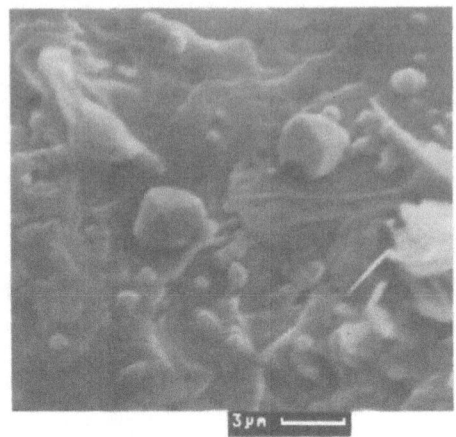

Fig.9 Effect of annealing temperature on the X-ray diffraction patterns of Tl-Ca-Ba-Cu-O films. (A) 805±5°C (B) 815±5°C; at Tl_2O_3 0.4g; annealing time 15 min., $Ca_1Ba_{0.75}Cu_{1.5}Ox$ target.

Fig.10 Typical surface morphology of Tl-Ca-Ba-Cu-O films.

The typical surface morphology of Tl-Ca-Ba-Cu-O films is shown in Fig. 10. It indicates that the plate-like grain and pyramid particle are formed. The EDAX analysis shows that the former is Tl-Ca-Ba-Cu-O compound and latter is Tl oxide. It means that the free Tl oxide forms on the film surface.

In summary, the superconducting Tl-Ca-Ba-Cu-O thin films (0.6~2μm) have been prepared by diffusion and interaction between sputtered Ca-Ba-Cu-O film on MgO or YSZ substrates and Tl_2O_3 vapor in the flowing O_2 furnace by the indicated three annealing ways. The high Ca content in the Ca-Ba-Cu-O film and high treatment temperature favor formation of higher Tc(zero) superconductor. By far, the best superconductivity of the films is Tc(onset)=120K and Tc(zero)=78K. The superconducting phase is highly oriented $Tl_1Ca_1Ba_2Cu_2Ox$. Study to establish optimum treatment condition of Ca-Ba-Cu-O films is in progress.

REFERENCES

1. Z.Z. Sheng and A.M. Herman, Nature 332:138(1988).
2. M.Nakao, R. Yuasa, M. Nemoto, H. Kuwahara, H. Mukaida and A. Mizukami, Jpn. J. Appl. Phys. 27:L849(1988).
3. W.Y. Lee, V.Y. Lee, J. Salern, T.C. Huang, R. Savoy, D.C. Bullock and S.S.P. Parkin, Appl. Phys. Lett. 53:329(1988).
4. T.C. Huang, W.Y. Lee, V.Y. Lee and R. Karimi, Jpn. J. Appl. Phys. 27: L1498(1988).
5. D.S. Ginley, J.F. Kwak, R.P. Hellmer, R.J. Baughman, E.L. Venturini and B. Morosin, Appl. Phys. Lett. 53:406(1988).
6. J.F. Kwak, E.L. Venturini, R.J. Baughman, B. Morosin and D.S. Ginley, Physica C 156:103(1988).
7. I. Shih and C.X. Qiu, Appl. Phys. Lett. 53:523(1988).

8. S.X. Dou, H.K. Liu, A.J. Bourdillon, N.X. Tan, J.P. Zhou and C.C.Sorrell, Mod. Phys. Lett. B 2:875(1988).

9. CRC Handbook of Chemistry and Physics, Ed. R.C. Weast, 58th edition (1978).

10. R.J. Lin, Y.C. Chen, J.H. Kung and P.T. Wu, Materials Research Society Symposium Proceedings, 99:319(1987).

11. R.J. Lin, J.H. Kung and P.T. Wu, Physica C 153-155:796(1988).

12. R.J. Lin, J.H. Kung and P.T. Wu, Proceedings of MRS International Meeting on Advanced Materials, Tokyo, Japan(1988).

13. A.O. Sandborg, R.B. Shen and S.G. Maegdlin, The EDAX EDITOR 10(3):11 (1980).

14. A.K. Ganguli, G.N. Subbanna and C.N.R. Rao, Physica C 156:116(1988).

15. R.M. Hazen, L.W. Finger, R.J. Angel, C.T. Prewitt, N.L. Ross, C.G. Hadidiacos, P.J. Heaney, D.R. Veblen, Z.Z. Sheng, A.El Ali and A.M. Hermann, Phys. Rev. Lett. 60:1657(1988).

16. Superconductivity News, 2(1):4(1988).

17. R. Beyers, S.S.P. Parkin, V.Y. Lee, A.I. Nazzal, R. Savoy, G. Gorman, T.C. Huang and S. LaPlaca, Appl. Phys. Lett. 53:432(1988).

TL-BASED SUPERCONDUCTING FILMS BY SPUTTERING USING A SINGLE TARGET

S. H. Liou

Department of Physics and Astronomy
University of Nebraska-Lincoln
Lincoln, Nebraska 68588-0111

M. Hong, A. R. Kortan, J. Kwo, D. D. Bacon
C. H. Chen, R. C. Farrow, and G. S. Grader

AT&T Bell Laboratories
Murray Hill, New Jersey 07974

ABSTRACT

We have prepared superconducting Tl-Ba-Ca-Cu-O films on MgO(100) and SrTiO$_3$(100) substrates by diode sputtering using a single composite oxide target. Films containing primarily Tl$_2$Ba$_2$Ca$_1$Cu$_2$O$_8$ phase have a T$_c$(R=0) at 102K and a transport J$_c$ of 10^4A/cm^2 at 90K. For Tl$_2$Ba$_2$Ca$_2$Cu$_3$O$_{10}$ films, T$_c$(R=0)'s are at 116K and transport J$_c$'s are as high as 10^5A/cm^2 at 100K. Both types of films are strongly textured with the c-axis perpendicular to the film plane. The rocking curve of the Tl$_2$Ba$_2$Ca$_2$Cu$_3$O$_{10}$ films is 0.22° wide. The film surfaces are porous as revealed by scanning electron microscopy (SEM) micrographs. Microstructures of the Tl$_2$Ba$_2$Ca$_2$Cu$_3$O$_{10}$ films are studied by transmission electron microscopy (TEM).

INTRODUCTION

The discovery of the high T$_c$ oxide superconductors has generated a great deal of excitement because of the scientific interest as well as many potential applications[1]. At present, the Tl-based superconductors offer the highest T$_c$(R=0) (up to 125K)[2]. As a consequence, intensive thin-film efforts have been devoted to these new high T$_c$ superconducting oxides. The film activities include electron-beam evaporation[3], sputtering using a single target[4-6] as well as multiple targets[7], and laser evaporation[8]. The ability to produce films with T$_c$(R=0) above 100K (up to 120K) and further to control the phase formation (Tl$_2$Ba$_2$Ca$_1$Cu$_2$O$_8$ or Tl$_2$Ba$_2$Ca$_2$Cu$_3$O$_{10}$) in films[6] has been clearly demonstrated. High-quality films are important for fundamental studies and may also be used in many practical applications, e.g. in the area of microelectronics and sensors.

In this paper, we discuss the preparation and characterization of Tl-based superconducting films. Films 0.2 to 0.4 μm thick have been prepared on both MgO(100) and $SrTiO_3(100)$ substrates. The temperatures and the duration of the post annealing were found to be crucial in obtaining films with $Tl_2Ba_2Ca_1Cu_2O_8$(2212) or $Tl_2Ba_2Ca_2Cu_3O_{10}$(2223) phase.

FILM GROWTH

Films were prepared by dc diode sputtering using a single composite oxide target. For the reasons of safety, film deposition was confined in a liquid nitrogen cooled stainless can 7.5 cm in diameter and 10 cm in length, thus eliminating the spread of Tl-compound over the rest of the chamber. The sputtering targets were made by sintering a mixture of Tl_2O_3, $CaCO_3$, $BaCO_3$, and CuO powders with an appropriate ratio. The target composition used in this study is 2.3:2:2:3:x of Tl:Ba:Ca:Cu:O. The sputtering gas was pure Ar. The input power was 3 W for the target 2.5 cm in diameter. The Ar pressure was maintained at 80 mTorr during the deposition. The deposition rate was low around 3.5 nm/min. The substrate temperatures were varied from room temperature to 500°C. However, we have not found any significant effects on the superconducting properties of the films by varying the substrate temperature.

COMPOSITION, STRUCTURE, AND HEAT TREATMENT

The film composition was determined by Rutherford backscattering spectrometry (RBS) using 1.8 MeV He^+ ions. In this study, we used a thin film 0.03 μm thick for a more accurate determination of the film composition. A typical film composition is $Tl_2Ba_{1.86}Ca_{2.08}Cu_{3.04}O_x$.

The as-deposited films are insulating, and need post-deposition annealing to become superconducting. In order to prevent further loss of Tl or other elements, films were wrapped in a gold foil with pellets of compressed composite Tl-Ba-Ca-Cu-O powders and sealed in a quartz tube. The film structure depends very much on the heat treatment conditions such as the temperature and the annealing duration. We found that films grown on different substrates required different heat treatments to obtain the desirable phase, 2212 or 2223. For films deposited on MgO, annealing below 860°C, or for a very short period time at 870°C (~5 min), gave the films with a primary phase of $Tl_2Ba_2Ca_1Cu_2O_8$. When films with the same composition were heat treated at 870°C for 10-15 min, $Tl_2Ba_2Ca_2Cu_3O_{10}$ phase became the almost pure phase in the films. For films grown on $SrTiO_3$ substrates, a longer annealing time is needed to obtain the similar results.

The crystal structure of the films was studied by high resolution X-ray diffraction measurements using a 12 kW rotating Cu anode source equipped with a triple-axis four-circle diffractometer. A pair of Ge (111) monochromator and analysis crystals were used to provide a spatial resolution of 8 x 10^{-4} $Å^{-1}$ for the parallel scans and 3 x 10^{-5} $Å^{-1}$ for the rocking scans. The X-ray diffraction pattern for the film deposited on MgO and annealed at 870°C for 5 min is shown in Figure 1(a). The sharp periodic peaks observed in the θ-2θ scan are the (00ℓ) peaks of 2212 phase with the lattice constant c = 29.41 Å. The unsymmetrical broadening of the diffraction peaks indicates that there are many defective structures in the film.

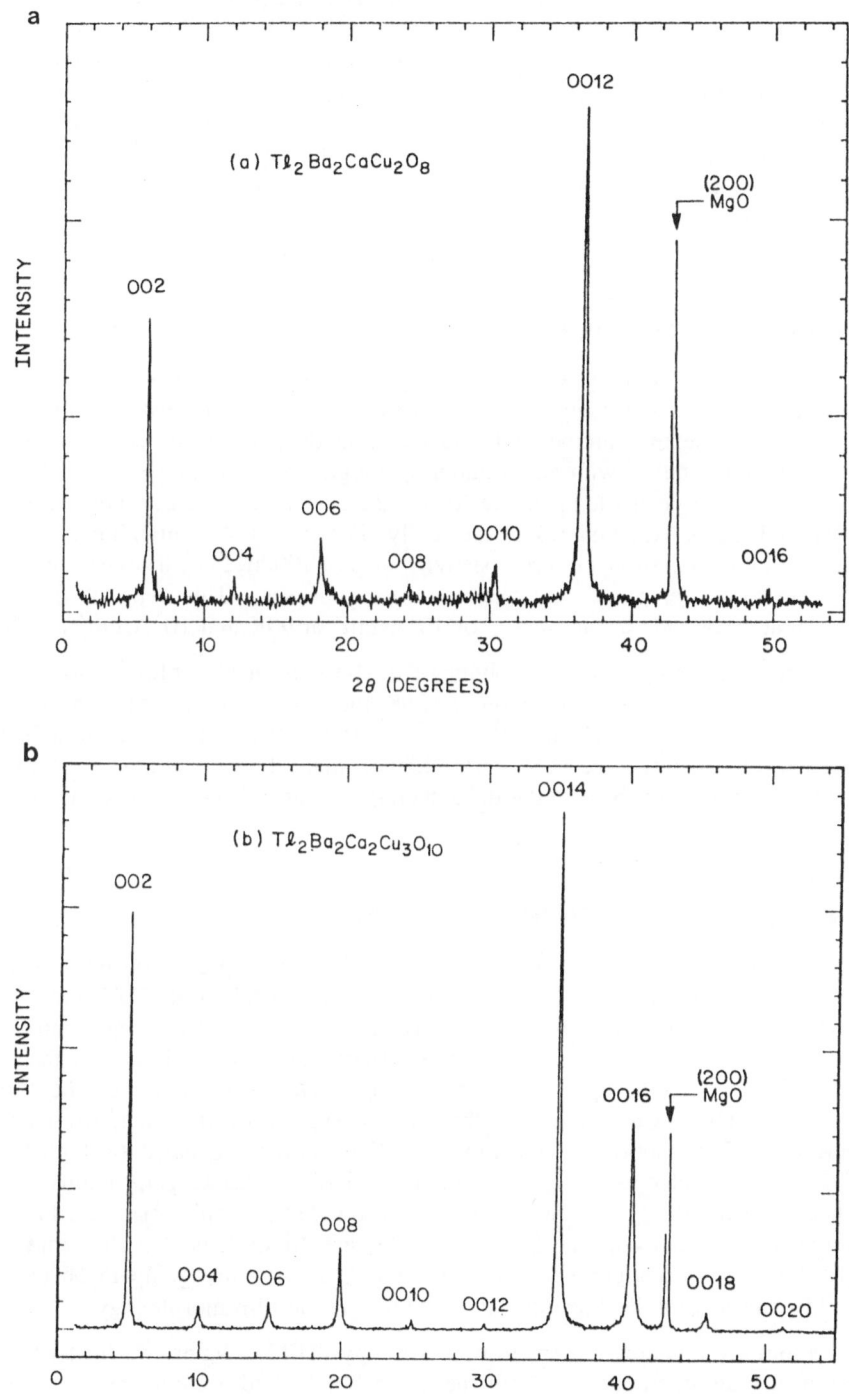

Figure 1   X-ray θ-2θ scans normal to the film plane for (a) Tl-Ba-Ca-Cu-O film containing primarily 2212 phase on MgO(100), and (b) 2223 phase film on MgO(100).

The X-ray diffraction pattern for the film deposited on MgO and annealed at 870°C for 10 min is shown in Figure 1(b). The sharp periodic peaks observed are assigned to be the (00ℓ)'s of the highest T_c phase $Tl_2Ba_2Ca_2Cu_3O_{10}$ with the lattice constant c = 35.647 Å. No second phase, down to the 1% level, was found as clearly shown in Figure 1 (b). The amount of misalignment of the ordered grains was measured by a rocking curve through the (0014) diffraction peak of the $Tl_2Ba_2Ca_2Cu_3O_{10}$ phase. The width of the rocking curve of the film is 0.22 °, and the rocking curve was centered on the same position as the (200) peak of MgO as shown in Figure 2.

SUPERCONDUCTING PROPERTIES

The superconducting and transport properties were measured by the standard four-point measurement using a dc method by switching the polarization of the applied current during the measurement. Critical current densities were measured in the van der Pauw configuration with and without a lithographic patterning. Figure 3 shows the data of resistance versus temperature for the 2212 and 2223 films, whose structures are shown in Figures 1(a) and (b), respectively. For the films containing 2212 phase, a relatively high room-temperature resistivity of 500-1000$\mu\Omega$-cm has been observed due to the existence of some non-superconducting phases. This type of films have a T_c(R=0) at 102K, and a transport J_c of 10^4A/cm^2 at 90K and 10^5A/cm^2 at 77K.

The room temperature resistivity of $Tl_2Ba_2Ca_2Cu_3O_{10}$ films is low about 170 $\mu\Omega$-cm. These films have a T_c onset at 125K and a T_c(R=0) at 116K. The J_c's, at zero magnetic field, are as high as 10^5A/cm^2 at 100K. We found that the J_c's in films deposited on SrTiO$_3$ are, in general, higher than those in films on MgO. Film morphologies may attribute to the difference in J_c's as will be discussed later.

MORPHOLOGY AND MICROSTRUCTURE

The morphology of the films was studied by scanning electron microscopy. The micrographs of the films on MgO and SrTiO$_3$ after annealing at 870°C for 10 min are shown in Figures 4(a) and (b), respectively. The films grown on MgO tend to have a larger grain size than those on SrTiO$_3$. There are minor phases in the form of separated particles on top of the film surface. The composition of these particles varies from Tl- to Ca-rich. Also, the particle shapes are different, from needle- to sphere-like. The majority of the films is 2223 phase as revealed by both the X-ray and transmission electron diffraction studies. The $Tl_2Ba_2Ca_2Cu_3O_{10}$ grains are weakly connected, particularly in the films grown on MgO. This may attribute to the difference in J_c's between the films on SrTiO$_3$ and MgO. It is also clear that the films on both substrates are porous. Therefore, the J_c is certainly going to be increased if the films can wet on the substrates thus improving the film morphology.

From the transmission electron microscopy (TEM) studies, we have observed an incommensurate component in both the a and b-axis, and a commensurate component along the c-axis. The diffraction spots from these components are sharper than those reported for a bulk ceramic sample containing 2212 phase[9]. A high-resolution lattice image obtained along the [100] zone axis is shown in Figure 5. The films grown on MgO show no intergrowth of perovskite layers with a different layer-thickness, which has been commonly observed in the Bi- and Tl-based bulk superconductors[10,11].

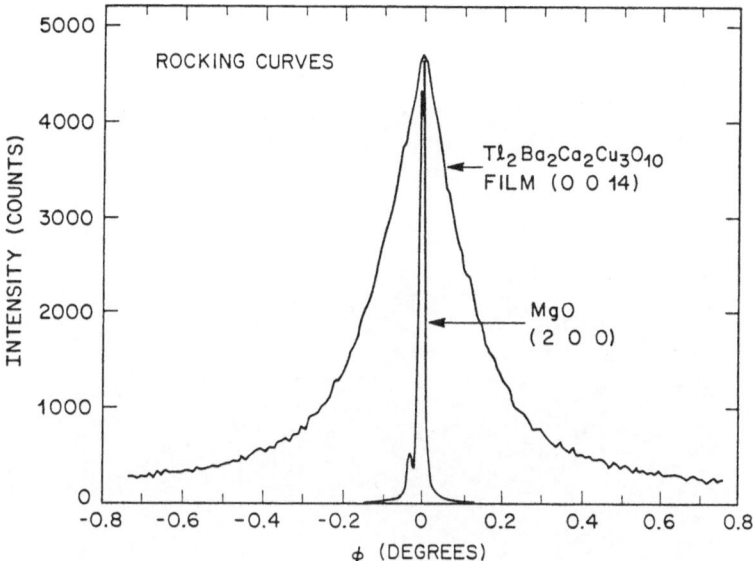

Figure 2   Rocking curve of (0014) of 2223 phase on MgO(100). (The center on
the same position is the (200) peak of the substrate.)

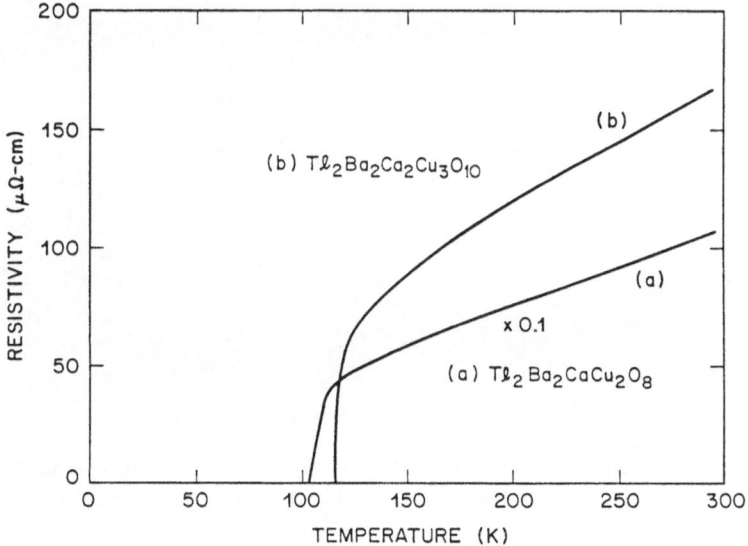

Figure 3   Resistivity versus temperature curves (a) for the same film as in
Fig.1(a), and (b) for the same film as in Fig.1(b).

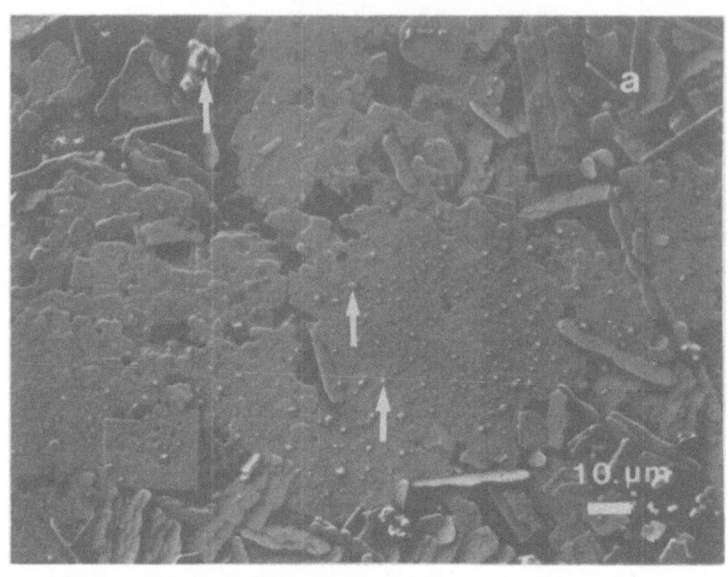

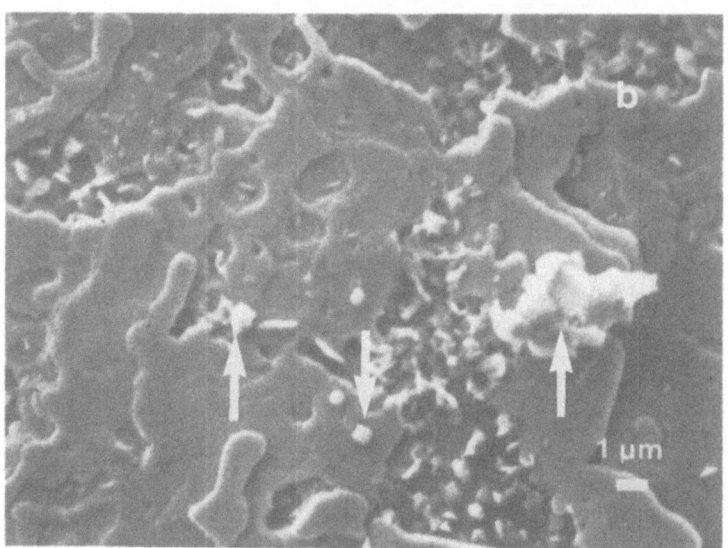

Figure 4   Scanning electron micrographs of 2223 films on (a) MgO(100) and (b)SrTiO$_3$(100). (Some of the Ca-rich portions were indicated by the white arrows.)

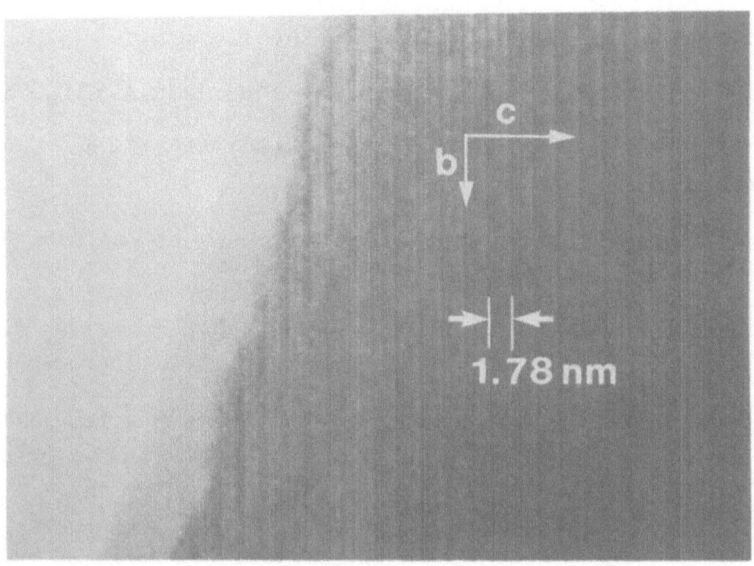

Figure 5   High-resolution TEM lattice images obtained along [100] zone-axis.

SUMMARY

We have routinely prepared superconducting films in the Tl-based system with pure 2212 or 2223 phase by using a simple dc diode sputtering technique and employing a single composite oxide target. The films on both MgO and SrTiO$_3$ are strongly textured with the c-axis normal to the film plane. Full width at half maximum (FWHM) of the rocking curve at (0014) of the 2223 films grown on MgO is as narrow as 0.22°. Using TEM we have also observed an incommensurate superlattice in the Tl$_2$Ba$_2$Ca$_2$Cu$_3$O$_{10}$ films. The 2223 films have a T$_c$ onset at 125K and T$_c$(R=0) at 116K. The transport J$_c$ is as high as 10^5A/cm^2 at 100K. This value can be further enhanced when the film morphology is optimized.

REFERENCES

[1] see papers presented in this conference proceedings.

[2] S.S.P. Parkin, V.Y. Lee, E.M. Engler, A.I. Nazzal, T.C. Huang, G. Gorman, R. Savoy, and R. Beyers, Phys. Rev. Lett. **60**, 2539 (1988).

[3] D.S. Ginley, J.F. Kwak, R.P. Hellmer, R.J. Baughman, E.L. Venturini, and B. Morosin, Appl. Phys. Lett, **53**, 406 (1988).

[4] W.Y. Lee, V.Y. Lee, J. Salem, T.C. Huang, R. Savoy, D.C. Bullock, and S.S.P. Parkin, Appl. Phys. Lett. **53**, 329 (1988).

[5] Yo. Ichikawa, H. Adachi, K. Setsune, S. Hatta, K. Hirochi, and K. Wasa, Appl. Phys. Lett. **53**, 919 (1988).

[6] M. Hong, S.H. Liou, d.D. Bacon, G.S. Grader, J. Kwo, A.R. Kortan, and B.A. Davidson, Appl. Phys. Lett. **53**, 2102 (1988).

[7] J.H. Kang, R.T. Kampwirth, and K.E. Gray, Phys. Lett. **A 131**, 208 (1988).

[8] S.H. Liou, N.J. Ianno, B. Johs, D. Thompson, D. Meyer, John A. Woollam, paper presented in this conference proceedings.

[9] J.D. Fitz Gerald, R.L. Withers, J.G. Thompson, J.S. Anderson, and B.G. Hyde, Phys. Rev. Lett. **60**, 2797 (1988).

[10] C.H. Chen, D.J. Werder, S.H. Liou, H.S. Chen, and M. Hong, Phys. Rev. **B37**, 9834 (1988).

[11] K.K. Fung, Y.L. Zhang, S.S. Xie, and Y.Q. Zhon, Phys. Rev. **B38**, 5028 (1988).

STRUCTURE AND PROPERTIES OF MAGNETRON SPUTTERED

SUPERCONDUCTING Y-Ba-Cu-O FILMS

M.E.Golovchanskii, S.N.Ermolov, N.A.Kislov,
O.P.Kostyleva, L.S.Kokhanchik, V.A.Marchenko,
A.V.Nikulov, and V.Zh.Rosenflants

Institute of Problems of Microelectronics Technology
and High Purity Materials, USSR Academy of Sciences
142432 Chernogolovka, Moscow District, USSR

INTRODUCTION

The paper reports on the structural properties and crystallization of amorphous $YBa_2Cu_3O_7$ films deposited on cold substrates as well as the properties of critical current through the boundary of a large grain with a fine-grained medium.

1. CRYSTALLIZATION OF AMORPHOUS $YBa_2Cu_3O_7$ FILMS

The small coherence length $\zeta \leqslant 35 \overset{\circ}{A}$ [1] in HTS $YBa_2Cu_3O_7$ prevents fabrication of Josephson junctions by the conventional methods. The known successful attempts [2] take advantage of the Josephson tunneling occurring on grain boundaries in these materials. It is of interest therefore to study the structure of $YBa_2Cu_3O_7$ grain films and possible control of the film structure and superconductor critical temperature T_c. So far high values of T_c have been obtained mainly on epitaxial (single crystal) films deposited on $SrTiO_3$ and MgO substrates [3]. On grain films T_c is significantly lower, sometimes by tens of Kelvins, which is due to the film–substrate interaction giving rise to parasitic phases in the film [4]. There is evidence in literature indicating that the influence of noble metal sublayers permits of an increase in T_c which is yet considerably lower than that of epitaxial films.

Below it is shown that sufficiently rapid thermal treatment and the use of an Ag sublayer on relatively cheap and commercially available ZrO_2 substrates allow obtaining polycrystalline films with T_c=86-88 K. We have studied the structure of $YBa_2Cu_3O_7$ films after crystallization from the amorphous state. The possibility is demonstrated of growing large (up to 200 μm) single crystals on films.

The starting films were made by magnetron sputtering onto a cold substrate from single crystal zirconia stabilized by yttria with a [100] axis perpendicular to the plane.

$YBa_{2-y}Cu_{3-z}O_{7-x}$ (y=0-0.5; z=0-0.3) films 1 μm thick have been used. The Ag interfacial layer was 0.16 μm. After sputtering the films have an amorphous structure. Auger-analysis of the film surface revealed an excess (by $\sim$40%) Ba concentration decreasing to a bulk value at a depth of $\sim$150 A. Variation of the time of film air storage between sputtering and analysis from 4 h to two weeks does not affect the distribution profile.

For crystallization the films were annealed in air at 400–890°C followed by phase reflection electron diffractometry, the layer to be analyzed being $\sim$50 A thick. The electron diffraction patterns were interpreted using our data obtained for $BaCuO_2$, Y_2BaCuO_5, $YCuO_5$, $YBa_3Cu_2O_5$ standards and those available in literature [5] on interatomic distances in Y-Ba-Cu-O compounds. The electron diffraction patterns of 30 films investigated reveal a large set of lines that is not always subject to interpretation. The most common structures are listed in Table 1.

Table 1.

Annealing mode	400°C 1–36 h	500°C 1 h	650°C 1 h	670°C 2 h
Phases	Y_2BaCuO_5 Y_2BaO_4	Y_2BaCuO_5 Y_2BaO_4	$BaCuO_2$ Y_2BaO_4	$BaCuO_2$ Y_2BaCuO_5 CuO

Annealing mode	670°C 8 h	750°C 1 h	850–890°C 0.5–1 h
Phases	$BaCuO_2$ $YBa_2Cu_3O_7$	$BaCuO_2$ Y_2BaCuO_5 $Y_2Cu_2O_5$	$YBa_2Cu_3O_7$ Y_2BaO_4 CuO

Auger-analysis shows that thermal treatment does not affect the excess Ba concentration in the outer film layer, no carbon signal being visible (carbon sensitivity $\sim$0.2% at.). The electron diffraction patterns are also free from oxycarbonate lines. This suggests that carbon is weakly absorbed by films under conventional conditions and its excess content reported elsewhere [6] is largely due to the incomplete synthesis using $BaCO_3$ as a starting component in target fabrication. It should be noted that the samples do not exhibit oxycarbonates with the use of $Ba(OH)_2 \times 8H_2O$ instead of $BaCO_3$ [5].

Crystallization at $T \leqslant 750$°C does not give rise to superconductivity, the resistivity of the films (without an Ag layer) $R_\square \geqslant 1$ MΩ, i.e. the $YBa_2Cu_3O_7$ conducting phase is present in the film to a small extent, if at all, and does not form a continuous grid. The

Y_2BaCuO_5 and $YBaO_4$ phases formed at low annealing temperatures (400–500°C) display higher (1300°C and 1400°C, respectively) stability temperatures [5] as compared to $YBa_2Cu_3O_7$ (~980°C). The presence of the former phases is likely to hamper the formation of $YBa_2Cu_3O_7$, its growth requiring high temperatures with the times of annealing in use (see Table 1). This may explain why films sputtered on cold substrates demand annealing within 900°C to attain superconductivity, whilst films sputtered on hot (600°C) substrates possess the structure desired [7]. The size of the grain therewith $d \leqslant 0.5 \ \mu m$.

In the second run of experiments we used films stabilized by air heating at 450°C for 1 h. The high temperature mode in the oxygen atmosphere was chosen with regard to two circumstances, namely, a possible reduction in the stay time of the sample at $T > 850°C$ involving the interlayer interaction and the kinetics of film–atmosphere oxygen exchange. From [8] it follows that the time required for attaining equilibrium oxygen concentration in an $YBa_2Cu_3O_7$ film 1 μm thick increases with decreasing temperature and comes to 15 min. at 450°C. Heating to 900–1040°C was done for ~10 min. and cooling down to 100°C for ~60 min. without holding at a maximum temperature. Over the range of 850–1040°C the heating rate was 50°C/min., the cooling rate 80°C/min Such thermal treatment allows fabrication of films with an onset of transition (R=0.9 R_N) at T=91–92 K. Fig.1 shows T_{co} corresponding to R=0. Film critical current $j_c(78 \ K) \approx 10^3$ A/cm^2. When holding the films at a maximum temperature for 20–40 min. T_{co}=86–88 K is achieved at lower (900–940°C) temperatures, but j_c is appreciably lower. The films exhibit a ceramic structure (Fig.2) with plate-like grains. The grain thickness is by an order of magnitude less than the diameter. Increasing temperature results in an increase of the size of the grains (Fig.1) and a change from their random orientation to arrangement of the grain plane parallel to the substrate. The metal grain composition corresponds to stoichiometric $YBa_2Cu_3O_x$. Heating to 980°C leads to abrupt refining of the structure caused by peritectic decay of $YBa_2Cu_3O_x$. Since no foreign elements are detected in the grains, this temperature corresponds to the peritectic horizontal for pure $YBa_2Cu_3O_x$ which enables the data available in literature to be improved [5]. A fine-grained layer with a composition similar to that of Y_2BaCuO_5 is often observed under $YBa_2Cu_3O_7$ grains (Fig.2b), i.e. high temperature crystallization of films with an yttrium-enriched starting composition is attended with vertical phase separation. This permits reducing the requirements on the accuracy of film composition. The dependence of the signal intensity of Y, Ba, Cu, Ag on electron probe energy (5,10,15 keV) in microanalysis, i.e. with increase in depth of the layer to be analyzed, shows that silver remains mainly between the substrate and the metal oxide film.

Heating above 980°C and subsequent crystallization on cooling give rise to single crystals 5–200 μm in size against the background of the fine-grained structure (Fig.2d). With similar morphology two types of crystals were observed, $YBa_2Cu_3O_x$ and $BaCuO_2$, the films remaining superconducting (Fig.1).

Thus, the annealing techniques described enables fabrication of superconducting $YBa_2Cu_3O_7$ films with T_{co}=86–88 K and different controlled structures on ZrO_2 with an Ag sublayer.

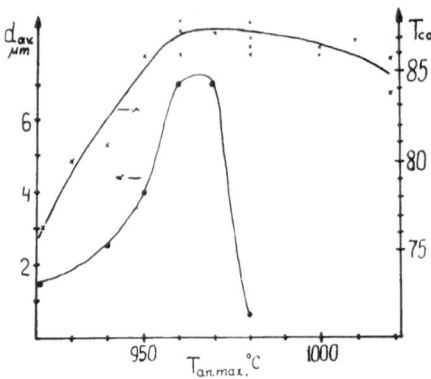

Fig.1. Critical temperature T_{co} and average grain size d_{av} against maximum temperature of thermal treatment $T_{an.max.}$

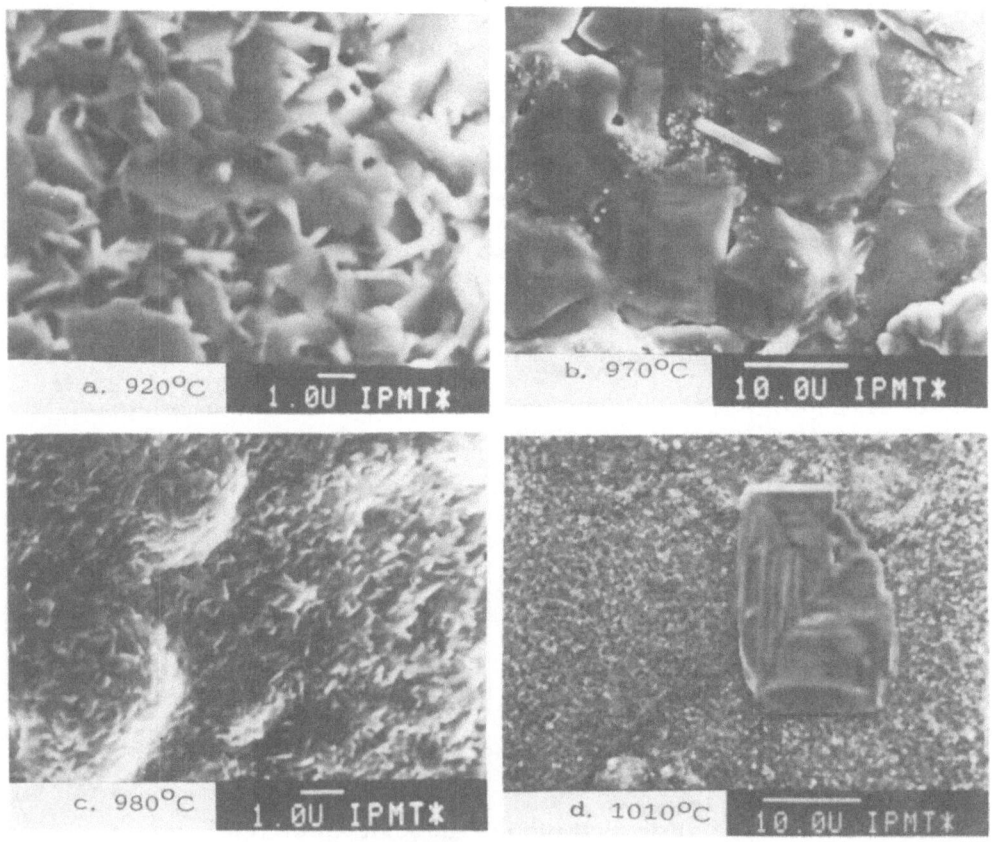

Fig.2. Film microstructure against $T_{an.max.}$

2. CRITICAL CURRENT ACROSS GRAIN BOUNDARY

Josephson current through a single grain boundary was first investigated in [2]. However, the form of the magnetic field dependence of critical current $j_c(H)$ and the value of oscillation period obtained differ from those expected for a josephson junction of the "sandwich" type. The authors must have used a system of two Josephson junctions rather than an individual one.

We investigated structures obtained by means of laser patterning of $YBa_2Cu_3O_7$ films annealed at $1040^\circ C$, $T_c = 86 = 88$ K, $j_c(78$ K$) \leqslant 10^3$ A/cm^2, $j_c(4.2$ K$) \leqslant 1.5 \cdot 10^4$ A/cm^2. A bridge was made from rectangular crystals by laser cutting across the film. The cut was 10-20 μm wide. The critical current density calculated for the bridge cross-section was also close to j_c of the starting film and used to be both greater and less than j_c. A voltage level of 0.2 V was used to determine j_c.

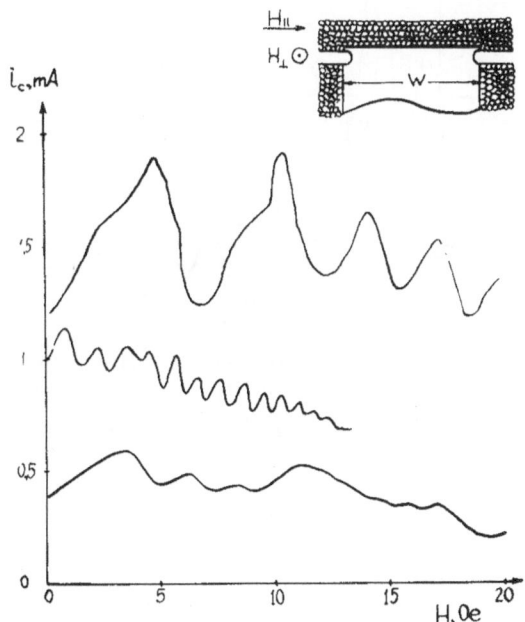

Fig. 3. Plot of $i_c(H)$ for a bridge with W=45 m T=77.4 K; upper curve — H||, middle curve — H⊥, lower curve — after recovery annealing at $500^\circ C$ for 0.5 h. The insert illustrates the structure geometry.

The temperature dependence of critical current in a zero magnetic field for the bridges is identical to that of the starting films in the vicinity of critical temperature $i_c \sim (T_c-T)^{3/2}$. At low temperatures the critical current increases linearly as the temperature decreases. Figs. 3 and 4 show the $i_c(H)$ dependences for two bridges formed by crystallites 40 and 45 μm wide on a film 4 μm thick. The form of the dependences resembles the Fraunhofer diffraction pattern characteristic of Josephson junctions of the "sandwich" type. The oscillations are not regular in character which may be associated with the inhomogeneity of critical current density over the junction area. Fig. 3 shows the plot of $i_c(H)$ for various orientations of magnetic field with reference to the film surface. It is readily seen that with magnetic field perpendicular to the film plane the oscillation

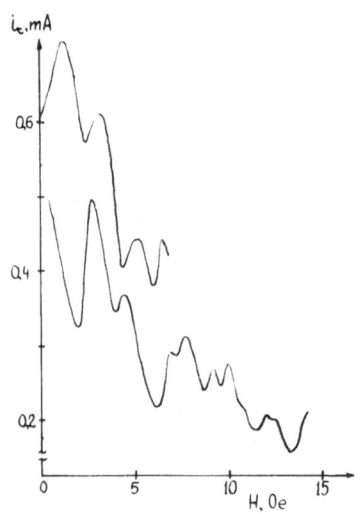

Fig. 4. Plot of $i_c(H)$ for a bridge with W=45 m: T=77.4 K. After cooling at $H_{cool.}$=0 – lower curve, $H_{cool.}$=6 Oe – upper curve.

period is smaller than with parallel orientation. Eight periods of the $i_c(H)$ plot with H perpendicular to the film fall between the first two maxima of $i_c(H)$ with H parallel to the film. This value is close to the ratio between the face sizes of rectangular crystallites. The intermaximum distance H at perpendicular orientation H varies from 1 to 2.5 Oe. Assuming that the width of the contact across the field W is equal to the bridge width, we may estimate the thickness of the contact area d= ϕ_0 W^{-1} H^{-1}=0.2–0.5 μm which is by the order of magnitude close to 2λ_L [9] (ϕ_0 is the flux quantum, λ_L the London penetration depth). Thus, the behaviour of the contact of a large single crystal with a fine-grained conglomerate is analogous to that of a wide Josephson junction.

The plot of $i_c(H)$ is not a single-valued magnetic field function. The plots of $i_c(H)$ do not coincide with increase and subsequent decrease of the field. $i_c(H)$ is also affected by the field in which the sample is cooled. When it is cooled in a weak magnetic field, $i_c(H)$ differs only slightly from the dependence obtained upon cooling in zero field but it is shifted in the x direction with respect to the latter (Fig. 4) towards the cooling field. This may be due to intragrain magnetic flux trapping (the magnetic flux and the cooling field having the same direction). The outer field of the trapped moment induces an additional r e v e r s e field in grain boundaries, i.e. in the regions with a small critical current [10]. Therefore, the maximum of Josephson critical current is achieved when external magnetic field compensates the outer field of the grain trapped moment.

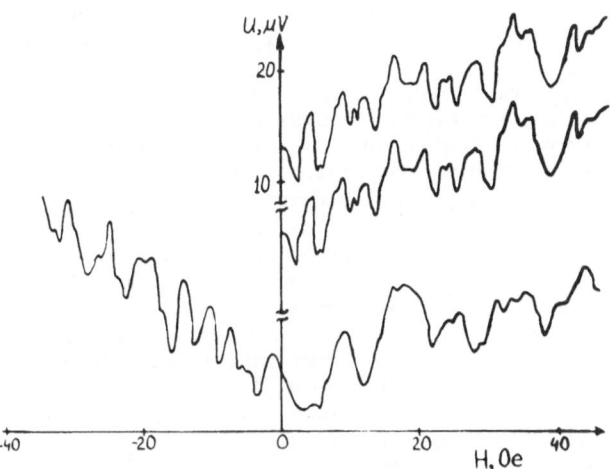

Fig. 5. Voltage U(H) on the bridge at I=2.5 mA as a function of magnetic history, the field predecreased from H=45 Oe to H=-35 Oe (lower curve) and to H=0 Oe (upper curves); T=57 K. The upper curves are vertically shifted.

Within such an approach the variation of $i_c(H)$ caused by application of a strong field may be accounted for by the inhomogeneity of the outer fields over the junction width. The plots of $i_c(H)$ are practically identical at the same magnetic history (Fig. 5, upper curves).

The value of critical current decreases with time. After a month of air storage the critical current decreased from 2.5 mA at T=77.4 K to $i_c < 10$ μA. Annealing at 500°C in an oxygen atmosphere recovers junction critical current. Fig. 3 shows original $i_c(H)$ and that of the bridge obtained by annealing recovery of i_c. Though the plots do not coincide, the interextremum distance differs only slightly which points to the recovery of the original junction.

REFERENCES

1. T.K.Worthington, W.J.Gallagler, and T.R.Dinger., "Anisotropic Nature of High-Temperature Superconductivity in Single-Crystal $YBa_2Cu_3O_{7-x}$", Phys. Rev. Lett. <u>59</u>, 10:1160 (1987).

2. P.Chaudhari, J.Mannhart, D.Dimos, C.C.Tsuei, J.Chi, M.M.Oprysko, and M.Sheuermann, "Direct Measurement of the Superconducting Properties of Single Grain Boundaries in $YBa_2Cu_3O_7$", Phys. Rev. Lett. <u>60</u>, 16:1653 (1988).

3. A.I.Golovashkin, E.B.Yakimov, S.I.Krasnosvobodtsev, and E.V.Pechen, "HTS Single Crystal Films with a Perovskite Structure", Pisma v ZhETF. <u>47</u>, 3:157 (1988).

4. A.Perrin, Z.Z.Li, O.Pena, J.Padiou, and M.Sergent, "d.c. Sputtering Elaboration of Thin Films of the High-T_c Superconductor $YBa_2Cu_3O_{7-x}$: Evidence for Strong Film-Substrate Interactions", Revue Phys. Appl. <u>23</u>, 3:257 (1988).

5. D.M.De Leeuw, C.A.H.A.Mutsaers, G.P.J.Geelen, H.C.A.Smoorenburg, and C.Langereis, "Compounds and Phase Compatibilities in the System Y_2O_3-(BaO)-SrO-CuO at $950^{o}C$", Physica. <u>C 152</u>, 5:508 (1988).

6. A.Z.Avdeev, V.T.Babayev, N.B.Brandt, A.V.Volkozub, M.B.Guseva, A.R.Kaul, V.V.Moshchalkov, H.Ramos, O.V.Snigirev, O.Yu.Sokol, Yu.D.Tretyakov, V.V.Khanin, and V.V.Khvostov, "Chemical Composition and Magnetic Properties of Yttrium-Based HTS ceramics", in: "Problemy vysokotemperaturnoi sverkhprovodimosti. Informatsionnye materialy", part II, UrO AN SSSR, Sverdlovsk (1987).

7. T.Terashima, K.Iijima, K.Yamamoto, Y.Bando, and H.Mazaki, "Single-crystal $YBa_2Cu_3O_{7-x}$ Thin Films by Activated Reactive Evaporation", Jpn. J. Appl. Phys. <u>27</u>, 1, 2:L91 (1988).

8. A.Davidson, A.Palevski, M.J.Brady, R.B.Laibowitz, R.Koch, M.Scheuermann, and C.C.Chi, "In situ Resistance of $YBa_2Cu_3O_x$ Films during Anneal", Appl. Phys. Lett. <u>52</u>, 2:157 (1988).

9. Y.Isikawa, K.Mori, K.Kobayashi, and K.Sato, "Magnetic Anisotropy in a High-T_c Superconductor $YBa_2Cu_3O_{7-}$ Single Crystal below T_c", Jpn. J. Appl. Phys. <u>27</u>, 3, pt.2:L403 (1988).

10. L.S.Kokhanchik, V.A.Marchenko, T.V.Nikiforova, and A.V.Nikulov, "Critical Current in $YBa_2Cu_3O_x$ in Low Magnetic Fields", Fizika nizkih temperatur. <u>14</u>, 8:872 (1988).

EFFECT OF ANNEALING CONDITIONS ON THE STRUCTURAL AND SUPERCONDUCTING PROPERTIES OF Y-Ba-Cu-O FILMS

J. Ryu[a], Y. Huang[a], C. Vittoria[a], D. F. Ryder, Jr.[b], J. Marzik[c], R. Benfer[c] and W. Spurgeon[c]

a: Northeastern University, Center for Electromagn. Research, Boston, MA 02115
b: Tufts University, Laboratory for Materials and Interfaces, Medford, MA 02155
c: Army Materials Technology Laboratory, Watertown, MA 02912

ABSTRACT

The effect of various annealing treatments on ion beam sputter-deposited $Y_1Ba_2Cu_yO_{7-x}$ (y=3 and 5) amorphous thin films has been studied using scanning electron microscopy (SEM), x-ray diffraction and dc resistivity measurements. SEM analysis showed that the morphology of the films were clearly divided into three categories: an equiaxial grain structure, a well oriented grid-like microstructure and a textured platelet structure. The equiaxial grain structure (random orientation according to x-ray diffraction measurements) was obtained when the film was annealed at slow heating and cooling rates in an oxygen atmosphere. The grid-like microstructure (b/c-axes oriented normal to the plane) was obtained in Cu-rich composition (y=5) by rapid thermal annealing. A textured platelet microstructure (c-axis oriented normal to the plane) was obtained using slow heating rates and by carefully controlling the gas environment at the temperature region near the orthorhombic-tetragonal transition. Thus, we have demonstrated the ability to control the morphology and orientation of the film by varying the annealing process.

INTRODUCTION

The discovery of superconductivity above liquid nitrogen temperature in oxide ceramics has ignited intense research on these materials throughout the world. For planar device applications, such as microwave and millimeter wave components, high power electromagnetics, and high speed electronic devices, it is essential to produce thin superconducting films. Furthermore, it is now well known that the superconducting properties of these materials are highly anisotropic.[1] Therefore, it is important to be able to control the orientation and morphology of these films. Although preferentially oriented polycrystalline films[2,3] and single crystalline films[4,5] have been reported by several groups, it is not yet clear what the determining factors are in controlling film orientation. In the present study, we report the annealing processes by which we have been able to produce thin films having several unique morphologies and orientations.

Futhermore, we report here the effect of microstructure on the superconducting, properties, wet chemical etching rate, and the stability of this film to the laboratory environment.

EXPERIMENTAL PROCEDURE

Films of YBaCuO were deposited by using an ion beam sputtering technique with sintered single oxide targets. The nominal compositions of the targets were Y=1, Ba=2.3 and Cu=3.8 or 6. The compositions of films measured by electron probe microanalysis (EPMA) were $Y_1Ba_2Cu_3O_{7-x}$ and $Y_1Ba_2Cu_5O_{7-x}$ using $YBa_{2.3}Cu_{3.8}O_{7-x}$ and $Y_1Ba_{2.3}Cu_6O_{7-x}$ targets, respectively. Usually, a 1μm thick film was deposited on single crystalline (100) oriented cubic zirconia (YSZ), $SrTiO_3$ or MgO substrates in 1hr. All films were deposited without externally heating the substrate. As-

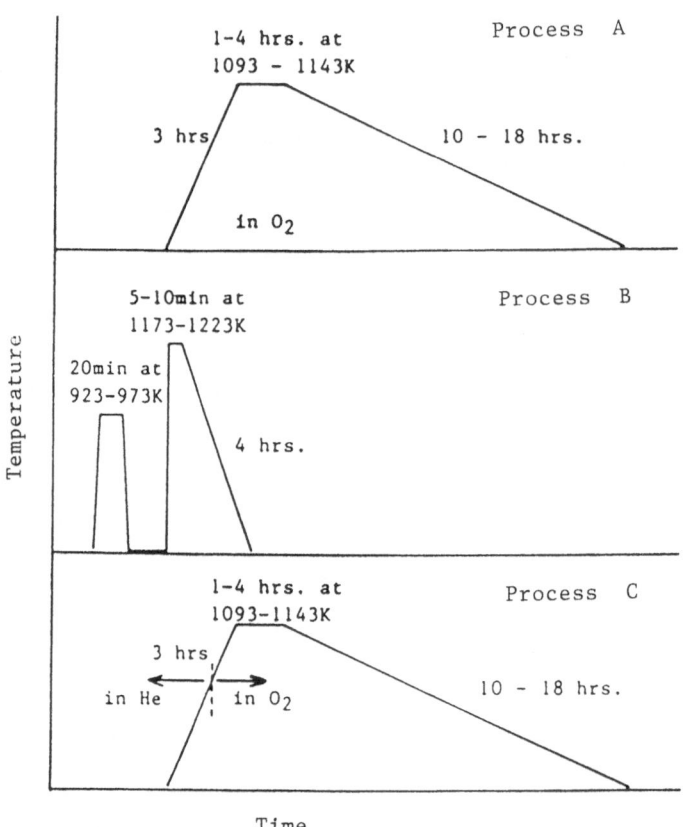

Fig. 1. Annealing procedures used in this research to produce films having a unique morphology and orientation. Note that the temperature-time cycle of process C is identical to that of process A. However, the heating cycle was performed under He gas in process C. Process B invloved three steps: in-situ rapid thermal annealing in a low pressure O_2; rapid heating in a He atmosphere; and followed by furnace cooling in O_2.

deposited films had an amorphous structure. The films were then annealed using various heating and cooling cycles, as shown in Fig. 1. In process A, the films were heated under flowing oxygen (20sccm) in a tube furnace to 1093K-1143K in 3hrs and held at this temperature for 1-4hrs and subsequently cooled to room temperature in 10-18hrs.

Process B involved three steps. In step one, the films were rapidly heated using a quartz lamp to temperatures ranging from 923K to 973K for 20 min in a 200-500 millitorr oxygen atmosphere and then cooled to room temperature. In step two, the films were heated from ambient temperature to 1173K-1223K in 30sec in a helium atmosphere, held at this temperature for 5-10min. In step 3, the films were cooled in an oxygen atmosphere to 473K over 4 hr. The temperature cycle of process C was identical to that of process A. However, in process C, the heating cycle was performed under helium gas in order to suppress the orthorhombic-tetragonal transition. The atmosphere was then switched to oxygen for the high temperature hold and subsequent cool down. Surface profillometry, SEM, and thin film x-ray diffraction were used to determine surface roughness, morphology, and orientation of the film, respectively. The dc-resistance of the film as a function of temperature was measured by using a standard four probe technique with pressed indium contacts. The current was 0.1mA in this measurement. For the stability study, the resistance of the film was remeasured after exposure to air for one month. In the chemical etching rate study, the films were etched using a 1:60 mixture of phosphoric acid and DI water.

RESULTS AND DISCUSSIONS

The SEM micrographs and x-ray diffraction pattern of the films are shown in Fig. 2 and Fig. 3, respectively. A random oriented equiaxial grain structure with grain size around $1\mu m$ was observed in films annealed using process A (slow heating and cooling in O_2). Films annealed in this process always exhibited a smooth, shiny surface (surface roughness less than 40nm). The texture was independent of the substrate material and slight variations in the Cu concentration. The changes in the film structure during the annealing cycle are complex with several factors to be considered including the amorphous-crystalline transition, the orthorhombic-tetragonal transition, and crystallite nucleation and growth. For process A in which the heating and cooling rates are relatively slow, it is likely that the nucleation rate is much higher than the crystallite growth rate, resulting in a fine grain, randomly oriented structure.

A well oriented grid-like structure with grain size around 1-2mm was observed from the film annealed using process B. The initial rapid heating (process B, step 1) was found to be critical to the formation of this microstructure. In addition, a successful deposition was only achieved using a $SrTiO_3$ substrate. Films deposited on YSZ and MgO tended to peel away from the substrate during the annealing cycle. Generally, process B was more successful for the films with higher Cu concentration. The surface roughness was determined to be less than 180nm. X-ray diffraction results (Fig. 3b) indicated that the films showed preferred orientation with strong indications of c-axis oriented normal to the plane and weaker orientation normal to the b-axis. Further work is required to determine whether there are two types of grains oriented normal to the (001) and (010) directions, respectively or there exists a single preferred orientation. In the Cu rich film, the existance of a secondary phase, such as CuO and $BaCuO_2$ was observed (as marked (*) in Fig. 3b). Further, in some films annealed by this process, we noticed a presence of an amorphous phase, even after heating at 1193K for 5min. While the average composition of the film was $YBa_2Cu_5O_{7-x}$, as determined by EPMA, the composition of grid area only was $YBa_2Cu_3O_{7-x}$. This result indicated that Cu rich films with the grid-like structure consisted of two phases - stoichiometric composition grid (see G in Fig. 2b), and Cu-rich matrix (see M in Fig. 2c). It has been reported elsewhere[5] that a c-axis oriented epitaxial film of YBaCuO has been produced on $SrTiO_3$ by heating the substrate during deposition at a temperature of 923K. Based on our results and others[4,5], we believe that the well oriented tetragonal crystallites are nucleated at the film-substrate interface during the first step rapid thermal heating at 923K-973K for 20min. These fine crystallites at the interface rapidly grow during the second rapid heating at 1173K-1223K. As a result, stoichiometric composition (as determined by EPMA) grid structure and Cu rich composition islands are formed.

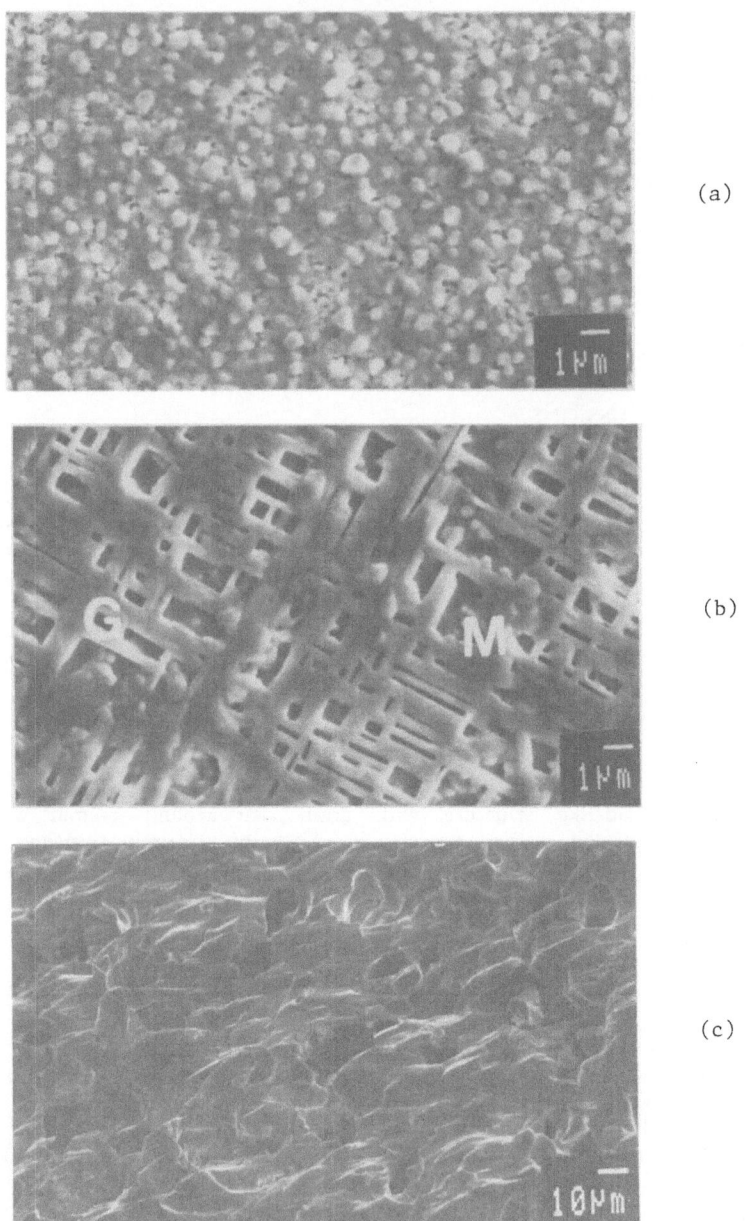

Fig. 2. The SEM micrographs show surface morphology of the YBaCuO thin film annealed various processes: (a) film with equiaxial grain structure on YSZ substrate; (b) film with grid-like structure on $SrTiO_3$ substrate. Note that it consists of two phases – the stoichiometeric composition grid (G) and matric (M); (c) platelet microstructure on YSZ substrate.

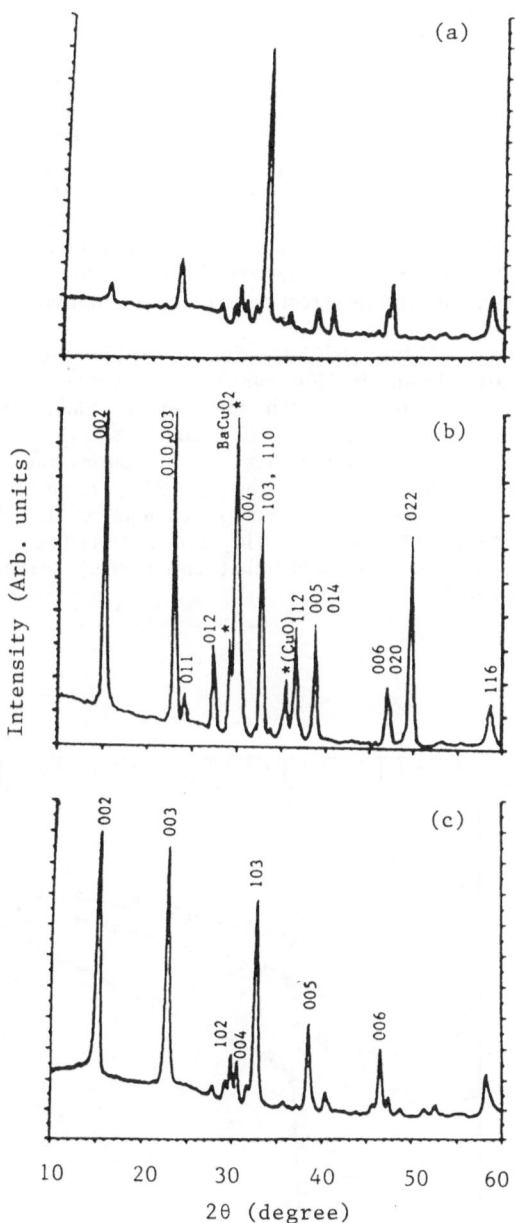

Fig. 3. Thin film x-ray diffraction pattern of the films with various microstructures: (a) equiaxial grain structure – random orientation; (b) grid-like structure – b/c-axes normal to the plane. The peaks from secondary phase are marked (*); (c) platelet microstructure – c-axis oriented normal to the plane.

In addition, taking into consideration that an excessive amount of Cu enhances the formation of the grid-like structure, the growth rate of the grid may be controlled by the diffusion rate of Cu ion during rapid thermal annealing.

A well oriented platelet structure was obtained in films on the three substrates investigated using process C. At temperatures around 973K-1073K, the annealing gas was switched abruptly from helium to oxygen. The helium gas was introduced during heating in order to increase the orthorhombic-tetragonal transition temperature . The surface roughness of the film was less than 300nm, and the platelet-like grains were approximately 20μm in size (Fig. 2c). The film showed strong preferred orientation with the plane of the film normal to the c-axis as shown by x-ray diffraction (Fig. 3C). It was found that the helium atmosphere during the heating step was critical to the formation of the platelet microstructure. Study on the film-substrate interface using cross section transmission electron microscopy is underway in order to understand the mechanism of nucleation of the crystallite and its effect on the film orientation.

Temperature dependence of the resistance for the films prepared by aforementioned annealing processes are shown in Fig. 4. The resistance (y-axis) is normalized to the room temperature resistance of each film. A T_c onset for film B (grid-like microstructure) and film C (platelet) is the same at 94K, but the zero resistance is observed at 74K and 84K, respectively. A T_c onset and T_c zero for film A (equiaxial grains) is observed at 80K and at 60K, respectively. It is noted that the slope in the resistance curve for film A changes at 180 K, and a slight drop in resistance is observed at 105 K. The broad transition and changes in slope ($d\rho/dK$) in the normal state may imply that cation (Cu) ordering is not uniform in the randomly oriented film.[6]

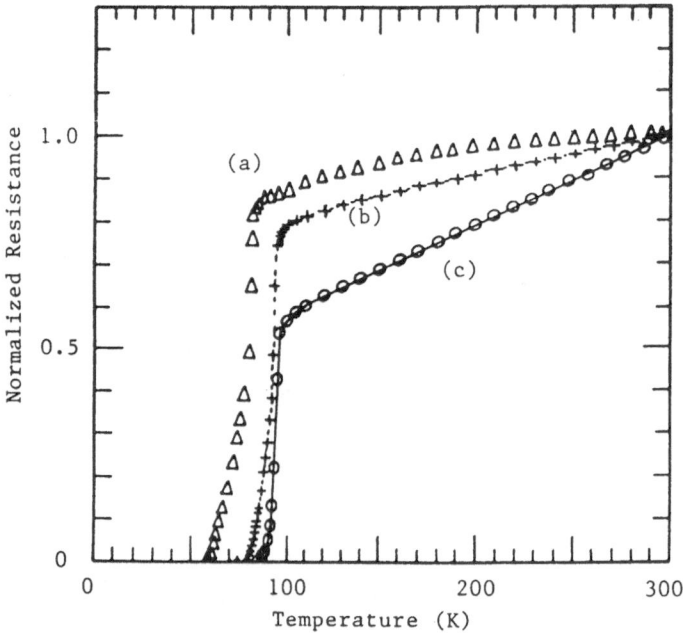

Fig. 4. Temperature dependence of the resistance for the film with various morphology and orientations. The resistance is normalized to room temperature resistance (r/r_{300}) of each film: (a) randomly oriented film; (b) b/c-axes normal to the plane; (c) c-axis normal to the plane.

Table 1. Chemical etching rate, microstructural and superconducting properties of the films with various orientations.

films	structure	room temp. resistivity ($m\Omega$–cm)	T_c on set (K)	T_c zero (K)	etch rate[a] (μm/min)
A[b]	equiaxial grains, random oriented	0.63	80	60	0.154
B[c]	grid-like, b/c-axes oriented	1.07	94	74	0.909
C[b]	platelet, c-axis oriented	0.21	94	84	0.309

a: Film was etched in a 1:60 mixture of phosphoric acid and DI water at room temperature.
b: On YSZ substrate.
c: On $SrTiO_3$ substrate.

In films with oriented microstructures, there was no degradation in T_c onset, but room temperature resistance increased 5-10% after exposure to air for one month. However, the chemical etching rate (in dilute phosphoric acid) of the oriented film is 2-3 times higher than that of random orientation film. Chemical etching rate, microstructural, and superconducting properties of the films are summarized in Table 1.

CONCLUSION

High quality thin films of the YBaCuO superconductor have been produced on YSZ, $SrTiO_3$ and MgO substrates via ion beam sputter-deposition and annealing. The morphology and orientation of the film have been successfully controlled by employing unique annealing processes. Preliminary results indicate that the orientation and morphology of the films are determined primarily by the nucleation rate of crystallites at the film-substrate interface during the initial stage of heating, and the growth rate (or temperature) of these crystallites during high temperature heating. In any case, the lattice matching between the film and the substrate is a key factor in this solid state phase transition. Additional work is required to define the mechanism of nucleation and crystallite growth, and its effect on film orientation. The observed T_c zero resistance is 60K, 74K, and 84K for the film with equiaxial grains, b/c-axis oriented grid-like structure and c-axis oriented platelet microstructure, respectively.

REFERENCES

1. S. W. Tozer, A. W. Kleinsasser, T. Penney, D. Kaiser, and F. Holtzberg, Measurement of Anisotropic resistivity and Hall Constant for Single-Crystal $YBa_2Cu_3O_{7-x}$, Phys. Rev. Lett. 59:1768 (1987).
2 J. Kwo, T. C. Hsieh, R. M. Fleming, M. Hong, S. H. Liou, B. A. Davidson, and L. C. Feldman, Structure and Superconducting Properties of Orientation-Ordered

$B_1Ba_2Cu_3O_{7-x}$ Films Prepared by Molecular-beam Epitaxy, <u>Phys. Rev.</u> B 36:4039 (1987).

3. K. Setsune, T. Kamada, H. Adachi, and K. Wasa, Epitaxial Y-Ba-Cu-O thin films prepared by rf-magnetron sputtering, <u>J. Appl. Phys.</u> 64:1318 (1988).

4. T. Terashima, K. Iijima, K. Yamamoto, Y. Bando, and H. Mazaki, Single-Crystal $YBa_2Cu_3O_{7-x}$ Thin Films by Activated Reactive Evaporation, <u>Jpn. J. Appl. Phys.</u> 27:L51 (1988).

5. J. Fujita, T. Yoshitake, A. Kamijo, T. Satoh, and H. Igarashi, Preferentially oriented epitaxial Y-Ba-Cu-O films prepared by the ion beam sputtering method, <u>J. Appl. Phys.</u> 64:1292 (1988).

6. G. F. Dionne, Transition-Metal Oxide Superconductivity, <u>Technical Report 802,</u> ESD-TR-277, Lincoln Laboratory, Lexington, Ma (1987)

SYNTHESIS OF $Bi_2Sr_2Ca_nCu_{n+1}O_x$ FILMS

BY MULTILAYER DEPOSITION TECHNIQUE

Junich Sato*, Masatsugu Kaise, Keikichi Nakamura and
Keiichi Ogawa

National Research Institute for Metals
1-2-1 Sengen, Tsukuba 305, Japan

Introduction

Soon after the discovery of high Tc superconductors in
Bi-Sr-Ca-Cu -O systems[1], i.e., 80 K and 110 K phases, it was
realized that the 110 K mono phase was extremely difficult to
be synthesized by the conventional solid state reaction.
Frequent intergrowth of many phases, i.e. $Bi_2Sr_2Ca_nCu_{n+1}O_x$
(n=0-6) are observed[2]. This suggests that the free energy of
these phases has no sharp minimum against the number of CaCu
layers sandwitched with two neighboring Bi_2O_2 layers. The
sequential deposition technique is one of the most promising
methods to obtain non equilibrium or metastable material having
layered structure. We have attempted to synthesize the phases
with n=0,1,2 or 3. In this paper we shall report on our
sequential vapor deposition technique, characterization of the
resultant multilayered films by X-ray diffraction and by SEM
and finally their superconducting transion.

Experimental

Our sequential vapor deposition apparatus is illustrated
schematically in Figure 1. The deposition technique used was
two target reactive rf and dc sputtering. Targets were metal
Bi (dc) and sintered oxide Sr-Ca-Cu-O (rf) with a cationic
ratio 2:2:2 in the above order. The power was 5 W for Bi and
140 W for Sr-Ca-Cu-O. The reactive gas was a mixture of Ar
and 10 to 30 % O_2, and the total pressure during the
sputtering was 1 to 1.2 Pa. The substrate was a cleaved MgO
with the surface parallel to (100). The substrate temperature
was varied in a range from RT to 700° C. The sputtering time
for one layer unit Bi-O/Sr-Ca-Cu-O was usually 28 to 46 sec,
which was repeated 100 times. To aid oxidation of the deposited
film, the oxygen shower was blown to the film. Some films were
post annealed in air at 810 or 850°C for 1 h. The resultant
films were characterized by the conventional θ-2θ scan of the
Cu Kα X-ray and by taking the SEM image. The electrical
resistivity was measured by the conventional four prove method.

---

* On leave from Metal Research Laboratory, Hitach Cable Ltd.,
1500 Kawajiricho, Hitachi 317, Japan

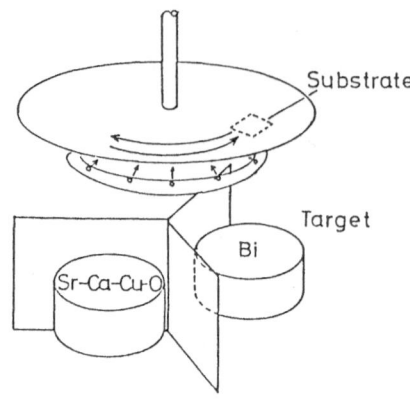

Figure 1. Schematic illustration of sequential vapor deposition apparatus, i.e., two for dc and one for rf sputtering.

Results

Figure 2 shows the X-ray diffraction pattern of the multilayered film, deposited onto the MgO substrate kept at RT and repeated 10 times. A series of zeroth order reflection peaks, i.e., N=11, 12,....,15 were observed and the repeating distance was calculated to be 315 Å. This is, of course, in agreement with the value calculated from the thickness of the film divided by the number of repeats, i.e., 10.

Let us examine the effects of the substrate temperature (Figure 3). The deposition period of each layer was reduced in the present film by the factor of 10 from that shown in Figure 2. When the substrate temperature Ts was 250°C, the period was found to be 28.4 A. In addition to the superlattice

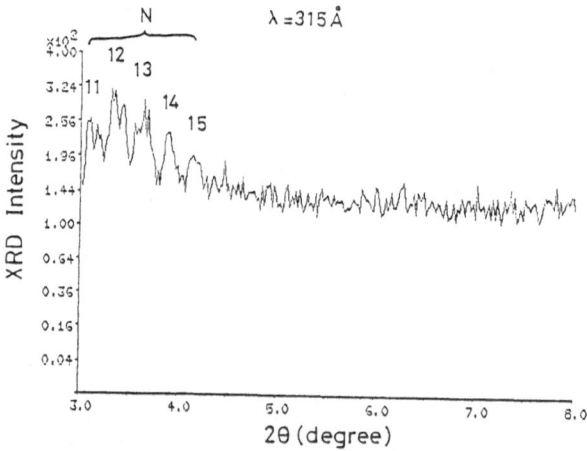

Figure 2. X-ray diffraction pattern of the multilayered film. A series of zeroth order reflections are numbered.

reflection, i.e., $2\theta = 3.0°$, we have a halo possibly arising from the amorphous phase in the as-deposited film.

   As Ts increased, the superlattice reflection peaks became broad and shifted to lower angles. It can be seen from Figure 3(c) that some crystllization is occurring when Ts is raised to 450°C. So far we were not able to identify these diffraction peaks yet.

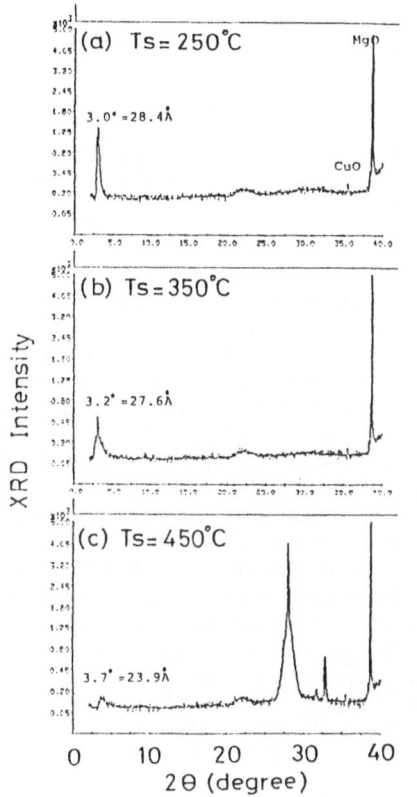

Figure 3. Effect of substrate temperature on the structure of multilayered films.

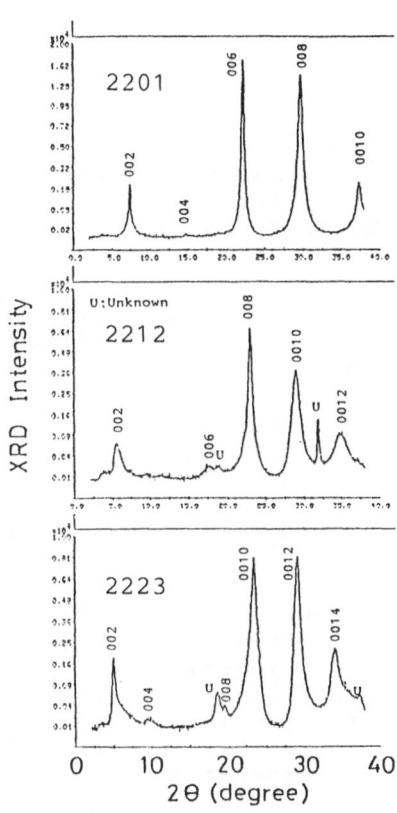

Figure 4. X-ray diffraction patterns of the multilayered films simulated to the 2201, 2212 and 2223 phases, respectively. The substrate temperature was 650°C.

   Figure 4 shows the X-ray diffraction patterns arising from the multilayered films deposited at Ts=650°C and simulated to the 2201, 2212 and 2223 phases. All diffraction peaks can be indexed by (00ℓ); the c planes of the present film are dominantly parallel to the substrate plane. It is noted that there are some unidentified peakes as well. These patterns are almost similar to the recently published results[3] except that the present patterns show no peaks arising from CuO phase.

255

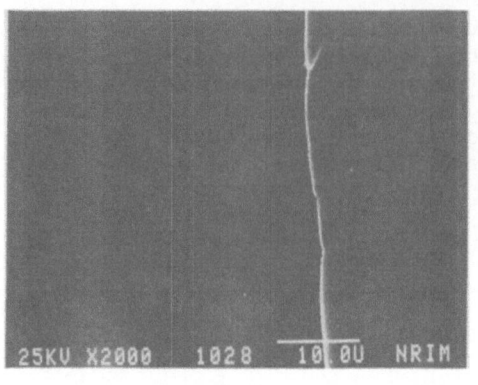 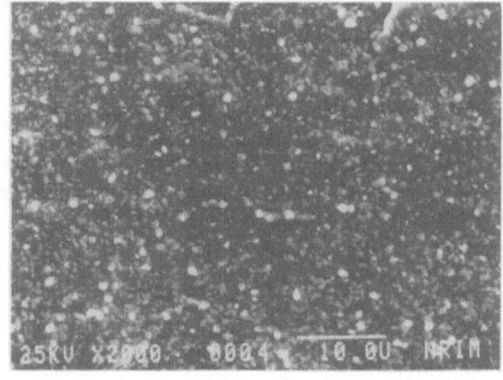

(a)   (b)

Figure 5. SEM images for the surface of the multilayered film (Ts+650°C) (a) and of the film sputtered from a single composite Bi-Sr-Ca-Cu target (Ts=650°C) (b).

The SEM image is shown for the surface of the multilayered film deposited on the substrate kept at Ts=650°C. The surface is smooth indeed. The bright line running nearly vertically in Figure 5(a) corresponds to a cleavage step on the MgO substrate. For the sake of comparison, we show the SEM image of the Bi-Sr-Ca-Cu-O film deposited from the single target onto the substrate kept at 650°C. Here we see many fine grains in the film, in contrast with the smooth surface of the multilayered one.

The resistivity vs temperature curves are shown in Figure 6 for the films of Fgure 4. All the curves show the semiconductor-like behavior, unexpected from the structure of Bi-Sr-Ca-Cu-O shown in Figure 4. When the films are annealed at 810°C for 1 h in air, they became superconducting (Figure 7). There is some sign of superconducting transition at 110 K for both films. The 2201 film was found to evaporate completely during the same heat treatment at 810°C. It is noted incidentally that the 2201 film has the Bi content more than the other films.

Discussions

It appears somewhat strange that the present multilayered films with a correct period corresponding to $Bi_2Sr_2Ca_nCu_{n+1}O_x$

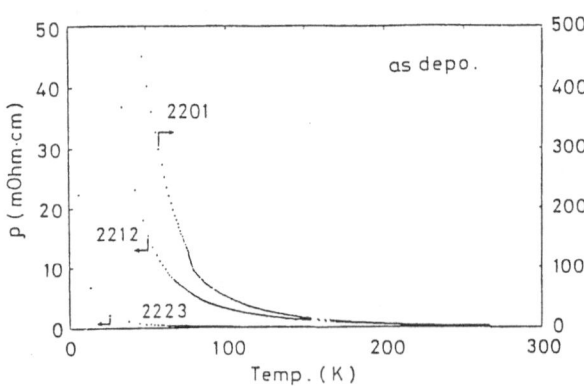

Figure 6. Resistivity vs temperature curves for the as-deposited films (Ts=650°C). These films were simulated to the 2201,2212 and 2223 phases, respectively.

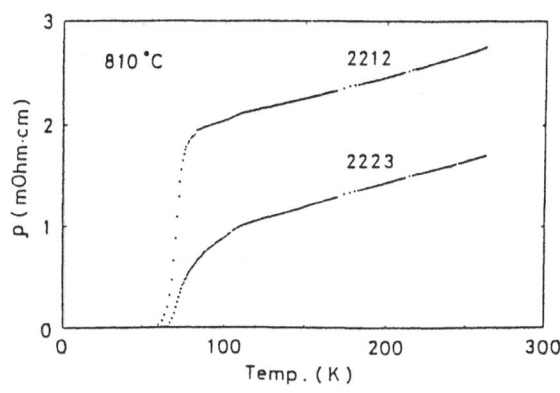

Figure 7. Resistivity vs temperature curves for the corresponding films post annealed at 810°C.

Figure 8. X-ray diffraction patterns of the multilayered film simulated to the 2234 phase.

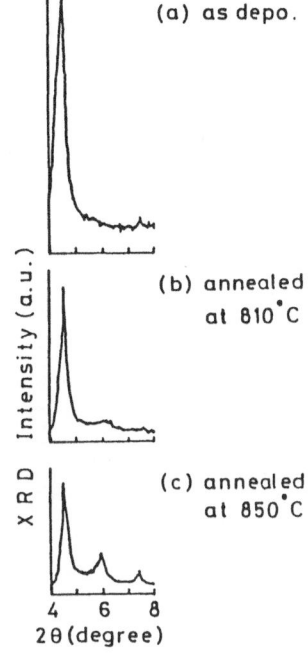

behaves like a semiconductor (Figures 4 and 6). It is likely that the chemical composition and chemical order are not quite right in the Sr-Ca-Cu-O layer. After annealing the as-deposited films at 810°C, the films became superconducting as shown in Figure 7. The X-ray diffraction pattern for the annealed sample having 2234 structure showed dominant 2234 peaks with a small amount of 2212 and 2201 peaks as shown in Figure 8. No additional peaks arising from impurities such as CuO were not observed. This indicates that the multilayered films deposited onto MgO substrate kept at 650° C have very stable layered structure. By annealing the layered films having incorrect chemical composition can only decompose into similar phases having different periods. However, the resistivity results shown in Figure 6 indicate that the chemical disorder seems to be remained in the majority phase.

References

1) H. Maeda, Y. Tanaka, M. Fukutomi and T. Asano, A new high Tc oxide superconductor without a rare earth element, Jpn. J. Appl. Phys., 27:L209(1988)
2) S. Ikeda, H. Ichinose, T. Kimura, T. Matsumoto, H. Maeda, Y. Ishida and K. Ogawa, TEM studies of intergrowth in BiSrCaCu O and high Tc superconducting phase, Jpn. J. Appl. Phys., 27:L999 (1988).
3) H. Adachi, S. Kohiki, K. Setsune, T. Matsuyu and K. Wasa, Formation of superconducting Bi-Sr-Ca-Cu-O thin films with controlled c-axis lattice spacings by multi-target sputtering, Jpn. J. Appl. Phys., 27:L1883(1988).

STUDY OF HIGH QUALITY $YBa_2Cu_3O_{7-y}$ THIN FILMS

B. R. Zhao, C. W. Yuan, J. Gao, Y. Z. Zhang, P. Out[*],
Y. Y. Zhao, Y. M. Yang, P. Xu and L. Li

Institute of Physics, Academia Sinica, Beijing

ABSTRACT

The high quality $YBa_2Cu_3O_{7-y}$ thin films were prepared by
RF magnetron sputtering and ion beam sputtering methods with post
annealing procedure. Zero resistance temperature ($T_{R=0}$) and
critical current density J_c (at 77 K) are 91 K and 1.3×10^5
A/cm^2 respectively. X-ray diffraction showed three types of
orientation for these thin films: a-axis orientation, c-axis
orientation and mixed a-axis and c-axis orientations. For a-axis
and c-axis orientation thin films, anisotropic values were
obtained in the measurement of $J_c(H)$ under high magnetiac field.
It is possible that the anisotropic flux pinning force might exist
in the oriented thin films.

INTRODUCTION

Since the discovery of high T_c oxide superconductors, many
research works have been carried out for these materials because
of the scientific importance and the potential technological
applications. It is well known that thin films of this kind of
materials should be very useful to realize and develope the
applications of superconductivity. Up to date, thin films of
several important materials have been studied; the Y-Ba-Cu-O,
Bi-Ca-Sr-Cu-O and Tl-Ba-Ca-Cu-O. Many methods have been used to
prepare the Y-Ba-Cu-O thin films, i.e., the RF (magnetron)
sputtering, DC (magnetron) sputtering, ion beam sputtering, e-beam
evaporation and MBE etc. High quality thin films were obtained by
some groups[1-3]. But the critical current density is still a
problem to be solved. Up to now, we know that the critical current
density of the thin films of this kind of material depends upon
the preferential orientation and microstructure. Due to the two
dimentional characteristic of the orthorombic perovskite structure
and the (a-b) plane plays an important role to determine the
superconductivity, the c-axis orientation of $YBa_2Cu_3O_{7-y}$ is
expected for obtaining high current densities of the film. So the
epitaxial growth of $YBa_2Cu_3O_{7-y}$ on single crystal (100)
$SrTiO_3$ substrate is been used. The experimental

* Permanent address: Twenty University, The Netherlands

results showed that the ideal epitaxial growth of the film not only depends on the substrates but also on the substrate temperature, the composition of the film and the post annealing conditions. So it is not easy to find an optimum growth procedure for the thin film samples. So far, only few groups in the world reported c-axis orientation of $YBa_2Cu_3O_{(3,3)-y}$ thin films with J_c (77 K) at an order of 10^6 A/cm^2.

On the other hand, the inhomogeneity and weak links in the microstructure of the samples, leads to the low limit of electron transpotation. The superconducting current has to flow through the Josephson effect like tunnelings (SIS) and proximity effect-tunnelings (S-N-S, S-I-N-S)[4-7]. Generally, the critical current density of the order of 10^6 A/cm^2 is only obtained in a small part of the film. This is very different from the conventional superconductors, for example, the intermetallic compounds A15 Nb_3Sn, Nb_3Ge etc. So at present it is still necessary to improve the preparation methods and to study the microstructure and physical properties of the thin films.

In this report, we will describe the RF magnetron sputtering (RMS) and ion beam sputtering (IBS) method for preparing $YBa_2Cu_3O_{7-y}$ thin films and give systematic results obtained up to date. Some results are discussed in detail.

EXPERIMENTAL

The $YBa_2Cu_3O_{7-y}$ thin films were made by RF magnetron sputtering and ion beam sputtering methods. We have reported the former method previously[8,9]. The schematic diagram of the ion beam sputtering system is shown in Fig. 1. The target was

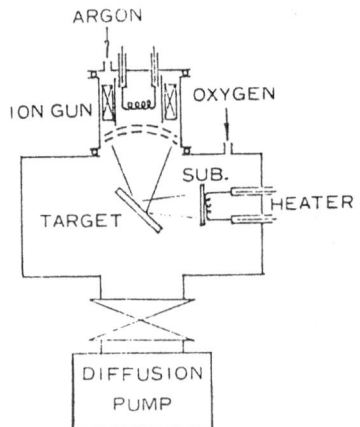

Fig. 1. The schematic diagram of the ion beam sputtering system.

mounted 45° to the ion beam. The distance between target and ion beam sourse is 20 cm. The ion beam can be focussed to 1.5-5.5 cm in diameter, for this case, 2 cm diameter of ion beam was used with which, the energy was 800-1200 eV, current was 100-150 mA. (100) $SrTiO_3$ single crystal was used as the substrate. The system was pumped by using the diffusion pump. The background pressure was $1-2 \times 10^{-5}$ torr and the pressure of sputtering gas was $3-5 \times 10^{-4}$ torr with ratio of Ar to $O_2 = 4$. The deposition temperature was 400-450°C, deposition rate was 200-260 A/min., the total thickness of the film is about 1 μm. The as deposited thin films were amorphous or highly disordered, post annealing treatment is necessary to convert it to orthorhombic perovskite structure. Two-step annealing procedure was performed; 600-700°C for 2-4 hr. followed by heating at 850-900°C for 1-2 hr. then slowly cooled down within the furnace to room temperature.

The films were characterized as following: X-ray diffraction was used to determine the structure of the films. The chemical composition was analysed by inductively coupled plasma atomic emission spectroscopy (ICPAES) and x-ray fluorescence spectroscopy (XRFS). The scanning Auger analysis was used to determine the depth profile of the film, in special to study the diffusion of the film and substrate at the interface. The normal state resistivity, critical temperature, critical current density J_c and $J_c(H)$, were measured by the standard four probe method.

RESULT AND DISCUSSION

The phase structure and physical parameters of the films are shown in Fig. 2 and listed in Table 1. We can see from Fig. 2 and Table 1 that most of the films are a-axis oriented. This is reasonable to understand from the knowledge of epitaxial films, since the lattice parameter of $SrTiO_3$ (a= 3.90 A) is nearly

Table 1. The physical parameters of the $YBa_2Cu_3O_{7-y}$ thin films.

Sample No.	$T_D(°C)$	Orientation	$\rho_{300K}/\rho_{Tconset}$	$T_{R=0}(K)$	J_c at 77 K (A/cm^2)
S-408	400	a>>c	2.0	87	1.4×10^4
S-414	400	a>>c	2.5	88	2.5×10^4
IB-108	425	a~c	2.8	90	4.5×10^4
IB-114	425	a~c	3.0	90	3.6×10^4
S-541	400	c>>a	3.1	90	1.3×10^5

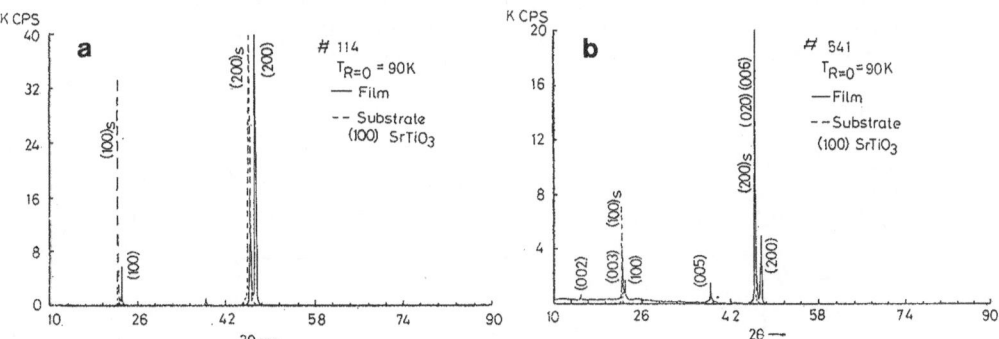

Fig. 2. The phase structure of the thin films, (a) a-axis oriented film, (b) c-axis oriented film.

equals to the b-axis and c-axis/3 of $YBa_2Cu_3O_{7-y}$, from the free energy point of view, the $YBa_2Cu_3O_{7-y}$ thin films tend to grow on (100) $SrTiO_3$ substrate by epitaxy along the b-axis and c-axis, i.e. it is easy to grow the film with a-axis orientation. The c-axis oriented thin films could be grown only when the composition of the film was controlled to "123" and the post annealing treatment was performed carefully. The experimental results show that when the composition deviates from "123" very much, serious interdiffusion between film and substrate has been observed, namely, the thickness of the diffusion layer becomes wider. It is easy to understand from Fig. 3. Fig. 3(a) shows the Auger profile of sample for which the composition deviates from "123" within 2%. $T_{R=0}$ = 90 K. The width of diffusion layer is much narrower than that of sample S-123 (Fig. 3 (b)) for which the composition is Ba rich.

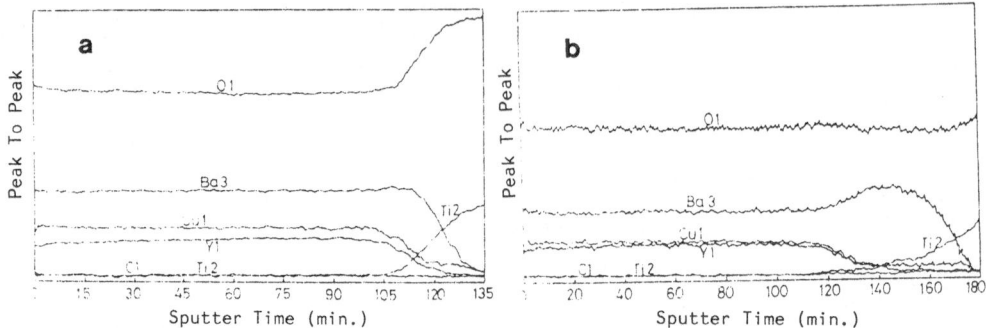

Fig. 3 The Auger profile analysis of the thin films (a) Ba rich film, (b) "123" composition film.

As mentioned above, the microstructure is also a basic factor for determining the critical current density. It is well known that, the metallic oxide superconductor is very different from the intermetallic superconductor. For the latter, one can easily obtain J_c at the order of 10^5 A/cm^2 due to its isotropic structure and perfect body effect of superconductivity. While the metal oxide superconductor is the quasi-2-dimensional strucutre the critical current density depends upon the orientation of the sample strongly. On the other hand, this kind of compound is also a glass-type material[10]. (although for this material it is hard to say is an intrinsic property). Various types of boundaries lead to the weak links between individual superconducting areas. From I-V curves and J_c-T correlations, several types of weak links were recognized, such as the Josephson type junction, proximity effect (S-N-S, S-I-N-S) junction etc[6].

For our samples, the weak links existed in films could be analysed from resistivity vs temperture measurments. From Fig. 4 we can see that for the samples showing very strong metallic behavior $\rho_{Tconset} \sim 100-150$ $\mu\Omega\cdot$cm, $\rho_{300K}/\rho_{Tconset} \gtrsim 2.5$ even show no intercepts at 0 K when the R-T curves were extrapolated

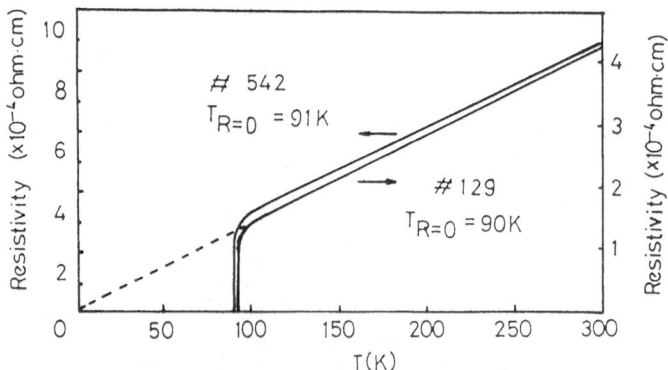

Fig. 4. The resistivity vs temperature curve of the films.

to 0 K. This result shows rather highly ordered structure for
"123" perovskite crystal in the films, so that the measured
critical current density J_c which is still less than the
magnitude of 10^6 A/cm^2 should be attributed to the weak links
existed between the highly ordered structure areas.

J_c vs magnetic field measurements can also be used to
evaluate the quality of the thin films, because the weak links are
easy to be destroyed under low magnetic field. Fig. 5 shows the
measurement of J_c vs magnetic field at 77 K. It can be seen that
for both a-axis and c-axis oriented thin films, J_c decreased
very rapidly in the low magnetic field, especialy in the range of
< 0.5 T, irrespective of the direction of the applied magnetic
field. This should be attributed by the fact that the weak links
were destroyed by low magnetic field.

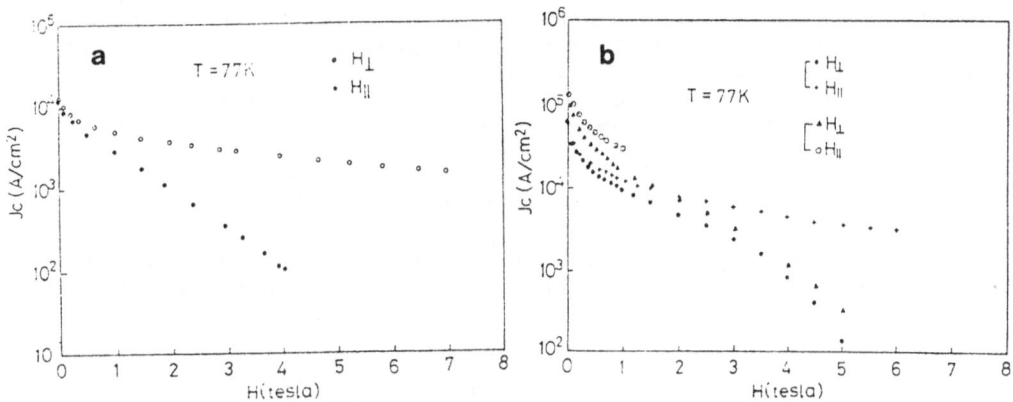

Fig. 5. J_c vs magnetic field (a) a-axis oriented film, (b)
c-axis oriented film.

When the magnetic field increases, the decrease of J_c became
slower and also depended on the direction of the applied magnetic
field and the orientation of the films; for a-axis oriented thin
films, J_c decreased more quickly when the magnetic field was
applied parallel to the film, while in the vertical field, J_c
decreased slowly (Fig. 5(a)). In the case of c-axis orented thin
films, the reverse results were observed, i.e., in the parallel
magnetic field, J_c decreased with field slowly, while in
vertical field the J_c decreased more quickly as shown in Fig.
5(b). These results can be explained with the anisotropic
structure of $YBa_2Cu_3O_{7-y}$ and flux pinning mechanism.

As mentioned above, the (a-b) plane plays an essential role
for superconductivity in $YBa_2Cu_3O_{7-y}$, so it is reasonable to
understand that the influence of magnetic field on superconductivity of $YBa_2Cu_3O_{7-y}$ are weaker when the magnetic
field was parallel to (a-b) plane. So in the case when the
magnetic field was applied perpendicular to a-axis oriented films
or parallel to c-axis oriented films, the magnetic field are
parallel to (a-b) plane, so J_c decreased with magnetic field
slowly. On the other hand, in the high magnetic field, this J_c
behavior should be also associated to the anisotropic flux pinning
mechanism.

We believe that for oriented thin films, some connected defect planes occur easily. The inter-connected imperfect (a-b) planes may be one kind of them, These plane defects are parallel to (a-b) plane. For a-axis oriented thin films, (a-b) planes are mostly perpendicular to the film, while the orientations of b-axis and c-axis may be random in the plane of the films. So in between the oriented (a-b) planes, many plane defects may exist. (Among these plane defects, imperfect (a-b) planes should be included). So that in both c-axis and a-axis oriented thin films, plane defects (which are usualy parallel to (a-b) planes) easily occur and become plane pinning center. When c-axis oriented film in the parallel magnetic field, and a-axis oriented film in the vertical field, these plane defects all act as a strong pinning force, subsequently, an higher $J_c(H)$ was obtained.

SUMMARY

For high quality (especialy High J_c) $YBa_2Cu_3O_{7-x}$ thin films, epitaxial growth is necessary, for which, either c-axis or a-axis orientation thin films could be grown, the weak links shown in the microstructure of the film can be decreased to the lowest level. Also, for oriented thin films, the plane flux pinning centers may exist which could act as a strong anisotropic flux pinning force in high magnetic field.

ACKNOWLEDGEMENT

The work was sponsered by the Chinese National Natural Science Foundation and the National Center for R & D on Superconductivity.

REFERENCE

(1) M. R. Beasley of the Stanford Thin Film Group. Physica, 148 B (1987) 191.
(2) B. Y. Bando, T. Terashima, K. Iijima, K. Yamamoto, J. Hirata and H. Mazahi, Proceeding of 20th. Conference On Solid State Devices and Materials, Tokyko, (1988) 419.
(3) P. M. Mankiewich, J. H. Scofield, W. J. Skocpol, R. E. Howard, A. H. Dayem and E. Good, Appl. Phys. Lett. 51(21) (1987)
(4) K. Moriwaki, Y. Enomoto and T. Murakami, Jpn. J. Appl. Phys. 4 (1987) L521.
(5) S. B. Ogale, D. Dijkkamp, T. Venkalesan, X. D. Wu and A. Inman, preprint.
(6) C. W. Yuan, B. R. Zhao, Y. Z. Zhang, Y. Y. Zhao, Y. Lu, H. S. Wang, Y. H. Shi, J. Gao and L. Li, to be published in J. Appl, Phys. (1988).
(7) J. W. C. de Vries, M. A. Gijs, G. M. Stollman, T. S. Baller and G. N. A. van Veen, J. Appl. Phys. 64(1>) (1988) 426.
(8) B. R. Zhao, L. Li, Y. Lu, H. S. Wang, Y. H. Shi, Y. Y. Zhaom C. W. Yuan and J. Gao, International J. of Modern Phys. B Vol. 1 No. 2 (1988) 561.
(9) B. R. Zhao, H. S. Wang, Y. Lu, Y. H. Shi, C. W. Yuan, Y. Y. Zhao and J. Gao, Chinese Phys. Lett. Vol.4, No. 12 (1988).
(10) K. A. Muller and J. G. Bednorz, Science, Sept. (1987).

DEPOSITION OF TlBaCaCuO THIN FILMS

C.X. Qiu and I. Shih

Electrical Engineering Dept., McGill University
3480 University St., Montreal, P.Q., H3A 2A7, Canada

C. Moreau, B. Champagne and S. Dallaire

IMRI, National Research Council
Boucherville, P.Q., Canada

INTRODUCTION

For the high T_c system of TlBaCaCuO[1], it is specially interesting to investigate the formation of films from sputtered BaCaCuO[2]. In such experiments, a Tl-free target of BaCaCuO is used for the film deposition and therefore the contamination problem of the vacuum system can be avoided. Using such rf sputtered BaCaCuO films, it is necessary to carry out oxidation and Tl diffusion to form the final compound. In this paper, some of the experimental results on the deposition of BaCaCuO thin films are first described. Variation of composition in the BaCaCuO films with the heat treatment is then presented.

EXPERIMENTAL RESULTS

The deposition of the BaCaCuO films were carried out in a system with an rf magnetron sputtering gun. The target, having a diameter of 5 cm, was prepared by sintering weighted amounts of $BaCO_3$, $CaCO_3$ and CuO at 850°C and pressing the sintered material. Polished ZrO_2 and glass slides were used as the substrates during the deposition which were mounted on a water-cooled Al holder. The distance between the substrates and the target was 4 cm and the incident rf power was maintained at 80 W. A mixture of Ar and 5% O_2 was used for the sputtering with a pressure of 2 mtorr. Glass slides were first used for the deposition to prepare films for the thickness determination. Fig. 1 shows the variation of the film thickness (measured with a commercial stylus) with the distance from the projection of the target center. It is seen that the maximum thickness is obtained at about 2 cm from the center. Observation of the films showed a shiny and dark surface in the region from 1 to 3.5 cm. In the region from 0.4 to 1 cm from the center, the film is brown and is semi-transparent. In the central region, the film is shiny and dark again. In order to see the difference of composition in these regions, the elemental concentration of the film was determined by electron probe micro analyzer (EPMA). Fig. 2 shows the elemental distribution of Ba, Ca and Cu. In the region from 2 to 4 cm, the concentration of Ba, Ca and Cu are fairly constant and the ratio of Ba/Ca is about 1.2. As one travels towards the center, the Cu concentration increases and that for Ba and Ca decrease. The Cu concentration reaches a maximum of 32% at about 0.9 cm and then decreases to 21% and remains at this value until one reaches the central region of the substrate. It is also seen that there is a corresponding abrupt change in the Ba concentration. The variation of the Ca concentration in the region is seen to be small. Such negative correlation of Cu and Ba has been observed on heat treated films of TlBaCaCuO prepared by a multilayer deposition method[3]. Using the rf magnetron sputtering, it is seen that there is a region (2 to 4 cm from the center) where the concentration for the constituent elements is essential constant, allowing one to prepare films for subsequent observation. However, it is also noted that the Ba/Ca ratio is greater than 1 even when the nominal Ba/Ca ratio in the target is equal to 1.

From our previous experiments on the thin films of TlBaCaCuO, it has been observed that the electrical properties of the material is very sensitive to the conditions used during the heat treatment. It is possible that the variation of electrical properties is directly relevant to the stoichiometry. In the present work, the variation of composition with the heat treatment has been specifically studied. Films of BaCaCuO deposited on ZrO_2 substrates were used for composition experiments. The composition of as-deposited BaCaCuO films on polished monocrystalline ZrO_2 substrates was first determined by EPMA and the corresponding composition was determined after heat treatment at different temperatures. The treatments were made in a horizontal furnace with a quartz tube. During the experiments, O_2 was allowed to flow through the tube with a flow rate of about 50 cc/min. The results obtained are summarized in Table 1. Before the heat treatment, the Ba/Ca ratio of the films is 1.3 and the Cu/Ca ratio is about 1.7. After the treatment at 785°C, the Ba/Ca ratio is seen to reduce to 1.15 and to 1.22 for the Cu/Ca ratio. The ratio is seen to reduce further to Ba/Ca = 0.99 and Cu/Ca = 1.16 as the treatment temperature is increased to 815°C. After the treatment at an even higher temperature of 845°C, the ratios are Ba/Ca = 0.89 and Cu/Ca = 0.93. From the above results, it is evident that there is a considerable loss of both Ba and Cu in the rf sputtered BaCaCuO films during the heat treatment at temperatures above 785°C. The elemental loss resulted in reduced Ba/Ca and Cu/Ca ratios. From these results, it is clear that it is not only important to sputter the BaCaCuO films from a target with appropriate composition, it is also important to select appropriate heat treatment conditions in order to reduce the variation of composition due to the un-even loss of the constitutient elements in the films.

Table 1   Variation of Relative Composition in BaCaCuO Films With Heat Treatment.

Sample No.	Before H.T.		After H.T.		Treatment Conditions
	Ba/Ca	Cu/Ca	Ba/Ca	Cu/Ca	
TP12-6	1.28	1.70	1.15	1.22	785°C, 5 min
TP14-2	1.26	1.63	0.99	1.16	815°C, 5 min
TP12-4	1.31	1.74	0.89	0.93	845°C, 5 min
TP14-3[a]	1.33	1.74	1.33	1.54	815°C, 5 min

[a] Presintered $Ba_2Ca_2Cu_3O_x$ present in the treatment chamber.

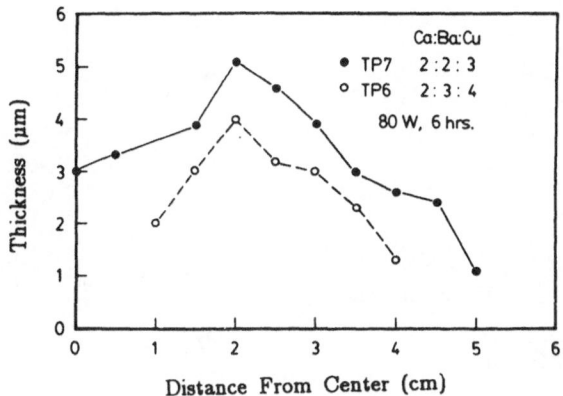

Fig. 1 Variation of thickness with distance from the projection of
target center for two as-deposited BaCaCuO films.

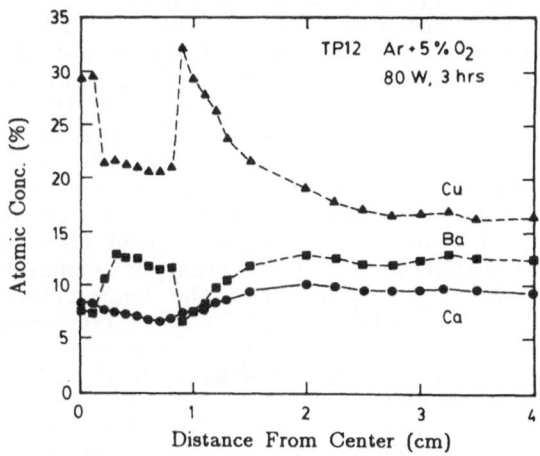

Fig. 2 Variation of composition of the rf-sputtered BaCaCuO film with
distance showing an essentially constant distribution for the
working region from 2 to 4 cm from the center.

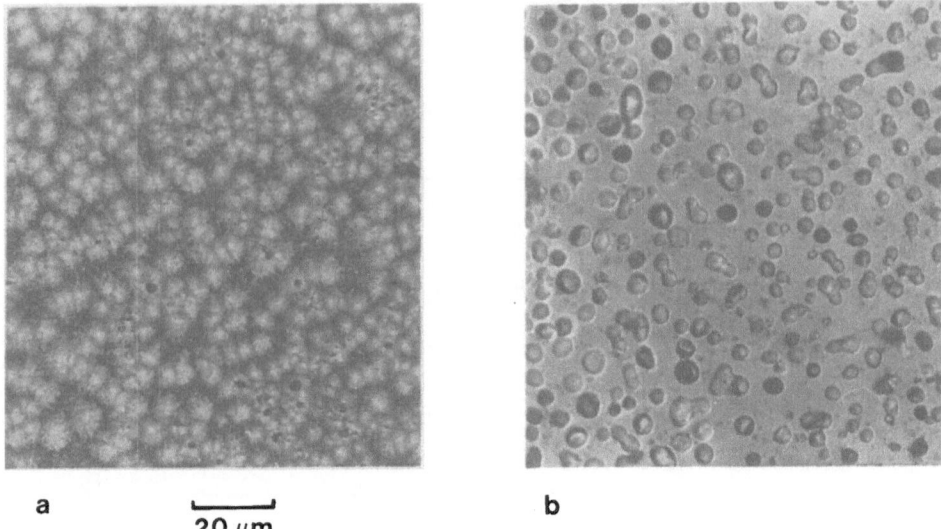

a $\overline{20\ \mu m}$ b

Fig. 3 (a) Photograph of an rf sputtered BaCaCuO film treated at 815°C showing dark and bright areas. Ba/Ca<1 for dark areas and Ba/Ca>1 for bright areas, and (b) photograph of a treated BaCaCuO film originally covered with a 100 Å thick Cu layer.

In order to minimize the un-even loss of the constitutient elements during the treatment, about 0.5 gm of presintered $Ba_2Ca_2Cu_3O_x$ was placed in a small quartz boat and the boat located near the films in an up-stream position during the treatment. Composition of a sample TP14-3 treated with the presence of the presintered BaCaCuO was then obtained and the results are given in Table 1. From Table 1, it is seen that after the 5 minute treatment at 815°C for TP14-3, the Ba/Ca ratio is not changed and the Cu/Ca ratio decreases from 1.74 to 1.54. The above results suggested that the presence of the presintered bulk sample is effect in preserving the required stoichiometry of the deposited BaCaCuO films. The surface morphology of the heat treated film with the presence of presintered BaCaCuO material was observed and a optical micrograph is shown in Fig. 3 (a). It is seen after the heat treatment, the film consists of many small bright areas which are separated by dark materials. This morphology is different from the one treated under similar conditions without the presintered BaCaCuO material which showed a flat surface. EPMA results showed the bright areas are Ba-rich with Ba/Ca<1 and the dark areas are Ba deficient with Ba/Ca<1. Similar segregation effect was also found for the treated films originally covered with a layer of Cu (about 100 Å). Fig. 3(b) shows the top view of such a film after a short treatment at 800°C in O_2.

In a previous work on the TlBaCaCuO multilayer films, it has been observed that the formation of this compound is complex and a lateral segregation effect occurred during the heat treatment and Tl diffusion process. It is interesting to determine whether a similar segregation effect is present in the rf sputtered films after the Tl diffusion. The observation was made with a sample deposited on a ZrO_2 substrate with a

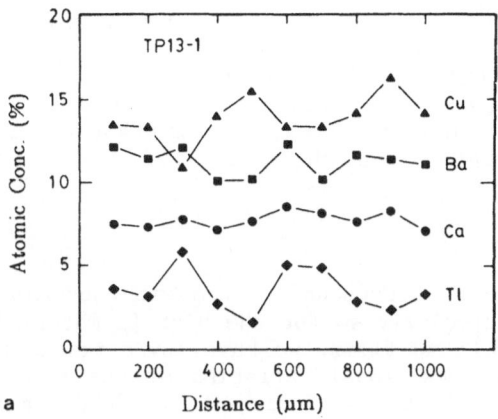

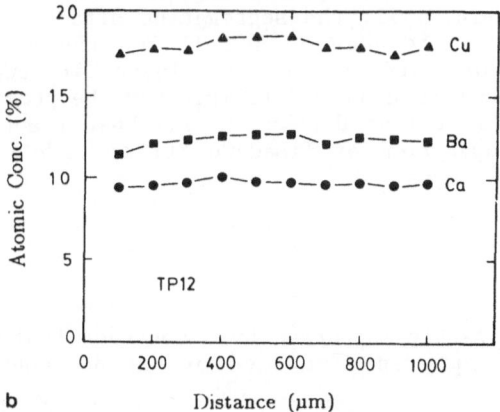

Fig. 4 (a) Compositional results from EPMA for an as-deposited BaCaCuO
film, and (b) results for a similar film after a short heat
treatment. It is seen there is a lateral segregation effect and
a positive correlation between Ba and Tl.

Tl diffusion time of about 1 hr and an as-deposited film on a glass substrate. The composition was determined along a line with a step of 100 microns and the results obtained are shown in Fig. 4. Fig. 4 (a) shows the compositional results of the as-deposited film on the glass substrate (TP12). One can see that the elemental distribution is quite uniform, with a deviation of about ±0.5%. The composition results for the sample after the Tl diffusion are shown in Fig. 4 (b), where one can see that there is a large variation of Cu concentration with the distance (±2.5 at.%). A large variation is also seen for the Ba and Tl distribution. For Ca, the variation is relatively small. In Fig. 4(b), it is also noted that there is a positive correlation between Ba and Tl and a negative correlation between Cu and Tl. In areas with high Cu concentration, the Ba concentration is also large. Such effect has been observed in the TlBaCaCuO films prepared by the multilayer deposition method[3]. The causes for the lateral segregation of elements have not been established for the present samples. It could be due to random crystallization of an unknown phase with a composition which is different from the initial stoichiometry.

DISCUSSION

The formation of a compound system with several elements is often complex. This is especially so for the high T_c material TlBaCaCuO. Using the rf sputtered BaCaCuO films, we have investigated the effect of heat treatment on the compositional variation of the films. For the BaCaCuO films without Tl, a short heat treatment at 800°C can result in a severe lateral elemental segregation. The segregation occurred even with the presence of presintered bulk BaCaCuO material. The resulted films consist of two apparent main regions, one with Ba/Ca>1 and the other with Ba/Ca<1. The size of the segregation region is several microns. For the treated films containing Tl, the segregation effect is also severe. The positive correlation between Ba and Tl and the negative correlation between Cu and Tl observed for the multilayer TlBaCaCuO films was also observed for the treated sputtered films. From the present results, it is thus evident that a refined deposition and heat treatment procedure is needed for the preparation of TlBaCaCuO films with uniform composition distribution.

ACKNOWLEDGEMENT

This work is partly supported by a contract from National Research Council through Supply and Services of Canada (contract No. 31950-8-0004/01-SS).

References

1 Z.Z. Sheng and A.M. Herman, Nature, **332**, 138 (1988).

2 C.X. Qiu and I. Shih, Applied Physics Letters, **53**, 1122 (1988).

3 M.W. Denhoff, P. Grant, C.X. Qiu, I. Shih, P.T. Robinson, I. Sproule and D. Mitchell, submitted to Applied Physics Letters.

METALORGANIC CHEMICAL VAPOR DEPOSITION OF HIGHLY TEXTURED

SUPERCONDUCTING $YBa_2Cu_3O_{7-x}$ FILMS

K. Zhang, B. S. Kwak, E. P. Boyd, A. C. Wright and A. Erbil[a]

School of Physics
Georgia Institute of Technology
Atlanta, Georgia 30332

ABSTRACT

C-axis textured single phase superconducting $YBa_2Cu_3O_{7-x}$ films have been successfully grown on the yttria-stabilized zirconia (100) substrates by using the metalorganic chemical vapor deposition technique. After the post annealing the films deposited on the yttria-stabilized zirconia substrates exhibited a highly textured x-ray pattern with c-axis perpendicular to the substrate surface. These films show an onset superconducting transition temperature of 93K with the resistance becoming zero at 84K. The films deposited on sapphire show a semiconducting feature in the normal state with a broad superconducting transition at much lower temperatures (10-40K). In order to gain some insight onto the growth process, we also studied the mass transport of metalorganic compounds as a function of source temperature, flow rate and reactor pressure.

INTRODUCTION

Since the discovery of high T_c superconductivity in the Y-Ba-Cu-O system,[1,2] several deposition techniques have succeeded in making the superconducting $YBa_2Cu_3O_{7-x}$ films with transition temperatures above liquid nitrogen temperature.[3-12] Among the thin film deposition techniques, one of the most important ones is the metalorganic chemical vapor deposition (MOCVD), which is currently the dominant method in depositing device quality III-V and II-VI compounds.[13] The method has the advantage of making the thin films in a low vacuum environment and can easily be scaled to production level. We report here our recent research on high temperature superconducting $YBa_2Cu_3O_{7-x}$ films deposited by MOCVD. The results show that after post-annealing, highly c-axis oriented single phase superconducting $YBa_2Cu_3O_{7-x}$ films can be obtained reproducibly on the yttria stabilized zirconia (YSZ) (100) substrates with T_c (R = 0) at 84K. Under similar processing conditions, the films deposited on sapphire ($1\bar{1}02$) also show a single phase $YBa_2Cu_3O_{7-x}$ structure with a very broad superconducting transition reaching the zero resistance at much lower temperature (10-40K).

a) To whom all correspondence should be addressed.

EXPERIMENT

Fig. 1 is a schematic diagram of the MOCVD system used for the deposition of the thin films. Metal β-diketonate complexes of $Y(C_{11}H_{19}O_2)_3$, $Ba(C_{11}H_{19}O_2)_2$ and $Cu(C_{11}H_{19}O_2)_2$ were used as the metalorganic precursors, which will be referred to as $Y(thd)_3$, $Ba(thd)_2$ and $Cu(thd)_2$. Each individual precursor source was placed in a separate stainless steel tube wrapped with heating tapes. An auxiliary outlet was introduced to avoid the deposition from the initial surge and unsteady gas flow. Argon was used as the carrier gas. By carefully controlling the gas flow rates and the source temperatures, the metalorganic precursors were evaporated and carried by argon gas to the reactor after premixing with oxygen gas, which

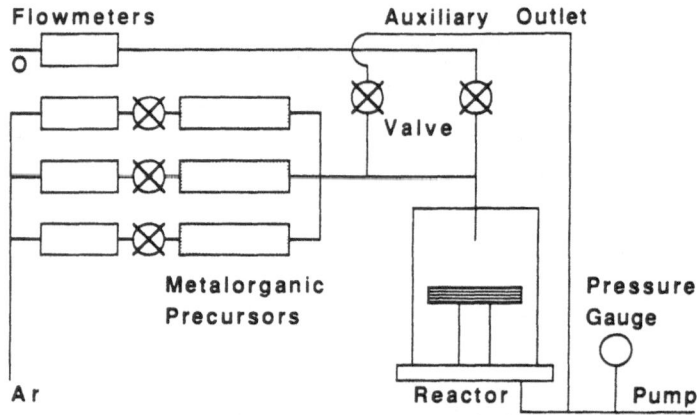

Fig. 1. Schematic diagram of the metalorganic chemical vapor deposition system used for the deposition of the thin films.

was at a flow rate of 1000 sccm. A warm wall vertical reactor made of stainless steel was used with a resistive heating stage for the deposition. The temperature of the susceptor was set at 650°C.

STUDY OF THE DEPOSITION PROCESS

In order to gain some insight into the deposition of $YBa_2Cu_3O_{7-x}$ superconducting films, first we investigated the deposition of Y_2O_3, $BaCO_3$ and CuO films by MOCVD. These films have been successfully deposited on fused quartz substrates. The Y_2O_3 and $BaCO_3$ films deposited on fused quartz substrates were transparent and had shiny surfaces. The CuO films show dark brown color generally with rougher surface under the flow rates and source temperatures investigated. Fig. 2 shows the x-ray diffraction patterns of the Y_2O_3, $BaCO_3$ and CuO thin films obtained from the respective metal β-diketonates when a single component is sent to the reactor at a time. As can be seen, they all show sharp and pronounced diffraction peaks indicating a polycrystalline structure on the quartz substrates.

Fig. 3 shows the mass transport rates as a function of carrier gas flow rates for $Cu(thd)_2$, $Ba(thd)_2$ and $Y(thd)_3$. The pressure was 50 torr on the sources and in the reactor with the Y and Cu compounds set at 160°C and the Ba compound set at 300°C. These experiments show that the mass transport rates are not very sensitive to the carrier gas flow rates for Y and Ba but for Cu the mass transport increases strongly with increasing flow rate. Above 500 sccm, the mass transport is quite insensitive to a change in the flow rate even for Cu. It should be noted that we need a

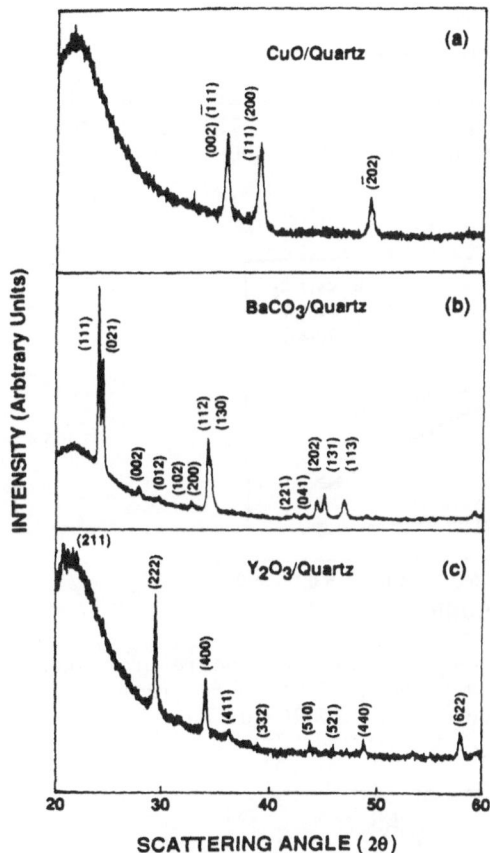

Fig. 2. X-ray diffraction patterns for the films of a) CuO, b) BaCO$_3$ and c) Y$_2$O$_3$ deposited on quartz.

temperature for the Ba precursor which is almost double the temperature needed for Cu and Y to obtain a comparable mass transport. Fig. 4 shows the mass transport rates as a function of reactor pressure. Again, we note that the Ba and Y precursors do not vary much in the pressure range studied in comparison to the more rapid variation for Cu. Fig. 5 shows the mass transport rates in the log scale as a function of the inverse source temperature for the Y, Ba and Cu precursor at a reactor pressure of 50 torr. It is clear from the figure that the Cu and Y precursors are much more volatile in comparison to the Ba precursor and the order of the volatility for these three compounds follows the corresponding ionic radius quite well.[14]

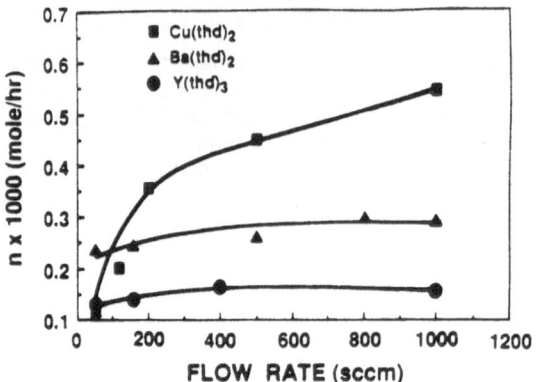

Figure 3. The mass transport rates as a function of carrier gas flow rates for $Cu(thd)_2$, $Ba(thd)_2$ and $Y(thd)_3$. The source temperatures were at 160°C for Y and Cu, and at 300°C for Ba.

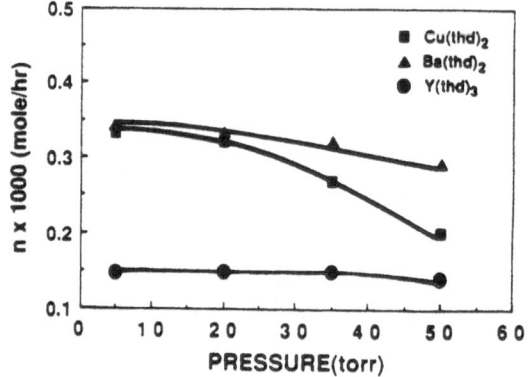

Figure 4. The mass transport rates as a function of the reactor pressure for $Cu(thd)_2$, $Ba(thd)_2$ and $Y(thd)_3$. The source temperatures were at 160°C for Y and Cu, and at 300°C for Ba.

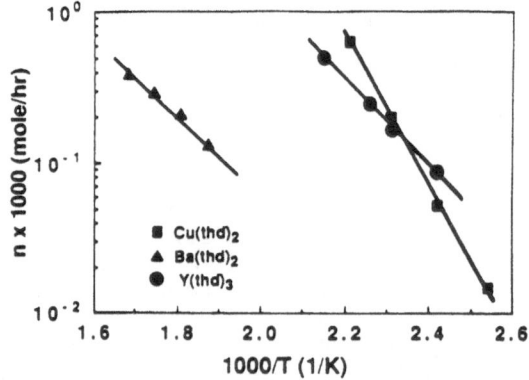

Figure 5. The mass transport rates in the log scale as a function of the inverse source temperature for $Cu(thd)_2$, $Ba(thd)_2$ and $Y(thd)_3$ at a reactor pressure of 50 torr. The flow rates were 125 sccm for Y, 1000 sccm for Ba and 160 sccm for Cu precursors.

The mass transport studies clearly show that among the deposition parameters the control of the source temperature is the most critical because the mass transport rate depends strongly on the source temperature. Based on these preliminary mass transport studies, the source temperatures were set at $160^{\circ}C$, $300^{\circ}C$ and $170^{\circ}C$ for the Y, Ba and Cu compounds, respectively, to deposit Y-Ba-Cu-O films. The flow rates were 125 sccm for Y, 1000 sccm for Ba and 160 sccm for Cu with the susceptor temperature set at $650^{\circ}C$. The reactor pressure was 50 torr. The partial pressures of the Y, Ba and Cu metalorganic precursors were estimated to be 3, 11 and 10 millitorrs, respectively. After one hour of deposition at a rate of 10 μm/h, the films were cooled to room temperature under oxygen flow. The as-deposited films on both YSZ and sapphire were black and had very high resistances (10-1000 KΩ) compared to the films after post annealing (10 ⁓ 500 Ω). The optimum post annealing conditions under oxygen flow were $950^{\circ}C$ for 30 minutes for the films deposited on YSZ substrates, and $895^{\circ}C$ for 15 minutes for the films deposited on sapphire. After the high temperature anneals the samples were cooled slowly (about $4^{\circ}C$/min) to ambient temperature.

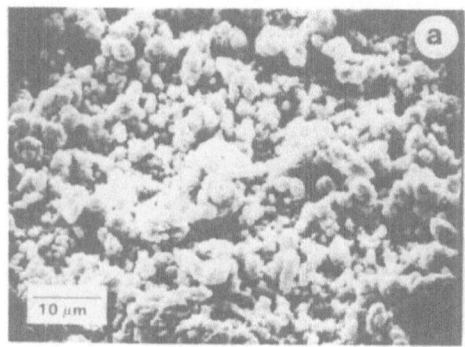 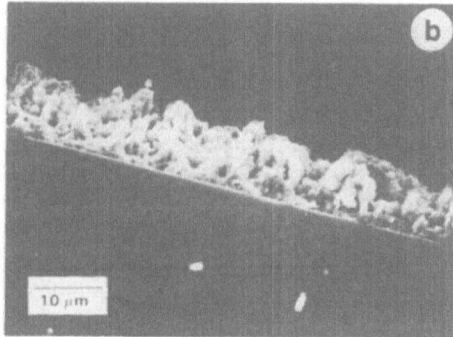

Figure 6.   SEM micrographs of (a) the surface and (b) the cross-section for an as-deposited film of YSZ.

The structure of the films was analyzed by using x-ray diffraction and the surface morphology was examined by scanning electron microscopy (SEM). For x-ray diffraction, a digitized horizontal diffractometer from Siemens with Ni-filtered Cu $K\alpha$ radiation was used. Four-terminal low-frequency ac resistivity measurements were made with indium pressure contacts.

Fig. 6(a) and (b) show surface and cross-section SEM micrographs, respectively, of an as-deposited film on YSZ substrate. As can be seen, the surface is fairly rough and the film consists of an aggregation of micron-size particles. The film seems to be highly porous, possibly due to homogeneous nucleation in the gas phase. Fig. 7(a) and (b) show surface and cross-section SEM micrographs, respectively, of the same film after post annealing at $950^{\circ}C$ for 30 minutes. The post annealing greatly improves the surface morphology and eliminates the porosity, providing a dense film. No cracking or peeling was observed in the films examined with thicknesses as high as 10 μm. Similar results were obtained for the films deposited on sapphire ($1\bar{1}02$) substrates.

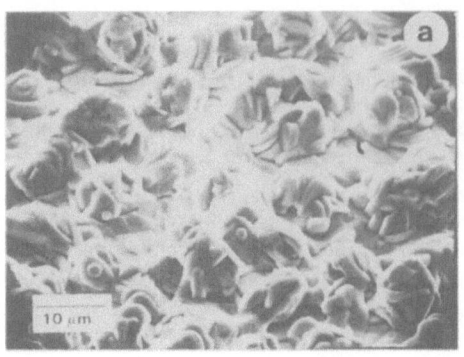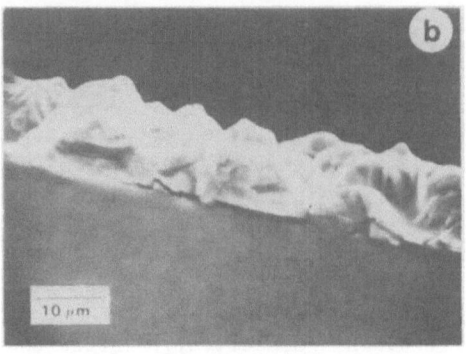

Figure 7. SEM micrographs of (a) the surface and (b) the cross-section for the film deposited on YSZ after post annealing at 950°C for 30 minutes.

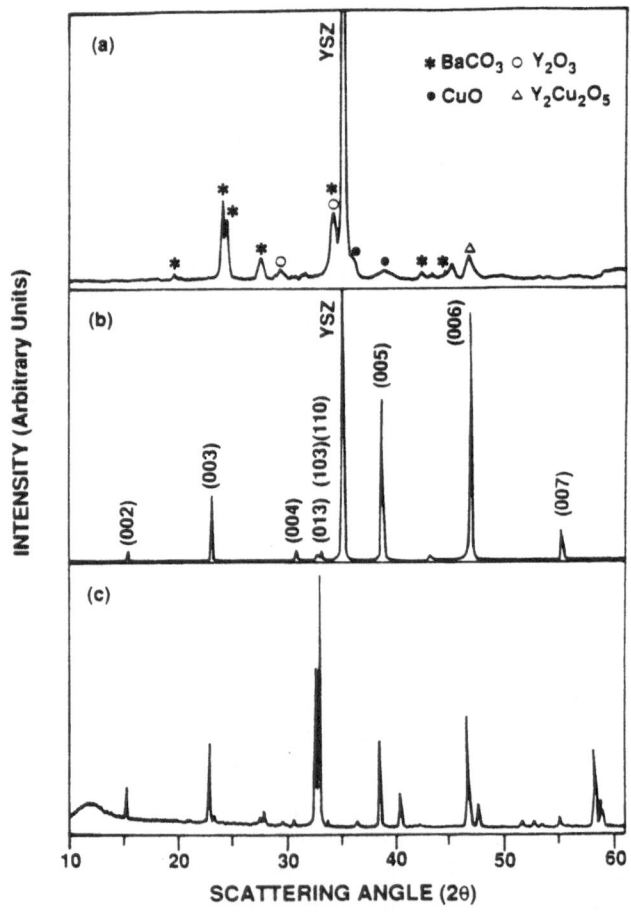

Figure 8. X-ray diffraction patterns for the samples (a) as-deposited on YSZ, (b) post-annealed at 950°C for 30 minutes, and (c) the powdered bulk $YBa_2Cu_3O_{7-x}$ superconductor.

The x-ray diffraction pattern of an as-deposited film on YSZ is shown in Fig. 8(a). This pattern shows that the as-deposited film is a mixture of yttrium oxide, copper oxide and barium carbonate and yttrium copper oxide. Electron microprobe analysis provides evidence that the mixture is homogeneous at submicron level. Similar results were obtained for the films deposited on sapphire. In Fig. 8(b), the diffraction pattern shows that, after post annealing, the film on YSZ has a single phase orthorhombic $YBa_2Cu_3O_{7-x}$ structure. This conclusion is drawn by comparing the thin film pattern in Fig. 8(b) to the pattern in Fig. 8(c) obtained from a powdered bulk superconducting $YBa_2Cu_3O_{7-x}$ sample. The pattern for the film shows that all the (00ℓ) lines have very pronounced intensity, indicating a highly textured structure with the c-axis oriented perpendicular to the substrate surface. It is remarkable that films with thicknesses as large as 10 μm can be grown almost completely textured. Highly textured films are important because of the large critical currents they can carry.[3,11] In Fig. 8(b), there are two extra peaks, which cannot be associated with the $YBa_2Cu_3O_{7-x}$ structure. The peak at 35° comes from the YSZ substrate. The peak at 43° can be attributed to $BaZrO_3$ compound forming at the interface due to the reaction between the film and the substrate.[15] Fig. 9 shows the x-ray diffraction pattern of a film deposited on a sapphire ($\bar{1}102$) after post-annealing at 895°C for 15 minutes. This pattern also shows a well crystallized $YBa_2Cu_3O_{7-x}$ phase in addition to a small amount of CuO.

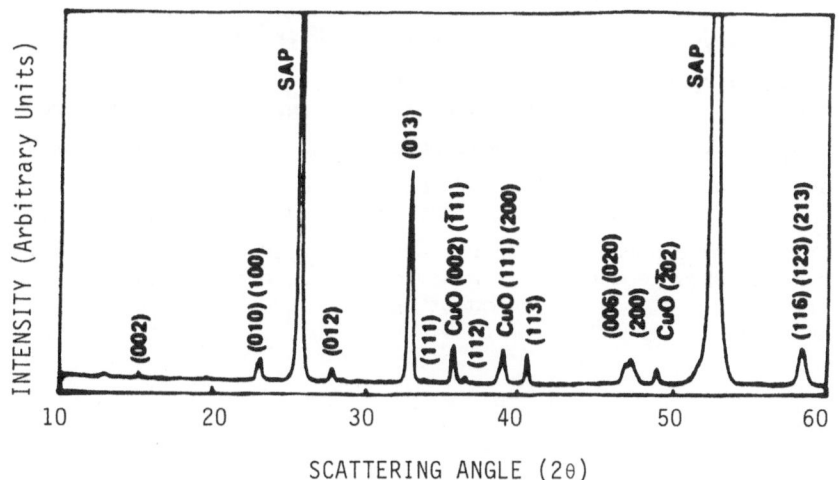

Figure 9.   X-ray diffraction pattern of a $YBa_2Cu_3O_{7-x}$ film deposited on a sapphire ($\bar{1}102$) after post annealing at 895°C for 15 minutes.

Fig. 10 shows the result of the resistivity measurements as a function of temperature for films deposited on YSZ and sapphire. The contact resistances were 20 Ω and 200 Ω for films on YSZ and sapphire, respectively. The film on YSZ in Fig. 10 shows a sharp transition with an onset of 93 K, zero resistance at 84 K, and a width (10-90%) of 4 K. The resistivities of the films on YSZ substrates are about 4.5×10^4 $\mu\Omega$-cm at room temperature. The resistivity of the film on YSZ decreases linearly from room temperature down to the transition onset with a resistivity ratio ρ_{300K}/ρ_{95K} of 2.6 indicating good metallic behavior in the normal state. The resistivity of

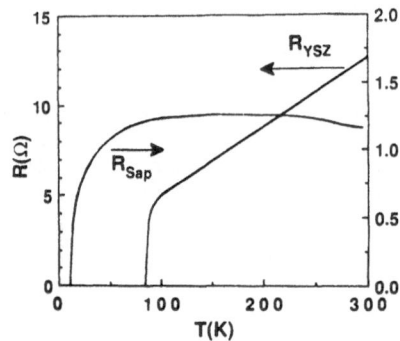

Figure 10. The result of resistivity measurements as a function of temperature for films deposited on YSZ and sapphire.

the film deposited on sapphire exhibits semiconducting behavior in the normal state. This film has a resistance peak value just above the superconducting transition. The zero resistance temperature is as low as 10 K indicating a strong interaction between the sapphire substrate and the film.

In summary, highly textured single phase $YBa_2Cu_3O_{7-x}$ thin films have been obtained by the MOCVD technique on YSZ (100) substrates with thicknesses as large as 10 μm. SEM shows a dense recrystallized homogeneous $YBa_2Cu_3O_{7-x}$ film after post annealing at 950°C for 30 minutes. X-ray diffraction indicates a mixture of metal oxides and carbonates for the as-deposited films and a c-axis highly oriented single phase $YBa_2Cu_3O_{7-x}$ with orthorhombic structure for the post annealed ones. Resistivity measurements show good metallic behavior in the normal state for the film deposited on YSZ with a superconducting onset temperature of 93 K and a zero resistance temperature of 84 K.

ACKNOWLEDGEMENTS

Authors would like to thank J. Johnson and G. Freeman for the SEM analysis, J. Cagle for his assistance in x-ray diffraction, and A. Zangwill for useful discussions. This work is partially supported by U. S. Air Force, Department of Energy, Texas Instruments, Sloan Foundation and Georgia Institute of Technology.

REFERENCES

(1)   J. G. Bednorz and K. A. Muller, Z. Phys. B 64, 189 (1986).
(2)   M. K. Wu, L. R. Ashburn, C. J. Torng, P. H. Hor, R. L. Meng, L. Gao, Z. J. Huang, Y. Q. Wang and C. W. Chu, Phys. Rev. Lett. 58, 908 (1987).

(3)   P. Chaudhari, R. H. Koch, R. B. Laibowitz, T. G. Mcguire, and R. J. Gambino, Phys. Rev. Lett. $\underline{58}$, 2684 (1987).

(4)   M. Naito, R. H. Hammond, B. Oh, M. Hahn, J. W. P. Hsu, P. Rosenthal, A. Marshall, M. R. Beasley, A. Kapitulnik, and T. H. Geballe, J. Mater. Res. $\underline{2}$, 713 (1987).

(5)   R. M. Silver, A. B. Berezin, M. Wendman, and A. L. de Lozanne, Appl. Phys. Lett. $\underline{52}$, 2174 (1988).

(6)   P. Berberich, J. Tate, W. Dietsche, and H. Kinder, Appl. Phys. Lett. $\underline{53}$, 925 (1988).

(7)   A. Inam, M. S. Hegde, X. D. Wu, T. Venkatesan, P. England, P. F. Miceli, E. W. Chase, C. C. Chang, J. M. Tarascon, and J. B. Wachtman, Appl. Phys. Lett. $\underline{53}$, 908 (1988).

(8)   Y. Enomato, T. Murakami, M. Suzuki, and K. Moriwaki, Jpn. J. Appl. Phys. $\underline{26}$, L1248 (1987).

(9)   K. Char, A. D. Kent, A. Kapitulnik, M. R. Beasley, and T. H. Geballe, Appl. Phys. Lett. $\underline{51}$, 1370 (1987).

(10) J. Kwo, T. C. Hsich, R. H. Fleming, M. Hong, S. H. Liou, B. A. Davidson, and L. C. Feldman, Phys. Rev. B $\underline{36}$, 4089 (1987).

(11) S. Shibata, T. Kitagawa, H. Okazaki, and T. Kimura, Jpn. J. Appl. Phys. $\underline{27}$, L646 (1988).

(12) Z. L. Bao, F. R. Wang, Q. P. Kiang, S. Z. Wang, Z. Y. Ye, K. Wu, C. Y. Li, and D. L. Yin, Appl. Phys. Lett. $\underline{51}$, 946 (1987).

(13) P. D. Dapkus, Ann. Rev. Mater. Sci. $\underline{12}$, 243 (1982).

(14) R. E. Sievers and J. E. Sadlowski, Science $\underline{201}$, 217 (1972).

(15) M. J. Cima, J. S. Schneider, S. C. Peterson, and W. Coblenz, Appl. Phys. Lett. $\underline{53}$, 710 (1988).

METAL-ORGANICS FOR CVD OF HT$_c$S

C.I.M.A. Spee and A. Mackor

TNO Institute of Applied Chemistry
P.O. Box 108
3700 AC Zeist, The Netherlands

INTRODUCTION

Chemical vapor deposition CVD has a high potential for preparing thin films of HT$_c$S (YBa$_2$Cu$_3$O$_7$ and Bi-Sr-Ca-Cu-O or Tl-Ca-Ba-Cu-O) on substrates (SrTiO$_3$, ZrO$_2$, MgO or other) by using suitable, volatile metal-organic precursors. As in semiconductor manufacturing, one may expect to establish mild process conditions and to obtain well-controlled microstructure, composition and product properties of the deposited layer in MO-CVD. A particular advantage of CVD over sputtering is its combination of preparation and shaping, e.g. on non-flat substrates. Very recently some groups have shown considerable progress and promise of MO-CVD for the preparation of thin films of Y-Ba-Cu-O [1-9] and Bi-Sr-Ca-Cu-O thin films [10-12].

The investigated HT$_c$S/substrate combinations were:
YBa$_2$Cu$_3$ O$_{7-x}$/SrTiO$_3$ [6,9], ZrO$_2$(:Y) [2-5,7,8,10], MgO [1,9], sapphire [7] or alumina [7] and Bi-Sr-Ca-Cu-O/MgO [11] or YSZ [12].

The following aspects of MO-CVD of HT$_c$S seem to be of great interest:
* with respect to the metal-organic precursors: 1. Selection; 2. Purity; 3. Volatility (vapor pressures); 4. Stability (thermochemistry);
* with respect to the CVD process: 5. Evaporation conditions of MO precursor; 6. Deposition conditions;
* with respect to film quality: 7. Film annealing (temperatures, O$_2$); 8. Film orientation; 9. Film substrate interactions, and 10. Superconductivity.

The results so far show the following:
<u>metal β-diketonates of Y, Cu and the alkaline earth metals Ca, Sr and Ba</u> have been found to be volatile at elevated temperatures.

In particular, some derivatives have been used of pentane-2,4-dione (acetylacetone Hacac), which forms complexes of the following type

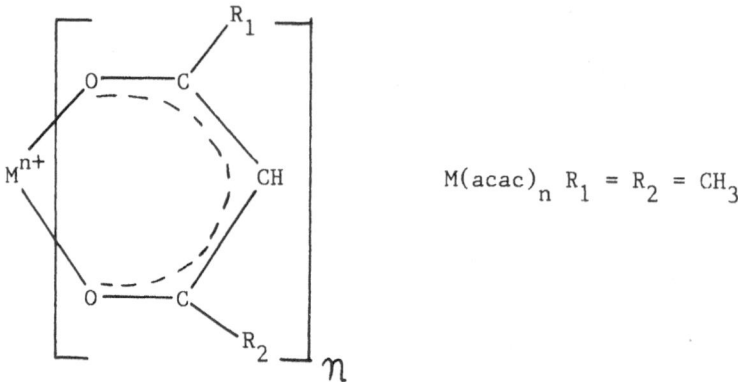

$$M(acac)_n \quad R_1 = R_2 = CH_3$$

With bulky R_1 and R_2 substituents, e.g. $R_1 = R_2$ = tert.-butyl, $-C(CH_3)_3$, one arrives at the 2,2,6,6-tetramethylheptane-3,5-dione (Htmhd or Hthd) ligand = dipivaloylmethane (Hdpm). Metal complexes with this ligand have received most attention [1,2,6,9], while some work has concentrated on complexes of Cu and Ba with the 1,1,1,5,5,5-hexafluoropentane-2,4-dione (Hhfa) ligand [2,9] ($R_1 = R_2 = CF_3$) and for Cu also on its complex with unsubstituted acetylacetone Cu $(acac)_2$.

From this and previous research on the volatility and stability of metal β-diketonate complexes it has become clear that for preparing Y-Ba-Cu-O the most difficult metal to handle is <u>barium</u>. While for <u>Y(thd)</u>$_3$ evaporation (sublimation) temperatures between 125° and 160° C have been reported [1,6,9] and similar temperatures for the various <u>copper</u>(II) bis-(β-diketonate) complexes, Ba(thd)$_2$ is volatile above 200 °C and evaporation (sublimation) temperatures of 240° and 253 °C have been reported [1,6] for its MO-CVD to give BaO deposits. The decomposition of Ba(hfa)$_2$ however at similar temperatures is restricted by decomposition above 260 °C and it has been reported to yield BaF$_2$ [9], which in itself is also a precursor for BaO in HT$_c$S [13].

Other problems with commercially obtained metal β-diketonates are the presence of adhering solvent or free ligand, e.g. Hthd. As these complexes may coordinatively be unsaturated, they may attract oxygen- or nitrogen-containing molecules during preparation, which may contaminate them upon crystallization or even enter the crystal lattice to give e.g. hydrates or other solvation effects. Problems of this kind may have been encountered by Nakamori et al. [10], who report instability of Ba(thd)$_2$ during evaporation and as a result, changing composition of the gas source with deposition time and therefore changes in deposition rate and film decomposition. For example, stepwise hydrolysis may take place to finally yield:

$$Ba(thd)_2 + H_2O \rightleftharpoons Ba(OH)_2 + 2 \; Hthd.$$

Also, the high evaporation temperatures impose high material demands on the MO-CVD equipment. Therefore materials with lower evaporation temperatures and higher thermal stabilities are desirable. The two lighter alkaline earth metals, <u>calcium</u> and <u>strontium</u>, which are presently used in HT_cS, are correspondingly less problematic, which is reflected in lower evaporation temperatures of their thd complexes, reported at 177° and 227 °C, respectively [12].

For <u>bismuth</u>, several options are open. A volatile and rather stable organometallic precursor seems to be triphenylbismuth $BiPh_3$. Other groups have used bismuth alkoxide, i.c. triethoxybismuth $Bi(OC_2H_5)_3$, which however is moisture sensitive [12].

Although <u>thallium</u> β-diketonates are easily accessible and also volatile, the toxicity of thallium compounds, especially in the gas phase, makes their MO-CVD less attractive. The TNO programme, which is outlined in the following, does not include their study at the present time. However, extensive study of the copper and the above-mentioned alkaline earth bis(β-diketonates) will yield valuable information on the preparation of Tl-Ca-Ba-Cu-O. In view of the problems, encountered here and elsewhere, in the handling of MO-CVD precursors, the TNO programme is concentrating in its first phase on these precursors (selection, synthesis and purification, volatility - vapor pressures, stability-thermochemistry). Three classes of precursors are under investigation: metal β-diketonates, metal alkoxides and organometallic precursors with groups like n-alkyl, phenyl, cyclopentadienyl, etc. These precursors must have a vapor pressure of at least 1 torr below 200 °C, without decomposition. Two series of metal β-diketonates are being studied: metal hfa complexes $M(hfa)_n$ (M = Cu, Sr or Ba, n=2; Y, n=3) and the same metals as metal-thd complexes $M(thd)_n$. As a volatile bismuth source, Ph_3Bi may be used up to 180 °C, or $Bi(OEt)_3$. The second step consists of measurements of vapor pressure versus temperature on these precursors. Next is the study of the thermochemical behavior. In particular, the nature of the metal oxide or other deposit and the volatile residues from the organic ligands are a subject of investigation.

For the actual deposition of pure and mixed metal oxides on the substrate ($SrTiO_3$, MgO or other substrate), a CVD reactor is in the final stage of construction.

Sofar, the deposition results in the literature are promising; e.g. Yamane et al. [6] report for $YBa_2Cu_3O_{7-x}$, grown on (100) $SrTiO_3$ by MO-CVD, using the thd complexes, a T_c (R=0) at 84 K and $J_c = 2x10^4$ A/cm^2 at

77K. The films were epitaxial, with the c-axis oriented perpendicular to the substrate surface. For Bi-Sr-Ca-Cu-O on MgO, preliminary results with $T_c(R=0)$ of 78 K have been reported, also with the c-axis perpendicular to the substrate. In these cases the deposition temperatures were 900° and 910 °C, respectively. Post-deposition treatment with oxygen was carried out by slow cooling in an oxygen atmosphere. It is clear from the published work that the MO-CVD procedure has to be improved on the many (ten) points which have been mentioned in this Introduction. Here we report our considerations and initial results.

RESULTS AND DISCUSSION

1. Composition, coordination and association of the metal β-diketonates

The choice of precursors and reactor conditions for MO-CVD is determined by the following requirements:
- availability of a well-defined precursor (composition), giving a
- stable composition of the vapor during the evaporation process,
so that the deposit will be homogeneous in composition, microstructure and properties.

As mentioned in the Introduction, the metal β-diketonates are coordinatively unsaturated, and they will attract oxygen- or nitrogen-containing molecules as extra ligands, such as water, alcohol, pyridine or even non-polar ones, such as benzene, to give stronger or weaker bonds with the metal complex, depending on the nature of the metal ion and the ligand [14].

Yttrium complexes: Although the complex $Y(acac)_3$ is known in the literature [15], like the other rare earth (RE) acetylacetonates, it is probably hygroscopic and it must be kept in a water-free atmosphere. Without special care, using ethanol/water mixtures, one obtains either the trihydrate $Y(acac)_3 \cdot 3H_2O$ (from 60% ethanol) or the monohydrate $Y(acac)_3 \cdot H_2O$ (from 95% ethanol). For the trihydrate the crystal structure is known and it shows an eight-coordinated yttrium ion with three acetylacetonate rings and two water molecules16 [16]. The third water molecule is used for hydrogen bonding of the complex molecules in pairs. Although for the monohydrate we do not know its crystal structure, it is comparable to that of other $RE(acac)_3 \cdot H_2O$ complexes, as shown by powder diffraction [17] and infrared spectroscopy [18]. These compounds appear as dimers, which may be another complication for MO-CVD. By using bulky substituents in the β-diketonate ligand, like tertiary butyl $-C(CH_3)_3$, one may attempt to suppress these interactions and thus obtain a more volatile material and a more constant evaporation rate. Here, several strategies can be followed, among which are:

- rigid drying and removal of all volatile materials from the solid precursor,
- dissolution of the precursor in an organic liquid and removal of all
  impurities from the solution, or
- structural modification of the ligand or the use of other types of ligands.

It is along these lines that we are trying to find suitable precursors
and evaporation conditions.

Alkaline earth complexes (Ca, Sr, Ba): For Ca(acac)$_2$·3H$_2$O the crystal
structure is known [19]; two water molecules are coordinating to give a
hexacoordinate structure. We may assume similar effects in the other
alkaline earth β-diketonates.

Copper(II) complexes: The β-diketonates of Cu^{II} are mononuclear, with
square planar coordination [14]. To what extent further coordination may
interfere with straightforward MO-CVD applications is presently under
investigation in our group (the complexes may contain one or more water
molecules in the solid state).

Thallium(I) complexes: For Tl(hfa) two structural units are present in
the crystal, one is monomeric and the other is a dimer, in which the
four oxygen atoms from the ligands together with two Tl atoms form an
octahedron. Moreover, two such units are linked by weak oxygen bridges
to form a linear polymeric chain, which is easily broken since the compound is quite volatile [20].

Bismuth complexes: Very little is known about bismuth β-diketonates.
Very recently, at least one commercial company [21] offers Bi(thd)$_3$ for
sale (m.p. 114-116 °C; dec. 295 °C; b.p. 150°C/0.05 mm).

EXPERIMENTAL

Vapor pressure measurements are performed by a static method using an
isoteniscope [22,23] (Fig. 1). In this method, the sample is brought
into a sample bulb. If necessary, this can be performed under nitrogen.
While the sample is cooled by a dry ice/acetone bath (-78.5 °C), the
system is brought under a vacuum of less than 0.001 torr. After the system is closed from the vacuum pump, mercury is introduced from a reser-
voir into a U-tube closing off a small volume, containing the sample,
from the remainder of the system. The mercury lock and sample bulb are
then brought to the desired temperature in a thermostat bath. Vapor
pressure is measured by a Pirani manometer by introducing the same pressure as in the sample bulb. This is done by the inlet of nitrogen in

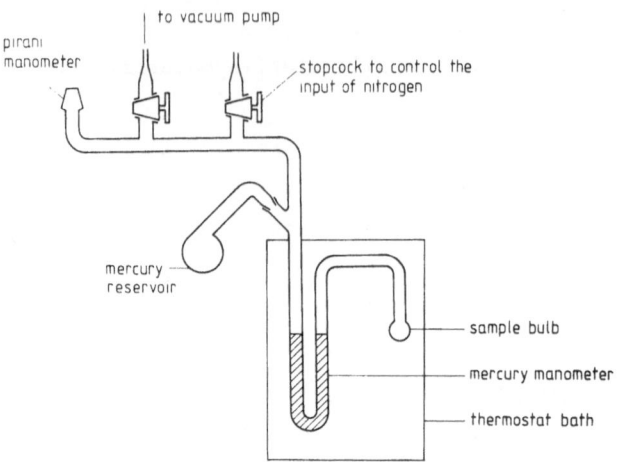

Figure 1. Isoteniscope.

such a quantity as to keep the mercury level in the lock in the same position at both sides of the U-tube. Limitations of this method are a minimum vapor pressure of 0.1 torr, which can be measured, and a maximum temperature of approximately 180 °C because of increasing p(Hg).

The thermochemical behaviour of the single precursors is studied by thermogravimetric analysis (TGA), differential scanning calorimetry (DSC), and in a small CVD-like apparatus (see Fig. 2).

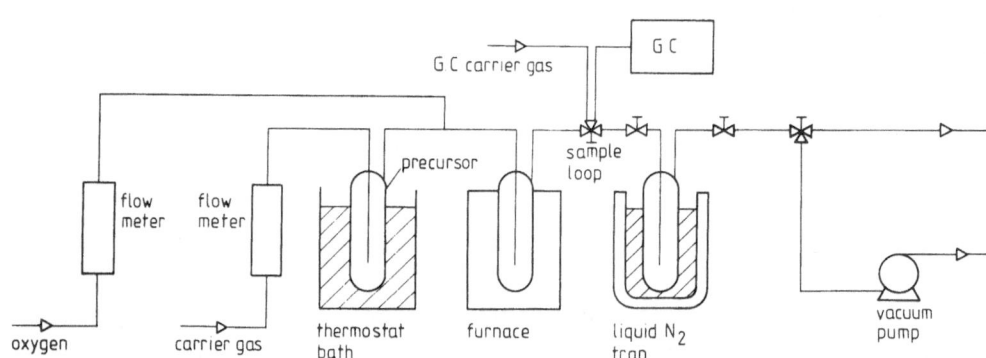

Figure 2. Apparatus for study of thermochemistry of HT_cS precursors.

In this apparatus, the precursors are placed in a thermostatted eva- porator and thus vapor is drawn along the tube to the furnace by a stream of nitrogen. Before entering the furnace, oxygen is added ($N_2:O_2$ = 1:1). The gases resulting from the thermolysis, are analyzed by means of a gas chromatograph or condensed in a trap for further analysis by GC and/or GCMS. Deposits in the reaction tube are examined by elemental analysis techniques (AAS, ICP).

The CVD apparatus is shown in Figures 3 and 4. It consists of a gas-handling system, evaporators and a mixing chamber, where all gases are mixed prior to entering the reactor. All pipes from the evaporators to the reactor can be heated in order to hold the vapor pressures set by

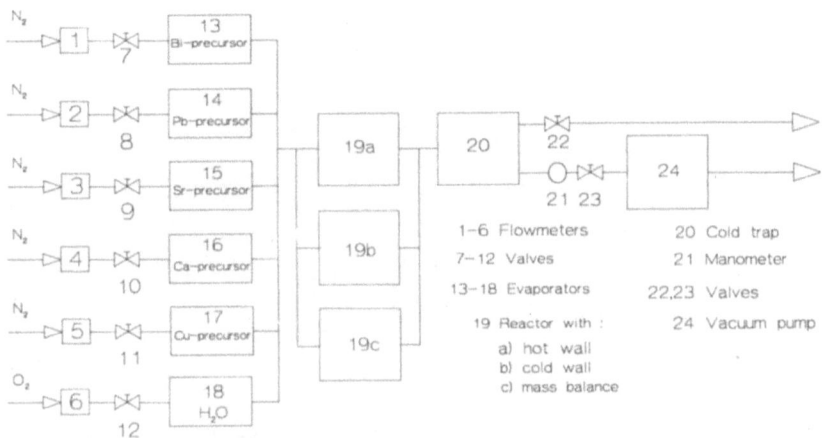

Figure 3. Deposition arrangement shown here for deposition of Bi-Sr-Ca-Cu oxides.

Figure 4. (LP)MOCVD-apparatus under construction.

the evaporators. Reactor configuration can be chosen out of three options, of which the first two have a horizontal reactor arrangement and the third one a vertical arrangement: a) substrate + reaction chamber are heated by a furnace (hot-wall configuration); b) substrate is inductively heated (cold-wall configuration); c) an arrangement, whereby kinetic measurements may be performed (Figure 5) by continuous measurement of mass differences during deposition.

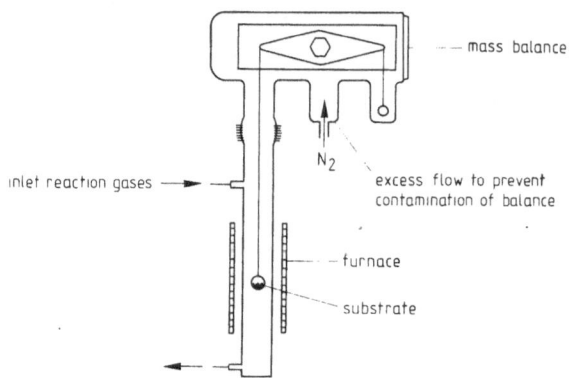

Figure 5. Mass balance arrangement.

PRELIMINARY RESULTS

Since our HT_cS programme has started very recently, only a few experimental results are presently available, besides the construction and testing of equipment.

<u>Vapor pressure measurements</u>

Vapor pressures have been measured for three copper precursors, $Cu(acac)_2$, 1, $Cu(thd)_2$, 2 and $Cu(hfa)_2$, 3. Starting with a commercial sample of 1, already at low temperatures a considerable vapor pressure has been found, which increases only slightly upon further raising the temperature. We also note that at room temperature 1 has the characteristic smell of the free ligand. Therefore, we conclude from both observations that the sample of 1 as obtained contains free acetylacetone as an impurity. This has been removed by heating 1 at 80 °C for 30 minutes at a pressure of 0.0008 torr, prior to the vapor pressure determinations. After this purification, the vapor pressures were determined as shown in Figure 6, series 2, *e.g.* as ca. 3 torr at 150 °C (after correction for the vapor pressure of mercury, as also shown in Figure 6).

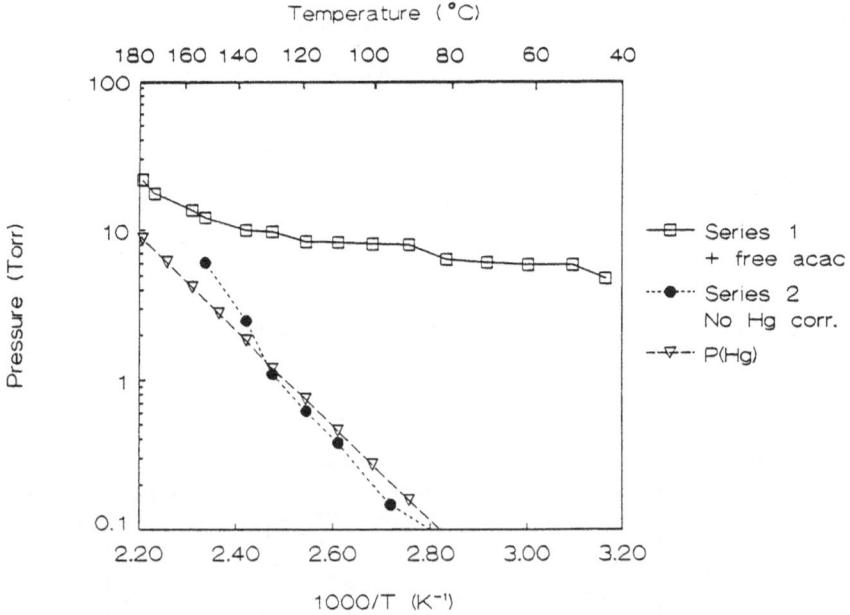

Figure 6. Vapor pressure measurements on Cu(acac)$_2$.

For commercial samples of 2, and 3, the measurements of vapor pressure do not indicate the presence of the free ligand (Figure 7). However, 3 did contain some toluene. Purification was done as for 1.

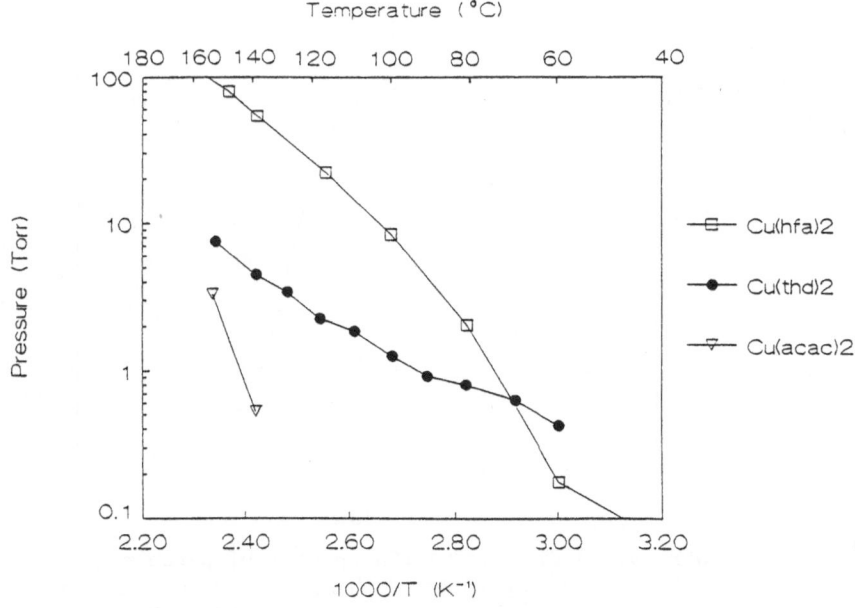

Figure 7. Vapor pressure measurements on Cu(acac)$_2$, Cu(thd)$_2$ and Cu(hfa)$_2$.

Other groups have also recently reported on similar measurements on precursors for CVD of $YBa_2Cu_3O_7$ or on preliminary CVD results cf. Introduction. The fact that the vapor pressures of the metal β-diketonates, as determined on practical samples, are in many (all?) cases larger than those of a pure sample, has not yet been noted in the literature.

Thermochemistry

Preliminary DTA and TGA measurements on some copper(II) β-diketonates [1, 3 and $Cu(tfa)_2$ *, 4] showed different features. For anhydrous 1 (cf.

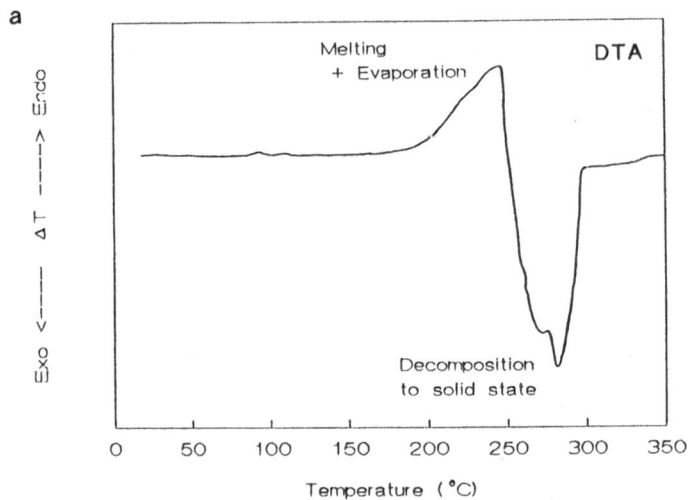

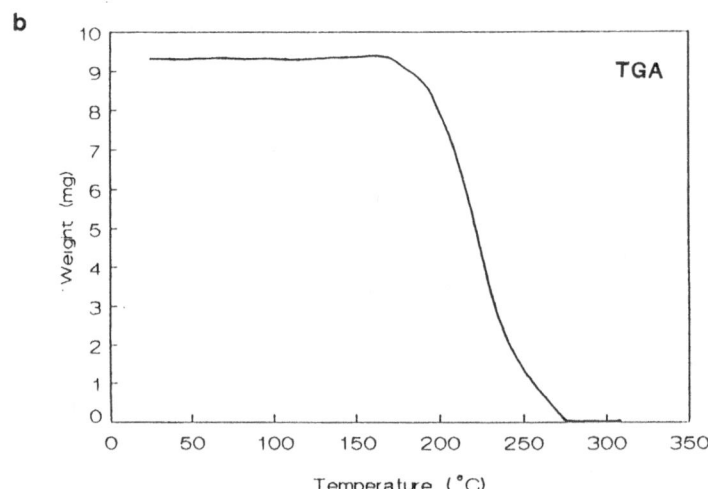

Figure 8. a) DTA and b) TGA spectra of $Cu(acac)_2$. Heating rate 6 °C/min.

---

*Htfa = 1,1,1-trifluoro-2,4 pentanedione.

Figure 8), we observe sublimation from 200 °C upward. Above 250 °C, a strongly exothermic decomposition occurs up to 300 °C, where no more weight losses occur. This thermochemical behaviour is different from that which has been observed by Murray et al. [25]. These authors only observe the evaporation and not the decomposition reaction. This difference may be due to a difference in heating rates and/or air flows. However, Jasim et al. [24] also report cupric oxide formation upon decomposition of 1 in air, which is consistent with other observations, e.g. ref. 2. For 3, we observe a phase transition at 177 °C and melting around 200 °C, followed by increasing evaporation without decomposition at higher temperatures (up to 270 °C), as shown in Figure 9. Compound 2 sublimes from 175 °C and melts at 198 °C, which is consistent with Fig. 10.

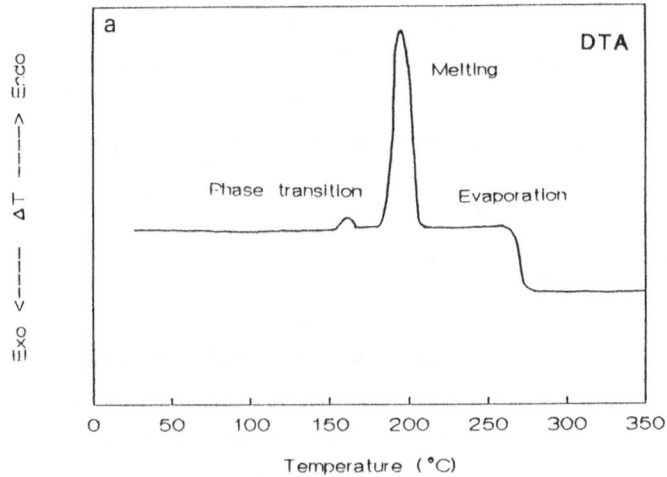

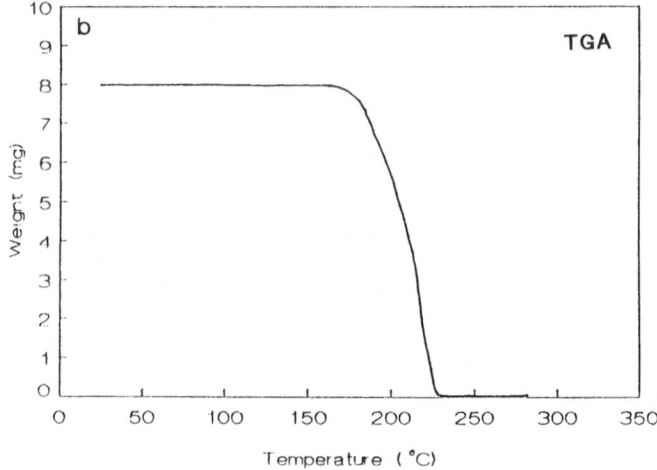

Figure 9. a) DTA and b) TGA spectra of Cu(thd)$_2$. Heating rate 6 °C/min.

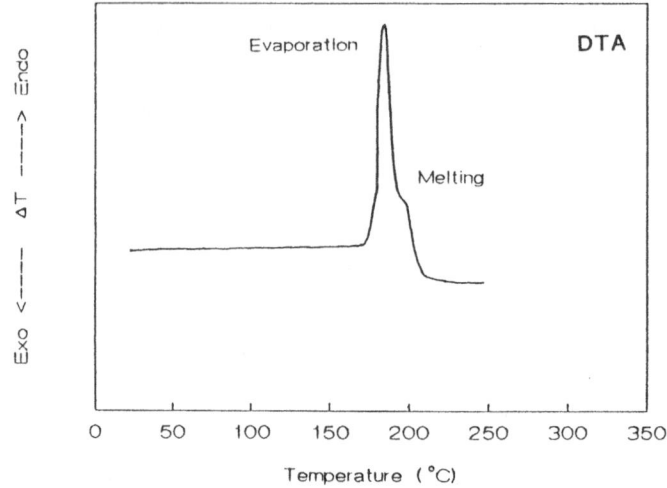

Figure 10. DTA spectra of Cu(tfa)$_2$. Heating rate 6 °C/min.

The first results with the evaporation/decomposition equipment of Figure 2, using solid Cu(thd)$_2$, show the formation of carbon-contaminated CuO. The carbon content has been lowered from 20% to 0.7% by adding oxygen ($N_2/O_2 = 1/1$) to the gas mixture in the reaction chamber.

ACKNOWLEDGMENT

We wish to acknowledge valuable contributions by Mr. K. Timmer (purification), by Mr. J.L. Linden and Mrs. J.G. Kraaykamp (vapor pressure measurements), Mrs. E.A. van der Zouwen-Assink and Mr. G.J. de Koning (thermochemistry) and discussions with Dr.Ir. G. Hakvoort and Prof. Dr. J. Schoonman - Technical University Delft and our colleague Dr. H.A. Meinema. This study is financially supported by the Netherlands National R&D Programme on HT$_c$S.

REFERENCES

1. A.D. Berry, D.K. Gaskill, R.T. Holm, E.J. Cukauskas, R. Kaplan, and R.L. Henry, Formation of high T$_c$ superconducting films by organometallic chemical vapor deposition, Appl. Phys. Lett. 52: 1743 (1988).
2. H. Suhr, Ch. Oehr, H. Holzschuh, F. Schmaderer, G. Wahl, Th. Kruck, and A. Kinnen, Thermal and plasma enhanced CVD of HT$_c$-superconductors, Physica C 153-155: 784 (1988).

3. C. Gonzalez-O., H. Schachner, H. Tippmann, and F.J. Trojer, Superconducting thin films of Y-Ba-Cu-oxide by sol-gel and CVD methods, <u>Physica</u> C 153-155: 1042 (1988).

4. H. Yamane, H. Kurosawa, H. Iwasaki, N. Kobayashi, and Y. Muto, Temperature Dependence of Electrical Resistivity for Y-Ba-Cu-O films prepared by CVD, <u>J. Ceram. Soc. Jpn Int. Ed.</u> 96: 776 (1988).

5. H. Yamane, H. Kurosawa, H. Iwasaki, N. Kobayashi, and Y. Muto, T_c of c-Axis-Oriented Y-Ba-Cu-O films prepared by CVD, <u>Jap. J. Appl. Phys.</u> 27: L1275 (1988).

6. H. Yamane, H. Masumoto, T. Hirai, H. Iwasaki, K. Watanabe, N. Kobayashi, and Y. Muto, Y-Ba-Cu-O superconducting films prepared on $SrTiO_3$ substrates by chemical vapor deposition, <u>Appl. Phys. Lett.</u>, submitted for publication.

7. H. Kurosawa, H. Yamane, H. Masumoto, and T. Hirai, Preparation of the c-Axis-Oriented Films of $YBa_2Cu_3O_{7-x}$ by Chemical Vapor Deposition, <u>Sci. Lett.</u>, in the press.

8. H. Yamane, H. Kurosawa, and T. Hirai, Preparation of $YBa_2Cu_3O_{7-x}$ films by Chemical Vapor Deposition, <u>Chem. Lett.</u> 939 (1988).

9. K. Shinohara, F. Munakata, and M. Yamanaka, (a) High-T_c superconductor prepared by chemical vapor deposition, <u>Proc. First Int. Symp. Superconductors Superconductivity</u>, Nagoya, PTF-7 (1988) and (b) Preparation of Y-Ba-Cu-O Superconducting Thin Film by Chemical Vapor Deposition, <u>Jap. J. Appl. Phys.</u> 27: L1683 (1988).

10. T. Nakamori, H. Abe, T. Kanamori, and S. Shibata, Superconducting Y-Ba-Cu-O Oxide Films by OMCVD, <u>Jap. J. Appl. Phys.</u> 27: L1265 (1988).

11. H. Yamane, H. Kurosawa, T. Hirai, H. Iwasaki, N. Kobayashi, and Y. Muto, Formation of Bismuth Strontium Calcium Copper Oxide Superconducting Films by Chemical Vapor Deposition, <u>Jap. J. Appl. Phys.</u> 27: L0000 (1988).

12. H. Yamane, H. Kurosawa, and T. Hirai, Preparation of Bi-Sr-Ca-Cu-O Films by Chemical Vapor Deposition with Metal Chelate and Alkoxide, <u>Chem. Lett.</u>, submitted for publication.

13. E.M. Logothetis, R.E. Soltis, R.M. Ager, W. Win, C.J. McEwan, K. Chang, J.T. Chen, T. Kushida, and L.E. Wenger, Deposition and characterization of superconducting YBaCuO films, <u>Physica C</u>, 153-155:1439 (1988).

14. A.R. Siedle, Diketones and related ligands, in: Comprehensive Coordination Chemistry, R.D. Gillard and J.A. McCleverty (eds.), Pergamon Press, Oxford (1987), Vol. 2, p. 365.

15. M.F. Richardson, W.F. Wagner, and D.E. Sands, Anhydrous and Hydrated Rare Earth Acetylacetonates and Their Infrared Spectra, Inorg. Chem. 7: 2495 (1968).

16. J.A. Cunningham, D.E. Sands, and W.F. Wagner, The Crystal and Molecular Structure of Yttrium Acetylacetonate Trihydrate, Inorg. Chem. 6: 499 (1967).

17. J.A. Cunningham, D.E. Sands, W.F. Wagner, and M.F. Richardson, The Crystal and Molecular Structure of Ytterbium Acetylacetonate Monohydrate, Inorg. Chem. 8: 22 (1969).

18. M.F. Richardson, W.F. Wagner, and D.E. Sands, Anhydrous and Hydrated Rare Earth Acetylacetonates and Their Infrared Spectra, Inorg. Chem. 7: 2495 (1968).

19. J.J. Sahbari and M.M. Olmstead, Structure of cis-bis(acetylacetonato)diaquacalcium monohydrate $[Ca(C_5H_7O_2)_2(H_2O)_2].H_2O$, Acta Cryst. C 39: 208 (1983).

20. S. Tachiyashiki, H. Nakayama, R. Kurodo, S. Sato, and Y. Saito, Hexafluoroacetylacetonatothallium (I), Acta Cryst. B 31: 1483 (1975).

21. Strem Chemicals, Inc.: Newburyport, Maryland, USA.

22. A. Weissberger, Physical Methods, Interscience, New York (1960), Part 1, p. 434.

23. A. Smith, and A.W.C. Menzies, Studies in vapor pressure: III. Static methods for determining the vapor pressures of solids and liquids, J. Amer. Chem. Soc. 32: 1412 (1910).

24. F. Jasim, and I. Hamid, Thermoanalysis and catalytic study of transition metal acetylacetonates, Thermochim. Acta 93: 65 (1985).

25. J.P. Murray, and J.O. Hill, DSC determination of the sublimation enthalpy of tris(2,4-pentanedionato)cobalt(III) and bis(2,4-pentanedionato)nickel(II) and -copper(II), Thermochim. Acta 109: 383 (1987).

ORGANOMETALLIC CHEMICAL VAPOR DEPOSITION OF SUPERCONDUCTING

YBaCuO FILMS AND POST-DEPOSITION PROCESSING

J. Zhao, H. O. Marcy, L. M. Tonge, B. W. Wessels, T. J.
Marks, and C. R. Kannewurf

Materials Research Center
Northwestern University
Evanston, IL 60208

INTRODUCTION

Since the recent discovery of materials that exhibit superconductivity above 90K, there has been much interest in the preparation of thin films and coatings of these materials for numerous electronic applications. To date, most of the effort in thin film preparation has centered on physical vapor deposition techniques. However, chemical vapor deposition (CVD), although not yet widely investigated for preparation of these materials, may offer several important advantages. These advantages could include simplified deposition apparatus, excellent film uniformity and compositional control, high deposition rates, and the ability to coat complex shapes. As an example, these advantages suggest that CVD would be particularly well-suited for the large-scale preparation of wires and ribbons. Despite the attractions of CVD for the fabrication of high T_c superconducting films, very few experiments have actually been reported. This situation reflects, in large part, a paucity of suitably volatile, stable barium sources. However, recent reports have shown that metal-organic complexes, especially β-diketonate complexes, are suitable sources of barium. Thus, Ba(dpm)$_2$ (dpm = dipivaloylmethanate) has been successfully used in the CVD of YBa$_2$Cu$_3$O$_{7-\delta}$ films.[1-3] Indeed, films with the onset of superconductivity at 90K and zero resistance as high as 80K have been recently prepared.[4] While these deposition experiments utilizing Ba(dpm)$_2$ have yielded encouraging preliminary results, there are alternative Ba precursors which may yield films with improved properties at lower deposition temperatures and/or with higher deposition rates and yields. A strategy for increasing the vapor pressure of a precursor is to form complexes that involve bulky fluorocarbon ligands.[5,6] The introduction of fluorinated precursors also provides a source of BaF$_2$ which has been previously shown to promote epitaxy in other types of YBa$_2$Cu$_3$O$_{7-\delta}$ deposition processes and ultimately leads to sharpening of the transition to the superconducting state.[7]

Using the above strategy, we recently reported the organometallic chemical vapor deposition (OMCVD) of high-T_c superconducting YBa$_2$Cu$_3$O$_{7-\delta}$ films using the fluorocarbon-based precursor Ba(fod)$_2$ (fod = heptafluorodimethyloctanedionate).[8] Films were prepared that exhibited the onset of superconductivity at 90K and zero resistance by 66K. In this paper, we report additional recent experiments on YBa$_2$Cu$_3$O$_{7-\delta}$ OMCVD, including the effects of post-deposition annealing and how it relates to increasing the zero-resistance temperature.

EXPERIMENTAL

OMCVD at low pressure was carried out in a horizontal quartz reactor having separate parallel quartz inlet tubes for the introduction of the precursors. The carbon susceptor on which the substrate was placed was heated by an infrared lamp. Deposition was carried out at a system pressure of 5 torr with an argon carrier gas transporting each precursor to the substrate region. For yttrium and copper precursors, the complexes $Y(dpm)_3$ and $Cu(acac)_2$ (acac = acetylacetonate) were used. For barium, $Ba(fod)_2$, prepared for this work by the reaction of $Ba(OH)_2$ and Hfod, was used. The $Ba(fod)_2$ was found to be more volatile than $Ba(dpm)_2$ and can be transported at lower temperatures. Decomposition of the $Ba(fod)_2$ during deposition was also less severe. Source temperatures of 150°, 100°, and 170°C were employed for the $Cu(acac)_2$, $Y(dpm)_3$, and $Ba(fod)_2$, respectively. Oxidants studied included H_2O, $O_2 + H_2O$, and tetrahydrofuran (THF). Deposition was carried out at a substrate temperature of 700°C–850°C. Substrates investigated included $SrTiO_3$, MgO, and yttrium-stabilized zirconia (YSZ). Film thicknesses were 0.5-2.0 μm and typical deposition rates were 10-30 nm min^{-1}. The superconducting phase is obtained after a high temperature anneal above 900°

RESULTS AND DISCUSSION

Fig. 1 shows a typical X-ray diffraction pattern of an OMCVD-derived film prior to annealing. The X-ray pattern indicates that the as-deposited films are essentially amorphous. Only traces of crystalline BaF_2 amd Y_2O_3 are observed. The as-deposited films are dark brown and have a featureless morphology as determined by optical microscopy. Such films are also highly resistive. In order to obtain superconducting material, the films were annealed in flowing oxygen at temperatures in excess of 850°C. Initially, experimental samples were annealed for 10 hours at 600°C, 1.5 hours at 900°C, and 10 min. at 950°C, followed by slow cooling in an O_2 stream. The long anneal at 600°C was carried out to remove any carbonaceous materials from the as-deposited films. The subsequent annealing steps were necessary to form the $YBa_2Cu_3O_{7-\delta}$ phase with the appropriate oxygen content. From the X-ray diffraction measurements, the annealed film was determined to be a single ortho-rhombic phase with a lattice constant in the $\underline{c}$ direction of 11.68 ± 0.03Å. This value of the lattice constant indicates that δ is less than 0.3 and that the films thus have the appropriate oxygen concentration.[9] Fig. 2 shows the resistance versus temperature dependence of an OMCVD film deposited on MgO after post-deposition annealing using the aforementioned conditions. As is characteristic of suitably annealed samples, charge transport at temperatures above 100K is metallic. A sharp onset of the transition to the superconducting state is observed at 90K with zero resistance being obtained at 66K. However, films deposited on $SrTiO_3$ exhibit a broad transition to the superconducting state with the best films deposited on this substrate having a zero resistance at 30K. The strong dependence of critical temperature on substrate indicates that substrate-film interactions are playing a major role in determining the superconducting properties, presumably as a result of interdiffusion and interfacial reactions.[10] Such reactions are believed to occur during the high temperature annealing step.

To eliminate substrate-superconductor reactions, rapid thermal annealing of the as-deposited films has been investigated. Fig. 3a shows a sample deposited on YSZ that was annealed according to the following protocol: 10 hours at 600°C, 1.5 hours at 900°C, and 10 min at 960°, followed by slow cooling. For this sample, zero resistance is observed at 60K. If a high temperature, rapid annealing process is utilized by initially heating to 870°C in $O_2 + H_2O$, subsequently annealing in pure O_2

296

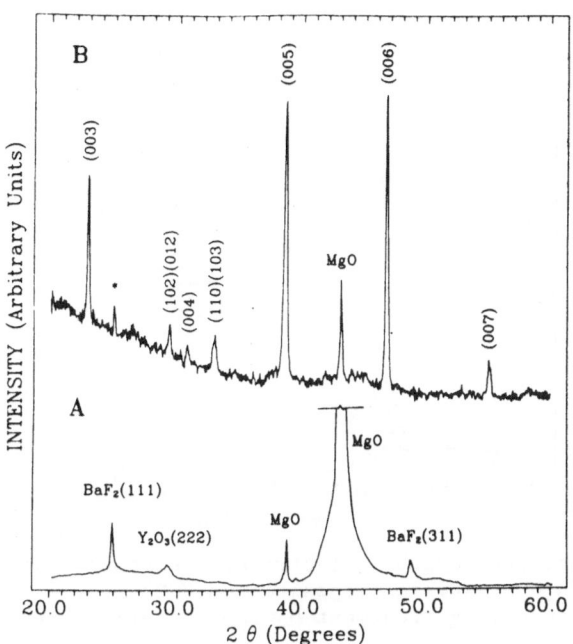

Fig. 1. X-ray diffraction pattern of an OMCVD film prepared at
700°C a) prior to annealing and b) after annealing. The *
denotes BaF$_2$

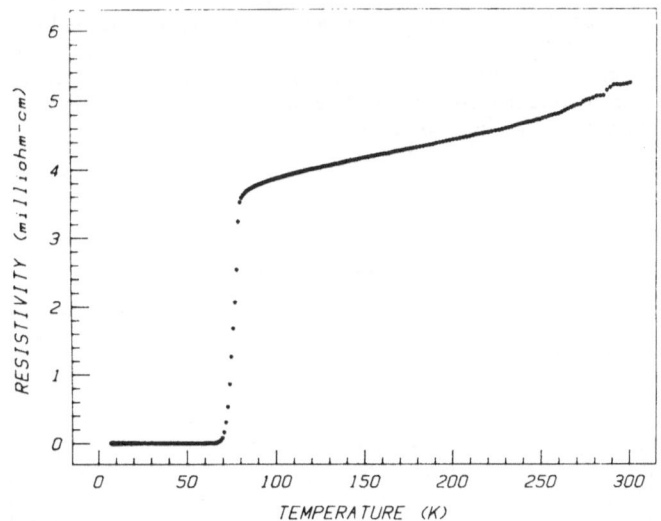

Fig. 2. Variable-temperature four-probe electrical resistivity data
for a YBa$_2$Cu$_3$O$_{7-\delta}$ film deposited on [100] MgO substrate.

at 870°C for 1.5 hours, and then heating to 980°C for one second, films
with improved superconducting properties are obtained as indicated in
Fig. 3b. For this sample, which was deposited on YSZ, zero resistance
is observed at 70K. It should be noted that if the rapid thermal
annealing step is not used, the samples are not superconducting. Moreover, if the sample is subjected to a thirty-second anneal instead of a

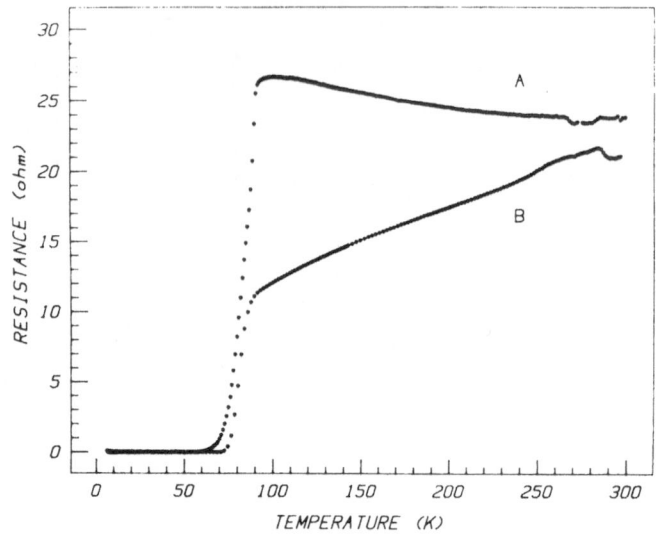

Fig. 3. Variable temperature electrical resistance data for
YBa$_2$Cu$_3$O$_{7-\delta}$ samples deposited on YSZ after a) a standard
post-annealing treatment b) rapid thermal annealing for one
second.

one-second anneal, considerable grain growth occurs with grains as large
as 40 μm in length and 3 μm in width being observed. Such samples show
an onset to the superconducting state at 90K and a narrowed 10%-90%
transition width of 8K, although the material never does become fully
superconducting. This presumably results from the formation of a
discontinuous grain structure. It is apparent that annealing processes
need to be further optimized if superconducting films with the desired
transition widths are to be obtained.

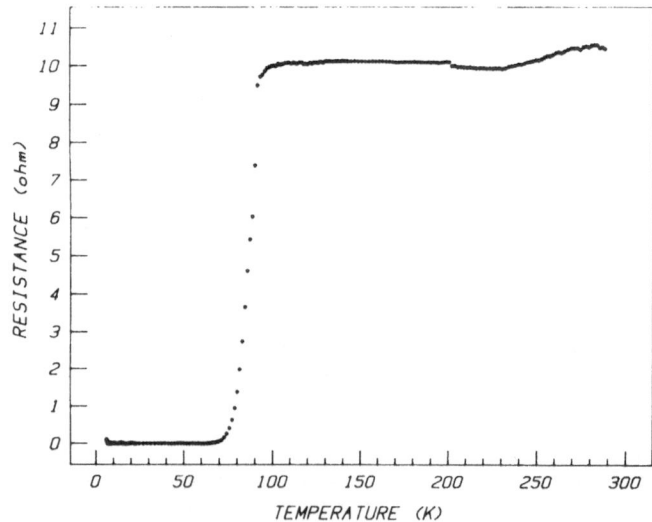

Fig. 4.   Variable temperature electrical resistance data for
YBa$_2$Cu$_3$O$_{7-\delta}$ films deposited on YSZ where THF was used as the
oxidant.

It was also noted in our studies that the required post-deposition treatment depends on the oxidant used during chemical vapor deposition. While for films deposited using H_2O, annealing, at 600°C for 6 hours prior to the high temperature step is sufficient, for films deposited using THF as the oxidant, low temperature anneals at 700°C of 10 hours in duration are required. Nevertheless, films with suitable superconducting properties can be obtained after high temperature annealing, as indicated in Fig. 4.

SUMMARY

Films of $YBa_2Cu_3O_{7-\delta}$ have been prepared by organometallic vapor deposition using a barium fluorocarbon-based precursor. Films deposited on YSZ and MgO have yielded the best electrical properties with a zero-resistance transition temperature of 70K being observed. Post-deposition annealing of the as-deposited films is required to obtain the super-conducting phase. Rapid thermal annealing leads to films with improved electrical properties.

ACKNOWLEDGMENTS

This research was supported by the National Science Foundation through the Northwestern Materials Research Center (grant DMR8520280) and in part by the Office of Naval Research. We thank Mr. X. K. Wong for helpful discussions.

REFERENCES

1. A. D. Berry, D. K. Gaskill, R. T. Holm, E. J. Cukauskas, R. Kaplan, and R. L. Henry, Appl. Phys. Lett. 52:1743 (1988).
2. H. Abe, T. Tsuruoka, and T. Nakamori, Jap. J. Appl. Phys. 27:L1473 (1988).
3. J. Zhao, K. Dahmen, H. O. Marcy, L. M. Tonge, B. W.Wessels, T. J. Marks and C. R. Kannewurf, Solid State Commun., in press.
4. H. Yamane, H. Masumoto, T. Hirai, H. Iwasaki, K. Watanabe, N. Kobayashi, Y. Muto, and H. Kurosawa, Appl. Phys. Lett. 53:1548 (1988).
5. R. E. Sievers and J. E. Sadlowski, Science 201:217 (1978) and references therein.
6. R. Belcher, C. R. Cranley, J. R. Majer, W. I. Stepton, and P. C. Uden, Anal. Chim. Acta 60:109 (1972).
7. G. Gupta, R. Jagannathan, E. I. Cooper, E. A. Geiss, J. I. Landman, and B. W. Hussey, Appl. Phys. Lett. 52:2077 (1988).
8. J. Zhao, K. Dahmen, H. O. Marcy, L. M. Tonge, T. J. Marks, B. W. Wessels, and C. R. Kannewurf, Appl. Phys. Lett. 53:1750 (1988).
9. R. J. Cava, B. Batlogg, C. H. Chen, E. A. Reitman, S. M. Zahurak, and D. Werder, Phys. Rev. B 36:5719 (1987).
10. H. Nakajima, S. Yamaguchi, K. Iwasaki, H. Morita, H. Fujori, and Y. Kujino, Appl. Phys. Lett. 53:1437 (1988).

MELT GROWTH OF Bi–Sr–Ca–Cu–O SUPERCONDUCTING

SHEETS AND FILAMENTS

T. F. Ciszek and C. D. Evans

Solar Energy Research Institute
1617 Cole Blvd., Golden, CO 80401, USA

ABSTRACT

We utilized a liquid-phase process for the solidification of high-T_c superconducting sheets and filaments from a melted mixture of the oxide powders Bi_2O_3, SrO, CaO, and CuO in approximately 1:2:2:2 molar ratios. Alumina sheets up to 13-mm wide and various filament materials with diameters ranging from 100 μm to 2.5 mm were coated with thin Bi-Sr-Ca-Cu-O layers by passage through the liquid mixed oxides and subsequent solidification. Coating rates between 2 and 1800 mm/min were used, and film thicknesses between 5 μm and 500 μm were attained. After annealing at 875°C for several hours, the coatings exhibited a superconducting transition onset near 80 K. We also report on a Bi-Sr-Ca-Cu-O wire casting technique.

INTRODUCTION

Most previous work in the Bi-Sr-Ca-Cu-O superconducting materials system has concentrated on sinter reacting pressed pellets of the source oxides and carbonates at relatively low temperatures near 860°C or on a variety of thin film deposition techniques at < 860°C. We present a method that utilizes high (>1100°C) reaction temperatures to completely liquefy the mixed oxide powders (no carbonates are used). Advantages of the high temperature melt process that we reported earlier include the ability to make dense specimens and to obtain single crystal superconducting platelets up to about 4-mm in width[1]. From analysis of the real, χ', and imaginary, χ'', components of ac magnetic susceptibility[2], along with resistivity measurements, we also found that there appears to be virtually lossless intergrain coupling and good connectivity in these Bi-Sr-Ca-Cu-O superconductors formed from the liquid phase as compared to those formed by sintering. The superconducting phase predominantly obtained by this process has composition $Bi_2Sr_2Ca_{0.8}Cu_2O_8$ (determined by electron-beam microprobe analysis of single crystals) and an orthorhombic/pseudotetragonal crystal structure with a $\approx$ 0.54 nm, b $\approx$ 0.54 nm, and c $\approx$ 3.06 nm (determined by x-ray diffraction and TEM selected-area diffraction)[1].

Here, we present a simple method for sheet and filament formation based on the liquid-phase process. Some of the geometries that have been produced by this technique are shown in the photograph of Figure 1. A 13-mm-wide

Fig. 1. Photograph of a 13-mm-wide sheet, 0.8 and 1.3-mm diameter rods, and a 100-μm-diameter wire coated with Bi-Sr-Ca-Cu-O.

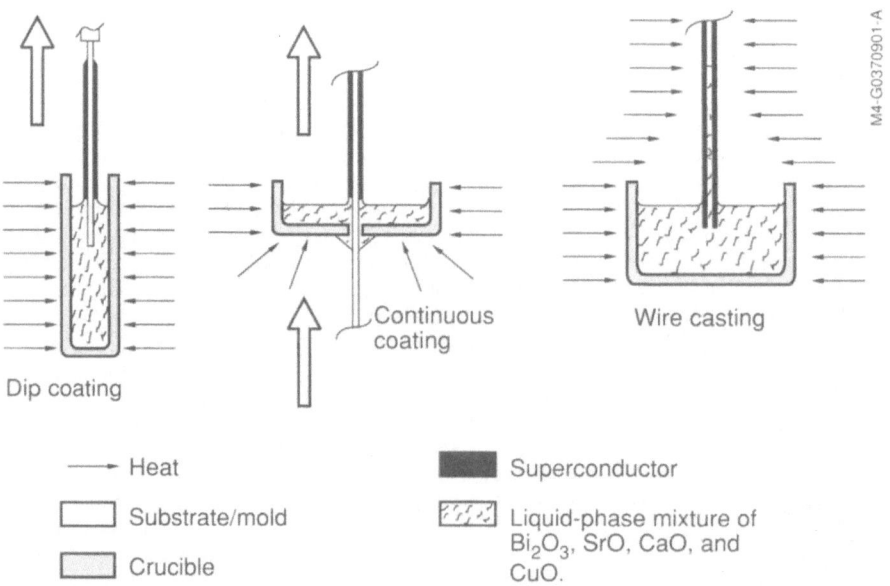

Fig. 2. Schematics of the processes used for dip coating (left), continuous coating (middle), and wire casting (right).

sheet, 0.8 and 1.3-mm-diameter rods, and a 100-μm-diameter wire are depicted. Wire diameters as small as 25 μm have also been obtained.

EXPERIMENTAL

Schematics of the techniques we used to coat sheets, rods, and wires appear in Figure 2. The dip-coating technique (Figure 2a) was used to make short coated lengths of sheet, rod, and wire geometries. Speeds between 2 and 1800 mm/min were used to withdraw the coated substrates. Only finite lengths can be made by this process, the limit being determined by the crucible depth. Figure 2b shows a continuous process, where the substrate is fed into the crucible, coated, and then passes out. We have applied this technique to wires and rods, but not sheets as yet. Figure 2c shows a casting process. We used hollow alumina cylinders and drew the liquid oxide mixture up into the warm tube where it was solidified when removed from the heated environment. This, like the process in Figure 2a, is a finite-length technique.

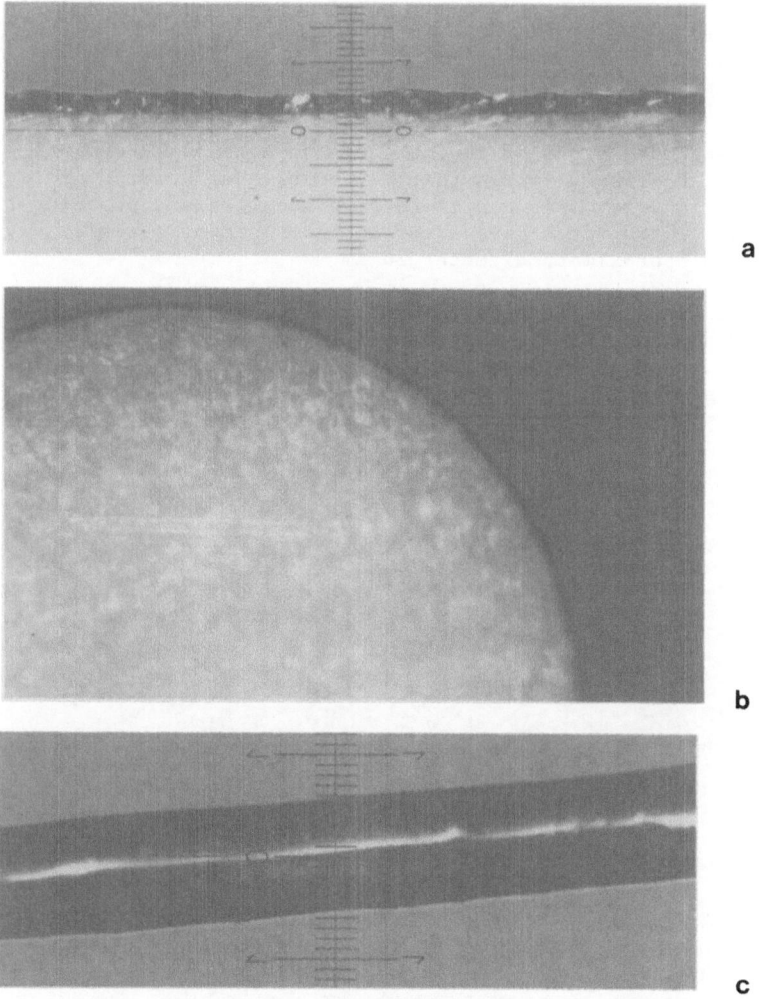

Fig. 3. Cross sections of an alumina sheet (a) and a 0.8-mm diameter alumina rod (b) and a plan view of a 100-μm diameter Pt wire (c) coated with Bi-Sr-Ca-Cu-O. The smallest scale division in a and c is 10 μm.

In all cases, crucibles composed of 99.8% alumina were used. The starting materials were Bi_2O_3, SrO, CaO, and CuO with ACS-grade or higher purity. A typical feed composition consisted of molar ratios Bi_2O_3:SrO:CaO: CuO = 1:2:2:2. In some cases, PbO was also added and the molar ratios were Bi_2O_3:PbO:SrO:CaO:CuO = 3:2:6:6:8. The oxide powders were ground and mixed before placement in the crucible. After heating to ≈1150°C and reaction of the oxides, the substrates were inserted in the liquid for coating. Alumina, Pt, and other metallic wires were used as substrates. In most cases, the coated substrates were not superconducting at 77 K in the as-grown state, but required post-growth annealing for ≥2 hours at 865-875°C followed by cooling at ≤30°C/hour.

RESULTS AND DISCUSSION

Coating thicknesses in the range 5 μm to several hundred μm were obtained. Thickness depended on substrate pulling speed, residence time of the substrate in the liquid, temperature of the substrate and temperature of the liquid. Cross sections of a coated alumina sheet and a 0.8-mm diameter coated alumina rod are shown in the photomicrographs of figures 3a and 3b respectively. The coating on the sheet is approximately 30-μm thick and the one on the rod is 6 to 8-μm thick. A 100-μm diameter Pt wire coated with a 5-μm Bi-Sr-Ca-Cu-O layer is shown in Figure 3c (plan view).

The surface morphology of the coated alumina sheets consists of interlocking blade-like crystals that may be several hundred microns long (Figure 4), or in some cases a mixture of blade-like and platelet-like crystals. The coated wires have a surface composed of platelet crystals (Figure 5) with approximate diameter 20 μm.

We used a tunnel-diode oscillator technique for characterizing the temperature dependence of the superconducting transition in the coated layers[3]. The resonant frequency f of a tunnel-diode LC oscillator circuit, with its coil encircling the superconductor sample, is monitored as the

Fig. 4. Surface morphology of Bi-Sr-Ca-Cu-O coated alumina sheets. The smallest scale division is 10 μm.

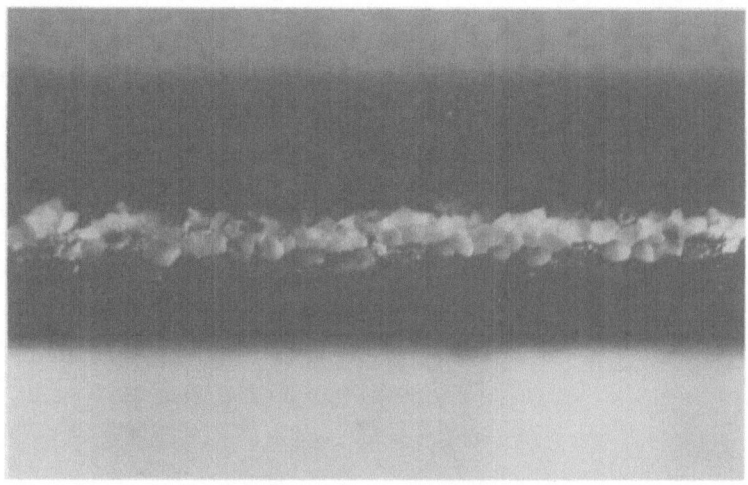

Fig. 5.   Crystal platlet morphology on the surface of a Bi-Sr-Ca-Cu-O coated, 250-μm diameter Pt wire.

sample temperature T is lowered through the superconducting transition.   When the sample becomes superconducting, flux exclusion occurs due to the Meissner effect.   This alters the inductance L of the circuit and hence f changes.   A plot of f vs. T characterizes the transition.   Figure 6 shows the transition (qualitatively similar to the real part χ' of magnetic susceptibility) for an annealed sheet and Figure 7 shows the transition for a 1.5-mm diameter cast wire containing Pb.

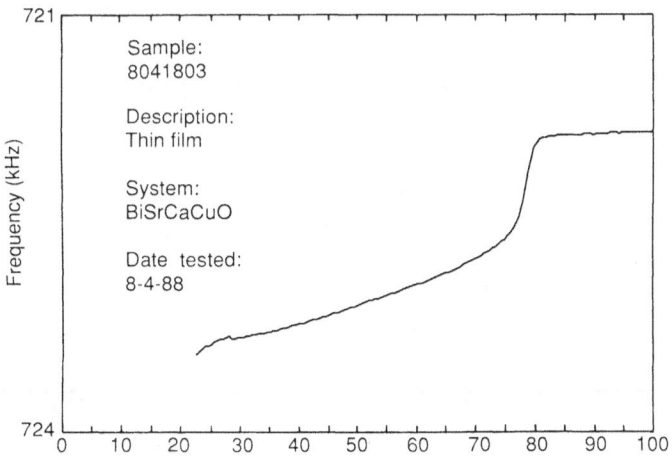

Fig. 6.   Flux-exclusion transition, by the tunnel diode oscillator technique, for an annealed Bi-Sr-Ca-Cu-O sheet coating.

The temperature dependence of resistance for a 200-μm thick Bi-Sr-Ca-Cu-O coating layer was monitored by measuring the voltage drop as a function of temperature while a constant current flows through a bar-shaped specimen cut from the sheet. A dc 4-probe configuration was used and leads were attached with silver paste. The result is shown in Figure 8.

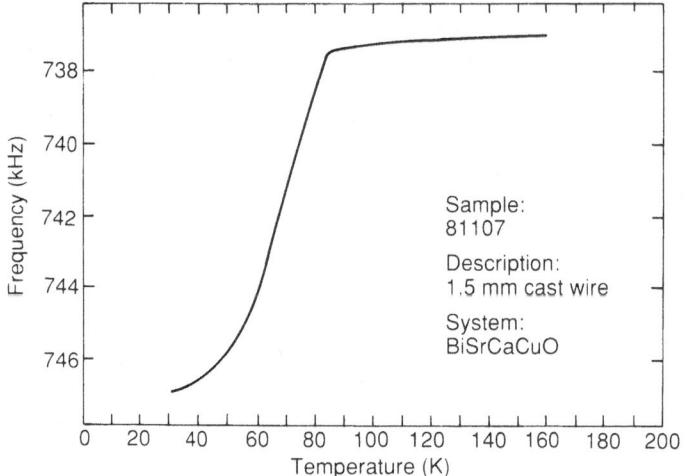

Fig. 7. Flux-exclusion transition, by the tunnel diode oscillator technique, for an annealed Bi-Pb-Sr-Ca-Cu-O cast wire.

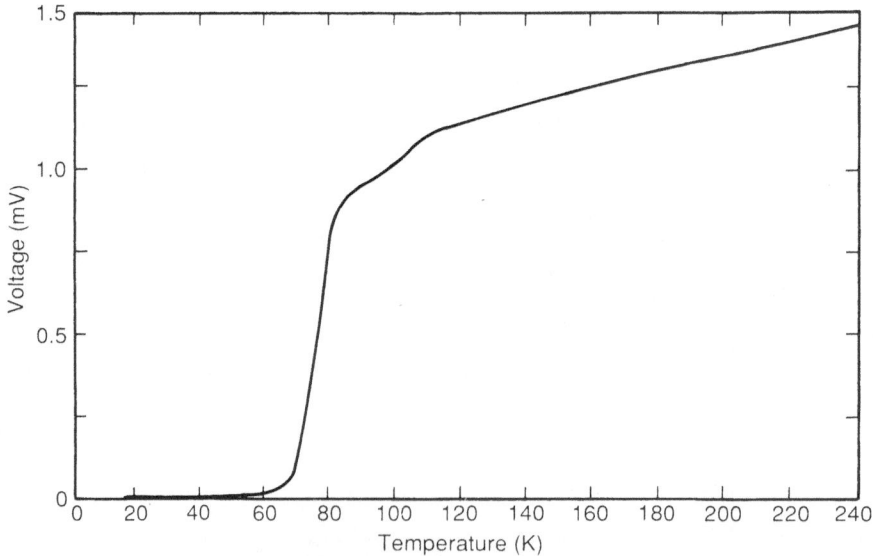

Fig. 8. Voltage drop vs. temperature, under constant current, for a segment of the same sheet coating that exhibited flux exclusion in Fig. 6.

306

SUMMARY AND CONCLUSIONS

Three methods were used for liquid-phase growth of Bi/Sr/Ca/Cu/O sheets and filaments: dip coating, continuous coating, and melt casting. Coating speeds up to 1.8 m/min were achieved to date. Deposited thicknesses between 5 μm and 500 μm were attained, on substrates ranging from 25-μm to 3-mm in thickness. Alumina substrates were used to form sheets and rods. Platinum substrates were used to form thin wires. While continuous coating was demonstrated for short lengths, considerable work remains to be done before the technique is reliable. The observed transition onset temperatures are in the vicinity of 80 K. An optimized annealing procedure may be necessary to raise this to higher temperatures.

ACKNOWLEDGMENTS

The authors are appreciative of the excellent technical assistance that was provided by Terry Schuyler during this investigation. This work was partially supported by the Office of Energy Storage and Distribution of the US Department of Energy.

REFERENCES

1. T.F. Ciszek, J.P. Goral, C.D. Evans, and H. Katayama-Yoshida, J. of Crystal Growth **91** (1988) 312.

2. R.B. Goldfarb, T.F. Ciszek, and C.D. Evans, J. Appl. Phys. **64** (1988) 5914.

3. T.F. Ciszek and E. Tarsa, elsewhere in this publication.

STRUCTURAL AND ELECTRICAL PROPERTIES OF SCREEN PRINTED AND SPRAY PYROLIZED HIGH T_c SUPERCONDUCTING FILMS

K.L. Chopra, P. Pramanik, B. Roul, S. Chakraborty, T.K. Dey,
S. Srinivasan, S.K. Ghatak and D. Bhattacharya

Indian Institute of Technology, Kharagpur 721302, India

ABSTRACT

Good superconducting screen printed films and tapes of various bismuth
and rare earth based cuprates have been prepared by using inks made from submicron
powders synthesized by a novel coprecipitation process developed by us. Further,
spray pyrolized superconducting films of bismuth based cuprate on YSZ substrates
have been synthesized. The films are largely single phasic and show highly oriented
growth. All films exhibit sharp superconducting transition temperatures. The
critical current density depends on processing parameters and can be enhanced
by densification using additives like CuO and Ag_2O. In Y-Ba-Cu-O materials,
Miessner effect as high as 70% is obtained.

INTRODUCTION

Sputtered and evaporated films of Y and Bi based cuprates with high T_c
have been extensively reported in literature[1,2]. As compared with sophisticated
physical vapour deposition techniques, such chemical techniques as spray pyrolysis
and screen printing are simpler and inexpensive and can be used to prepare films
for a variety of electronic applications. But, as reported in literature [3-6], such
films generally have a low T_c and a broad transition because of chemical and
metallurgical reactions with the substrate at high processing temperatures which
are essential for sintering. Thus either low processing temperature or some exotic
inert substrates need to be used.

One of the methods of lowering sintering temperature and time is to use
very fine particles, obtained by precipitation techniques, or thick and thin films
of the materials. Most of the coprecipitation processes reported in the literature[7]
are actually successive precipitation processes. Coprecipitation has however
been successful to a limited extent by citrate, oxalate and carbonate routes.
The metal-oxalate/citrate chelates have a small solubility in aqueous media
and hence metal ions in excess of stoichiometric requirements have to be used
during coprecipitation for obtaining desired metal ion ratios in the precipitates.
The precipitates formed by the carbonate method suffer from the drawback
that the nascent metal oxides and carbon dioxide, generated during calcination,
react back with each other thus necessitating higher processing temperatures.

We have prepared good superconducting screen printed films and tapes of
various cuprates using inks made from submicron powders synthesised by a novel
coprecipitation process. Also, spray pyrolized superconducting films of Bi based
cuprates on YSZ substrates have been synthesized. This paper reports on the
preparation, structure and properties of these films.

EXPERIMENTAL

Screen Printed Films

Submicron sized particles of the ceramic oxides M_1-M_2-M_3-Cu-O (where M_1 = Y/Bi/Tl/Gd/Sc/Sm, M_2/M_3 = alkaline earth metals) were prepared by a novel coprecipitation process. Aqueous solutions of the metal ions, mixed in appropriate ratios, were buffered by triethylamine. This solution was added to an alcoholic solution of 0.02M triethylammoniumoxalate solution which yielded a bluish coloured precipitate. The precipitates were filtered, washed, dried at 120°C and then calcined at 820°C. The particle sizes of the calcined masses are less than 0.5 μ. These powders were used to prepare inks for screen printing of thick films. An alternative technique based on coprecipation of hydroxides of the various constituents of rare earth cuprates has been developed [8]. This yields materials with significantly improved superconducting properties. The technique is being studied for the preparation of thin films.

Inks for screen printing of the films were prepared by ball milling the calcined powders with 0.1% ethylcellulose in isopropylalcohol for 24 hours. The alcohol in the ball milled suspension was allowed to evaporate and as the suspension became thicker, progressive amounts of terpeneol were added to the mass to maintain viscosity of the paste. The ink was then screeen printed onto the alumina substrate through a 300 mesh screen. The films were then dried at 120°C. For Bi-Ca-SrCu-O films, the films were heated at 850°C in air for 24 hours and the films were then slowly cooled to room temperture in the furnace.

In the case of $YBa_2Cu_3O_x$ films, the ink was prepared the same way as mentioned above except that 1% of a BPSG glass was added to the mixture before ball milling. The glass acts as a barrier layer in preventing migration of alumina from the substrate into the Y-Ba-Cu-O film. Green films of Y-Ba-Cu-O were heated at 900°C for 10 hours under oxygen and were then cooled @ 1°C/min to room temperature.

Screen printed films on such substrate as alumina and silver were found to be stable and strongly adherent as determined by a scratch test.

Spray Pyrolyzed Films

Stock solutions of nitrates of constituent ions were prepared by dissolving Bi_2O_3, $CaCO_3$, $SrCO_3$ and CuO in minimum volume of nitric acid and then diluting them to required volume with 70:30 methanol-water mixture. The stock solutions contained about 5 wt% of the metal ion nitrates. Methanol was used to assist in faster evaportion of the solvent during spray deposition on the substrate. An atomizer driven by Ar/N_2 gas was used to spray, from a distance of 12 cm, the alcoholic solution of the mixture of the nitrates containing Bi^{3+},Ca^{2+},Sr^{2+} and Cu^{2+} in the ratio 2:1:2:2 onto a YSZ substrate heated to 350°C. Typical deposition rate was 1 μ/min. Films are dull in appeerence when the first few layers are deposited which on progressive thickening become glossy. The as deposited films were then pyrolyzed by heating the substrate to 500°C. The films were sintered by heating at 2°C/min to 780°C in air which was raised rapidly at 10°C/min to 840°C. After holding the film at 840°C for 5 minutes, the substrate was quench cooled in air to 600°C, followed by slow cooling in the furnace to room temperature. The films were black and strongly adherent to the substrates as seen by scratch test.

RESULTS AND DISCUSSION

Screen printed superconducting films of various bismuth and rare earth based cuprates have been prepared. Films of Y-Ba-Cu-O show the orthorhombic structure. The SEM micrograph (fig. 1) indicate extensive columnar growth of sintered grains. EDAX analysis does not indicate the presence of any alumina in the film

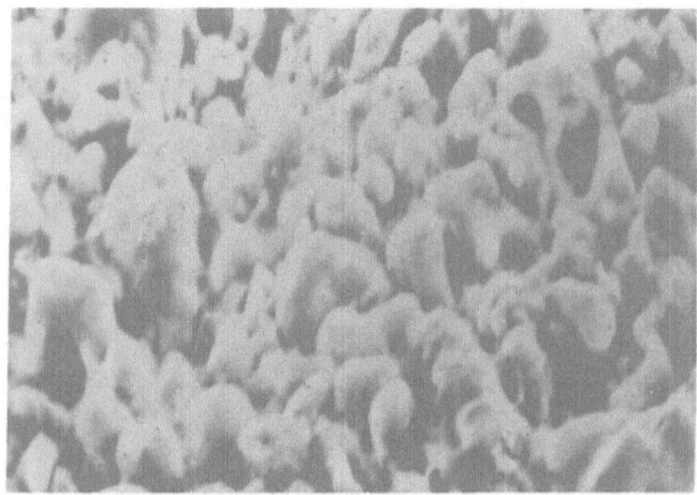

Fig. 1 SEM micrograph of screen printed $YBa_2Cu_3O_x$
film showing columnar growth

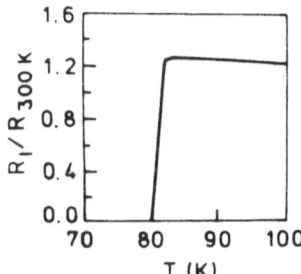

Fig. 2 Temperature dependence of resistivity of screen
printed $YBa_2Cu_3O_x$ film

so that there is no diffusion from the substrate into the film. The porosity of
the films as measured from the micrograph was 10%. The films show onset of
superconducting transition at 82 K with a transition width of 2 K (fig. 2).

The X-ray diffraction pattern of screen printed Bi-Ca-Sr-Cu-O films show
tetragonal structure with extensive platelet type of crystal growth as seen in
the SEM micrograph (fig. 3). The films exhibit a sharp onset of superconductivity
at 90 K (fig. 4), however the resistance reaches zero at 66 K. In order to establish
the role of diffusion of the substrate into the film, EDAX analysis and chemical
mapping of the cross section of the fractured film/substrate interface were undertaken.
Micrographs in figs. 5 and 6 show that Bi diffuses into alumina substrate. There
is, however, no evidence of diffusion of Al into the film material. It should also
be noted that no evidence of diffusion of Ca, Sr or Cu into alumina substrate
could be seen. We conclude that the low T_c is due to nonoptimal processing
conditions rather than the reaction of the film material with the substrate.

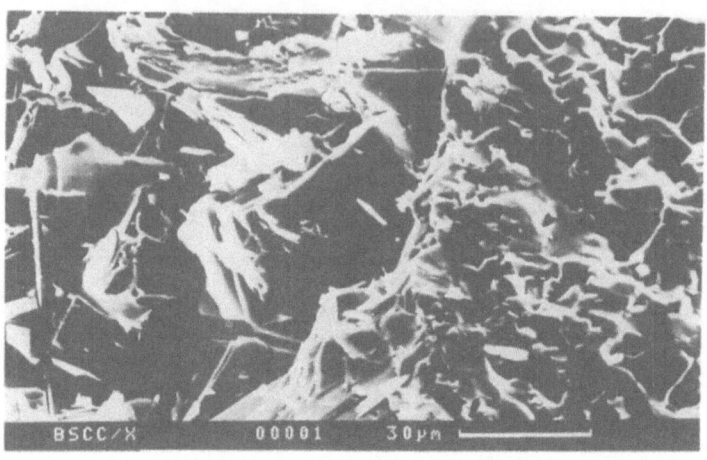

Fig. 3 SEM micrograph of screen printed $Bi_2(Ca,Sr)_3Cu_2O_x$ film

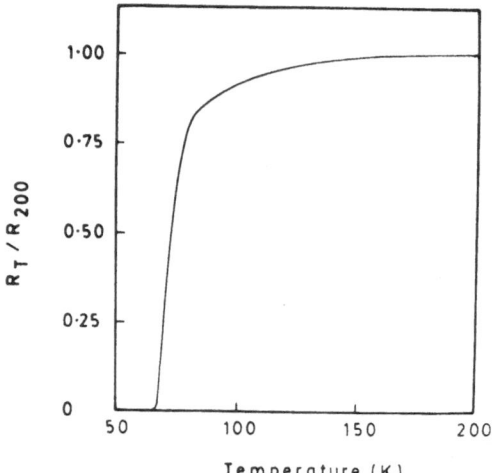

Fig. 4 Temperature dependence of resistivity of screen printed $Bi_2(Ca,Sr)_3Cu_2O_x$ film.

Spray pyrolyzed films from a solution corresponding to the composition $Bi_2CaSr_2Cu_2O_x$ on YSZ showed a nominal composition (at %) Cu : 23.7, Ca : 9 Sr : 19.2, Bi : 30.6 and Zr : 17.5. Note that the composition of the film differs significantly from the composition of the sprayed solution, possibly due to the difference in vapour pressures of the constituents. The X-ray diffraction pattern (fig. 7) shows the structure of the film to be tetragonal with a=b=3.832 A and c=30.78 A. The grains are oriented with c-axis normal to the film. No diffraction lines due to unreacted compounds were detectable. Extensive platelet type of crystal growth is seen in fig. 8. The presence of ZrO_2 in the film is easily discernible in the chemical map of the distribution of Zr (fig. 9). It appears that Zr is dispersed throughout and does not segregate at any interface or grain boundaries. EDAX analysis indicates that Zr is in the form of ZrO_2 precipitates. This observation indicates that nitric acid in the spray solution has reacted with the ZrO_2 in the YSZ

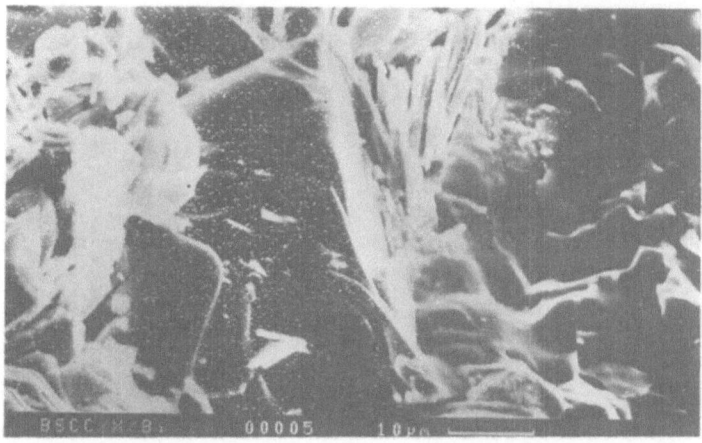

Fig. 5 Chemical map of film/substrate interface of screen printed $Bi_2(Ca,Sr)_3Cu_2O_x$ film showing extent of diffusion of Bi into alumina substrate

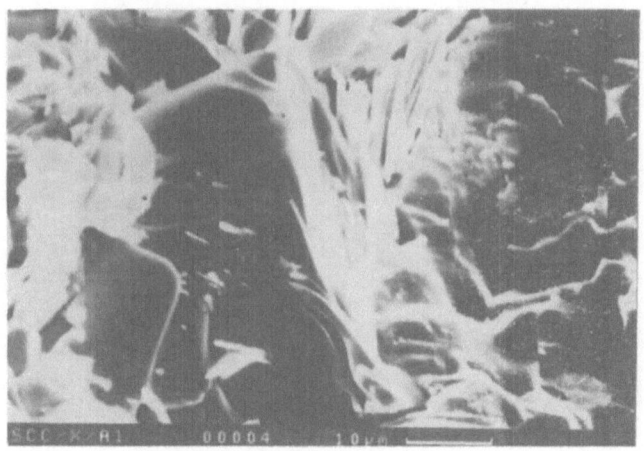

Fig. 6 Chemical map of Al at the film/substrate interface of screen printed $Bi_2(Ca,Sr)_3Cu_2O_x$ film

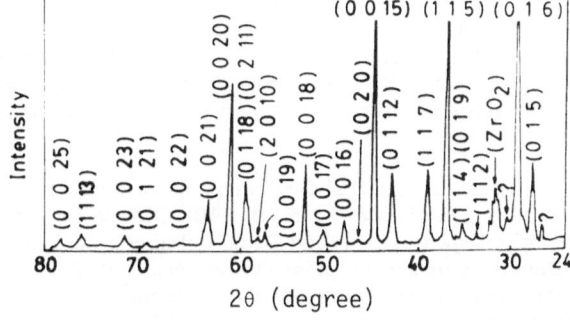

Fig. 7 X-ray diffraction pattern of spray pyrolyzed $Bi_2CaSr_2Cu_2O_x$ film on YSZ substrate

Fig. 8 SEM micrograph of spray pyrolyzed $Bi_2CaSr_2Cu_2O_x$ film on YSZ substrate

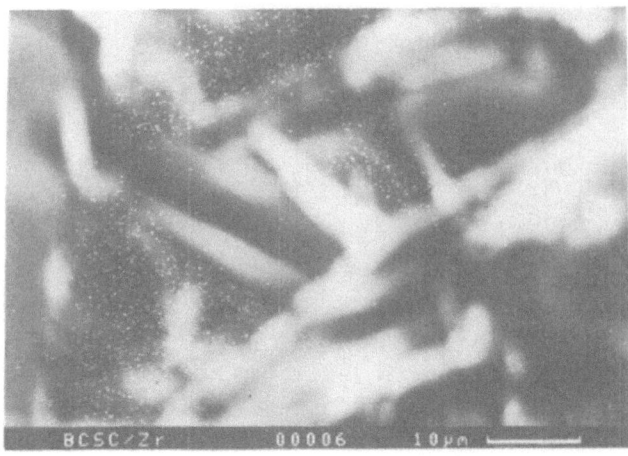

Fig. 9 Chemical map of Zr on the surface of spray pyrolyzed pyrolyzed $Bi_2CaSr_2Cu_2O_x$ film on YSZ substrate

substrate to form $Zr(NO_3)_4$ which is then deposited in the film. During sintering, $Zr(NO_3)_4$ disscociates to form ZrO_2 precipitates. It should be noted that in the case of Al_2O_3 substrates reaction with HNO_3 results in extensive diffusion of $Al(NO_3)_3$ and thus it is not possible to obtain superconducting films on alumina substrates.

Spray pyrolyzed films of Bi-Ca-Sr-Cu-O show sharp superconductive onset at 92 K, the zero resistance being reached at 77 K (fig. 10). This is the first observation of the zero resistance of Bi based spray pyrolyzed films being reached above 77 K. The critical current densities of screen printed and spray pyrolyzed films are dependant on process parameters and spacing between measuring electrodes. Typical current densities of large area samples range from 10 to 100 A/cm^2. With increasing densities of films obtained by addition of additives like Ag_2O and CuO, the current density increases sharply.

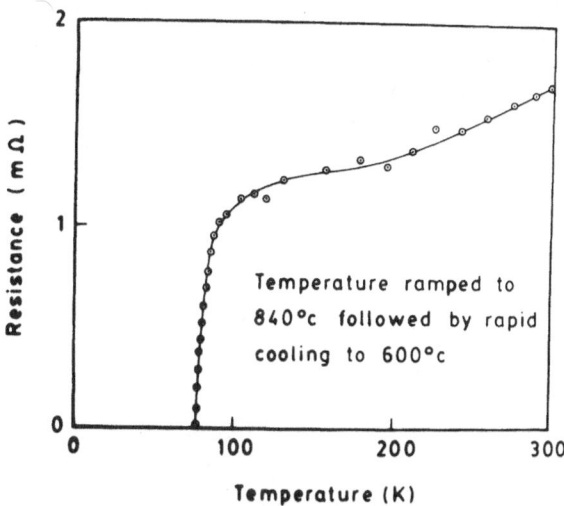

Fig. 10 Temperature dependence of resistivity of spray
pyrolyzed $Bi_2CaSr_2Cu_2O_x$ film on YSZ substrate

Magnetization behaviour of coprecipitated superconducting materials show
typical type II superconducting behaviour with low H_{c1}. Flux expulsion studies
show a large Miessner effect as large as 70% at 77 K in $YBa_2Cu_3O_x$ material.
In case of Bi based materials, the corresponding effect is 10%

CONCLUSIONS

1. Our coprecipitation technique allows preparation of multicomponent cuprates
   of Y,Bi,Gd,Tl,Sc,Sm based materials mixed homogeneously on microscopic
   scale with quantitative stoichiometry.
2. The processing temperatures and time of the coprecipitated powders are
   significantly lower than those prepred by conventional processes.
3. The precipitates can be used to make bulk, screen printed films and tapes
   with sharp (1 K) superconducting transition temperatures.
4. It should be possible to control the coprecipitation process to yield thin
   films by ion-by-ion deposition process.

REFERENCES

1. Rice, C.E., Levi, A.F.J., Fleming, R.M., Marsh, P., Baldwin, K.W., Anzlowar,
   M., White, A.E., Short, K.T., Nakahara, S. and Stormer, H.L., 1988, Preparation
   of superconducting thin film of calcium strontium bismuth copper oxides by
   coevaporation, Appl. Phys. Lett., 52, 1828
2. Liou, S.H., Hong, M., Kwo, J., Davidson, B.A., Chen, H.S., Nakahara, S.,
   Boone, T. and Felder, R.J., 1988, Y-Ba-Cu-O films by rf magnetron sputtering
   using single composite targets : superconducting and structural properties,
   Appl. Phys. Lett., in press
3. Bansal, N.P., Simmons, R.N. and Farrell, D.E., 1988, High T_C screen printed
   $YBa_2Cu_3O_{7-x}$: effect of substrate material, Appl. Phys. Lett., 53, 603.
4. Tabuchi, J. and Utsumi, K., 1988, Preparation of superconducting Y-Ba-
   Cu-O thick films with preferred c-axis orientation by a screen printing method,
   Appl. Phys. Lett., 53, 606
5. Sacchi, M., Sirolti, F., Morten, B. and Prudenziati, M., 1988, High T_C superconductivity in Y-Ba-Cu-O screen printed films, Appl. Phys. Lett., 53, 1110
6. Vaslow, D.F., Dieckman, G.H., Elli, D.D., Ellis, A.B., Holmes, D.S., Lefkow,
   A., MacGregor, M., Nordman, J.E., Petras, M.F. and Yang, Y., 1988, Super-

conducting Bi-Ca-Sr-Cu oxide films by spray pyrolysis of metal acetates, Appl. Phys. Lett., 53, 324.

7.   Schrodt, D.J., Osofsky, M.S., Bender, B.A., Lawrence, S.H., Lechter, W.L., LeTorneau, V., Skelton, E.F., Qadri, S.B., Wolf, S.A., Toth, L.E., Singh, A.K., Hoff, H.A., Richards, L.E. and Pande, C.S., 1988, Properties of Bi-Ca-Sr-Cu-O High T_c superconductors by coprecipitation processing, Solid State Commun., 67, 871.

8.   Our unpublished results.

SPRAY-DEPOSITED SUPERCONDUCTING Bi-Sr-Ca-Cu-O THIN FILMS

Roman Sobolewski[1,2] and Witold Kula[1]

Institute of Physics,[1] Polish Academy of Sciences
PL-02668 Warszawa, Poland

Department of Electrical Engineering,[2] University of Rochester
Rochester, NY 14627, U. S. A.

INTRODUCTION

Since the first successful deposition of the $YBa_2Cu_3O_{7-y}$ (hereafter referred as to YBCO) superconducting thin-film[1] with properties approaching those of the bulk material, a wide spectrum of methods and techniques (both high-vacuum and chemical) have been successfully implemented for fabrication of thin film high-T_c superconductors.[2] Currently, high-vacuum deposition at elevated temperatures (at least 650 °C) appears to have a decisive advantage in growing epitaxial or textured films with high transport-current densities. On the other hand chemical methods give a precise control over the stoichiometry of the deposited material and produce the best-quality granular films.

In general, chemical deposition consists of dissolving proper amounts of organic or inorganic compounds of the elements (e.g., yttrium, barium, and copper nitrates for a YBCO film) in a solvent, and then spin-coating[3,4] or spraying[5,6] the solution of precursors onto a suitable substrate. After the solvent evaporates the film is pyrolized and subsequently heat-treated at high temperatures to form a desired superconducting phase. Despite the fact that this technique is quite crude as compared to high-vacuum technologies, it is an excellent method for testing variations on the compositions of the superconductor studied, is extremely simple, inexpensive, and results in very high quality granular films. Chemically deposited films can also be easily patterned either by laser[7] or electron beam[8] induced pyrolysis or by spraying through a stencil mask which is placed in close contact with the substrate.[9]

In this paper we report our progress on the fabrication and characterization of Bi-Sr-Ca-Cu-O (BSCCO) thin films deposited and patterned using a chemical spray technique previously developed for YBCO films.[9] Our interest is mainly focused on properties of films of different stoichiometries that exhibit superconducting transition around 80 K. We show that although superconducting BSCCO films are easy to make (certainly the procedure is less involved than that for YBCO films), fabricating the film with a narrow superconducting transition requires a very precise control over the temperature of the annealing process.

FABRICATION PROCEDURE

Film deposition

We started our deposition process by separately diluting each of the nitrate precursors in water to a standard 0.1 M concentration of the appropriate ion. Next, proper amounts were mixed to produce a desired composition. Three different starting compositions were tested: $Bi_1Sr_1Ca_1Cu_2O_x$ (referred hereafter as 1:1:1:2) - a "generic" composition in which superconductivity in BSCCO was first discovered,[10] $Bi_4Sr_3Ca_3Cu_4O_x$ (4:3:3:4), and $Bi_2Sr_2Ca_1Cu_2O_x$ (2:2:1:2). The latter two are known[11,12] to be responsible for the 80-K superconducting phase.

The final mixture was sprayed onto a heated (180 − 300°C) substrate using a fine air brush with clean air as the carrier gas. Using a simple spray-gun we obtained fairly uniform films over the area of about 1 cm.[2] For film patterning we covered the substrate with a thin (~15 μm) stencil mask, made from nickel electroformed on a photolithographically patterned stainless steel plate.[13] In order to assure the proper contact between the mask and the substrate and to achieve a full compatibility with the evaporation process, a special mask-substrate fixture was designed. For spraying convenience we put the fixture vertically in the front of the spray-gun nozzle. The distance between the substrate and the spray nozzle was about 4 cm.

Spraying rate (fixed by the pressure of the gas carrier) and the substrate temperature played a most important role in obtaining uniform, well defined structures. We observed that, unlike in the YBCO spraying process,[9] the increase of the substrate temperature during the deposition resulted in very uniform films with small grains and well defined edges. Thus, our standard procedure was the following. We started the deposition process at the substrate temperature of 180 °C and continued spraying for 4 min., slowly rising the temperature to about 300 °C. Starting the spraying at a relatively low temperature was the key in obtaining a good adhesion of the BSCCO film to the substrate. Depending on the stencil mask used, we could reliably pattern structures with smallest features of the order of 50 μm. In most cases we fabricated simple strips 150 − 700 μm wide, about 4 mm long, and 1 − 2 μm thick.

Unpatterned films could be obtained with the substrate temperature as low as 150 °C, however, they were not as uniform in thickness as those deposited according to the procedure described above.

Several crystalline substrates, such as yttrium stabilized ZrO_2, MgO, $LiNbO_3$, and Al_2O_3 were tried. However, only films deposited on MgO exhibited superconducting properties. Films on ZrO_2 and $LiNbO_3$ reacted (violently in the case of ZrO_2) with the substrate during annealing, while films on Al_2O_3 melted above 820°C and were semiconducting.

Heat treatment

The sprayed films or structures were heat-treated in a tube furnace in air and no special procedure was applied for the pyrolysis process. In all cases the procedure was the same with the annealing temperature being the only variable (we always started with the furnace preheated to a desired annealing temperature). After the initial rapid heating (1 min), the sample was annealed for 5 min., and then slowly cooled down with the furnace to room temperature (2 – 3 hours) in flowing air. Substituting oxygen for air during the annealing and cool-down did not affect film properties. The influence of annealing temperature on the superconducting transition is presented in the next Section.

Once removed from the furnace, the films were studied under high-power optical and scanning electron microscopes. We found that although they were granular, the grains were quite small (about 1 – 3 μm) and the film surface was much smoother and uniform than the surface of YBCO pyrolytic films. Our best films were not sensitive to water and could easily withstand many cooling cycles with no apparent degradation of their superconducting properties.

ELECTRICAL PROPERTIES

Superconducting transition

Resistance vs. temperature [R(T)] dependences for 700-μm-wide strips annealed at different temperatures are shown in Figs. 1, 2 and 3. All measurements were performed in a standard four-probe arrangement, with current and voltage leads soldered directly to silver contact pads evaporated on the films precleaned by glow discharge. The sample temperature was monitored by a thermally grounded 100 Ω platinum resistor, as well as a calibrated Allen-Bradley resistor.

Figure 1 presents a collection of R(T) curves for films with the starting composition 1:1:1:2. It is clear that the annealing temperature played the major role in superconducting properties of 1:1:1:2 films. Annealing at 830 °C produced a material with a semiconductor-type behavior above the transition. The film apparently consisted of mainly the $Bi_2Sr_2Cu_2O_x$ phase characterized by a single Cu-O plane and very low T_c.[14] Increasing the annealing temperature we observed substantial improvement in the superconducting transition. However, the only films exhibiting sharp drop in resistance at T_c, without a resistive "tail" below the transition, were the ones annealed at temperatures between 895 and 907 °C. Above 907 °C films melted, nevertheless, they remained superconducting at temperatures below 77 K.

Similar behavior (presented in Fig. 2) was observed for films with the starting composition 4:3:3:4. Films annealed at 830 °C were semiconducting above the transition and they did not superconduct even at liquid helium temperatures. The range of temperatures leading to the sharpest superconducting transition was narrower: films with no resistive "tail" on their R(T) curves were obtained for temperatures between 895 and 900 °C. Again, above 900 °C the material melted but was superconducting.

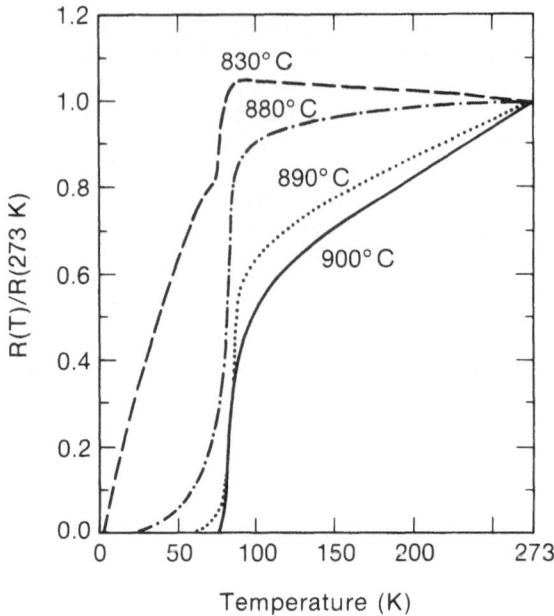

Figure 1.   Resistive transition for for BSCCO strips of the starting composition 1:1:1:2 annealed at different temperatures. The annealing time - 5 min.

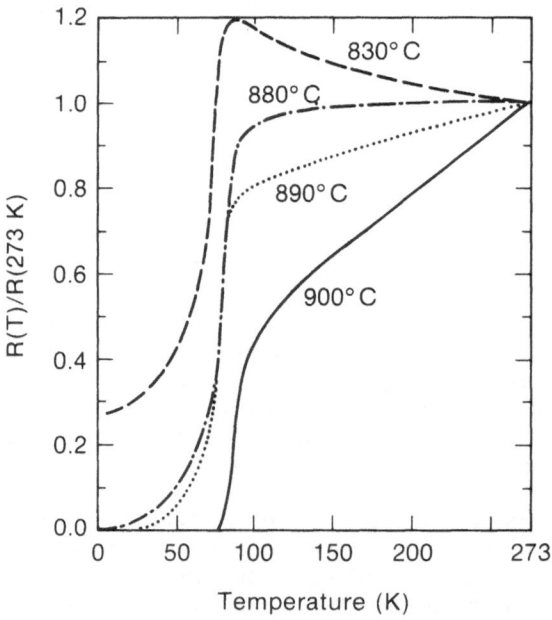

Figure 2.   Resistive transitions for BSCCO strips of the starting composition 4:3:3:4, annealed at different temperatures. The annealing time - 5 min.

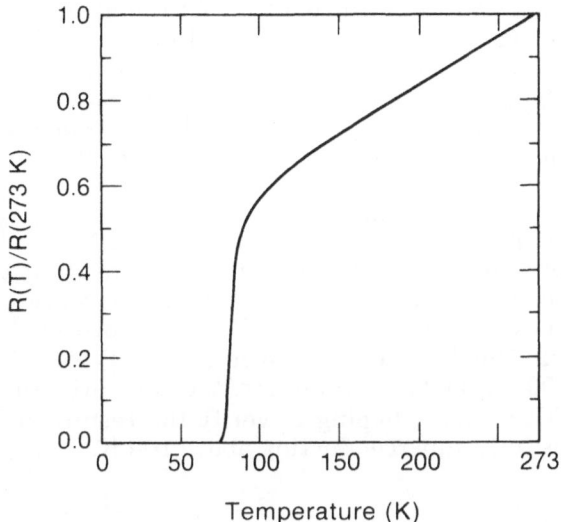

Figure 3.   Resistive transition for the BSCCO strip of the starting
composition 2:2:1:2 annealed at 900 °C. The annealing time - 5
min.

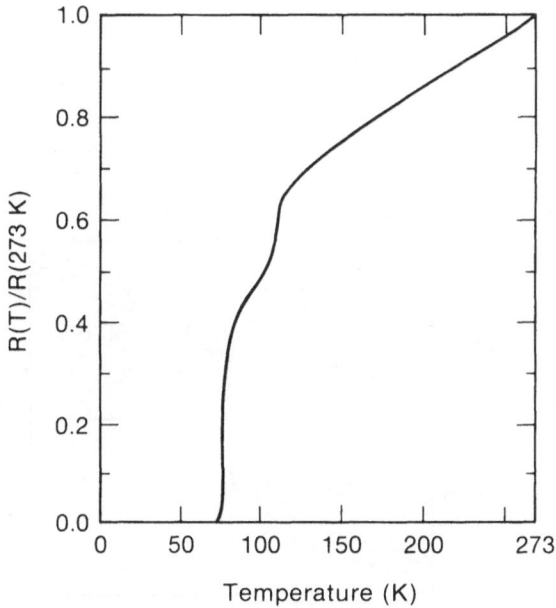

Figure 4.   Resistive transition for the BSCCO strip of the starting composition 1:1:1:2 annealed at 890 °C for 16 hours. A drop in resistance
at approximately 110 K is about 30%.

We have also tested films of the starting composition 2:2:1:2 and observed that films annealed at temperatures below 840 °C exhibited purely semiconducitng R(T) characteristics down to 4.2 K. At the same time a sharp transition, as shown in Fig. 3, was measured for films annealed at temperatures between 895 and 900 °C, similarly to the 4:3:3:4 films. This finding led us to conclude that compositions 4:3:3:4 and 2:2:1:2 corresponded to the same 80-K superconducting phase within a solid solubility range of Ca and Sr, in agreement with observations presented by Satoh et al.[15]

The very short annealing time applied for these studies precluded formation of even traces of the 110-K phase (the 2:2:2:3 phase). Thus, in a separate series of tests we tried much longer annealing times in order to check if any of the starting compositions used in this work could lead eventually to formation of the higher T_c phase. We succeeded in the case of starting composition 1:1:1:2. Our best result (shown in Fig. 4) was obtained for a sample annealed at 890 °C for 16 hours in air. We currently are trying to synthesize Pb doped BSCCO films, hoping to verify the results of Nobumasa et al.[6] and to obtain films fully superconducting above 100 K.

Granular properties

As we mentioned in Sec. II, microscope studies of our films revealed their granular morphology. This finding was also supported by transport current measurements and studies of microwave properties of BSCCO films.[16] Our best films exhibited critical current densities above 10^4 A/cm^2 at 4.2 K, and about 100 A/cm^2 at 70 K. These values are much smaller than the expected, intrinsic values. They are, however, an order of magnitude larger than the best values obtained for the pyrolytic YBCO films of the same geometry.[9]

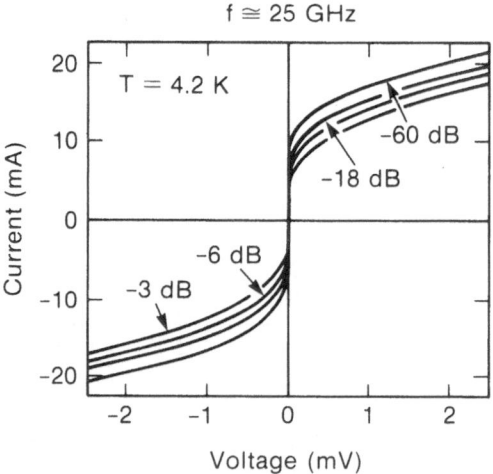

Figure 5.   Current vs. voltage characteristics for microwave-irradiated BSCCO strip of starting composition 1:1:1:2, annealed for 5 min. at 900 °C.

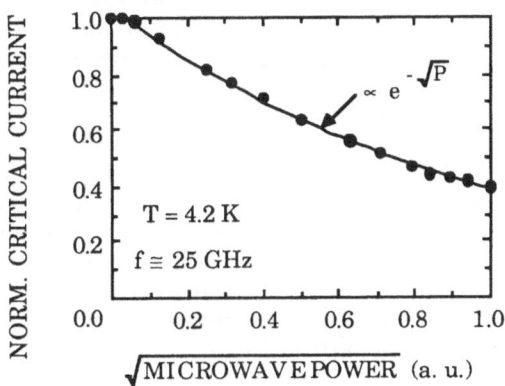

Figure 6. Experimentally measured normalized critical current as a function of the square root of applied microwave power (dots) for the $Bi_1Sr_1Ca_1Cu_2O_x$ strip. Solid line corresponds to simulation based on the granular-film model developed in Ref. 17.

Figure 5 presents a family of current-voltage (I–V) characteristics of a 150 μm wide $Bi_1Sr_1Ca_1Cu_2O_x$ strip irradiated by microwaves. We see that all I–V curves are smooth and exhibit a large positive curvature characteristic of strongly coupled weak-links (qualitatively the same behavior was observed for all temperatures up to the superconducting transition). Increasing the applied microwave power (800 mW max.) we were able to suppress the film I_c up to about 40% of its initial value.

The measured dependence of I_c on the square root of the incident microwave power ("microwave electric field") is shown in Fig. 6. We observe that, for moderate and large microwave powers, the I_c decrease is exponential but is not related directly to the incident power, as one would expect in the case of a simple heating. Instead, the observed behavior can be well explained in terms of a microwave response of the network of grain-boundary Josephson junctions interconnected by strongly superconducting grains.[17] According to this model, the I_c dependence on microwaves can be quantitatively simulated by the zero-order Bessel function averaged over many junctions with an distribution function appropriate for the film. For the distribution of the type of $p(t) = t/(t^2 + t_0^2)^{3/2}$ (a skewed distribution with the center around t_0; t is the junction coupling strength, and in a constant-voltage model corresponds to the barrier thickness) the integral can be evaluated analytically and we obtain a very simple dependence of I_c (solid line), presented in Fig. 6.

CONCLUSIONS

In summary, superconducting BSCCO thin films in a variety of stoichiometries were produced from nitrate precursors by spray-deposition technique. A main emphasis was placed on a systematic measurement of the influence of the annealing procedure on the shape of the R(T) characteristics. We proved that indeed the annealing temperature plays the crucial role in fabricating films exhibiting very narrow superconducting transitions. While most of our studies were performed on films annealed for a short period of time (5 min.) and characterized by the low-T_c (80 K) superconducting phase, we showed that long annealing (16 hours) can produce in the film of the same starting stoichiometry a substantial percentage of the high-T_c (110 K) phase. Our dc measurements were supplemented by high frequency experiments. We demonstrated that BSCCO microwave properties can be satisfactorily explained by properly modelling the granular nature of our films.

From the technological point of view, we have presented a fabrication process which results in granular films characterized by excellent superconducting properties, is convenient for patterning of simple structures, and eliminates any degradation of the film superconducting properties due to the patterning process. Despite the requirement of a very precise control of the annealing process, we have found the high-quality BSCCO films to be easier to produce and handle than the YBCO films.

ACKNOWLEDGMENT

The authors would like to thank P. Gierłowski for help and assistance in measurements. One of us (R. S.) thanks A. Kadin and T. Hsiang for many valuable discussions. This work was supported in part by the Polish Academy of Sciences under grant RPBP 01.9, and by the National Science Foundation grant ECS-8721247.

REFERENCES

1. P. Chaudhari, R. H. Koch, R. B. Laibowitz, T. R. McGuire, and R. J. Gambino, Critical-current measurements in epitaxial films of YBa$_2$Cu$_3$O$_{7-y}$ compound, *Phys. Rev. Lett.* 58:2684 (1987).

2. see e.g., Proc. of the International Conference on High-T_c Superconductivity and Materials and Mechanisms for Superconductivity (HTSC-M^{2}S, Interlaken 1988), *ed. by* J. Müller and J. L. Olsen, *Physica C*, 153-155 (1988); Proc. of the Applied Superconductivity Conference (ASC'88, San Francisco 1988), to appear in *IEEE Trans. Magn.* MAG-25 (1989).

3. C. E. Rice, R. B. van Dover, and G. J. Fisanick, Preparation of superconducting thin films of Ba$_2$YCu$_3$O$_7$ by a novel spin-on pyrolysis technique, *Appl. Phys. Lett.* 51:1842 (1987).

4. J. A. Agostinelli, G. R. Paz-Pujalt, and A. K. Mehrotra, Superconducting thin films in the Bi-Sr-Ca-Cu-O system by the decomposition of metallo-organic precursors, *Physica C* 156:208 (1988).

5. A. Gupta, G. Koren, E. A. Giess, N. R. Moore, E. J. M. O'Sullivan, and E. I. Cooper, YBa$_2$Cu$_3$O$_{7-y}$ thin films grown by a simple spray deposition technique, *Appl. Phys. Lett.* 52:163 (1988).

6. H. Nobumasa, K. Shimizu, Y. Kitano, and T. Kawai, Formation of a 100 K superconducting Bi(Pb)-Sr-Ca-Cu-O film by a spray pyrolysis, *Jpn. J. Appl. Phys.* 27:L1669 (1988).

7. A. Gupta and G. Koren, Direct writing of superconducting patterns of Y$_1$Ba$_2$Cu$_3$O$_{7-\delta}$, *Appl. Phys. Lett.* 52:665 (1988).

8. J. V. Mantese, A. B. Catalan, A. H. Hamdi, A. L. Micheli, and K. Studer-Rabeler, Use of electron beam lithography to selectively decompose metallo organics into patterned thin film superconductors *Appl. Phys. Lett.* 53:526 (1988).

9. W. Kula, R. Sobolewski, P. Gierłowski, S. J. Lewandowski, J. Konopka, and A. Graczyk, Simple patterning of spray deposited Y-Ba-Cu-O films, to appear in *Supercond. Sci & Technol.* (1988).

10. H. Maeda, Y. Tanaka, and T. Asano, A new high-T_c oxide superconductor without a rare earth element, *Jpn. J. Appl. Phys.* 27:L209 (1988).

11. J. M. Tarascon, Y. LePage, P. Barboux, B. G. Bagley, L. H. Greene, W. E. McKinnon, G. W. Hull, M. Giroud, and D. M. Hwang, Crystal substructure and physical preperties of the superconducting phase $Bi_4(Sr,Ca)_6Cu_4O_{16+x}$, *Phys. Rev. B* 37:9382 (1988).

12. H. W. Zandbergen, P. Groen, G. Van Tendeloo, J. Van Landuyt, and S. Amelinckx, Electron diffraction and electron microscopy of the high T_c superconductive phase in the Bi-Ca-Sr-Cu-O system, *Solid State Commun.* 66:397 (1988).

13. H. Bielska-Lewandowska, G. Jung, S. J. Lewandowski, and S. Wątkowski, Polish Patent No. 108 621 (1981); these stencil nickel masks are routinely used by us for the vacuum evaporation process.

14. C. Michel, M. Hervieu, M. M. Borel, A. Grandin, F. Deslandes, J. Provost, and B. Raveau, Superconductivity in the Bi-Sr-Cu-O system, *Z. Phys. B* 68:421 (1987).

15. T. Satoh, T. Yoshitake, Y. Kubo, and H. Igarashi, Composition dependence of superconducting properties in $Bi_2(Sr_{1-x}Ca_x)_{n+1}Cu_nO_y$ (n = 2, 3) thin films, *Appl. Phys. Lett.* 53:1213 (1988).

16. R. Sobolewski, J. Konopka, W. Kula, P. Gierłowski, A. Konopka, and S. J. Lewandowski, Fabrication and microwave properties of Y-Ba-Cu-O and Bi-Ca-Sr-Cu-O thin films, to appear in *IEEE Trans. Magn.* MAG-25 (1989).

17. R. Sobolewski, Modelling of the AC Josephson effect in granular high-T_c films, *Phys. Rev. B* - submitted for publication.

MICROSTRUCTURE AND CRITICAL CURRENTS OF $YBa_2Cu_3O_{7-x}$ THIN FILMS ON $SrTiO_3$ AND A NEW SUBSTRATE: $KTaO_3$*

R. Feenstra, J. D. Budai, D. K. Christen, M. F. Chisholm,
L. A. Boatner, M. D. Galloway, and D. B. Poker

Solid State Division, Oak Ridge National Laboratory
P. O. Box 2008, Oak Ridge, Tennessee 37831

INTRODUCTION AND SUMMARY

Superconducting thin films of high-T_C materials are of interest for use in fundamental research as well as for a variety of device applications. Epitaxial thin films of high-T_C superconductors have previously been produced using substrates such as $SrTiO_3$, cubic ZrO_2, MgO, and Al_2O_3 with the best results being reported for (100) surfaces of strontium titanate. In the present work, the properties of epitaxial films of $YBa_2Cu_3O_{7-x}$ (Y123) formed by coevaporation of Y, Cu, and BaF_2 onto (100) surfaces of $SrTiO_3$ are presented and compared with those found for identical films on a new substrate material: single crystal $KTaO_3$ or potassium tantalate. This material, like $SrTiO_3$, is also a cubic perovskite with a suitable lattice constant for promoting epitaxial growth. In Sections I and II of the present work, the details of the film preparation are outlined, along with the properties of the resulting high-T_C films formed on (100) and (110) surfaces of $SrTiO_3$. These results are similar to the findings by others and a critical dependency on the film microstructure and orientation is found. In Section III, the results obtained for the new substrate material $KTaO_3$ are presented. It is shown that the properties of Y123 films epitaxially grown on the (100) surface of $KTaO_3$ single crystals compare favorably with those of films formed on $SrTiO_3$. In Section IV more detailed comparisons of the microstructural properties of epitaxial films on the two substrate types are made. The results show that $KTaO_3$ represents a promising new substrate material for the epitaxial growth of high-T_C Y123 films.

I.   FILM PREPARATION AND EXPERIMENTAL DETAILS

The Y123 films investigated here were produced by a coevaporation of Y, Cu, and BaF_2,[1] followed by an ex-situ anneal in wet oxygen at 850°C. This method was found to be a reproducible technique for the fabrication of Y123

---
*Research sponsored by the Division of Materials Sciences, U.S. Department of Energy under contract no. DE-AC05-84OR21400 with Martin Marietta Energy Systems, Inc.

films. The films were deposited at a substrate temperature of 450°C in a background pressure of oxygen stabilized at 5×10^{-6} torr. The rates of the evaporators (e-beam guns for Y and Cu and resistive heating for BaF_2) were controlled by three separate feedback loops. In order to obtain the proper Y:Ba:Cu ratios, the evaporators were calibrated through compositional analyses of as-deposited films by means of Rutherford backscattering spectroscopy (RBS). After deposition, the films were slowly cooled in 100 torr of pure oxygen in a separate chamber, connected to the deposition chamber. A film thickness of 300 to 400 nm was chosen as a standard in this investigation.

In order to remove the fluorine and to form the Y123 compound, the films were annealed in oxygen bubbled through distilled water at room temperature. RBS analysis of the annealed films confirmed the substitution of fluorine by oxygen. Two annealing procedures were employed, which yielded distinct ratios of $a\perp$ to $c\perp$ domains in the final films. In the first procedure, the sample is heated to 850°C in 45 min in dry oxygen, reacted for 30 min in wet flowing oxygen at 850°C, then cooled to 550°C for a 30 min soak to enhance oxygen uptake, after which the sample is further cooled to room temperature. The cooling sequence takes place in dry oxygen. In the second procedure, the sample is first heated to a temperature between 650°C and 750°C in dry oxygen for approximately one hour, before the fluorine is replaced at 850°C following procedure 1. The second procedure yields films that exhibit a much larger $c\perp$ orientation than the first. After a partial anneal at 650 to 750°C, the films are still insulating (since the fluorine has not yet been replaced) and have an orange color. X-ray diffraction in a θ-2θ scan along the surface normal shows the appearance of Bragg peaks at angles corresponding to the 248 phase with the c-axis perpendicular to the substrate.[2,3] The 248 peaks are absent in off-axis diffraction scans. We also observe the growth of polycrystalline BaF_2 peaks and other peaks which were not uniquely identified. The formation of the 248 phase has previously been associated with the presence of fluorine in the film by Kwo et al.[4] and, as we will argue below, its occurrence could be important for reaching high critical current densities.

The electrical transport properties were measured using a standard four-probe technique with collinear spring-loaded contacts. The distance between the voltage contacts was 4 mm. The low-frequency AC measuring current densities were ~1 A/cm^2. In some cases, this leads to a small tail in the superconducting transition. For the resistive measurement of critical currents, the current density was increased by the formation of a 100 μm wide bridge (length 3 mm) in the films. The bridge was formed by a low-temperature oxygen implantation through a tungsten mask. With consecutive implantation doses of 10^{15} oxygen atoms/cm^2 at 240, 160 ,and 80 kV incident energy, the implanted regions became amorphous and insulating.[5] Thin films of gold were used to reduce the contact resistance of the current pads.

II. $YBa_2Cu_3O_{7-x}$ FILMS ON (100) AND (110) $SrTiO_3$

The $SrTiO_3$ substrates used in this study were purchased from Commercial Crystal Laboratories, Inc., Naples, Florida, as single-crystal disks with a high quality polished (100) or (110) surface. The crystals were cut into 2 or 3 mm wide rectangular samples along one of the in-plane symmetry directions. Most crystals showed narrow x-ray rocking curves (<0.1 degrees); in some cases, however, the crystals had a mosaic spread 3 to 4 times larger.

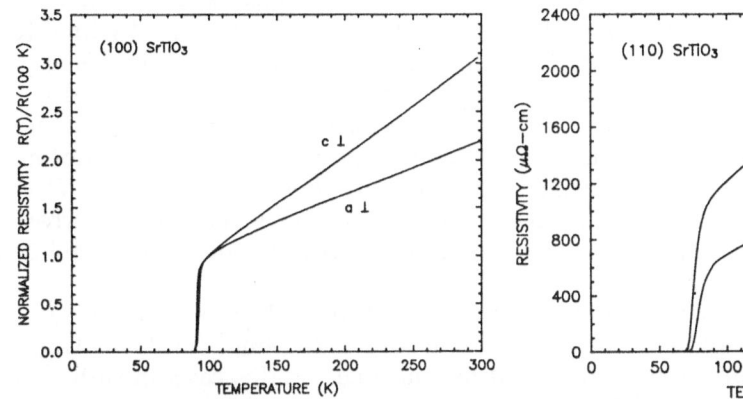

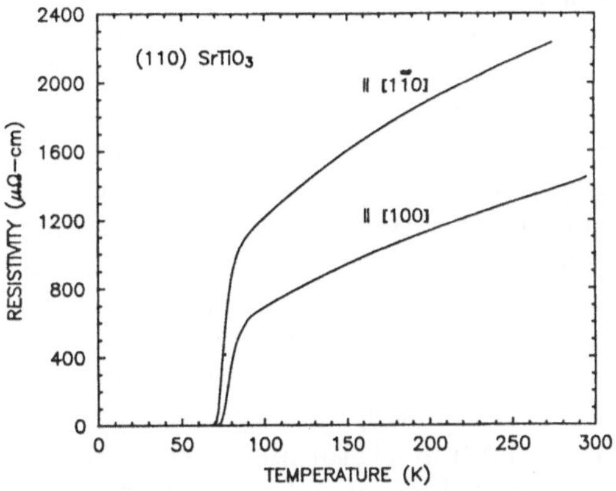

Fig. 1a. Normalized resistance as a function of temperature for stoichiometric Y123 films on (100) SrTiO₃ with the a-axis or the c-axis perpendicular to the substrate.

Fig. 1b. Resistivity as a function of temperature for stoichiometric [110]⊥ + [103]⊥ Y123 films on (110) SrTiO₃ measured along the in-plane [001] and [1$\bar{1}$0] direction.

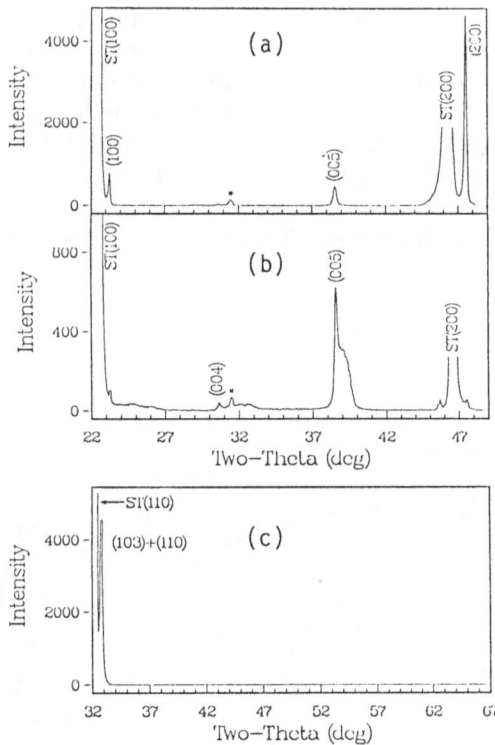

Fig. 2. X-ray diffraction data in a θ-2θ scan along the surface normal for (a) a largely a⊥ Y123 film and (b) c⊥ Y123 film on (100) SrTiO₃ and (c) [110]⊥ + [103]⊥ Y123 film on (110) SrTiO₃. The peaks marked with an asterisk (*) originate from impurity phases in the film.

In Fig. 1a, we show typical temperature dependences of the normalized resistance of films with a majority of a⊥ or c⊥ domains in the film. X-ray diffraction scans along the surface normal for these films are shown in Figs. 2a and b.

For c⊥ films, the resistivity at 300 K usually ranges between 250 and 350 μΩcm, while the resistance ratio R(300 K)/R(100 K) ranges between 2.8 and 3.4 with 0 K intercepts between ± 20 μΩcm. For some films, however, 300 K resistivities as high as 800 μΩcm were found with 0 K intercepts close to zero as well (see also Fig. 6 for KTaO₃). The a⊥ films have 300 K/100 K resistance ratios close to 2 with large positive intercepts. In addition, the temperature dependence of the normal state resistivity is slightly curved below 150 K. For all films deposited on (100) SrTiO₃ (and KTaO₃) with a cation ratio close to 1:2:3, we have observed a narrow superconducting transition in the vicinity of 90 K. This illustrates the reliability of the BaF₂ deposition method.

As shown in Fig. 1b, the situation for films on (110) SrTiO₃ is different. In an earlier report, Enomoto et al.[6] fabricated epitaxial Y123 films on (110) SrTiO₃ with the film's [110] direction along the surface normal. In this case, the a-b planes of the unit cell are perpendicular to the substrate and the c-axis is parallel to the substrate's [001] direction. In this [110]⊥ orientation, a large anisotropy exists in the transport properties of the film, i.e., a high resistivity (low critical currents) parallel to the [001] direction of the substrate and a low resistivity (high critical currents) parallel to [1̄10]. On axis x-ray scans such as Fig. 2c, combined with off axis scans reveal that our films on (110) SrTiO₃ contain [110]⊥ domains, as well as [103]⊥ domains in which the Cu-O planes make a 45° angle with the substrate. No a⊥ or c⊥domains nor peaks belonging to different phases are observed, however. SEM micrographs show that the film consists of many small grains that appear to grow out of the substrate plane. Thus, in these films, the current passes through numerous high angle grain boundaries with a small contact area. This is assumed to be the reason for the lower critical currents found in these films (Fig. 3), and the lower T_c's. Note in Fig. 1b that the films exhibit an anisotropy which is opposite to that of films containing only [110]⊥ domains.

Critical current densities for the films on (100) SrTiO₃ are presented in Fig. 3

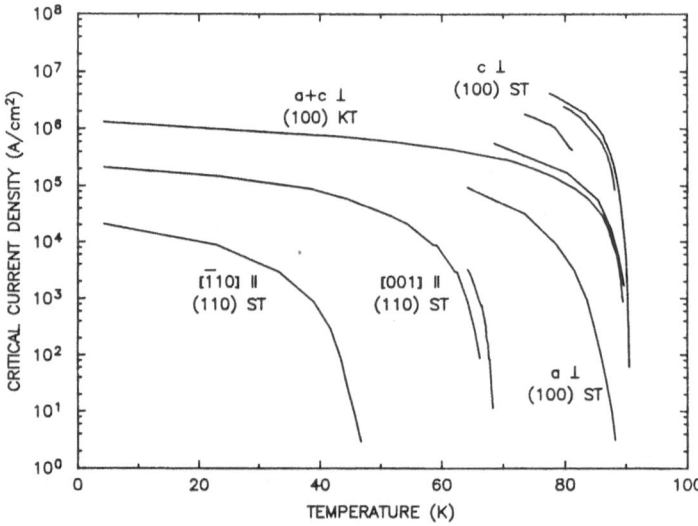

Fig. 3. Critical current density as a function of temperature for stoichiometric, epitaxial Y123 films on (100) SrTiO₃, (110) SrTiO₃, and (100) KTaO₃. The epitaxial orientations are indicated in the figure. For the films on (110) SrTiO₃, curves measured in two different in-plane lattice directions are presented in the figure.

as a function of temperature. A large anisotropy, observed for the first time by Chaudhari et al.,[7] exists between films with the Cu-O planes parallel or perpendicular to the substrate. For the first (c⊥ films), the highest current density (in zero applied field) we have measured is 4 MA/cm^2 at 77 K. Here the geometrical dimensions were used for the conversion to current density. The actual current through the 100 μm wide and 355 nm thick bridge was 1.43 A. At 85 K, J_c = 1 MA/cm^2. These high values were, in fact, measured on several samples that were all annealed according to the second procedure mentioned in Section I. The diffraction pattern (Fig. 2b) shows that the film consists of c⊥ Y123 domains with some shifted 248 peaks, that presumably remain from the 650–750°C anneal in dry oxygen. Char et al.[2] have associated these shifted peaks with the occurrence of additional Cu-O planes as stacking faults in the Y123 structure. At present, it is unclear what the effect of these stacking faults is on the transport properties of the films. For the a⊥ films on (100) SrTiO$_3$, critical current densities are typically 2 or 3 orders of magnitude lower than for the c⊥ films. In general, our findings for Y123 films on (100) SrTiO$_3$ are in good agreement with other published results.

III. YBa$_2$Cu$_3$O$_{7-x}$ FILMS ON (100) KTaO$_3$

To this point, the best film qualities, with the highest critical currents, have been obtained on (100) SrTiO$_3$ substrates.[8,9] With other substrates (e.g., cubic ZrO$_2$, Al$_2$O$_3$, Si/SiO$_2$, BaF$_2$, BaTiO$_3$, LiNbO$_3$, and MgO), interdiffusion across the film-substrate interface or the formation of a reaction layer[10] during the high-temperature anneal lead to a deterioration of the film's superconducting properties. In addition, for these other substrates, the lattice match between the film and substrate is usually significantly worse than in the case of (100) SrTiO$_3$.

In this section, we demonstrate that excellent Y123 films can also be grown on (100) surfaces of KTaO$_3$. Like SrTiO$_3$, potassium tantalate is a cubic perovskite. The lattice parameter a=0.399 nm matches reasonably well with the principal unit lengths of Y123 (c/3≅b=0.389 nm), although not quite as well as SrTiO$_3$ (a=0.391 nm). Epitaxial properties are, however, not entirely determined by the lattice match.

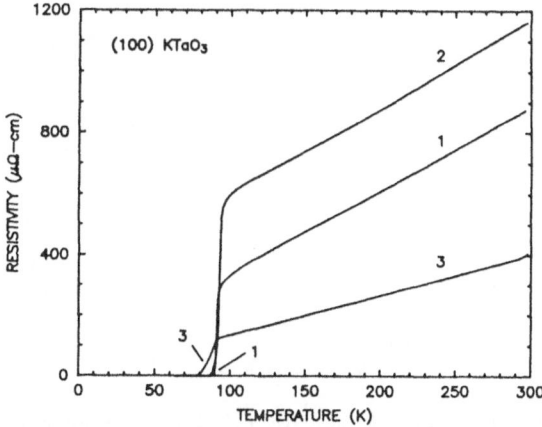

Fig. 4. Resistivity as a function of temperature for stoichiometric, mixed a⊥ + c⊥ Y123 films deposited on the (100) growth surface (curve 1) and a (100) cleaved surface (curve 2) of a KTaO$_3$ single crystal. Curve 3 belongs to a Cu-rich film (Y/Cu=0.27; Ba/Cu=0.50), containing the 248 phase.

The undoped, colorless single crystals of KTaO$_3$ ($5 \times 2 \times 1$ cm^3) used in this work were grown at ORNL using a technique similar to that described by Hannon.[11] Sections from the (100) growth surface or cleaved (100) surfaces from the interior of the crystal were used as substrates. The properties of Y123 films deposited on these faces were not always the identical, as illustrated by the resistivity vs. temperature data of Fig. 4. The normal state resistivity of the film on the growth surface (curve 1) is lower than that of the film on the cleaved surface (curve 2). In addition, curve 1 extrapolates to a resistivity close to zero at 0 K.

Based on prior experience with films on (100) SrTiO$_3$, this difference was expected to originate from different a⊥ to c⊥ fractions in the films. This interpretation was verified by the x-ray data (Fig. 5a,b) and by SEM micrographs of both films. The film on the growth surface was found to contain more c⊥ than a⊥ grains, whereas the film on the cleaved surface contained more a⊥ domains. The cause for this difference is unclear. Both films were produced in the same (stoichiometric) deposition batch and they were both annealed according to procedure 1 described earlier. Other sets of films on growth and cleaved surfaces showed similar behavior, but not as pronounced as that illustrated in Fig. 4.

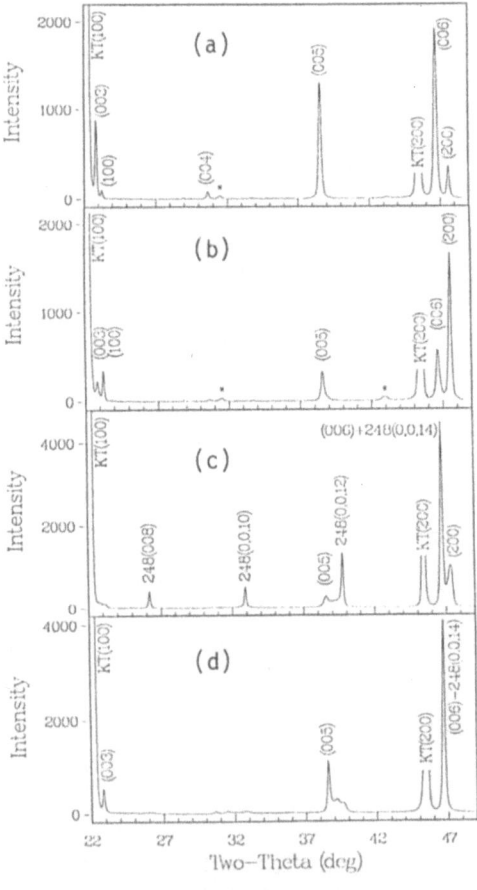

Fig. 5. X-ray diffraction patterns of stoichiometric Y123 films deposited on (a) the (100) growth surface, (b) a (100) cleaved surface of a KTaO$_3$ single crystal, (c) a Cu-rich film on (100) KTaO$_3$ (growth surface), containing the 248 phase, and (d) a stoichiometric Y123 film on a cleaved (100) surface annealed according to procedure 2 (see text). The peaks marked with an asterisk (*) originate from impurity phases in the film.

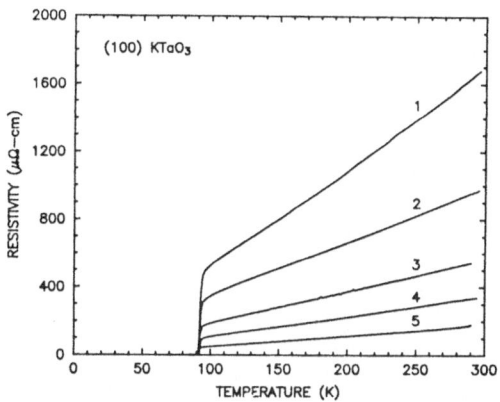

Fig. 6. Resistivity as a function of temperature for a series of c⊥ Y123 films deposited on (100) KTaO$_3$ and annealed according to procedure 2 (see text).

The critical current density as a function of temperature for the film on the growth surface is shown in Fig. 3. At 77 K j_c=0.15 MA/cm^2 and at 4.2 K j_c=1.3 MA/cm^2. For another film containing a larger fraction of c⊥ domains, j_c=0.3 MA/cm^2 at 77 K. These values are intermediate between those for c⊥ and a⊥ films on (100) SrTiO$_3$. In combination with the sharp superconducting transitions close to 90 K, this observation indicates that the superconducting properties of Y123 films on (100) KTaO$_3$ are not significantly affected by the high-temperature processing at 850°C. Thus KTaO$_3$ substrates provide an attractive, viable alternative for SrTiO$_3$.

In general, films on (100) KTaO$_3$ (cleaved or growth), annealed with procedure 1, exhibit larger fractions of c⊥ domains than films on (100) SrTiO$_3$. With the second annealing procedure, c⊥ films with only very small fractions of a⊥ domains are formed on either (100) KTaO$_3$ or SrTiO$_3$. As shown in Figs. 5d and 2b, both films contain the additional planar Cu-O stacking faults that are associated with the 248 phase. Usually, c⊥ films have a low resistivity with 0 K intercepts close to zero. In several (stoichiometric) films, however, we have observed much higher resistivities with temperature dependences that still extrapolate to the vicinity of the origin. For all curves in Fig. 6, the resistivity has been calculated using the geometrical dimensions of the film. The lowest of these curves (curve 5) has, in fact, been obtained for a 175 nm thick film. The other films are all approximately 355 nm thick (determined by means of RBS using the stopping power of Ba in Y123). For curve 1, the resistivity at 300 K is even higher than for films containing a majority of a⊥ domains. Variations of this magnitude were not observed for films annealed with procedure 1. As the resistivity perpendicular to the c-axis of single crystalline Y123 is a material property, the observed spread in resistivities implies that large differences can exist between the geometrical cross section of the film and the actual "effective" current carrying cross section. It seems that this variation is related to a variation in surface quality of the substrates.

Macroscopic domains of Y$_2$Ba$_4$Cu$_8$O$_{16-x}$ are formed in films with a Cu-rich metallic composition. This is shown in Fig. 5c, where sharp diffraction peaks are observed for c-⊥ domains of both the 123 and the 248 phase. The nominal composition of this particular film is Y$_1$Ba$_{1.9}$Cu$_{3.7}$O$_{7-x}$, i.e., close to the 1:2:4 cation ratio. The presence of the 248 phase in this film also follows[2,3] from the wider superconducting transition (curve 3 in Fig. 4), with zero resistance at 79 K rather than 90 K. As principal lattice parameters for the 248 phase we find a≅b=0.386(1) nm and c=2.7193(8) nm, in close agreement with refs. 2 and 3 for films on SrTiO$_3$.

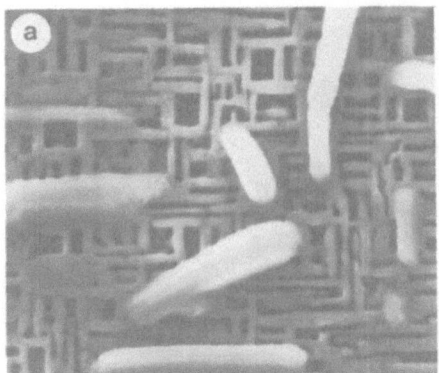
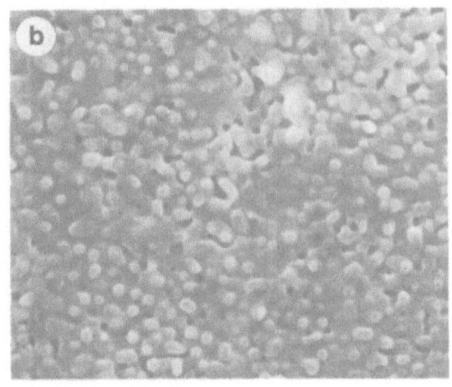

Fig. 7. Scanning electron micrographs of stoichiometric Y123 films, grown epitaxially on $SrTiO_3$ and $KTaO_3$, showing (a) a largely a⊥ film on (100) $SrTiO_3$, (b) a c⊥ film on (100) $KTaO_3$, and (c) a mixed a⊥ + c⊥ film on (100) $KTaO_3$.

IV. MICROSCOPICAL PROPERTIES

Figure 7 shows scanning electron micrographs of typical a⊥, c⊥, and mixed a+c⊥ films respectively on (100) $SrTiO_3$ and $KTaO_3$. Because of the similarity in epitaxial properties, the main features of these micrographs are the same for both substrate materials. The c⊥ film was formed on $KTaO_3$ and annealed with procedure 2. The a⊥ and the mixed films are on $SrTiO_3$ and $KTaO_3$ respectively, with both films annealed according to the first procedure. All three films have a Y:Ba:Cu ratio close to 1:2:3. The a⊥ grains form a rectangular network at the surface of the film. Due to an anisotropy in growth kinetics,[12] in these grains, the Y123 c-axis parallel to the substrate in the a⊥ domains points in the short dimension of the grain and the b-axis in the long dimension. The c⊥ domains grow as thin platelets that form plateaus parallel to the substrate. In the micrographs, they appear as a "pebble-grained" area. The light plates which seem to be lying on top of the surface were found to be substantially Y deficient and probably represent a barium-copper oxide. Note that the c⊥ film contains less of this impurity than the other films. Fewer Ba-Cu-O plates were also observed for films on (110) $SrTiO_3$.

The TEM cross section of a Y123 film/$KTaO_3$ substrate, shown in Fig. 8, reveals a number of interesting properties. First, it shows that the interface between the film and the substrate is well defined and that there is no evidence of a substantial reaction layer. Furthermore, the image shows that the c⊥ grains

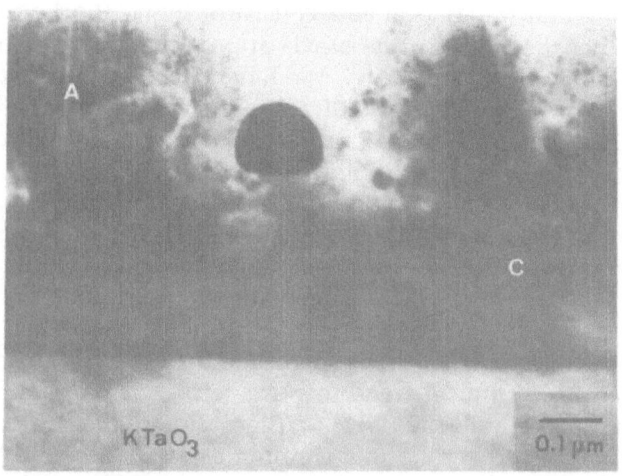

Fig. 8. TEM cross section of a mixed a⊥ + c⊥ Y123 film on the (100) growth surface of a KTaO₃ single crystal. The a⊥ and c⊥ domains are indicated in the figure.

are adjacent to the substrate with the a⊥ grains growing epitaxially on the c⊥ domains. The thickness of the c⊥ layer varies between 150 and 300 nm. This composition is consistent with the x-ray data and the SEM micrographs. In particular, for an a⊥ film, practically no c⊥ domains are observed at the surface (Fig. 7a) even though the x-ray diffraction pattern (Fig. 2a) contains non-negligible Y123 (00ℓ) reflections.

   This morphology allows for an interpretation of an interesting dissimilarity we have observed in θ-2θ scans through the (205) and (025) reflections of c⊥ grains in several Y123 films on (100) SrTiO₃ and KTaO₃. In Fig. 9a, such a scan is shown for a film on KTaO₃ that contains more a⊥ than c⊥ domains (curve 2 of

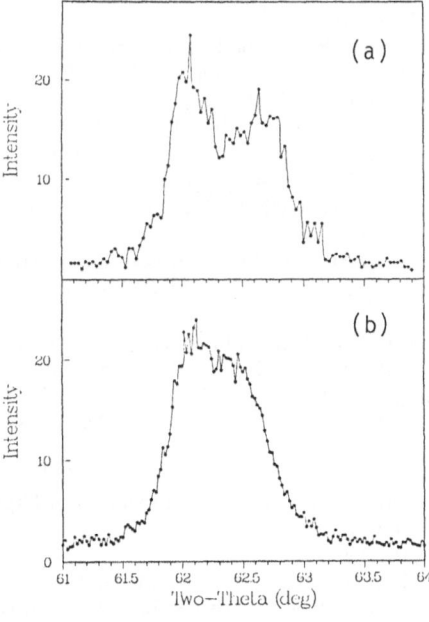

Fig. 9. Off axis θ-2θ x-ray scans through the (205) and (025) reflections of largely a⊥ Y123 films on (a) (100) KTaO₃ and (b) (100) SrTiO₃.

Fig. 4 is the corresponding $\rho(T)$ dependence). Two separate peaks can be distinguished, signifying that the c$\perp$ domains consist of two orthorhombic subdomains with either the a-axis or the b-axis approximately aligned with an in-plane <100> direction of the substrate. We have observed this pattern for all Y123 films on (100) KTaO$_3$. Typical lattice parameters for the c$\perp$ domains are a=0.383 nm, b=0.389 nm, and c=1.167 nm, in good agreement with values for bulk material. Similar values are also found for the a$\perp$ domains, although there may be slight differences between a$\perp$ and c$\perp$ within the same film.

For the film on (100) SrTiO$_3$, however, a single broadened peak is observed centered at a scattering angle intermediate between a and b (Fig. 9b) (separated (205 and (025) peaks are obtained for films with a larger fraction of c$\perp$ domains). There are several causes which may give rise to such a broadened diffraction peak. Possibly, the difference $|a-b|$ is smaller in the c$\perp$ layer. As an upper limit, we estimate $|a-b| < 0.004$ nm. A trend towards tetragonality could result from interdiffusion of Sr or Ti into the film.[13] Alternatively, the broadened diffraction peak might correspond to a continuous range of a and b lattice parameters in this layer, caused by a stress distribution or an inhomogeneous Sr or Ti content. The fact that we have not observed such effects for films on KTaO$_3$ is not conclusive for a comparison with SrTiO$_3$, but it does suggest a substantially reduced film-substrate interaction in the case of Y123 on KTaO$_3$.

In conclusion, we have shown that excellent Y123 film can be grown on (100) growth surfaces and cleaved surfaces of KTaO$_3$ by means of a fabrication process that includes a high-temperature anneal. The films exhibit sharp superconducting transitions with $T_c \cong 90$ K and are epitaxially oriented in three dimensions. The critical current densities measured thus far hold promise for high values which are comparable to those for films on (100) SrTiO$_3$.

REFERENCES

1. P. M. Mankiewich, J. H. Scofield, W. J. Skopcol, R. E. Howard, A. H. Dayem, and E. Good, Appl. Phys. Lett. 51:1753 (1987).
2. K. Char, M. Lee, R. W. Barton, A. F. Marshall, I. Bozovic, R. H. Hammond, M. R. Beasley, T. H. Geballe, A. Kapitulnik, and S. S. Laderman, Phys. Rev. B 38:834 (1988).
3. M. L. Mandlich, A. M. DeSantolo, R. M. Fleming, P. Marsh, S. Nakahara, S. Sunshine, J. Kwo, M. Hong, T. Boone, and T. Y. Kometani, Phys. Rev. B 38:5031 (1988).
4. J. Kwo, M. Hong, R. M. Fleming, A. F. Hebard, M. L. Mandlich, A. M. DeSantolo, B. A. Davidson, P. Marsh, and N. D. Hobbins, Appl. Phys. Lett. 52:1625 (1988).
5. G. J. Clark, F. K. LeGoues, A. D. Marwick, R. B. Laibowitz, and R. Koch, Appl. Phys. Lett. 51:1462 (1987).
6. Y. Enomoto, T. Murakawi, M. Suzuki, and K. Moriwaki, Jap. J. Appl. Phys. 26:L1248 (1987).
7. P. Chaudhari, R. H. Koch, R. B. Laibowitz, T. R. McGuire, and R. J. Gambino, Phys. Rev. Lett. 58:2684 (1987).
8. M. Naito, R. H. Hammond, B. Oh, M. R. Hahn, J. W. P. Hsu, P. Rosenthal, A. F. Marshall, M. R. Beasley, T. H. Geballe, and A. Kapitulnik, J. Mater. Res. 2:713 (1987).
9. T. Venkatesan, C. C. Chang, D. Dijkkamp, S. B. Ogale, E. W. Chase, L. A. Farrow, D. M. Hwang, P. F. Miceli, S. A. Schwarz, J. M. Tarascon, X. D. Wu, and A. Inam, J. Appl. Phys. 63:4591 (1988).
10. J. J. Cuomo, M. F. Chisholm, D. S. Lee, D. J. Mikalsen, P. B. Madakson, R. A. Roy, E. Giess, G. Scilla, in "Thin Film Processing and Characterization of High-Temperature Superconductors," AIP Conf. Proc. 165:141 (1988).
11. D. M. Hannon, Phys. Rev. 164:366 (1967).
12. D. L. Kaiser, F. Holtzberg, B. A. Scott, and T. R. McGuire, Appl. Phys. Lett. 51:1040 (1987); L. F. Schneemeyer, J. V. Wasczak, T. Siegrist, R. B. van Dover, L. W. Rupp, B. Batlogg, R. J. Cava, and D. W. Murphy, Nature 328:601 (1987).
13. M. F. Yan, W. W. Rhodes, and P. K. Gallagher, J. Appl. Phys. 63:821 (1988).

GROWTH OF HIGH-TEMPERATURE SUPERCONDUCTOR THIN FILMS ON

LANTHANUM ALUMINATE SUBSTRATES

R. W. Simon, A. E. Lee, C. E. Platt, K. P. Daly,
and J. A. Luine

TRW Space and Technology Group
Redondo Beach, CA 90278

C. B. Eom and P. A. Rosenthal

Applied Physics Department
Stanford University
Stanford, CA 94305

X. D. Wu and T. Venkatesan

Physics Department
Rutgers University
Piscataway, NJ 08854

ABSTRACT

We report on the use of the pseudo-perovskite crystalline
compound $LaAlO_3$ as a substrate for thin-film growth of high-
temperature superconductors. Lanthanum aluminate provides a
good lattice match to the basal plane of the superconducting
cuprates with its lattice constant of 3.792Å. In contrast with
strontium titanate, $LaAlO_3$ also serves as a useful substrate
for high-frequency applications with its low dielectric
constant (~16) and low microwave losses (5 x 10^{-6} loss tangent
at 4.2 K). In this paper we compare the properties of a number
of crystalline substrates for use with high-temperature
superconducting films.

We have tested the use of $LaAlO_3$ with thin films of erbium-
barium-copper-oxide and yttrium-barium-copper-oxide made by a
variety of techniques including multisource sputtering, single-
target sputtering, electron-beam evaporation, and pulsed laser
deposition. High-quality epitaxial films have been produced
using both single-step (in situ) and multi-step (post-anneal)
processing. Films on $LaAlO_3$ exhibit sharp resistive
transitions, high current density, strong c-axis orientation,
and bulk superconducting properties as evidenced by magnetic
susceptibility measurements. Substrate-film interaction
effects appear to be minimal for the $LaAlO_3$/1-2-3 system and
the thermal expansion properties of $LaAlO_3$ substrates are very
compatible with 1-2-3 films.

INTRODUCTION

The growth of high-quality thin films of high-temperature superconductors places unusually severe demands upon substrate materials because of high processing temperatures and the chemical reactivity and complex crystal structure of the superconducting compounds. Because of the anisotropic conductivity of the cuprate superconductors, optimum electrical performance has only been obtained in highly oriented film specimens. To date, epitaxial film growth on crystalline substrates of stronium titanate has produced the highest quality 1-2-3 superconductor films.

Strontium titanate has proven to be an advantageous substrate material for 1-2-3 thin films because of its close lattice match and its relative lack of chemical reactivity with the superconductor. Disadvantages of this substrate include its costliness, its unavailability in large wafers, and, most seriously, its highly undesirable high frequency properties. High dielectric losses and an extremely large dielectric constant make strontium titanate nearly useless as a substrate for many microwave electronic applications.

Therefore, considerable effort has been expended in developing alternative substrate materials for high-temperature superconductor thin film growth. The simple perovskite structure exhibited by strontium titanate is found in a large group of related oxide compounds with the ABO_3 formula. Depending upon the size and properties of the particular "A" and "B" ions in the compound, a variety of structural types occur as distortions from the cubic perovskite structure. These compounds occur in monoclinic, tetragonal, orthorhombic, or rhombahedral forms and often are polymorphic, exhibiting structural phase transitions as their temperature is varied. A variety of electrical properties occur in these materials as well; the ABO_3 perovskites include antiferroelectric, ferroelectric, pyroelectric, and piezoelectric materials.

From the standpoint of cuprate superconductor substrate use, many of the ABO_3 compounds have suitable crystalline properties. In many cases, the distortions from the cubic lattice are quite small and the slight rhombohedral or orthorhombic character of the lattice can be neglected. Thus a rather large number of compounds with pseudo-cubic lattice constants close to the 3.83Å a-axis parameter of the 1-2-3 compound may be able to support epitaxial growth of superconductor films. These materials must be evaluated in terms of high-frequency properties, chemical and structural stability during superconductor film growth, and ability to be grown economically in usable form. Among the compounds already investigated for this purpose are the cubic material $KTaO_3$[1], the orthorhomic compound $LaGaO_3$[2], and the rhombohedral compound investigated here, $LaAlO_3$[3].

SUBSTRATE PROPERTIES

Lanthanum aluminate deviates from a cubic structure because, at room temperature, its atomic planes meet an an angle of $90°6'$ rather than at right angles. Viewed as a cube, the unit cell for the material is 3.792Å on a side. This provides a lattice mismatch to 1-2-3 compounds of less than 1%.

LaAlO$_3$ is a relatively soft crystal (hardness 5-6 on the Mohs scale), can be clear or brownish due to F-centers, and congruently melts at 2110°C. The wafers used in these studies were grown by a commercial source[4] using flame fusion techniques. Crystal boules of 15mm x 15mm x 40mm dimensions were sectioned into small wafers polished on (100) surfaces. In the past, Czochralski growth techniques have also been successfully applied to this material[5].

Apart from the lattice constant matchup, the ideal substrate should also have a similar thermal expansion coefficient to the film in order that the film not be unduly stressed by thermal cycling during film growth and subsequent measurements. Figure 1 shows the lattice constants as a function of temperature for LaAlO$_3$[6], LaGaO$_3$[2], SrTiO$_3$[6], and 1-2-3[7]. Clearly all of these structurally related materials have similar thermal expansion properties below the orthorhombic-tetragonal phase transition in 1-2-3. The linear coefficients of thermal expansion are roughly 10 ppm/°C for all these compounds. These values are considerably higher than those for MgO, Si, and for KTaO$_3$, a cubic perovskite with a 3.99Å lattice constant. Thus the three substrates shown not only have lattic constants within 1% of one of the basal plane axes of the superconductor, but also have highly compatible coefficients of thermal expansion.

Our interest in developing LaAlO$_3$ for substrate use stems from its dielectric properties. We have measured the loss tangent of LaAlO$_3$[3] and have found it to be lower than that of

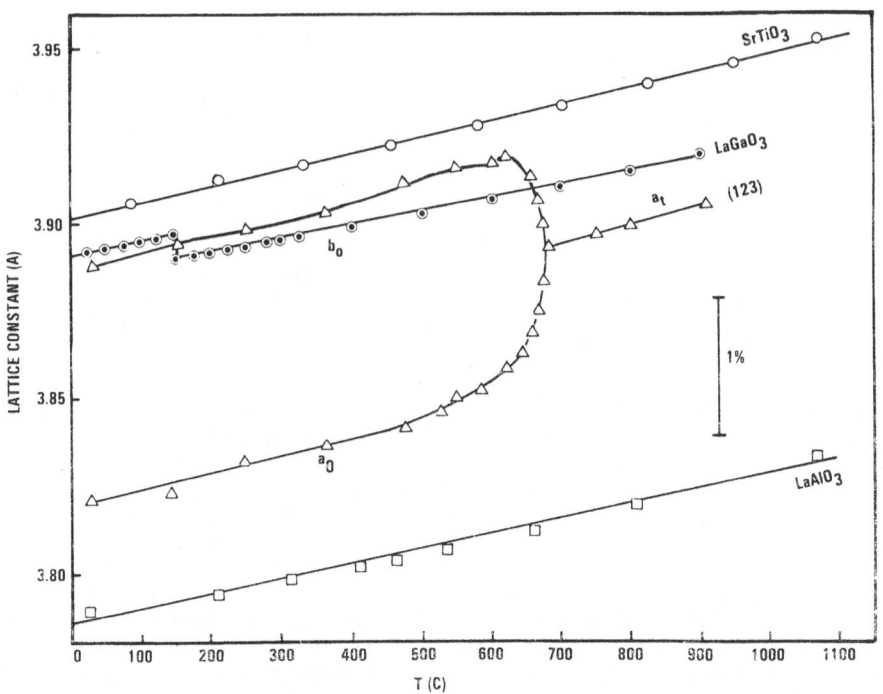

Figure 1 - Temperature Dependence of Lattice Constants of perovskite substrates and 1-2-3 superconductor.

other substrates currently in use for 1-2-3 films. Table I summarizes the room temperature values for a variety of substrates and lists the low temperature values we measured for LaAlO$_3$. In terms of dielectric constant, LaAlO$_3$ is markedly superior to the other perovskite substrates. Since the publication of our original dielectric measurements[3], we have used an enclosed suspended stripline measurement to measure the high frequency dielectric constant of LaAlO$_3$ and obtained a value of 16.8 at 52.6GHz. Table II summarizes the dielectric constant data for a variety of substrate materials[1,2,3]. By the criterion of dielectric performance, LaAlO$_3$ appears to be the most desirable substrate for superconductor film growth.

Table I. Dielectric Loss Tangents for
Some Potential Superconductor Film Substrates

Substrate	Loss Tangent ($\times 10^{-4}$)	
SrTiO$_3$	>200	
MgO	91	@ 1MHz
ZrO$_2$	54	@ 1MHz
LaGaO$_3$	18	@ 1MHz
LaAlO$_3$	5.8	@ 10GHz

Table II. Dielectric Constants for
Some Potential Superconductor Film Substrates

Substrate	Dielectric Constant	
Sapphire	8.6	
MgO	9.65	
Si	11.7	
LaAlO$_3$	16	
YSZ	25	
LaGaO$_3$	27	
KTaO$_3$	4000	
SrTiO$_3$	18,000	(4 K)

One other major consideration in the choice of a substrate for high-temperature superconductor films is the amount of chemical interaction between the film and the substrate. This issue is especially critical for films that undergo high-temperature post-annealing processes, a procedure common for 1-2-3 films and currently universal for the thallium-based superconductors. To investigate this issue, we performed Auger depth-profiling on erbium 1-2-3 films deposited on LaAlO$_3$ substrates. A compositional depth profile of a 500 nm film is shown in Fig. 2. To within the sensitivity of this measurement (about 2 atomic percent), there was no detected lanthanum or aluminum within the film until a region about 50 nm from the substrate. Similar profiling performed upon a film deposited on SrTiO$_3$ revealed detectable levels of strontium and especially titanium throughout the entire film. Deep within the film, differential sputtering effects have destroyed much of the depth resolution of the measurement and no detailed conclusions can be drawn about the extent of the nature of the interface or interaction region. We nonetheless can conclude that even in conjunction with high-temperature post-processing, chemical interactions between $LaAlO_3$ and 1-2-3 films do not appear to be a significant phenomenon.

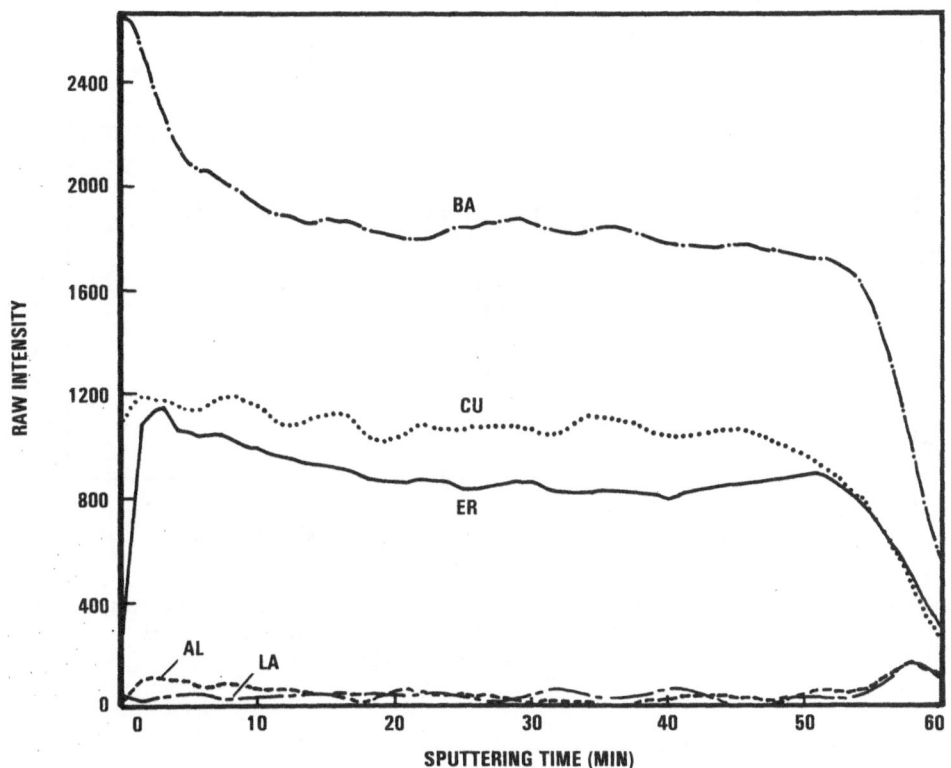

Figure 2 - Auger depth profiling into 500nm EBCO sputtered film on $LaAlO_3$.

SUPERCONDUCTOR FILMS

As reported in our earlier work[3], $LaAlO_3$ appears to be capable of supporting the same sort of epitaxial films growth that has been observed on $SrTiO_3$. We have deposited a number of erbium 1-2-3 films on $LaAlO_3$ substrates using a three-target sputtering technique presented elsewhere[8]. The best of these films are at least as good as our best films on $SrTiO_3$, exhibiting very sharp resistive transitions at 90 K (see Fig. 3a), high current density (2×10^6 amps/cm^2 at 4 K, 8×10^5 amps/cm^2 at 70 K), and highly oriented growth with the c-axis normal to the substrate surface. Figure 4 shows an x-ray diffraction spectrum for a 700nm EBCO sputtered film on $LaAlO_3$. The spectrum is dominated by (h,0,0) substrate peaks and (0,0,1) film peaks. The (1,1,0) peak -- indicative of unoriented grains -- is nearly absent, while the a-axis peaks are masked by the substrate peaks. Rocking curve data shows

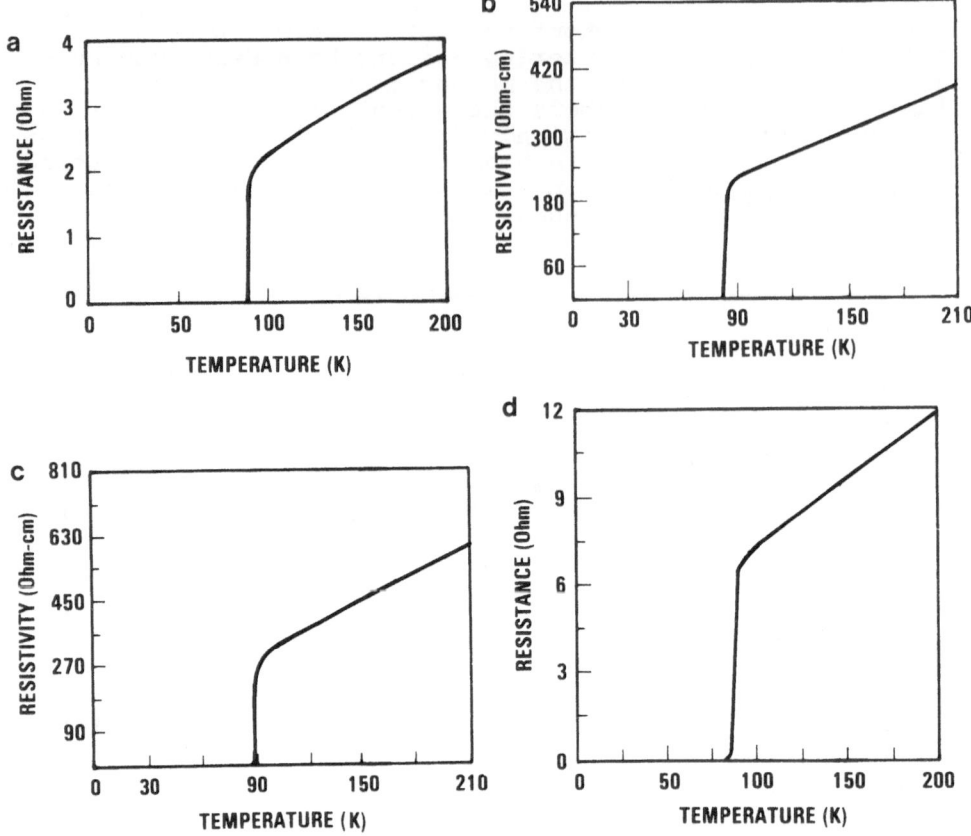

Figure 3 - Resistive transitions of 1-2-3 films on LaAlO$_3$ made by (a) multisource sputtering, (b) single-target sputtering, (c) electron-beam evaporation, and (d) pulsed laser deposition.

the film to be essentially 100% c-axis oriented although it is not single crystal in nature.

While the performance of these films was quite good, the current densities were not as high as the best numbers reported on SrTiO$_3$. An in situ growth process is under development in our laboratory; however, the co-sputtered films discussed above were produced by a post-annealing crystallization/oxidation process and their conduction properties are limited by grain-boundary effects. To learn more about superconductor film growth on LaAlO$_3$, a number of 1-2-3 films were produced elsewhere by several different techniques.

A film of yttrium 1-2-3 was grown in situ at Stanford University by a high-pressure single target sputtering process. The 350 nm film was deposited onto a LaAlO$_3$ substrate clamped to a 650°C block in a 5:1 Ar/O$_2$ gas mixture with a total pressure of 60 mT. The resultant film exhibited a 2 K-wide

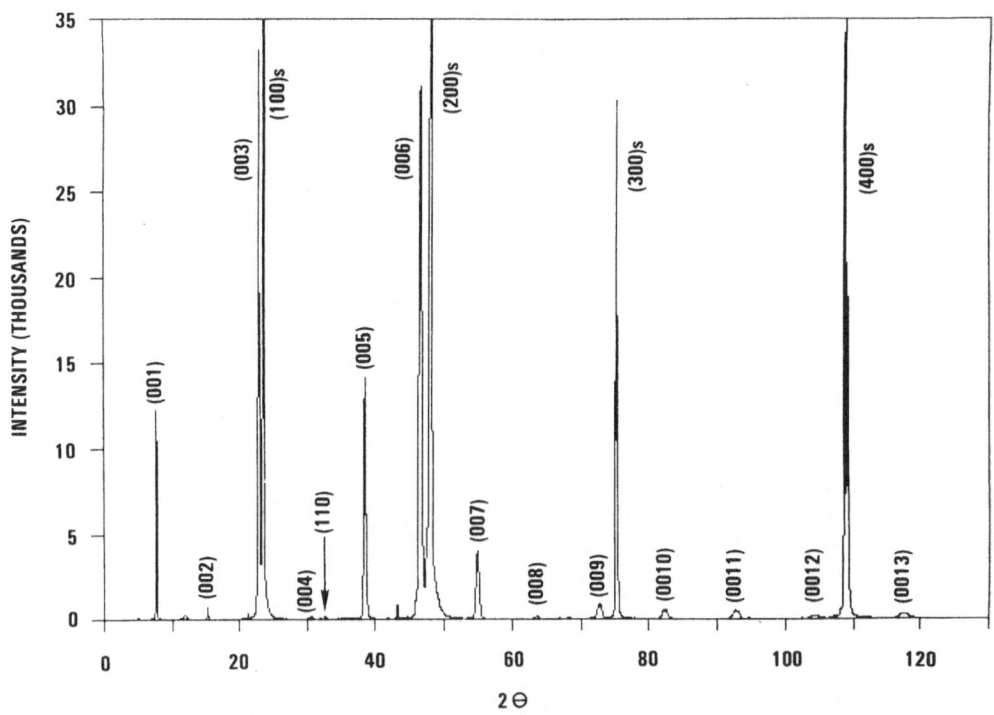

Figure 4 - X-ray diffraction spectrum for
EBCO film on LaAlO₃.

resistive transition with zero resistance at 84 K (Fig. 3b).
The film had a current density of 5×10^6 amps/cm^2 at 4 K. WDX
compositional analysis of the film showed it to be copper rich.
This was eventually traced to the bulk sputter target which ICP
analysis showed to have a cation ratio of 1:2:4 for
yttrium:barium:copper. These results are comparable to those
obtained on SrTiO₃ substrates in this system. X-ray analysis
revealed considerable amounts of a-axis growth on this
particular film. An SEM image of its surface is seen in Fig.
5; densely-packed a-axis grains dominate the morphology of this
film.

Another film was deposited at Stanford using the BaF₂
electron-beam evaporation process first developed at ATT Bell
Laboratories[9]. This 300 nm film exhibited a sharp resistive
transition at 89.4 K (Fig. 3c) and mostly c-axis growth
according to x-ray diffraction measurements. This one film had
a measured current density of 4×10^5 amps/cm^2 at 4.2 K.

The highest current density films have generally been those
with the densest morphology, ideally those devoid of current-
limiting grain boundaries. A technique that has thusfar
produced the smoothest films has been the pulsed laser
deposition technique developed by Bell Communications Research
and Rutgers University[10,11]. To test this technique with the
new LaAlO₃ substrates, a 100 nm film was deposited at Rutgers
using a 650°C substrate block and a 100 mT oxygen atmosphere.
The resultant film was extremely smooth (essentially
featureless on a 1 micron scale) and exhibited a 3 K-wide
resistive transition at 84 K (Fig. 3d). The somewhat low zero-

Figure 5 - SEM image of in situ sputtered film
of YBCO on LaAlO$_3$.

resistance temperature may be due to insufficient thermal
sinking of the substrate to the sample block. Despite the
slightly reduced T$_c$, this film was found to have a critical
current density of 1 x 10^7 amps/cm^2 at 4 K and 5 x 10^6 amps/cm^2
at 70 K. Current densities were measured in a vibrating sample
magnetometer using the Bean critical state model (in a system
cross-calibrated against transport measurements) and also by a
inductive technique using a coupled pancake coil[12]. Both
techniques gave the same results. These values are consistent
with films on SrTiO$_3$ produced by the Rutgers laser deposition
system.

While these results at other laboratories are preliminary -
- clearly no optimization of growth conditions for LaAlO$_3$ has
been performed -- it appears that the growth of 1-2-3 films on
LaAlO$_3$ is very similar to that seen on SrTiO$_3$. Critical
temperature, critical current density, and growth morphology
are similar for the new substrate. Such performance has also
recently been seen for the KTaO$_3$ substrate[1], but its dielectric
properties leave much to be desired. The LaGaO$_3$ material shows
greater promise. Its dielectric properties are not quite as
good as LaAlO$_3$, but are acceptable nonetheless. Of more
serious concern is a problem with macroscopic twin domains in

the substrate crystals which appear to change with thermal cycling above 150° C. This behavior is seen as a discontinuous jump in lattice constant in the thermal expansion data[2] in Fig. 1.

To summarize the current status of substrates for high-temperature superconductor film growth, the majority of high-quality films have been deposited upon $SrTiO_3$ substrates. The poor high-frequency performance and unlikelihood of the availability of large substrates make it unlikely that a superconductive electronics technology will be developed on this material. MgO has marginal dielectric loss properties and may not be able to produce films as good as those on the perovskite substrates. Silicon wafers, while they are the substrate of choice for conventional electronics, have serious problems with thermal mismatch with the superconductor as well as problems with the film growth itself. Thusfar, other materials such as sapphire, quartz, and cubic zirconia have shown only limited promise as substrates.

Therefore, it appears that a perovskite crystal such as $LaAlO_3$, $LaGaO_3$, or perhaps another material in the same class, may provide the best solution to the substrate problem. If the lattice match is truly critical, these materials can be engineered with mixing or doping to "tune" the lattice constant and stabilize the structure, as is done with yttrium doping in zirconium oxide. As an example, the basal plane lattice constant of the bismuth and thallium cuprate superconductors is 3.85A. An aluminate-galleate perovskite could match up exactly with these tetragonal superconductors. Such materials will undoubtedly be investigated in the future.

In meantime, both the galleate and the aluminate compound already show promise. $LaAlO_3$ has yet to be produced in the form of large area wafers; $LaGaO_3$ wafers have thusfar exhibited serious twinning problems that appear to limit the critical current density of films deposited on them[2,13]. These immediate problems may well be solved for each substrate. In the future, if both materials can be produced with equal quality and at equal cost, then the superior dielectric performance of $LaAlO_3$ may make it the preferred substrate for high-temperature superconductor films.

REFERENCES

1. R. Feenstra, L. A. Boatner, J. D. Budai, D. K. Christen, M. D. Galloway, D. B. Poker, submitted to _Appl. Phys. Lett_.
2. R. L. Sandstrom, E. A. Geiss, W. J. Gallagher, A. Segmuller, E. I. Cooper, M. F. Chisholm, A. Gupta, S. Shinde, R. B. Laibowitz, _Appl. Phys. Lett_. 53, 1874 (1988).
3. R. W. Simon, C. E. Platt, A. E. Lee, G. S. Lee, K. P. Daly, M. S. Wire, J. A. Luine, M. Urbanik, _Appl. Phys. Lett_. (to be published).
4. Commercial Crystal Laboratories, Naples, Florida 33942.
5. H. Fay, C. D. Brandle, _J. Phys. Chem. Solids, Suppl._, 1, 51 (1967).
6. D. Taylor, _Trans. & J. Br. Ceram. Soc_, 84, 181 (1985).
7. E. D. Specht, C. J. Sparks, A. G. Dhere, J. Brynestad, O. B. Cavin, D. M. Kroeger, _Phys. Rev_, B37, 7426 (1988).

8. R. W. Simon, C. E. Platt, A. E. Lee, K. P. Daly,
    M. K. Wagner, <u>J. Superconductivity</u> (to be published).
9. P. M. Mankiewich, J. H. Scofield, W. J. Skocpol,
    R. E. Howard, A. H. Dayem, E. Good, <u>Appl. Phys. Lett</u>. <u>51</u>,
    1753 (1987).
10. A. Inam, M. S. Hegde, X. D. Wu, T. Venkatesan, P. England,
    P. F. Miceli, E. W. Chase, C. C. Chang, J. M. Terascon, J.
    Wachtman, <u>Appl. Phys. Lett</u>. <u>53</u>, 908 (1988).
11. X. D. Wu, A. Inam, M. S. Hedge, T. Venkatesan, C. C.
    Chang, E. W. Chase, B. Wilkens, J. M. Tarascon, <u>Phys. Rev.</u>
    <u>B</u> (to be published).
12. J. H. Claassen, <u>IEEE Trans. Mag. 25</u>, (to be published).
13. P. M. Mankiewich, R. E. Howard, M. Nuss, D. B. Schwartz,
    B. Straugn, E. G. Burkhardt, T. Harvey, these proceedings.

SUPERCONDUCTING FILMS OF YBCO ON BARE SILICON

J. Tate, P. Berberich, W.Dietsche, and H. Kinder

Physik-Department E 10
Technische Universität München
8046 Garching, FRG

ABSTRACT

We have prepared thin films of superconducting $YBa_2Cu_3O_7$ on bare silicon (100) by thermal coevaporation of yttrium, barium and copper in an oxygen atmosphere. The highest temperature attained in the process is about 650° C, the temperature of the heated substrate. No post-deposition anneal is necessary to achieve superconductivity at 80 K and critical current densities, at 4.2 K, of 9×10^4 Acm^{-2}. In the best case so far, the YBCO film was superconducting at 85 K on bare Si after a short post anneal at 480° C. X-ray diffraction studies show that the best films are oriented with the c-axis perpendicular to the substrate, with no trace of reflections corresponding to other than those from $(00l)$ planes. We have also produced superconducting YBCO films on Si substrates with buffer layers of amorphous SiO_2 and amorphous Si_3N_4. These films are also largely c-axis oriented. The best superconducting characteristics were transition temperatures of $68 - 69$ K and critical current densities at 4.2 K of 3×10^3 Acm^{-2}.

INTRODUCTION

Thin films of the cuprate superconductors have been made on many different substrates, the most successful of which has undoubtedly been $SrTiO_3$. $YBa_2Cu_3O_7$ (YBCO) films have been grown epitaxially on this substrate, with critical currents greater than 10^6 A/cm^2 at 77 K [1] and transition temperatures above 90 K. However, $SrTiO_3$ is not a particularly useful substrate in many applications, and if thin film cuprate superconductors are to be widely useful in fundamental research, or are to be incorporated into integrated circuit technology, other substrates are necessary. Silicon, because of its widespread use in electronics, is one obvious choice.

The first attempts to grow superconducting thin films of YBCO on Si using the method so successful for $SrTiO_3$ (evaporation of the metals at low substrate temperatures, followed by annealing at $800 - 900°$ C in oxygen) failed because of the interdiffusion of YBCO and Si at the elevated temperatures. A lower temperature process was certainly desirable, particularly if it was to be compatible with current semiconductor technology. Several groups have reported superconducting as-deposited YBCO films on Si,[2-6] with the highest transition temperature to our knowledge, T_c (R=0), 85 K, being reported by our group. The methods are varied: dc sputtering[2,4] and laser evaporation[5] of a stoichiometric YBCO target, and e−beam[4] and thermal[6] evaporation of separate metal targets. The results of these experiments (see Table I) show that the superconducting, orthorhombic phase of YBCO can be

Table I. Recent results from groups reporting superconducting YBCO films on bare Si

Ref.	Method	T_{subst} (°C)	oxygen pressure (mbar)	T_c (K)	orientation
2	dc–sputtering	370	0.1 to 2*	56	polycrystalline, unoriented
3	e-beam, thermal coevaporation†	540	3×10^{-2}	68	polycrystalline, c- and b-axis ⊥ to substrate
4	dc-sputtering	650	4×10^{-4} *	76	polycrystalline, slight c-axis in plane
5	laser ablation	600	few $\times 10^{-3}$	67	c-axis ⊥ to substrate
6	thermal coevaporation	650	8×10^{-3}	85	c-axis ⊥ to substrate

* partial pressure of O_2 in Ar/O_2 mixture
† with plasma excited oxygen

produced at temperatures around 600° C, and the high temperature anneal can be avoided.

PREPARATION

We discuss here the characteristics of YBCO films on Si produced by thermal coevaporation of the constituent metals in an oxygen atmosphere. The evaporation rate of each metal is monitored by a quartz crystal oscillator and regulated by a feedback system. The substrate is located about 30 cm above the sources and is heated to about 650° C (as measured by a thermocouple pressed against the substrate surface) by a platinum wire heater. Oxygen is introduced close to the substrate so that the oxygen pressure is about 8×10^{-3} mbar in the vicinity of the substrate and 4×10^{-4} mbar in the remainder of the evaporation chamber. The evaporation rate is about 20 Ås^{-1}, (roughly the same as for laser ablation and faster than sputtering) and the films are typically $0.6 - 0.8$ μm thick. The films cool to room temperature in $1 - 10$ mbar of oxygen in about 40 minutes.

This procedure routinely produces metallic films ($\rho = 1$ mΩcm at 100 K) with superconducting transition temperatures of about $77 - 82$ K, as shown in Fig. 1. The best transition temperature to date is 85 K. Occasionally, we vary the post deposition treatment to include a post anneal at 480° C in flowing oxygen for 1 h. Usually, such a post anneal produces no improvement in the superconducting properties. We assume that sufficient oxygen is incorporated in the deposition phase.

CHARACTERIZATION

The composition of some of our films was determined by Rutherford backscattering to be within 10% of the nominal 1:2:3 ratio. No impurity phases were detectable by X-ray diffraction. Raman spectroscopy indicated the presence of $BaCuO_2$ in some films.

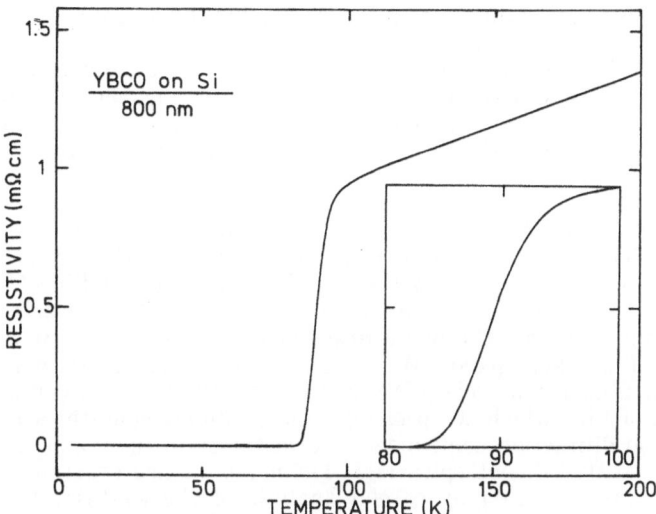

Fig. 1. Resistive transition of a YBCO film on Si. The inset shows an expanded scale in the region of the transition. (T_c = 81 K)

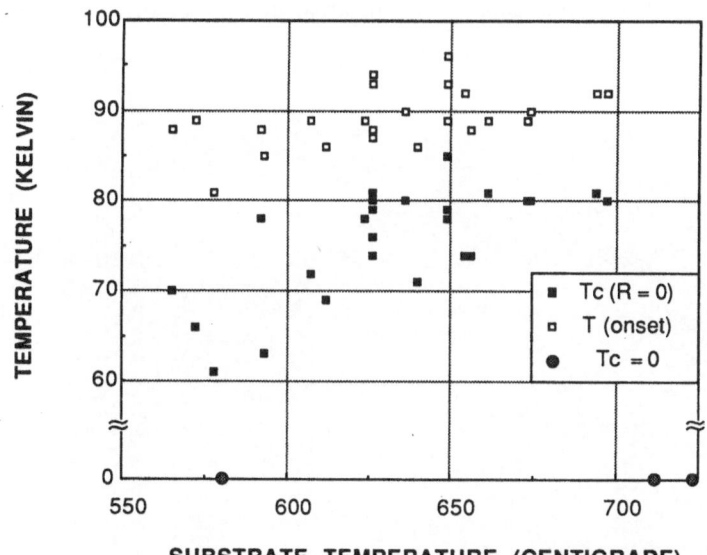

Fig. 2. Dependence of onset (T_o) and transition (T_c) temperatures on Si substrate temperature. Films prepared above T_s = 700° C were not superconducting.

We have investigated the effect of substrate temperature (T_s) on the films' properties. Fig. 2 shows that both the onset of superconductivity (T_o) and T_c itself tend to increase with increasing T_s until about $T_s \sim 700°$C. Films deposited on substrates above this temperature were not superconducting at 4.2 K.

We measured the critical current density (J_c) by noting the current required to produce a voltage of 1 μV across a constriction in a film. The constriction was either 100 μm or 500 μm wide, made by laser patterning or by a photolithographic process. Typically, J_c at 4.2 K for YBCO films on Si is $1 - 2 \times 10^4$ Acm^{-2}, with 9×10^4 Acm^{-2} the best result so far.

X-ray diffraction investigations reveal that the films are strongly oriented with the c-axis perpendicular to the Si substrate. In most cases there was no trace of other than (00l) reflections in a $\theta - 2\theta$ scan. This was true for films deposited on both (100) and (111) oriented Si. The extent of c-axis alignment was measured by rotating the film and substrate through a Bragg angle as the angle between the X-ray source and detector was kept fixed. We measured such rocking curves for the (005) peak in two of our films on Si. The FWHM was $\Delta\theta = 2.6°$, which can be compared to $\Delta\theta = 0.8°$ for a film which we prepared under similar conditions on a SrTiO$_3$ substrate. Electron diffraction confirmed the orientation of the c-axis and indicated that the films are textured, not epitaxial. The c-axis lattice constant for one film with $T_c = 67$ K was measured by careful analysis of the (0013) reflection. The result, c = 1.172 nm, is larger than the value for the bulk material, c = 1.168 nm. This is consistent with data published by Cava et al,[12] which related elongation of the c-parameter to reduced oxygen content and reduced T_c. An independent determination of the oxygen content will determine whether the elongation is due to insufficient oxygen content or to strains in the film.

Further elucidation of the nature of our films is provided by SEM and TEM studies. Fig. 3 shows a scanning electron micrograph of the surface of one of our films on Si. There is a smooth background matrix of 0.5 μm-sized YBCO crystallites, which, however, are not perfectly packed. This may account for the low critical current densities of these films compared to those on MgO or SrTiO$_3$. Also evident are micron-sized protrusions with an areal density of 0.1 μm^{-2}. Auger and

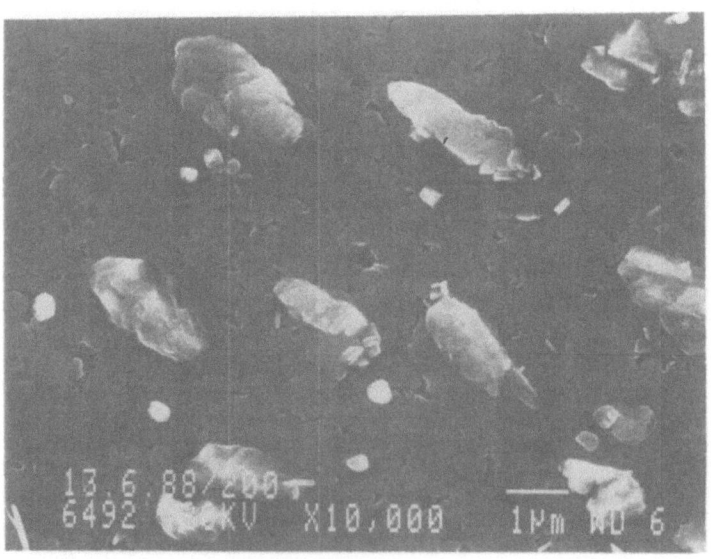

Fig. 3.   SEM image of the surface of an YBCO film on Si. The large grains are Cu rich.

EDX analyses indicate that these grains are Cu rich and Ba poor with respect to the background matrix. A TEM analysis of the cross section of the film indicated that the grains protrude about 0.1 μm above the surface.

TEM also reveals an amorphous SiO_2 layer approximately 10 nm thick between the Si substrate and the YBCO film. This is significantly thicker than the natural oxide layer on our Si substrates which we measured by ellipsometry to be 2 nm thick. To check whether the thicker oxide layer could have formed before evaporation of the metal film, we heated the Si substrate to 650° C in 8×10^{-3} mbar oxygen (the normal operating conditions) for 30 minutes without evaporating the film. Subsequent ellipsometry revealed that the oxide layer was no thicker than before. We therefore conclude that the 10–nm SiO_2 layer must have grown during evaporation of the film. At various points, the oxide layer is penetrated by grains which have grown into the Si substrate and the YBCO film. The grains are of the order of 0.5 μm in size and consist of a mixture of Si and Cu. Above the 10 nm oxide layer is a 20 nm thick crystalline layer which contains some Si as well as Y, Ba, Cu and O. Aside from this 30-nm region above the substrate, the bulk of the 600-nm film contains no Si and is rather uniform in composition.

In the light of the existence of the thin amorphous SiO_2 layer on the "bare" silicon substrates, it is interesting to compare the films prepared on "bare" Si and on Si with an 80–nm oxide layer grown by the manufacturer. At substrate temperatures up to about 650° C, the films on both substrates have very similar properties with the films on bare Si having higher T_c's by a few degrees and lower resistivity above T_c. However, at the higher substrate temperatures needed to produce $T_c = 80$ K and above for YBCO films on Si, we have found that the films on Si/SiO_2 have distinctly inferior properties. The transitions are broader and the critical currents are $1 - 2$ orders of magnitude smaller.

We also used hydrofluoric acid to etch away the natural oxide on the Si substrate before evaporating films. Such films have T_c's less than 50 K and J_c's of 200 Acm^{-2}.

We have also used our deposition method to produce YBCO films on Si coated with various buffer layers. Some work of this type has been reported,[5] [8-12] using buffer layers of ZrO_2, SiO_2 and various nitrides. The most successful buffer layer appears to be ZrO_2, which, when itself epitaxially grown, promotes epitaxial growth of YBCO films.[10] We used buffer layers of amorphous Si_3N_4 (150 nm) and amorphous SiO_2 (80 nm). The films prepared on these substrates at $T_s \sim 600–650°$ C also have their c-axes perpendicular to the substrate. In the X–ray diffractograms, the reflection at $2\theta = 32.85°$, which dominates in polycrystalline materials, is reduced to the size of the (004) reflection. The best YBCO films prepared on the amorphous buffer layers had T_c's up to 69 K and J_c's up to 3×10^3 Acm^{-2} at 4.2 K.

ACKNOWLEDGEMENTS

We wish to thank S. Schild and O. Eibl of Siemens for the TEM analysis, M. Scheib of Siemens for the lattice constant determination.

REFERENCES

1. B. Roas, G. Endres, and L. Schultz, Epitaxial growth of $YBa_2Cu_3O_{7-x}$ thin films by a laser ablation process, submitted to Appl. Phys. Lett.
2. R.J. Lin, J.H. Kung, and P.T. Wu, High T_c superconducting thin film fabrication by modified dc sputtering process, Physica C 153–155:796 (1988).
3. R.M. Silver, A.B. Berezin, M. Wendman, and A.L. de Lozanne, As-deposited superconducting Y-Ba-Cu-O thin films on Si, Al_2O_3, and $SrTiO_3$ substrates, Appl. Phys. Lett. 52:2174 (1988).
4. W.Y. Lee, J. Salem, V. Lee, T. Huang, R. Savoy, V. Deline, and J. Duran, High T_c $YBa_2Cu_3O_{7-x}$ thin films on Si substrates by dc magnetron sputtering from a stoichiometric oxide target, Appl. Phys. Lett. 52:2263 (1988).

5. T. Venkatesan, E.W. Chase, X.D. Wu, A. Inam, C.C. Chang, and F.K. Shokoohi, Superconducting $Y_1Ba_2Cu_3O_{7-x}$ films on Si, <u>Appl. Phys. Lett.</u> 53:243 (1988).

6. P. Berberich, J. Tate, W. Dietsche, and H. Kinder, Low-temperature preparation of superconducting $YBa_2Cu_3O_{7-x}$ films on Si, MgO and $SrTiO_3$ by thermal coevaporation, <u>Appl. Phys. Lett.</u> 53:925 (1988).

7. R.J. Cava, B. Batlogg, C.H. Chen, E.A. Rietmar, S.M. Zahurak, and D. Werder, Single-phase 60-K bulk superconductor in annealed $Ba_2YCu_3O_{7-x}$ ($0.3<x<0.4$) with correlated oxygen vacancies in the Cu-O chains, <u>Phys. Rev. B</u> 36:5719 (1987).

8. A. Mogro-Campero and L.G. Turner, Thin films of Y-Ba-Cu-O on silicon and silicon dioxide, <u>Appl. Phys. Lett.</u> 52:1185 (1988).

9. A. Mogro-Campero, L.G. Turner, E.L. Hall, and M.C. Burrell, Characterization of thin films of Y-Ba-Cu-O on oxidized silicon with a zirconia buffer layer, <u>Appl. Phys. Lett.</u> 52:2068 (1988).

10. G. Poullain, B. Mercey, H. Murray, and B. Raveau, High T_c "1 2 3" superconductive films on silicon substrates, <u>Mod. Phys. Lett.</u> B 2:523 (1988).

11. H. Myoren, Y. Nishiyama, H. Nasu, T. Imura, Y. Osaka, S. Yamanaka, and M. Hattori, Epitaxial growth of $Ba_2YCu_3O_x$ thin film on epitaxial $ZrO_2/Si(100)$, <u>Japn. Jour. Appl. Phys.</u> 27:L1068 (1988).

12. U. Poppe, J. Schubert, R.R. Arons, W. Evers, C.H. Freiburg, W. Reichert, K. Schmidt, W. Sybertz, and K. Urban, Direct production of crystalline superconducting thin films of $YBa_2Cu_3O_7$ by high-pressure oxygen sputtering, <u>Solid State Commun.</u> 66:661 (1988).

PEROVSKITE RELATED SUBSTRATES

FOR SUPERCONDUCTOR FILMS –

RARE EARTH ORTHOGALLATES

Roger F. Belt and Robert Uhrin

Airtron Division, Litton Systems Inc.
200 E. Hanover Avenue
Morris Plains, NJ 07950

INTRODUCTION

High quality single crystal thin films of the new oxidic high T_C superconductors are desirable for microwave, infrared, and Josephson junction devices. Initial attempts to grow films have utilized substrates of many available single crystals. All of the latter have been mixed oxides or other materials which do not possess the perovskite or a derivative type structure. $SrTiO_3$ has been the single exception and favorable high current capacity films have been deposited by several groups.[1-3] Unfortuntely, $SrTiO_3$ is grown by the flame fusion method and has some unfavorable characteristics of price, quality, size, and certain physical properties, e.g. high dielectric constant, high loss, lattice constant mismatch, or cleavage.

The rare earth orthogallates, $RGaO_3$ where R=La-Lu, all have a perovskite structure. Polycrystalline compositions were prepared and recognized[4] as early as 1954. Roth[5] gave preparation and X-ray data on many perovskites including a few containing gallium. The work of Geller[6] was again restricted to La, Pr, and Nd Polycrystals. An important step occurred in preparation of high pressure sintered[7] and finally small flux grown single crystals[8] of all rare earths. It was shown[9] that the larger size R ions give perovskites even at normal pressure. Ruse and Geller[10] grew small crystals of $NdGaO_3$ while those of $GdGaO_3$ were grown by Guitel[11]. To our knowledge, no one has grown large (>2 cm) crystals of $LaGaO_3$ or other gallates. In this paper we report the first results on this crystal and the growth of several other simple and mixed crystal gallates. We believe these materials offer good potential for substrates where single crystal superconductor films may deposit epitaxially.

EXPERIMENTAL

Starting components of rare earth oxides and gallium oxides were purchased from commercial suppliers. The La_2O_3 and Gd_2O_3

were 99.99 % purity and obtained from Aesar Chemical, Eagles Landing, NH. The Ga_2O_3 was 99.99 % purity and obtained from KBI Chemical Division of Billiton Witmetal, NY. The Nd_2O_3 was 99.999 % purity and obtained from Research Chemical of Phoenix, AZ. The oxides were used without further sintering or drying. The particle size of the powders was in the 1-5 μm range. For each particular crystal grown, the components were mixed in the stoichiometric 1:1 mole ratio for the orthogallates.

The melts were contained in iridium crucibles of 5-7.5 cm height x 5-7.5 cm diameter x 2 mm wall thickness. Crucibles were fabricated and purchased from Engelhard Corp. of Carteret, NJ in a purity of 99.99 %. Since seed crystals were not available for growth, the first boules were grown by using an iridium wire for nucleation. In this procedure, our preferred direction of growth appeared to be about 5-10° from [112] of the orthorhombic unit cell. This was repeated several times, but may not occur always. Once a single crystal of $LaGaO_3$ was obtained, it was oriented by X-ray Laue back reflection to obtain seed crystals of [100], [001], [110] and [112] orientations. These seeds were core drilled to a cylinder of 6 mm diameter x 2-4 cm long and used for subsequent growth of boules. It should be remarked that no prominent facets were revealed on the boule exteriors and shapes were cylindrical for all orientations.

All crystals were grown by the Czochralski method in a conventional station arrangement. The iridium crucibles were fitted with a lid which had an opening of 3-4 cm diameter. Our diameter control system was of the optical pyrometer type with appropriate melt level drop programmed into the electrical feedback. The RF heating consisted of either a 350 kHz tube type Lepel system or a 15 kHz Pillar solid state unit. Both of these were rated at 40 kW input capability and only 30-40 % of capacity was needed to obtain acceptable melts near 1750° C. The crystals were grown at a pull rate of 1-2 mm/hr and a rotation rate of 30-45 rpm. The radial and axial gradients were not measured specifically for our $LaGaO_3$. However, from similar geometrical arrangements in growing other garnets, e.g. $Gd_3Ga_5O_{12}$, we estimate about a 30-50°/cm radial gradient and 50-100°/cm axial gradient. The actual shape of the solid-liquid interface was convex and the full taper angle of a boule end pulled free of the melt was about 60°.

After a boule was grown, fabrication into substrates of several orientations was performed. The respective boule axis was achieved by the correct seed in growth. The boule was ground to a cylinder of 1.00 inch diameter on a Landis Tool outside diameter grinder. The cylinder was mounted on a fixture and X-ray oriented to ±0.25° of the desired direction on a Secasi diffractometer. The boule was then sliced on an STC inside diameter diamond wheel slicer with a rough cut wafer thickness of 0.030 inch. Wafers were then optically polished on one surface and epitaxially polished on the opposite surface. The latter steps included an initial lapping and removal of about 100 μm and finishing to parallel surfaces. The epitaxial surface is finished by lapping with 0.05 μm colloidal silica to remove an additional 25-75 μm. The resulting surface is scratch free with no underlying bulk damage. Finished wafers are 0.020 ± 0.002 inch thick.

RESULTS

Our initial work on Czochralski growth of $LaGaO_3$ was performed by melting stoichiometric 50-50 mole % compositions of La_2O_3 and Ga_2O_3. Stable melts were obtained with no apparent volatilization of Ga_2O_3 in spite of the large discrepancy between the melting points of components. Seeding of a crystal was achieved by use of an iridium wire. An uncorrected pyrometer reading of the melt temperature during seeding indicated that the melting point was about 1700° C or 50-100° C below that of Ga_2O_3. Crystal growth proceeded easily at a pull rate of about 1-2 mm/hr. A satisfactory boule was obtained with no cracking at a size of 2 cm diameter x 8 cm long. No faceting or other distinguishing features were found on the crystal. The color was not water white, but slightly yellow brown. The color may be attributed to either residual Fe^{3+}, estimated at 50-200 ppm, or small amounts of oxygen vacancies arising from trace losses of Ga_2O_3. Our first crystal was grown from no preferred orientation, but Laue X-ray examination showed that the boule axis was within 10° of [112] of the orthorhombic unit cell. X-ray diffractometer patterns were recorded from portions of the grown single crystal and the residual melt. These patterns were identical to each other and contained no other lines beyond those recorded by Geller[6] and Marezio[8]. However, we should note that all weak or apparently missing lines in their patterns were present in our patterns. This may be a result of the excellent crystal quality of our preparation. We also believe our data indicate that $LaGaO_3$ is congruently melting, but possibly not at 50-50 stoichiometry. The X-ray data confirm that the unit cell is orthorhombic at room temperature. The crystallographic transition to the rhombohedral structure at 875° C reported by Geller[6] was not examined in detail. It did not have any apparent effect on growth or quality.

The preliminary phase diagram of the La_2O_3-Ga_2O_3 system was investigated[12] and indicated that $LaGaO_3$ formed a compound. We performed an extensive literature search (to March 1988) and learned that no single crystals were grown previously. We also found that a thorough recent study of the phase diagram was made.[13] Mizuno's data show that $LaGaO_3$ is congruently melting at 1715° C. One phase transition from rhombohedral to orthorhombic is reported at 900° C and confirmed by high temperature X-ray data. The 25° C lattice constants of Mizuno[13] from polycrystalline $LaGaO_3$ agree with those of Marezio[8] with an interchange of a and b axes of the orthorhombic unit cell. Again our values are identical but show all very weak lines because of crystal quality. The 2:1 compound melting at 1704° C is apparently the only other compound in the system. This compound was indexed as monoclinic from a complete powder pattern. None of our $LaGaO_3$ has been found to contain any trace of the 2:1 in the pulled single crystals or residual melts.

For substrate utilization of $LaGaO_3$ single crystals, it is desirable to grow along certain preferred directions, e.g. the [100] or [010], [001], and [110] of the orthorhombic unit cell. The latter has been designated somewhat differently by workers, particularly in the interchange of the a and b axes. We prefer to adopt the convention of Geller[6] with c>b>a. The recognized space group of the structure is Pbnm-D$_{2h}^{16}$. In this space group which is in point group mmm, the projection symmetry is 2 mm along the major axes and m along (hko). Thus, Laue back reflection in itself is not too much help. However, the near identity of the $\underline{a}$ and $\underline{b}$ axial lengths introduces a pseudo four fold axis along [00$\bar{1}$]. Reference can be made to this direction. The positive identification of the other axes can be made by examination of high angle (hho) and (oko) reflections on a diffractometer. However, twins may interfere with this determination. Furthermore, (220) spacings are nearly identical to (002) in the unit cell and these again must be separated for positive assignment. Thus, this crystal and others of a similar crystal structure must be examined carefully for directional assignments. Our lattice constant data are essentially equal to those of previous workers.

The appearance of micro twins is common in the perovskite and related structures. We feel it is instructive to describe these in detail and point out some interesting features of our LaGaO$_3$ single crystals. Figure 1 is a photograph of a crystal grown along the [001] direction. The orientation was checked by X=ray examination of 2θ angles and intensities of possible (001) reflections where l = 2-10, and (kko) reflections where h=k=1-5. The checking was performed on one of several slices which were cut perpendicular to the boule axis. Each slice was polished prior to examination. Slices were then mounted on a diffractometer and planes parallel to the slices were examined along with a (111) cut standard silicon single crystal. Fig. 2a shows a surface slice of the single crystal after mechanical polishing with 20 µm Al$_2$O$_3$, 6 µm diamond, and 1 µm Al$_2$O$_3$. Twin planes course through the crystal and their diagonal traces are evident. The planes are parallel neatly to the [001] growth axis. The likely plane assignment is (hoo) and (oko). Fig. 2b is the same slice which has been etched in a 1:3 mixture of conc. H$_2$SO$_4$ and 30 % H$_2$O$_2$ at 160° C for 1 minute. This etchant is used commonly for GGG wafers and reveals damage, dislocations, and other artifacts in these substrates. The figure shows the random scratch damage due to sub-surface defects introduced by polishing, a few light digs, and no apparent defects at the twin boundaries. We take these data to imply that the crystal quality is excellent except for the twins. X-ray line widths on (ool) reflections confirm this since those measured for LaGaO$_3$ were only slightly larger than those for a nearly perfect and dislocation free (111) silicon single crystal planes.

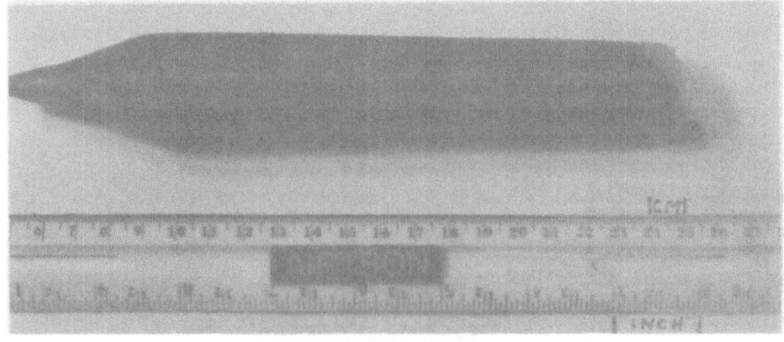

Figure 1.   Czochralski grown boule of LaGaO$_3$

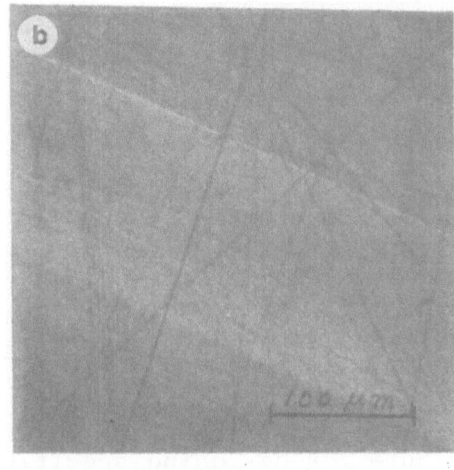

(a)   (b)

Figure 2.   Polished slice of (001) wafer of LaGaO$_3$
            in (a). Etched surface is shown is (b)

Fig. 3 is a transmitted light photograph of a 2.6 cm
diameter slice of LaGaO$_3$ cut from the same boule as above. The
following features are notable. The slice is divided into five
major different domains by poorly defined boundaries which run
vertically in the illustration. Within each major domain, the
twin planes all run parallel to each other. Some areas may be
free of twins or intersect further boundaries. The twins in
the adjacent major domains are at 90° to each other and both
sets are parallel to the [001] growth axis. It may be mentioned
that the introduction of this slice between crossed polarizers
does not change or extinguish any planes. We believe the twin
planes to be of type (hoo) and (oko). This type of twinning
has been described previously by several workers[14,15] and is
different from that of Ruse and Geller[10] observed on (110) planes
of NdGaO$_3$ and apparently revealed only under cross polarized
light.

The twin structures observed in our crystals appear to be
generated by strains induced during the crystal growth or
cooling. Properly LaGaO$_3$ must be classified as a ferroelastic
crystal, i.e. it is characterized by a net spontaneous strain
Es which can be reoriented experimentally or conceptually by
atomic motions required in changing the orientations[16]. The
criteria listed by Abrahams[16] are pertinent to those crystals
containing twins. They are recognizable by (1) optical
examination, since a twinned crystal with common c* but with a*
in some domains parallel to b* in other domains, does not
extinguish in polarized light (2) has (hol) and (okl) reflections
simultaneously present, and (3) the (600,060) reflections are
more than twice as broad as (600) or (060) alone. The ferroelastic crystal behavior of LaGaO$_3$ can be demonstrated by detwinning
under application of a compressive stress and a single crystal
with orientation abc can be transformed to one with orientation
bac.

Other Orthogallates

For some superconductor films, the X-ray lattice constants of $LaGaO_3$ may be too large for near ideal match. Convenient changes can be achieved by growing other pure rare earth gallates which contain the elements Ce-Gd. For example we have grown the $NdGaO_3$ in both [001] and [100] orientations. These crystals gave a highly colored purple-red boule since the Nd has many absorption bands from 350 to 900 nm. Our boules were about 2.5 cm diameter and 10 cm long in contrast to the one previous report of a 1 cm diameter crystal. We have also grown $GdGaO_3$ in a single crystal close to the [112] orientation. This crystal was not of good quality and had some inclusions of the garnet phase. Obviously the exact growth stoichiometry is important for attaining best quality. We have grown mixed crystals of $La_{1-x}Nd_xGaO_3$ where x = 0.2. This crystal grew without difficulty in an [001] orientation. It suggests that other solid solutions are indeed possible and opens an alternate procedure for designing specific substrate lattice constants to match films.

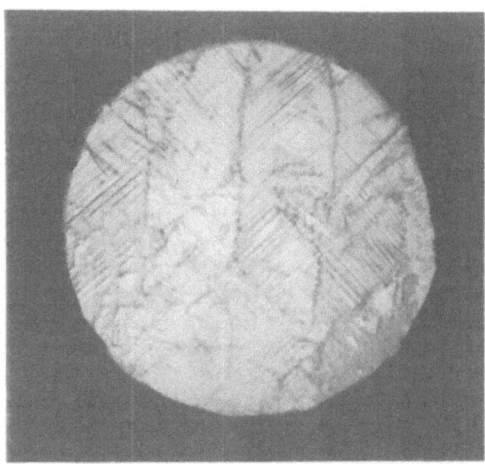

Figure 3. Transmitted light photograph of 1 inch diameter $LaGaO_3$ (001) polished slice. Twins and domains are clearly evident.

One of the critical physical properties of substrates is the dielectric constant and loss at microwave frequencies. We have measured several samples of the $LaGaO_3$, $NdGaO_3$ and the mixed system $La_{1-x}Nd_xGaO_3$. Samples were prepared by core drilling a cylindrical rod and placing it in a cavity. A standard procedure was followed according to ASTM Standard designation F131-86. All of our initial measurements were performed at 9.1 GHz and 300 K. For each of the samples, a dielectric constant of 25-28 and a loss tangent of no larger than 0.002 was measured. Additional data are being taken at 77K and will be reported in a separate communication.[17]

DISCUSSION

It has become abundantly clear that single crystal films of high perfection are most desirable for devices. The films generated by true epitaxy are likely to have the highest current carrying capacity, good stability, and fewest defects. The main criteria of epitaxy are crystal structure identity and lattice constant match. Past examples are Si on Si, garnets on $Gd_3Ga_5O_{12}$, and 3-5's on GaAs. To prevent serious strains and film cracking, the lattice match should be within ±0.3 %, although some heteroepitaxial films tolerate larger mismatch. Soon after the discovery of the $YBa_2Cu_3O_7$ superconductor, many workers grew films on currently available single crystal substrates. Table 1 gives a a list of these materials with melting points, lattice constants, and dielectric data. It can be seen from the table that the lattice constants offer no near matches to superconductors of of Table 3. However, the crystals do offer high melting points, a rather inert reactivity to very corrosive oxidic components which could lower T_c, and very favorable dielectric properties. Nearly all films grown on these substrates have been oriented polycrystals with varying electrical properties.

Table 1. Non-Perovskite Single Crystal Substrate Materials for Microwave Applications

Material	Melting Point(°C)	Crystal Structure	Lattice Constants a(A)	c(A)	ε at 9.1 GHz
Al_2O_3	2050	hexagonal	4.76	12.991	10
MgO	2400	cubic	4.21		9
$ZrO_2 \cdot Y_2O_3$	2800	cubic	5.23		10
$MgAl_2O_4$	2100	cubic	8.085		12
Si	1450	cubic	5.43		12
$Gd_3Ga_5O_{12}$	1800	cubic	12.383		15
BaO	1900	cubic	5.54		10
TiO_2	1950	tetragonal	4.60	2.960	150
SiO_2	575	hexagonal	4.91	5.394	5

A big step forward occurred when films were attempted on perovskite single crystal substrates. The most readily available material was $SrTiO_3$, but the growth method of flame fusion is not conducive to high perfection. It also suffers from high dielectric constant and loss. Table 2 compiles a list of some typical perovskites. Hundreds more are listed in a book[18], but unfortunately, 95 % of these are not congruently melting, melt too high to be grown in iridium, react with iridium, or have a deleterious effect on T_c of the superconductor. Of all the materials in Table 2, the $LaGaO_3$ (including other rare earth gallates) offers direct lattice match to most of the new superconductors.

Table 2. Perovskite Related Single Crystal Substrate Materials

Material	M.P. (°C)	Unit Cell	a(A)	b(A)	c(A)	ABO_3 (A)	ε
$SrTiO_3$	2000	c	3.905	----	----	3.905	150
$LaGaO_3$	1750	o	5.519	5.494·	7.770	3.92	27
$LaAlO_3$	2050	r	5.357	$\alpha=60°6'$		3.790	15
$BaSrNb_2O_6$	1450	t	12.430	----	3.913	3.693	25
$LiNbO_3$	1260	h	5.15	----	13.86	3.18	25
$LiTaO_3$	1560	h	5.15	----	13.77	3.16	30

For convenience we have listed in Table 3 the lattice constant data for most of the new oxidic superconductors. It should be realized that for the Tl compounds, the structures are found to be incommensurate, i.e. regions of varying axes are present and polytype phenomena occur as in other layered compounds, e.g. SiC. Thus, the reported c-axes may also be an integer n (where n = 1,2,3....) times that listed. We now ask what may be the possible epitaxial relationships between a substrate such as $LaGaO_3$ and a superconductor film, e.g. e.g. $BaY_2Cu_3O_7$. For clarity we have constructed Table 4 which gives calculated d lattice spacings for these respective structures. The last column summarizes several of the idealized relations. Not all of these have been investigated yet, and only one case has been documented[19], i.e. that of (003)F/(002)S. In this situation, the highest current carrying directions of $BaY_2Cu_3O_7$ films are lying in the a-b planes which deposited on (001) planes of $LaGaO_3$. However, the in-plane axial relations cannot be preserved simultaneously since there is a large discrepancy in magnitudes of the unit cell vectors. Therefore, it is likely that a and b axes of the film align along other directions such as 1/2 [110]. Obviously, further work has to be done to verify the exact nature of the epitaxy.

Table 3. Film Epitaxy for Oxidic Superconductors and Substrates

Material	T_c (°K)	Structure	a(A)	b(A)	c(A)
Superconductors					
$La_{1.85}Sr_{.15}CuO_4$	45	Tetrag	3.78	3.78	13.23
$YBa_2Cu_3O_7$	94	Ortho	3.856	3.870	11.666
$Bi_2CaSr_2Cu_2O_9$	90	Tetrag	3.817	3.817	30.6
		Ortho	5.41	5.44	30.8
$Tl_2Ba_2CuO_6$	90	Tetrag	3.87	3.87	23.24
$TlBa_2CaCu_2O_7$	91	Tetrag	3.85	3.85	12.75
$TlBa_2Ca_3Cu_4O_{11}$	125	Tetrag	3.85	3.85	19.01

One other important question arises in the use of $LaGaO_3$ substrates. Since most films are deposited at temperatures of 400-700° C, some diffusion of La or Ga into the superconductor may occur. The effect of La on T_C is clearly not significant. For Ga several studies have been attempted and the latest[20] gives rather ambiguous results. However, all of these studies were performed on bulk polycrystalline materials of $YBa_2(Cu_{1-x}Ga_x)_3O_y$ where x may reach a maximum of 0.1. For single crystal films, the T_C's reported[19] seem to be changed only slightly and film quality or preparation technique may control T_C rather than Ga diffusion.

Table 4. Epitaxial Relations For Substrate and Film

$LaGaO_3$ Substrate(A)	$YBa_2Cu_3O_7$ Film (A)	Near Lattice Matches, Film(F) on Substrate(S)
a = 5.482	a = 3.856	(100)F / (002)S
b = 5.526	b = 3.870	(100)F / (220)S
c = 7.780	c = 11.666	(010)F / (002)S
d_{100} = 5.482	d_{100} = 3.856	(010)F / (220)S
d_{010} = 5.526	d_{010} = 3.870	(110)F / (112)S
d_{001} = 7.780	d_{110} = 2.727	(003)F / (220)S
d_{220} = 3.892	d_{003} = 3.889	(003)F / (002)S
d_{002} = 3.890	d_{101} = 3.661	
d_{112} = 2.751	d_{002} = 5.833	

CONCLUSIONS

Single crystals of the perovskite $LaGaO_3$ were grown in one inch diameter boules. Substrates were prepared with quality suitable for high T_C single crystal superconductor films. The lattice constants are close to those of several film compositions and can be adjusted by other rare earths in mixed systems. Different orientations as (100), (001), (110) were grown, cut, and fabricated into polished wafers for evaluation. At least two twin structures of (100) and (110) type are present in the as grown crystals. Their effects on film properties must be evaluated. The orthogallates offer a dielectric constant around 25 and a loss of .002 at microwave frequencies. (9.1 GHz) These substrates are now available for use in various superconductor devices.

ACKNOWLEDGMENTS

The authors wish to thank Litton Industries for support of a portion of this research. Discussions with E. Giess of IBM Research Center were appreciated. Wafers were fabricated at the Airtron facility under the supervision of R. Hathaway. We thank Dr. J. Ings for several X-ray orientations.

REFERENCES

1. X.D. Wu, D. Dijkkamp, S.B. Ogale, A. Inam, E.W. Chase, P.F. Miceli, C.C. Chang, J.M. Tarascon, T. Venkatesan, Epitaxial Ordering of Superconductor Films on (001) $SrTiO_3$, Appl. Phys. Letters 58:861 (1987).
2. K. Char, A.D, Dent, A. Kapitulnik, M.R. Beasley, T.H. Geballe, Reactive Magnetron Sputtering of Thin Film $YBa_2Cu_3O_{7-x}$, Appl. Phys. Letters 51:1370 (1987).
3. J. Kwo, T.C. Hsieh, R.M. Fleming, M. Hong, S.H. Lion, B.A. Davidson, L.C. Feldman, Structural Properties of Orientation Ordered $YBa_2Cu_3O_7$ Films, Phys. Rev. 36:4039 (1987).
4. M.L. Keith and R. Roy, Structural Relations Among Double Oxides of Trivalent Elements, Am. Min. 39:1 (1954).
5. R.S. Roth, Classification of Perovskite and Other ABO_3 Compounds, J. Research Natl. Bur. Standards 58:75 (1957).
6. S. Geller, Crystallographic Studies of Perovskite-Like Compounds, Acta Cryst. 10:243 (1957).
7. M. Marezio, J.P. Remeika, and P.D. Dernier, High Pressure Synthesis of $YGaO_3$, $GdGaO_3$ and $YbGaO_3$, Mat. Res. Bull. 1:247 (1966).
8. Marezio, J.P. Remeika and P.D. Dernier, Rare Earth Ortho-gallates, Inorg. Chem 7:1337 (1968).
9. S. Geller, P.J. Curlander and G.F. Ruse, Perovskite-Like Rare Earth Gallium Oxides, Mat. Res. Bull. 9:637 (1974).
10. G.F. Ruse and S. Geller, Growth of Neodymium Gallium Oxide Crystals, J. Crystal Growth 29:305 (1975).
11. J.C. Guitel, M. Marezio and J. Mareschal, Single Crystal Synthesis of $GdGaO_3$, Mat. Res. Bull. 11:739 (1976).
12. S.J. Schneider, R.S. Roth and J.L. Waring, Phase Diagrams of $Ga_2O_3-R_2O_3$ Systems, J. Research Natl. Bur. Standards 65A:365 (1961).
13. M. Mizuno, T. Yamada and T. Ohtake, Phase Diagram of the System $Ga_2O_3-La_2O_3$, Yogyo-Kijokai Shi 93:295 (1985). In Japanese.
14. M. Marezio and P.D. Dernier, The Bond Lengths in $LaFeO_3$, Mat. Res. Bull. 6:23 (1971).
15. S.C. Abrahams, J.L. Bernstein and J.P. Remeika, Ferroelastic Transformation in $SmAlO_3$, Mat. Res. Bull 9:1613 (1974).
16. S.C. Abrahams and E.T. Keve, Structural Basis of Ferroelectricity, Ferroelectrics 2:129 (1971).
17. Andrew Felce and Roger F. Belt, Dielectric Measurements at X-Band on Substrates, Submitted to J. Appl. Phys.
18. F.S. Galasso, "Structure, Properties, and Preparation of Perovskite-Type Compounds", Pergamon Press, New York, 1969.
19. R.L. Sandstrom, E.A. Giess, W.J. Gallagher, A. Segmuller, E.I. Cooper, M.F. Chisholm, A. Gupta, S. Sharide and R.B. Laibowitz, $LaGaO_3$ Substrates for Epitaxial High T_C Thin Films, Appl. Phys. Letters, To be published.
20. Y. Xu, R.L. Sabatini, A.R. Moodenbaugh and M. Suenaga, Effect of Ga Addition to $YBa_2Cu_3O_7$, Phys. Rev. B38:7084 (1988).

PREPARATION OF SUPERCONDUCTING Y–Ba–Cu–O

THIN FILMS ON METALLIC SUBSTRATES

Masanori Ozaki, Nakahiro Harada, Shoji Akashita
and Jon-Chi Chang

Yokohama R & D Laboratories, The Furukawa Electric Co., Ltd.
2-4-3, Okano, Nishi-ku, Yokohama, 220 Japan

ABSTRACT

The YBCO thin films have been prepared on metallic substrates by the rf magnetron sputtering. Sputtered thin films have strongly oriented c-axis. After proper annealing, YBCO thin films with Tc,zero = 82K were obtained. Also, the magnetic field dependence and the strain dependence of the critical current density for YBCO thin films were investigated.

INTRODUCTION

Much effort has been dedicated to the preparation of high Tc superconducting thin films recently. Significantly high values, around 10^6 A/cm^2 at 77K, of the critical current densities have been successfully obtained for both the epitaxial grown films and polycrystalline films which were deposited on single crystal substrates, such as MgO, SrTiO$_3$ and sapphire, etc.[1][2][3][4][5][6] For wider applications of those high Tc superconducting oxides for wires and cables, electromagnetic shields, electronic devices, etc., it is necessary to develop the techniques which can make superconducting thin films onto flexible metallic substrates.

In this paper, the preparation of YBCO thin films on metallic substrates by using rf magnetron sputtering was reported. After heating at 700°C for 2 hours in flowing oxygen, the zero resistivity temperature of 82K was obtained for a YBCO thin film with a buffer layer. The magnetic field dependence and the strain dependence of the critical current densities of YBCO thin films were also investigated. The experiment and the results will be detailed below.

FILM PREPARATION

YBCO thin films were deposited onto the metallic substrates directly or on the buffer layer which has been coated on the substrate. Both YBCO films and buffer layers were deposited by using rf magnetron sputtering.

Pt, stainless, and Ni-alloy substrates were selected because of small differences in the coefficients of expansion and low misfit in the lattice parameters to YBCO films. The surface of the metallic substrates was polished by using electrolytic technique. However, the structure of the YBCO films deposited directly on the metallic substrates were polycrystalline with random orientation.

YBCO thin films with oriented c-axis

In order to obtain highly oriented YBCO thin films, multi-layers structures of YBCO/buffer layer/substrate were investigated. As a result, MgO and ZrO_2 films seem to be the most favorable buffer on metallic substrates for YBCO thin films. Typical X-ray diffraction (XRD) patterns are shown in Fig.1. (0, 0, 1) (1 = 1, 2, 3 ···) reflections of YBCO structure are clearly observed. The films are highly oriented with c-axis perpendicular to the film plane. The standard deviation angle of c-axis of the films are around 1.5°. The SEM micrograph of the surface of a c-axis oriented YBCO film is shown in Fig.2. The surface is smooth (glossy) and grains about 0.1~0.2μm in diameter can be observed.

EFFECT OF HEAT TREATMENT

The zero resistivity temperature Tc,zero of as-sputtered c-axis oriented YBCO films is below 77K. In order to improve the superconducting properties, the heat treatment was employed.

The X-ray diffraction patterns and the energy spectra of Auger electron of an as-sputtered YBCO thin film and that annealed at 700°C for 2 hours were shown in Fig.3 and Fig.4, respectively. The heat treatment was carried out in tube furnace with flowing oxygen. The interdiffusion of Ni element occurred and the YBCO structure was destroyed during the heating. It indicates that in order to prevent the interdiffusion, the lower annealing temperature is required for YBCO thin films deposited on metallic substrates with a buffer layer. For the smaller interdiffusion, the thicker buffer layer (MgO or ZrO_2) and the multi-buffer layers were investigated. However, thicker buffer layer always lower the orientation of YBCO thin films. Deposition of YBCO thin films on multi-buffer layers of ZrO_2/Ag/Ni-alloy substrate was investigated.

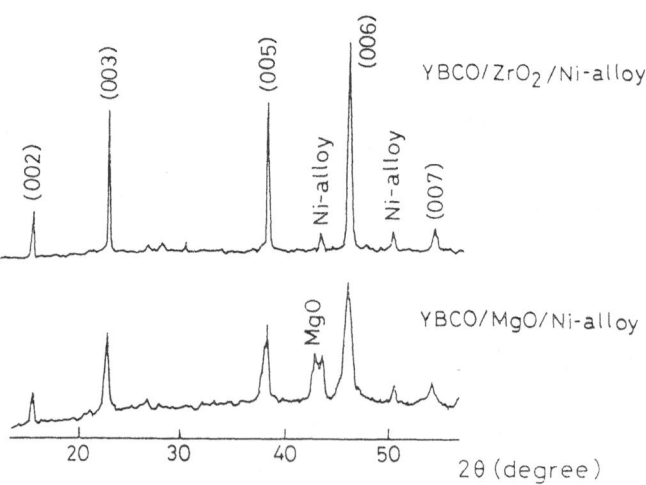

Fig.1 Typical X-ray diffraction patterns of YBCO films deposited on Ni-alloy substrates with a ZrC_2 or MgO buffer layer.

Fig.2 SEM micrograph of a YBCO thin film
with oriented c-axis.

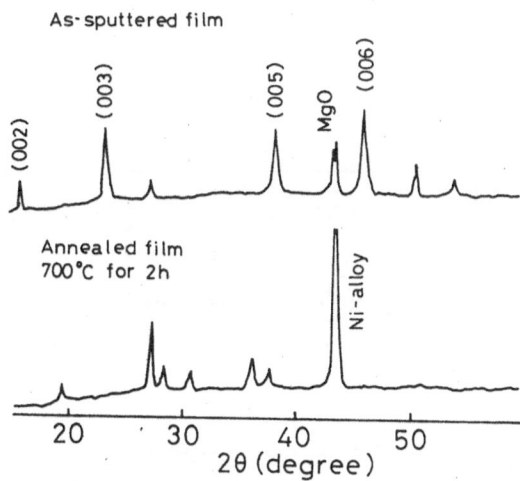

Fig.3 X-ray diffraction pattern of as-sputtered
YBCO thin film and that annealed at 700°C
for 2 hours.

Figure 5 shows the X-ray diffraction patterns of a 0.2µm thick YBCO thin film deposited on ZrO_2 layer (0.2µm)/Ag layer (0.4µm)/Ni-Alloy substrate. Strongly oriented c-axis was observed. This film was annealed at 700°C for 2 hours and the energy spectra of Auger electron of that were shown in Fig.6. The interdiffusion of Ag or Ni element did not occur. The multi-buffer layers seem to be more effective for YBCO films because of the smaller interdiffusion in ZrO_2. The c-axis XRD intensity ratio of these films as a function of annealing temperature was shown in Fig.7. Above 700°C heating temperature, the YBCO structure was destructed. The interdiffusion problem for YBCO films on metallic substrates during the heat treatment was more serious than that for the YBCO films deposited on single crystal substrates (Heating at 900°C was allowed)[6].

PROPERTIES OF YBCO THIN FILMS

Tc, Jc of the YBCO thin film

The zero resistivity temperature Tc,zero and the critical current density, Jc, were measured by using the conventional dc four-probe transport method. The temperature dependence of the resistance of YBCO thin films annealed at various temperature was shown in Fig.8. Tc,zero = 68K for an as-sputtered YBCO thin film with 0.2µm thickness and Tc,zero = 82K for that annealed at 700°C were obtained. Annealing at 800°C lower the Tc,zero to about 20K. The result of Jc measurement of the film annealed at 700°C and its SEM micrograph of the surface were shown in Fig.9. Here, we defined Jc_0 as the current density necessary to produce

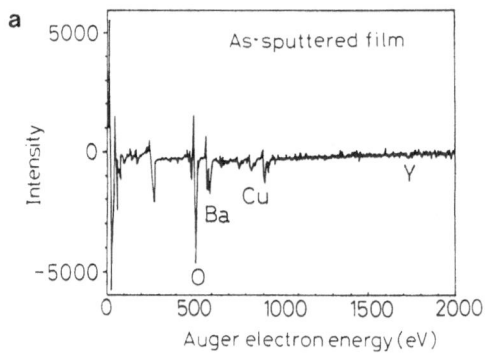

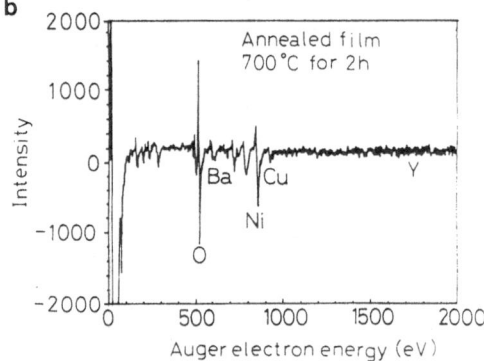

Fig.4 The energy spectra of Auger electron for
(a) an as-sputtered YBCO thin film.
(b) an annealed YBCO thin film (700°C for 2 hours).

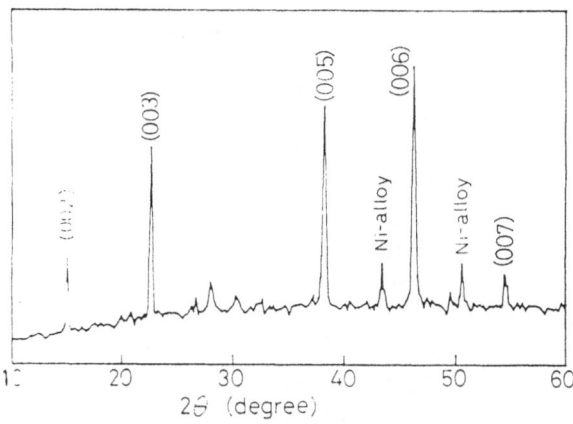

Fig.5　The X-ray diffraction pattern of
an as-sputtered YBCO thin film.

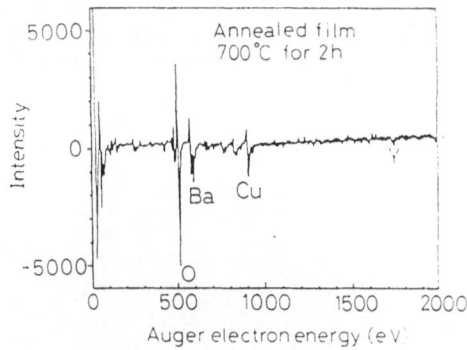

Fig.6　The energy spectra of Auger electron for
an annealed YBCO thin film
(700°C for 2 hours).

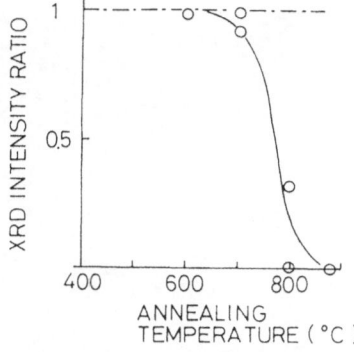

Fig.7　The c-axis XRD intensity ratio of
YBCO thin films as a function of
annealing temperature.

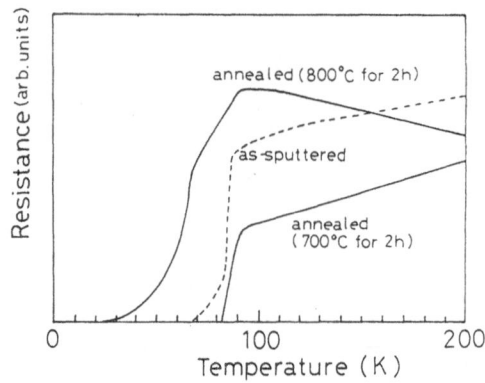

Fig.8　The temperature dependence of the resistance
of YBCO thin films.

an electric field equal to luV and $Jc_0 = 100A/cm^2$ at 77K was obtained.
The depth profile of this annealed YBCO thin film was analyzed by the
secondary ion mass spectrometry and the result was shown in Fig.10.　The
interdiffusion of Ni and Ag elements was repressed in ZrO_2 layer.

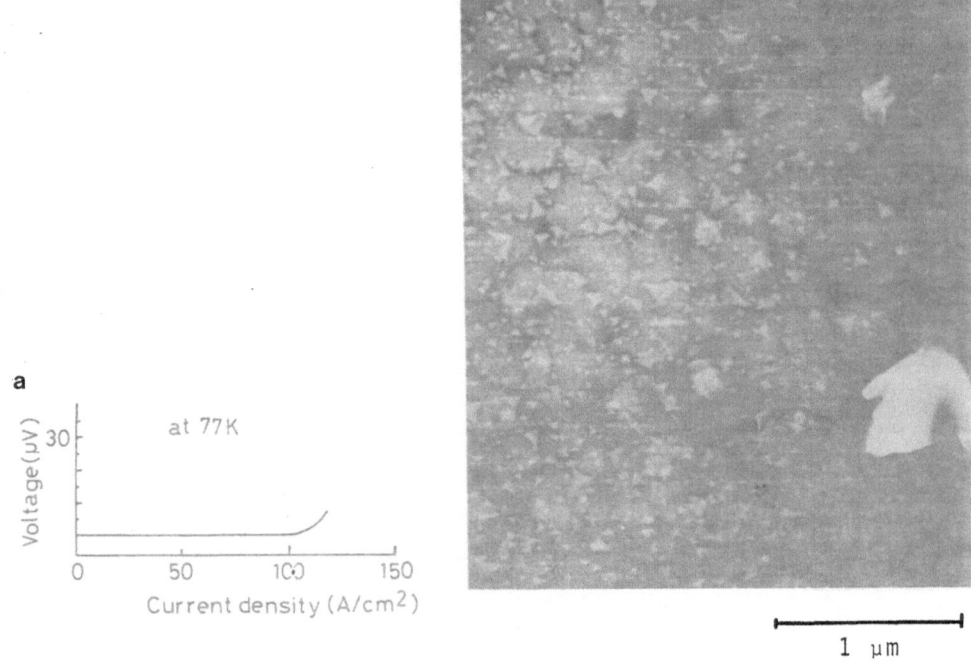

Fig.9　(a) Jc for a YBCO thin film at 77K.
(b) SEM micrograph of the YBCO thin film.

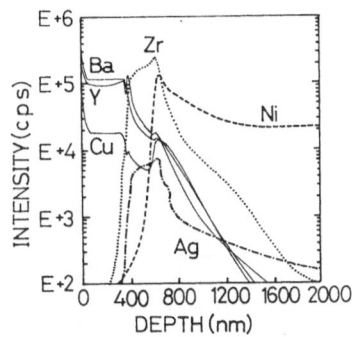

Fig.10 Depth profile of a YBCO thin film annealed at 700°C for 2 hours.

Jc–B property of the YBCO thin film

Figure 11 shows the magnetic field dependence of the critical current density for an annealed YBCO thin film at 4.2K. The direction of applied magnetic field is in the film plane. Jc decreases with the increase of a magnetic filed and reaches one tenth at one tesla.

Jc–strain property of the YBCO thin film

Jc was measured as the substrate was bended to various curvature. Figure 12 shows the strain dependence of Jc at 4.2K for a YBCO thin film. The critical current density remains constant up to 0.2% strain.

CONCLUSIONS

In summary, we have prepared the YBCO thin films with strongly

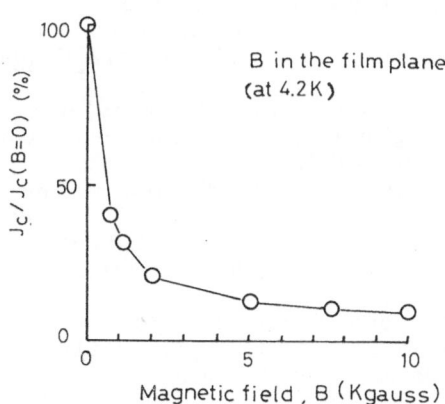

Fig.11 The magnetic field dependence of Jc at 4.2K for YBCO thin films deposited on metallic substrates.

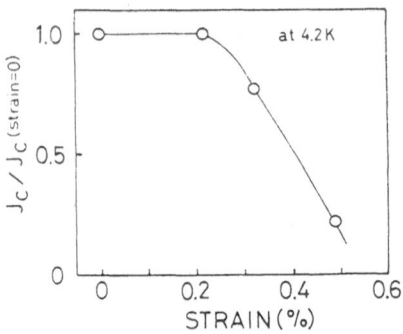

Fig.12 The strain dependence of Jc at 4.2K for YBCO thin films deposited on metallic substrates.

oriented c-axis on the metallic substrates by using the rf magnetron sputtering. After heating at 700°C for 2 hours, the YBCO thin films with Tc,zero = 82K were obtained. It is shown that the structure of multiple buffer layers can repress the interdiffusion of elements from the metallic substrate. The Jc, Jc-magnetic field dependence, and Jc-strain dependence were also investigated. Further research will be extended to improve the crystallinity and transport properties of YBCO thin films.

ACKNOWLEDGEMENTS

Authors wish to express great appreciations for fruitful technical discussions and funding bestowed by a group of Japanese electric power companies---Tokyo Electric Power Co., Tohoku Electric Power Co., Hokkaido Electric Power Co., and Electric Power Development Co.

REFERENCES

1. Y.Enomoto, T.Murakami, M.Suzuki and K.Moriwaki, Largely anisotropic superconducting critical current in epitaxially grown $Ba_2YCu_3O_{7-y}$ thin film, Jpn. J. Appl. Phys. 26: L1248 (1987).
2. J.L.Makous, L.Maritato and C.M.Falco, Superconducting and structural properties of sputtered thin films of $YBa_2Cu_3O_{7-y}$, Appl. Phys. Lett. 51: 2164 (1987).
3. K.Char, A.D.Kent, A.Kapitulnik, M.R.Beasley and T.H.Geballe, Reactive magnetron sputtering of thin-film superconductor $YBa_2Cu_3O_{7-x}$, Appl. Phys. Lett. 51: 1370 (1987).
4. R.B.Laibowitz, R.H.Koch, P.Chaudhari and R.J.Gambino, Thin superconducting oxide films, Phys. Rev. B, 35: 8821 (1987).
5. T.Terashima, K.Iijima, K.Yamamoto, Y.Bando and H.Mazaki, Single-Crystal $YBa_2Cu_3O_{7-x}$ thin films by activated reactive evaporation, Jpn. J. Appl. Phys. 27: L91 (1987).
6. J.Chang, S.Seo, A.Sayama, M.Matsui, K.Yamamoto and N.Harada, Transport properties of high Tc Y-Ba-Cu-O thin films made by rf magnetron sputtering", ISS'88 at Nagoya, Japan.

SUBSTRATE-HTcS THIN FILM INTERACTION STUDIES BY (S)TEM

P.P.J. Ramaekers°, D. Klepper° and A. Mackor*

°Centre for Technical Ceramics
P.O. Box 513, 5600 MB   Eindhoven
The Netherlands

*ITC-TNO
P.O. Box 108, 3700 AC Zeist
The Netherlands

ABSTRACT

This paper concerns with compatibility aspects between HT_cS thin films and their substrates. The influence of substrate-thin film interaction and thin film microstructure on the superconducting properties is discussed. In this respect, data based on (S)TEM observations are presented. It is concluded, that it is of the utmost importance to exactly control the growth and crystallisation process of a superconductor thin film.

INTRODUCTION

In the technology of thin film devices based on high-temperature (HT_cS) superconductors the choice and preparation of the substrate is one of the key factors. Both chemical interaction, thin film orientation and thin film microstructure are critical for the superconducting properties (Tc, Jc, etc.) and these can be strongly influenced by the compatibility of substrate and superconducting layer. Presently, we are studying chemical interactions and interdiffusion between $YBa_2Cu_3O_7$ thin layers and $SrTiO_3$ (100), Al_2O_3 (sapphire), ZrO_2(YSZ) and Si and between $BiCaSrCuO_x$ layers and $SrTiO_3$ (100) substrates. In addition we examine the effects of nitride, noble metal or $SrTiO_3$ intermediate layers as diffusion barriers. In this paper some results obtained by (S)TEM studies are presented.

Another key factor in the construction of hybrid HTcS devices is the choice of the thin film technology. In this paper we will summarize literature data on how the deposition method and the annealing procedure are related to the superconductor's microstructure and thus its superconducting behaviour.

EXPERIMENTAL

Thin film preparation and characterisation

$YBa_2Cu_3O_{7-x}$ thin films were deposited using UHV triple E-gun
deposition with BaF_2, Cu and Y as source materials. Oxygen was supplied
close to the substrate, a pressure of 10^{-7} mbar was maintained during
deposition. The experimental results in this paper report on 0.25 μm
thick layers on $SrTiO_3$ (100).
Correct stoichiometry was obtained by firing in a 1 bar O_2/H_2O mixture
at 850°C, as checked by Rutherford Backscattering. Overall thin film
orientation was determined by X-ray diffraction.
    Orientation of thin films was determined by X-ray diffraction and by
TEM selected area diffraction techniques.

TEM specimen preparation

    Small pieces of a wafer were cemented with Epox 812 with their thin
film faces together. The two opposite sides of the "sandwich" were
flatted along one of the major crystallographic axes with wet 600 grit
silicon carbide abrasive paper.
    One side was dimpled until about 25 microns with 6 micron diamond
paste and ending with 0.05 micron alumina (1).
With this face the sandwich was cemented using wax on a dimpler pin.
The specimen was thinned to 150 microns. A second dimple was made until
about 20 microns of material remained between the two.
    After dimpling and cleaning a copper ring (one hole grid) was fixed
to the specimen with epoxy. The specimen was placed in a Gatan duo ion
mill during about 8 hours at 6 kV and 0.5 mA. Specimens are observed in
a JEOL 2000 FX at 200 kV (2).

RESULTS AND DISCUSSION

Literature survey on substrate-HT_cS compatibility

    The first phenomenon to be considered when studying substrate-thin
film compatibility is the possibility of chemical reaction,
interdiffusion or segregation. Although many researchers have described
chemical interactions between substrate and HT_cS thin films, very few
elaborate studies using a variety of substrates have been performed.
Gurvitch and Fiory (3) report on a number of substrates and possible
buffer layers, giving a description of substrate-film reactions and
interdiffusivity problems. Dam et al. (4) describe results obtained on
thin films prepared by DC sputtering, substrates being Al_2O_3 (0001),
MgO (100) and (110), Yttria stabilized ZrO_2 (111), Si (100), Gadolinium
Gallium Garnet (100) and $SrTiO_3$ (100). Invariably the best results
(minimal interaction and deterioration of superconducting properties)
were found using monocrystalline $SrTiO_3$ substrates. In the case of
$SrTiO_3$ Gurvitch and Fiory (3) observe sharp resistive transitions but
mention poisoning of the superconductor by Ti, and loss of Cu from
$YBa_2Cu_3O_7$. Poisoning by Ti and also by Sr was found by Gavaler et al.
(5), whereas Agatsuma et al. (6) even found thin reaction layers of
Bariumtitaniumoxide on $SrTiO_3$ (100) substrates. In this case, however,
pastes containing non-reacted Y_2O_3, $BaCO_3$ and CuO powder were used,
which might result in direct reaction with the $SrTiO_3$.
    Also in the case of monocrystalline ZrO_2 substrates good results
have been reported. Gurvitch and Fiory (3) give a ranking where ZrO_2
takes first place and $SrTiO_3$ second. Degradation of the HT_cS thin film,

however, always takes place and progressively so with higher annealing temperatures and longer annealing times. That is why these authors recommend the use of a 300 A thin Ag or Nb buffer layer. ZrO_2 itself is also in the picture as a diffusion barrier in hybrid Si/or SiO_2/superconductor devices (7).

Important in this respect is the possibility that impurities are introduced into the HT_cS film during deposition or annealing, or coming from improper treatment of the substrate. Notorious is the inclusion of Ar^+ during sputtering, which, although in many cases partly removed by annealing can strongly decrease the critical current density. The quality of the polishing, cleaning, etching and drying procedure of the substrate to be covered will define the integrity of the interface and may also be a source of problems, not well realised by some research groups.

A second compatibility aspect of prime importance for the superconducting properties is the microstructure of the superconductor. The following microstructural features can be influenced by the substrate and by the deposition/annealing procedure: density, stoichiometry, crystallite orientation, presence of second phases, crystallite defects and lattice strains. In the context of this publication it is impossible to systematically treat all the relationships between these features and the thin film technology. Therefore we will refrain ourselves to a discussion of the most prominent effects described in the literature. The substrate's choice is not only important from the interaction point of view. Thin film orientation, lattice mismatch and strain, and macrostresses caused by thermal expansion differences should be also considered. For the example of $YBa_2Cu_3O_{7-x}$ deposition on monocrystalline $SrTiO_3$ many comments on the orientation dependence of the substrate and the deposition/annealing technology are known.

A (001) orientation of a $YBa_2Cu_3O_{7-x}$ thin film is one of the prerequisites for a high Jc. On $SrTiO_3$ this orientation is often preferentially obtained when starting with a (100) or (001) substrate. At least this is true for thinner films: the thicker the film, the more (h00) orientation is observed, although at the interface still (001) can be found (8).

Kentgens et al. (8) clearly indicate, for sputtered films, that a preferential c orientation is less advantageous for reasons of lattice mismatch. The fact that this orientation still appears for thinner films indicates that other factors have to be considered. Kentgens et al. postulate that a (001) $YBa_2Cu_3O_7$ orientation is more favourable for interfacial free energy reasons because of the different metal oxide surface structure, as compared to the (h00) orientation. Obviously, not only nucleation but also subsequent (or simultaneous) crystallization processes will determine the ultimate thin film orientation. Systematic experiments where growth temperature, deposition time, annealing characteristics (time, temperature, atmosphere) and perhaps also substrate preparation are varied under well defined conditions are needed to resolve this matter. Following the suggestion of Fujita et al. (9) that the lattice constants of $YBa_2Cu_3O_{7-x}$ have a temperature dependent and an oxygen deficiency dependent portion one can understand why it is so extremely important to be able to exactly control the growth process of a HT_cS thin film.

$YBa_2Cu_3O_{7-x}$ is an exciting material since its physical properties are strongly anisotropic. For instance the anisotropy in the thermal expansion coefficient (10) is one of the critical factors which determine the microstructural (and the superconducting) properties during thermal treatment. Differences in thermal expansion between substrate and HT_cS layer should also be considered for technological applications. In a recent paper Hashimoto et al. (11) presented some data (also temperature dependent) on this. Clearly the use of substrates with a large thermal expansion coefficient (MgO, SrTiO$_3$, YSZ) is favourable. A short annealing cycle at a sufficiently low temperature (preferably lower than the orthorhombic-tetragonal transition in YBa$_2$Cu$_3$O$_{7-x}$) is also recommended; this is also benificial for the reduction of substrate-film interactions.

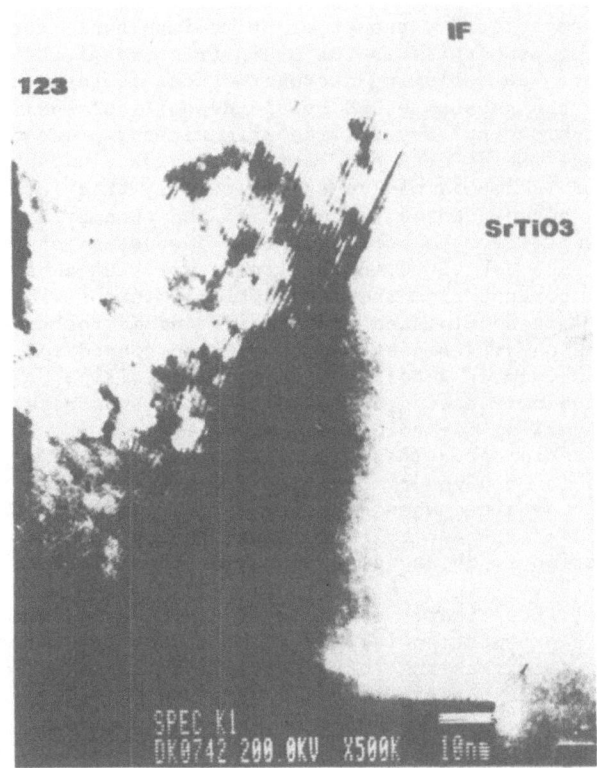

Figure 1.   (S)TEM Analysis Results

OBSERVATIONS

All YBa$_2$Cu$_3$O$_7$ thin films on (100) SrTiO$_3$ showed (001) preferential orientation near the substrate. In some cases also different textures, e.g. (h00) or (110) (see figure) have been determined.
Stacking faults and other defects in the HT$_c$S crystal structure have been regularly observed.
EDX analysis in the superconductor layer near the interfaces showed evidence of Sr and Ti impurities in a few cases, thus implying interdiffusion between substrate and thin film.

CONCLUSIONS

1. For the optimisation of superconducting properties of HT_cS thin
   films one should minimise substrate-thin film interaction. If on Si,
   SiO_2 or GaAs substrates annealing above 900 K is needed the use of a
   buffer layer is recommended. Also with other substrates, even with
   $SrTiO_3$ or YSZ, a diffusion barrier may be needed.
   A viable alternative is the use of a plasma assisted MBE (molecular
   beam epitaxy) or CBE (chemical beam epitaxy) process, where a high
   temperature annealing step can be avoided (12).
2. For the optimisation of superconducting properties of HT_cS thin
   films it is of the utmost importance to control the growth and
   crystallisation process during and after deposition. In this way
   microstructural control is possible.
3. Regular access to a (scanning) transmission electron microscope and
   experience with HT_cS thin film specimen preparation and handling is
   one of the prerequisites for the development of device technology
   based on high-temperature superconductors.

REFERENCES

1. L.R. Nazar, Proceedings of 46th Annual meeting of EMSA (1988) p.
   862.
2. M. Sarikaya, Proceedings of 46th Annual meeting of EMSA (1988) p.
   858.
3. M. Gurvitch and A.T. Fiory, Mat. Res. Symp. Proc., Vol. 99, (1988)
   297-301.
4. B. Dam, H.A.M. van Hal and C. Langereis, Europhys. Lett. 5 (1988)
   455-460.
5. J.R. Gavaler, A.I. Braginski, J. Telvacchio, M.A. Janocko, M.G.
   Forrester and J. Greggi, Extended Abstracts High-Temperature
   Superconductors II, April 5-9, 1988, Reno (Nevada), 193-196.
6. K. Agatsuma, T. Ohara, H. Tateishi, K. Kaiho, K. Ohkubo and H.
   Karasawa,
   Physica C 153-155 (1988) 814-815.
7. A. Mogro-Campero and L.G. Turner, Appl. Phys. Lett. 52 (1988)
   1185-1186.
8. A.P.M. Kentgens, A.K. Carim and B. Dam, J. Crystal Growth, accepted
   for publication.
9. J. Fujita, T. Yoshitake, A. Kamijo, T. Satoh and H. Igarashi,
   Extended Abstracts High-Temperature Superconductors II, April 5-9,
   1988, Reno (Nevada), 109-112.
10. W. Wong-ng and L.P. Cook, Adv. Ceram. Mater. 2 (1987) 624; H.M.
    O'Bryan and P.K.
    Galagher, ibidem 2 (1987) 640.
11. T. Hashimoto, K. Fueki, A. Kishi, T. Azumi and H. Koinuma, Japanese
    Journal of
    Applied Physics 27 (1988) L214-L216.
12. R. Cabanel, J.P. Hirtz, G. Garry, F. Hosseini Teherani and G.
    Creuzet, Proceedings Proceedings First International Symposium on
    Superconductivity, August 28-31, Nagoya, to be published.

CHARACTERIZATION OF ZrO_2 BUFFER LAYERS FOR SEQUENTIALLY

EVAPORATED Y-Ba-Cu-O ON Si AND Al_2O_3 SUBSTRATES

George J. Valco, Norman J. Rohrer, John J. Pouch[*],
Joseph D. Warner[*] and Kul B. Bhasin[*]

Department of Electrical Engineering
The Ohio State University
Columbus, Ohio 43210

[*]National Aeronautics and Space Administration
Lewis Research Center
Cleveland, Ohio 44135

INTRODUCTION

A large variety of techniques have been employed for the formation of thin films of the high temperature superconducting oxides on many different substrates. These efforts have been driven by the desire to investigate and develop electronic applications of these materials. Several of the substrates that have been used possess undesirable electronic properties, such as the large dielectric constant of $SrTiO_3$. For many applications it is desirable to obtain the superconducting films on silicon or, for high frequency applications, on substrates such as Al_2O_3 or GaAs. Unfortunately, the deleterious interactions between the films and these substrates or common dielectrics[1], such as silicon dioxide, result in unacceptable degradation of both the superconductor and the substrate. This has led to the investigation of the use of thin films of materials such as ZrO_2[2] or silver[3] as buffer layers between the superconductor and the substrate.

One of the techniques being employed to form thin films of high temperature superconductor involves the sequential evaporation of a multi-layer stack containing the constituents of the superconductor followed by annealing in oxygen. When performed by electron beam evaporation from a multi-hearth gun, this technique allows deposition of films with little spatial variation of the stoichiometry as all components of the film are evaporated from the same point in space. In addition, by controlling the thickness of the individually deposited layers, the stoichiometry of the film is easily adjusted. Films have been formed from a variety of starting materials in different combinations using this technique. These include Y, Ba and Cu metals[3,4,5], a combination of the metals and their oxides[5-7], a combination of metals and BaF_2[2,8] and a combination of oxides and BaF_2[9].

We have performed sequential evaporation of Cu, Y and BaF_2 to study the formation of superconducting films on $SrTiO_3$, MgO, yttrium stabilized ZrO_2 (YSZ), Si and sapphire substrates. For the silicon and sapphire substrates, we have used a thin film of ZrO_2 as a buffer layer. In addition, Auger electron spectroscopy (AES) with argon ion sputtering has been utilized to obtain depth profiles through the films and to study the interfacial interactions, particularly on silicon and sapphire substrates. A variety of thicknesses, compositions and annealing conditions have been used to form superconducting films and to prepare samples for the AES analysis. The results of these characterizations are presented here.

SAMPLE PREPARATION

Sample preparation involved deposition of the ZrO_2 buffer layer on those substrates which needed one followed by deposition of the multi-layered stack of the constituent elements. The multi-layer film was then converted into the YBaCuO compound through a multiple step annealing sequence. To allow electrical characterization, silver contacts were evaporated and heat treated. The details of each of these procedures follow.

Deposition of Films

Deposition of the Cu, Y and BaF_2 films was performed in a CHA Industries electron beam evaporator. The system is equipped with a four hearth gun, allowing deposition of the multi-layer stack without breaking vacuum. Thickness of the layers was controlled via an Inficon XTC thickness monitor and rate controller. The depositions were calibrated by measurements of step heights using a surface profilometer.

ZrO_2 for the buffer layers on the silicon and sapphire substrates was also performed in this evaporator. Samples were prepared with $0.2\mu m$ and $0.9\mu m$ thick buffer layers.

A cross sectional drawing of the structure of a typical as deposited film with buffer layer is shown on the left side of Figure 1. For a typical film, after deposition of the buffer layer, approximately 510Å of copper was deposited on the substrate. This was followed by an approximately 480Å thick layer of yttrium which was followed by an approximately 1710Å thick layer of barium fluoride. Thicknesses of the layers were adjusted to vary the composition. For most of our depositions, this multi-layered sequence was repeated four times for a total of twelve layers. For the thicknesses listed above, the film is characterized by a barium/yttrium atomic ratio of 1.98 and a copper/yttrium atomic ratio of 2.98. We have investigated the properties of films with barium/yttrium ratios ranging from 1.9 to 2.4 and copper/yttrium ratios ranging from 2.9 to 3.3 on most substrates and a larger range on $SrTiO_3$. We have used barium fluoride rather than elemental barium since barium fluoride is less reactive.

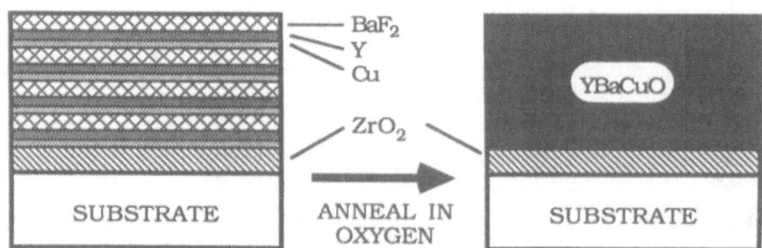

Figure 1.   On the left: Typical twelve layer structure of the as deposited film with ZrO_2 buffer layer. On the right: $YBa_2Cu_3O_{7-\delta}$ formed after annealing in oxygen.

Annealing

The deposited films were annealed in a hot wall, programable, quartz tube furnace. The furnace was preheated to the annealing temperature and purged with oxygen prior to inserting the samples. Temperatures ranged from 850°C to 900°C. The samples were usually pushed into the furnace with a slow 5min push. The duration of the anneals ranged from 15min to 3hr. The temperature was then ramped to 450°C at a rate of -2°C/min. The samples were held at 450°C for 6hr and then the temperature was ramped to room temperature at -1°C/min. During the high temperature portion of the anneal the ambient consisted of ultra high purity oxygen bubbled through room temperature water to assist in removal of fluorine from the films. Dry oxygen was used during all other portions of the annealing process.

Electrical Contacts

Ohmic contacts were formed on the films to allow measurement of their resistance as a function of temperature. Most of the samples were rectangular in shape with widths of approximately 5mm and lengths of approximately 1cm. The contacts for these samples were deposited by evaporation of $1\mu m$ of silver through shadow masks to produce four stripes across the width of the samples. For some irregularly shaped samples, shadow masks which produced four dots were used. The contacts were annealed in dry oxygen at 500°C for 1hr. They were placed in the tube furnace at room temperature and the temperature was ramped up at 20°C/min. At the end of the anneal the temperature was ramped to 250°C at a rate of -2°C/min and then to room temperature at -1°C/min.

CHARACTERIZATION

To allow measurement of the resistance of the films as a function of temperature, the samples were cooled in a closed cycle helium refrigerator. The samples were mounted onto a sample holder and gold ribbon bonds were made between the silver contacts and bonding posts. A four probe DC measurement was employed to determine the resistance. Measurements were usually started at room temperature and proceeded as the sample was cooled. A few samples were measured both while cooling and while heating, with the same results in both directions. Measurements were continued to well below the transition temperature for superconducting films or to approximately 10K for non-superconducting films.

Auger electron spectroscopy with argon ion sputtering was used to study the Y-Ba-Cu-O film and the film/substrate interface. Excitation was obtained through a primary electron beam of 3 keV energy. Sputtering for the depth profiles was performed with a 3 keV argon ion beam. In addition, scanning electron microscopy (SEM) was utilized to observe the morphology of the films.

RESULTS

Figure 2 shows an AES depth profile through a Y-Ba-Cu-O film on silicon with a $0.9\mu m$ thick ZrO_2 buffer layer. The film was deposited in twelve layers and was approximately $1.2\mu m$ thick prior to annealing. The sample was annealed at 850°C for 180 min. The film was dark grey in appearance and scanning electron microscopy showed it to be polycrystalline with randomly oriented grains. From the Auger profile, the superconducting film is seen to be quite uniform throughout its thickness, with no evidence of its multi-layer origin. The film is also free of fluorine to the limits of detection by AES. The interface between the film and the buffer layer,

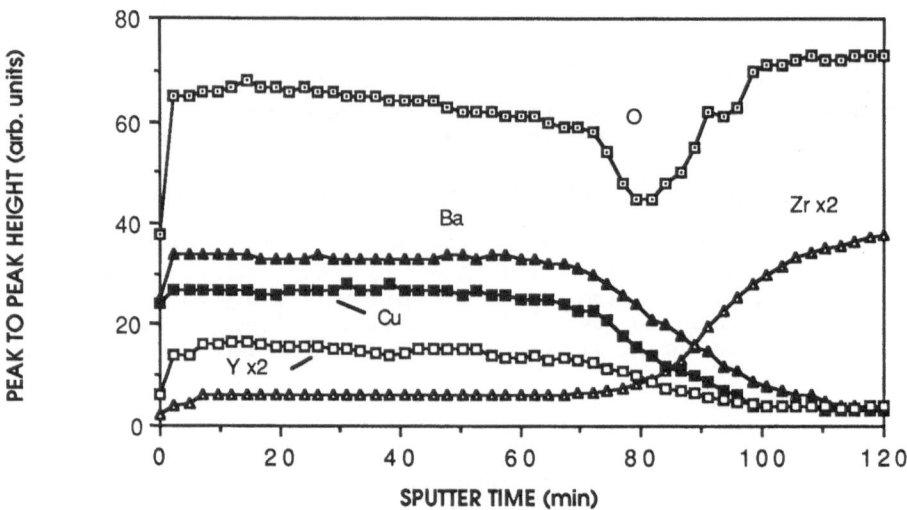

Figure 2. Auger depth profile through a Y-Ba-Cu-O film on silicon
with a 0.9μm buffer layer.

although quite broad, is well defined. The profile was stopped before
sputtering through the ZrO$_2$ layer but after the Y,Ba and Cu signals had
fallen to the background signal level. A survey spectrum was then obtained, to check for the presence of silicon from the substrate in the
buffer layer. None was observed.

    The width of the interface is probably a result of the long anneal at
850°C, however that anneal was needed to form the superconductor. This is
indicated in Figure 3, which shows the normalized resistance-temperature
(R-T) characteristic of two films similar to the one in Figure 2. Both of
these films have as deposited compositions of Ba/Y=2.00 and Cu/Y=3.01.

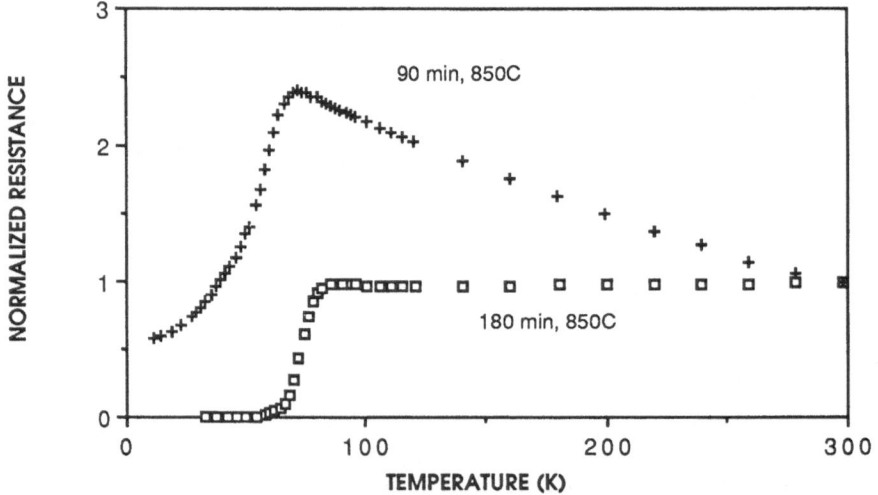

Figure 3. Normalized resistance as a function of temperature for
two Y-Ba-Cu-O films on silicon with a ZrO$_2$ buffer.

They were annealed at 850°C for 90 min and 180 min. The sample annealed for the shorter time has a semiconducting normal state characteristic and never achieved zero resistance. Its room temperature resistivity was also more than an order of magnitude larger than that of the sample annealed for 180 min. We have tried shorter anneals at 950°C, comparable to the procedure we use with $SrTiO_3$ substrates[8], but they also produced poor films on silicon.

An Auger depth profile through a film deposited directly on a YSZ substrate is shown in Figure 4. This sample was annealed at 900°C for 45 min and produced a superconducting film. With the exception of the larger Ba concentration (Ba/Y=2.3) this profile is very similar to that of the film with the thick ZrO_2 buffer layer on silicon.

A six layer deposition with a pre-anneal thickness of approximately $0.6\mu m$ was performed onto several silicon samples with a $0.2\mu m$ buffer layer. There were two reasons for this deposition. The first was to check the utility of a thin ZrO_2 buffer layer on silicon. The second was to reduce the sputtering time required to reach the interface and thereby decrease any artificial broadening of the interface due to the sputtering process. Three samples were annealed for 60, 120 and 180 min at 850°C. Upon removal of the samples from the furnace, it was apparent that extensive interaction had taken place between the film and the silicon as the film was not black and resembled a film deposited an silicon. The Auger profiles through the 60 min annealed and 120 min annealed films are shown in Figures 5 and 6 respectively. Neither the film/buffer or buffer/substrate interfaces are well defined. There appears to have been considerable interaction among all of the constituents of the film, buffer layer and substrate. In particular, barium is present in large amounts throughout the buffer layer and into the silicon, silicon is present throughout the buffer layer and appears to have accumulated at the surface of the film and zirconia extends through the film and into the substrate. Copper and yttrium both have long tails into the buffer layer for the 60 min sample and copper extends through the buffer layer and has accumulated at the silicon surface for the 120 min sample.

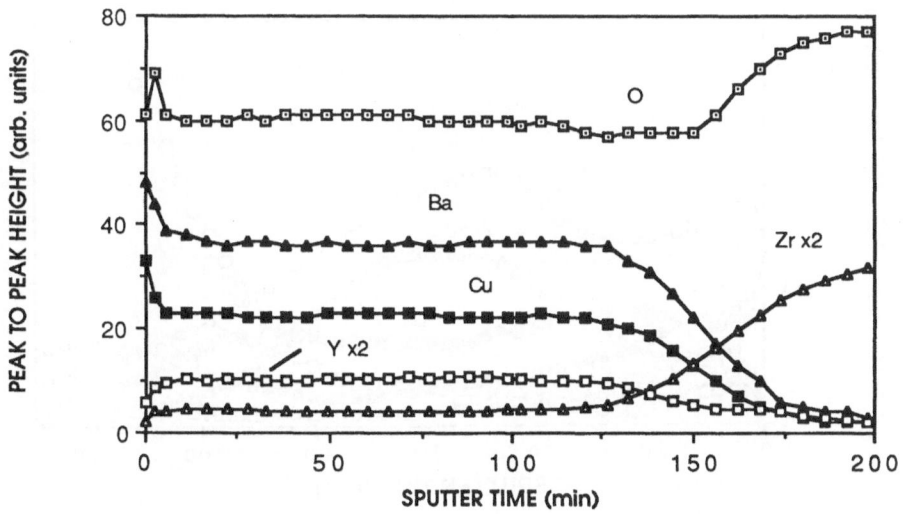

Figure 4. Auger depth profile through a Y-Ba-Cu-O film on a YSZ substrate.

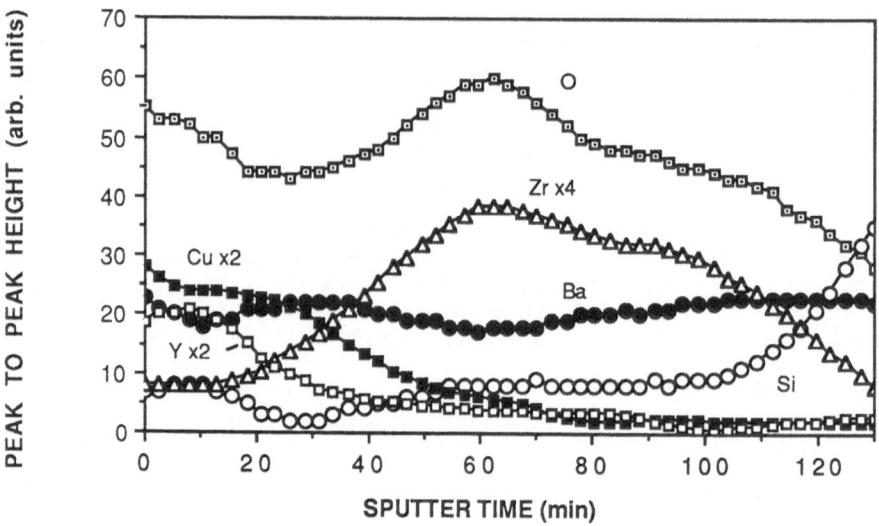

Figure 5. Auger depth profile through a Y-Ba-Cu-O film on silicon
with a 0.2μm buffer layer. Annealed 60 min at 850°C.

The same deposition, with 0.2μm ZrO_2 buffer layer, was performed onto
three sapphire substrates. They were also annealed for 60, 120 and 180 min
at 850°C. These samples were shiny, smooth and black when removed from the
furnace. Silver contacts were formed on the samples and their R-T characteristics were measured. They are plotted in Figure 7. There is a clear
improvement in the normal state characteristic and in the superconducting
transition with increasing annealing time, however, unlike the case for
silicon, even the sample annealed for 60 min became a superconductor.

Auger profiles for the samples annealed for 60 min and 180 min are
shown in Figures 8 and 9 respectively. These superconducting films are not
as uniform as the thicker films on the thicker buffer layer, but they are

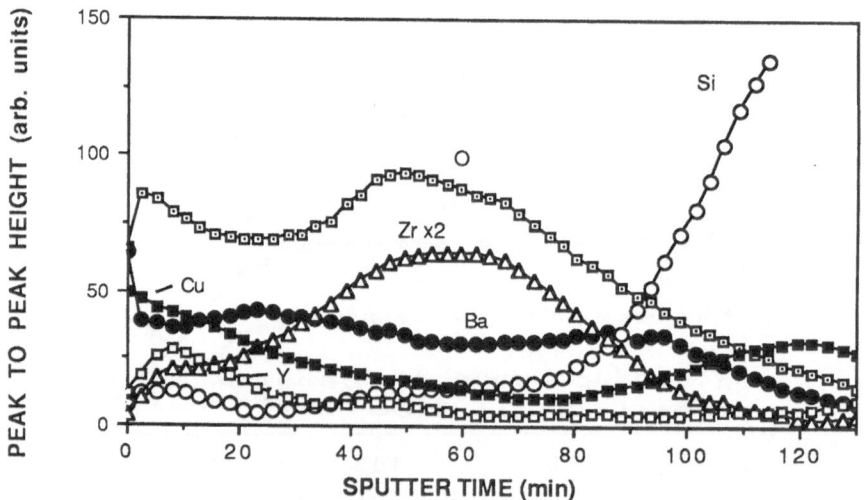

Figure 6. Auger depth profile through a Y-Ba-Cu-O film on silicon
with a 0.2μm buffer layer. Annealed 120 min at 850°C.

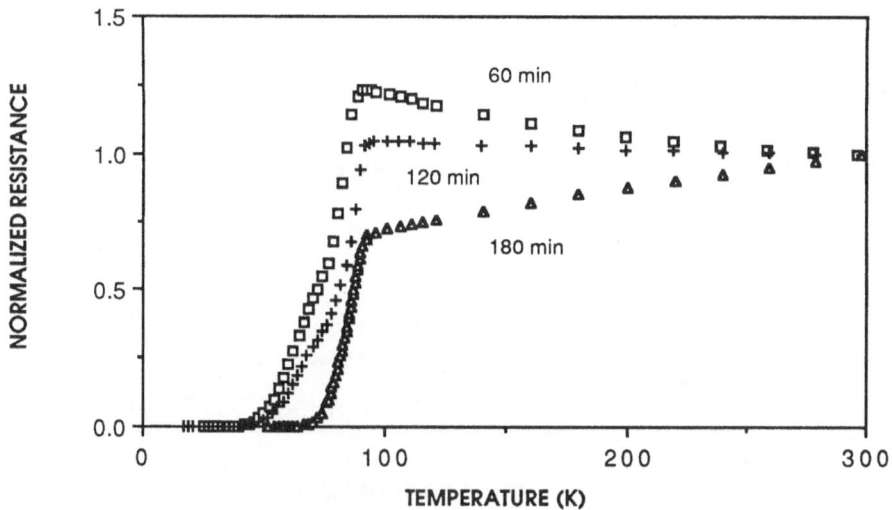

Figure 7. Normalized resistance as a function of temperature for three Y-Ba-Cu-O films on sapphire with a 0.2μm buffer layer.

greatly better than the films on silicon with the same thickness buffer layer. The interfaces on both sides of the buffer layer are reasonably well defined. There is significant penetration of barium into the buffer layer but it appears no worse for the 180 min film than for the 60 min film. Copper and yttrium also have tails into the buffer layer but they too are no worse with the longer anneal. The aluminum tail into the buffer layer may be an artifact due to the proximity of the barium and aluminum peaks in the Auger spectrum.

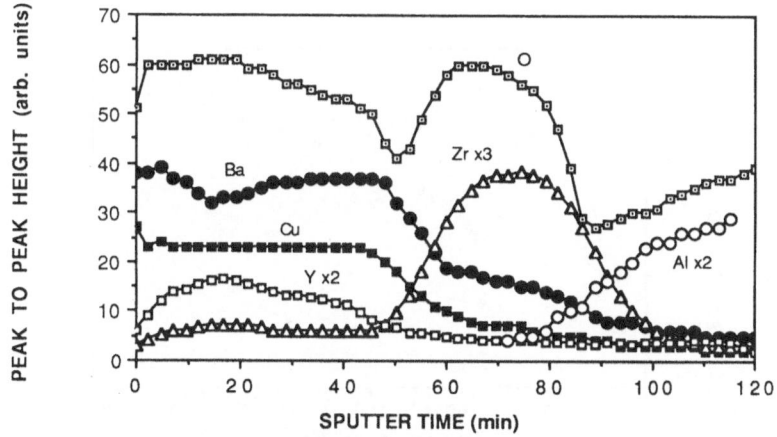

Figure 8. Auger depth profile through a Y-Ba-Cu-O film on sapphire with a 0.2μm buffer layer. Annealed 60 min at 850°C

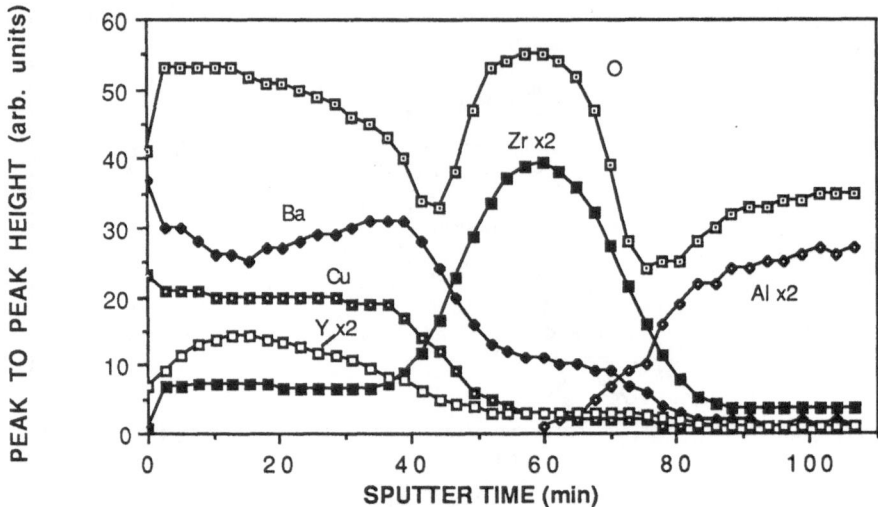

Figure 9. Auger depth profile through a Y–Ba–Cu–O film on sapphire with a 0.2μm buffer layer. Annealed 180 min at
850°C.

Figure 10 shows the resistance temperature characteristic of a
$YBa_2Cu_3O_{7-\delta}$ film on $SrTiO_3$. This film was deposited with a Ba/Y ratio of
2.25 and a Cu/Y ratio of 3.01 and was annealed at 900°C for 45 min. The
onset temperature was 93K, the 90% to 10% transition width was 3.6K and the
critical temperature was 85K. The transition temperature was found to
degrade only slightly over a period of one month. In earlier work[8], we

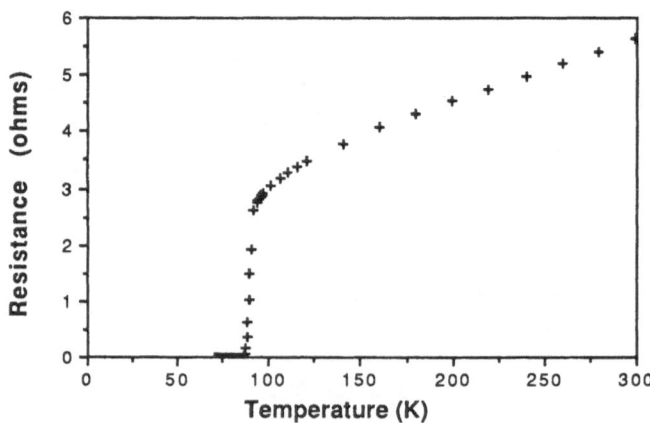

Figure 10.   Resistance as a function of temperature for a
Y–Ba–Cu–O film on $SrTiO_3$.

reported that for films on $SrTiO_3$, low Ba/Y ratios resulted in normal state resistance-temperature characteristics with a negative slope while films with Ba/Y greater than about 2.2 had metallic normal state characteristics. While we have not performed as thorough an investigation into the dependence on composition of the superconducting transition for silicon and sapphire substrates, the data we currently have show different behavior. For both silicon and sapphire substrates, when the film is annealed at 850°C for 180 min, the normal state resistance is metallic for lower Ba/Y ratios and tends to become more like a semiconductor as the Ba/Y ratio is increased. In addition, the as deposited compositions which produce the better critical temperatures appear to have significantly lower Ba/Y ratios than for films on $SrTiO_3$.

SUMMARY AND CONCLUSIONS

Multi-layer sequential evaporation of BaF_2, Y, and Cu layers has been employed for the formation of Y-Ba-Cu-O thin films. Auger electron spectroscopic depth profiles have been used to study the films formed on silicon and sapphire substrates coated with ZrO_2 buffer layers.

Films that were uniform throughout their thickness were formed on silicon when a thick buffer layer was used. The interface was broad but well defined and since the film and buffer layer were thick, a superconducting film could be formed with a 180 min anneal at 850°C. For the 0.2μm buffer layers on silicon, a 60 min anneal was sufficient to cause drastic intermixing of the constituents of the film, buffer layer and silicon. This intermixing proceeded to become worse for longer anneals.

The 0.2μm ZrO_2 buffer layers on sapphire, on the other hand, were sufficient to protect the superconductor from the substrate. Although the Auger depth profiles showed a significant amount of barium in the films, superconductivity was achieved for 60, 120 and 180 min anneals at 850°C. The amount of intermixing did not appear to become worse for the longer anneals and the electrical properties of the films improved.

The data generated during these investigations also suggest that the optimal as deposited composition for multi-layer formation of superconducting films on silicon and sapphire is different from that on $SrTiO_3$. Further investigations into this are being undertaken.

ACKNOWLEDGEMENTS

This research is supported by the National Aeronautics and Space Administration, Lewis Research Center under cooperative research agreement NCC 3-105.

REFERENCES

1. M. Gurvitch and A. T. Fiory, Preparation and Substrate Reactions of Superconducting Y-Ba-Cu-O Films, Appl. Phys. Lett. 51, 1027 (1987).
2. A. Mogro-Campero and L. G. Turner, Thin Films of Y-Ba-Cu-O on Silicon and Silicon Dioxide, Appl. Phys. Lett. 52, 1185 (1988).
3. K. Harada, N. Fujimori and S. Yazu, Y-Ba-Cu-O Thin Films on Si Substrate, Jpn. J. Appl. Phys. 27, L1524 (1988).
4. B-Y. Tsaur, M. S. Dilorio and A. J. Strauss, Preparation of Superconducting $YBa_2Cu_3O_x$ Thin Films by Oxygen Annealing of Multilayer Metal Films, Appl. Phys. Lett. 51, 858 (1987).

5. C. X. Qiu and I. Shih, Y–Ba–Cu–O Thin Films Prepared by a Multilayer Vacuum Method, Appl. Phys. Lett. 52, 587 (1988).

6. C–A. Chang, C. C. Tsuei, C. C. Chi and T. R. McGuire, Thin Film YBaCuO Superconductors Formed by $Cu/BaO/Y_2O_3$ Layer Structures, Appl. Phys. Lett. 52, 72 (1988).

7. Z. L. Bao, F. R. Wang, Q. D. Jiang, S. Z. Wang, Z. Y. Ye, K. Wu, C. Y. Li and D. L. Yin, YBaCuO Superconducting Thin Films With Zero Resistance at 84K by Multilayer Deposition, Appl. Phys. Lett. 51, 946 (1987).

8. G. J. Valco, N. J. Rohrer, J. D. Warner and K. B. Bhasin, Sequentially Evaporated Thin Y–Ba–Cu–O Superconductor Films: Composition and Processing Effects, American Institute of Physics Conference Proceedings, to be published.

9. N. Hess, L. R. Tessler, U. Dai and G. Deutscher, Preparation and Patterning of YBaCuO Thin Films Obtained by Sequential Deposition of $CuO_x/Y_2O_3/BaF_2$, Appl. Phys. Lett. 53, 698 (1988).

THE TRANSPORT BEHAVIOR OF THE YBaCuO SUPERCONDUCTING

THIN FILMS IN THE MAGNETIC FIELDS

Guo-Guang Zheng, Dong-Qi Li, Wei Wang, Jia-Qi Zheng, Wei-Yan Guan, and Pu-Fen Chen*

Institute of Physics and Surface Physics Laboratory
Chinese Academy of Sciences, Beijing, China
* Cryogenic Laboratory, Chinese Academy of Sciences
Beijing, China

INTRODUCTION

The transport behavior of the high Tc superconducting thin films is of importance both for practical reasons in applications and for fundamental reasons in relation to microstructures and measurement conditions. The critical current density is one of the most important superconducting parameters in determining the possible application of the new oxide superconductors. It is also important for understanding the physical properties of these materials in connecting with the various theories.[1-5]

In this paper we present the preparation of the YBaCuO superconducting thin films by electron beam evaporation and the critical current measurements as a function of temperatures and applied magnetic fields. The critical temperature of the sample was 84.4 K. The temperature dependence of the critical current Ic(T) showed a linear behavior except near Tc. At T<<Tc, the critical current Ic decreased rapidly with increasing the applied magnetic fields in the lower field region and then changed only a little. The critical current was irreversible while decreasing the fields and then the critical current Icl at zero applied field was much lower than the original value Ico. A peak of the critical current was observed in the Ic(H) curve. When the measuring current of the sample was slightly higher than Icl, the negative magnetoresistance phenomenon was also observed with sweeping the applied magnetic fields. The explanation was presented in terms of flux trapped in weak link networks in the films.

SAMPLE PREPARATION

The YBaCuO thin films were prepared by electron beam evaporation as previous description.[6] The starting materials Y_2O_3 (or Y), Ba, and Cu were deposited sequentially onto the BaF_2 substrate. The thickness of each layer was in between 7

and 30 nm and was controlled by a quartz crystal monitor
according to the stoichiometry of (123) phase. The total
thickness containing more than eight Y/Ba/Cu modulated periods
was 500 nm. The evaporation rate was 0.1-0.4 nm/sec. During
deposition the pressure was kept in 10^{-5} to 10^{-7} mbarr and the
substrate temperature was at room temperature.

From the depth profile of Auger electron spectrum a
composition modulation structure of the film was clearly
observed. It was noted that there is a small amount of
yttrium in the copper layer. Both the yttrium and copper
layers were separated in between the two barium layers. It
meant that the yttrium atom can diffuse into the copper layer
through the barium layer even in the as deposited films. The
barium layer was granular growth with many intervals and grain
boundaries, which provided a diffuse path for the yttrium
atom. After heat treatment at 600 °C for an hour in the
oxygen flowing the composition modulation structure disappeared. For sapphire substrate the barium diffused into the
substrate apparently. It was caused that the stoichiometric
of three metals in the film deviated from the (123) phase.
The single crystal substrate of BaF_2 which contains barium ion
might suppress the diffusion between the film and the substrate. The composition of the thin film would easily remain
unchanged. After the proper heat treatment at 850 °C for an
hour in oxygen flowing the sample we reported here showed a
critical temperature of 84.4 K and zero resistance temperature
of 80 K.

EXPERIMENT RESULTS AND DISCUSSION

In this work the measurements were made on a thin film
with a cross section area of 10^{-5} cm^2. The sample was mounted
in a vacuum sealed cryostat with a inner liquid helium bath.
The measuring leads were cold pressed onto the surface of the
film with indium balls. A four point method was used in our
measurements. The current-voltage characteristics were measured by standard techniques and registered in a X-Y recorder
with sensitivity of 5 μV/cm. The critical current Ic is
defined as the value of the current producing 1 μV across the
sample. The temperature was determined by a calibrated silicon diode thermometer.

Fig. 1 showed the transition curve of the sample we
studied. The measuring current was 1 μA. In the normal phase
the sample displayed a metallic behavior. The critical currents were first measured as a function of temperature when
the sample was first cooled down to below Tc. The critical
current at liquid helium temperature was 12000 A/cm^2. The
critical current Ic varied linearly with (1-T/Tc) except near
Tc as shown in Fig. 2. In the region close to Tc, the temperature dependence of the critical current has a nonlinear
tail. It can be fitted by a form Ic$\propto$(1-T/Tc)^n with n=3/2.
The result indicated that the thin films behaved as a three
dimensional percolating network of Josephson weak links. It
was similar with that of bulk samples pointed by Aponte[2]. The
SEM results showed that the film has an average size of the
grains of 0.1-0.3 μm. On the other hand, we can deduce the
grain size from a figure presented by Clem et al[7]. for a granular system. The average size of grains was given by

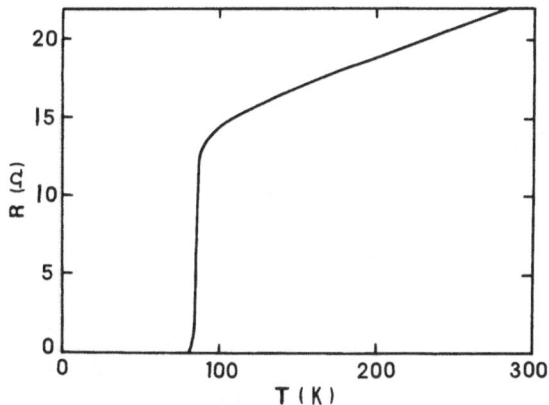

Fig. 1. Resistive transition curve of the YBaCuO
thin film.

$$a_0 \approx [2 \times 0.882/(1-Tx/Tc)]^{1/2} \times 10^{-3} \qquad (1)$$

where Tx was the crossover temperature at which the temperature dependence of the critical current changed from one
behavior to the other one. The unit of a_0 was in micrometer.
In our case Tx/Tc was around 0.81. So the size of grain in
our film was deduced to be 3 nm. It was much smaller than the
value observed from SEM. It might correspond to the characteristic size of domains between twin boundaries within the
grains. The percolating network might consisted of the twin

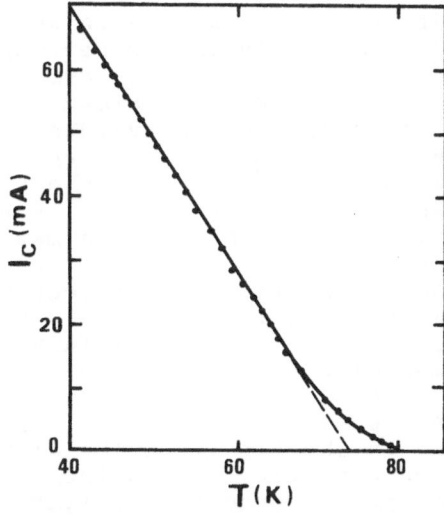

Fig. 2. The critical current Ic versus temperatures for the YBaCuO thin film at zero
field.

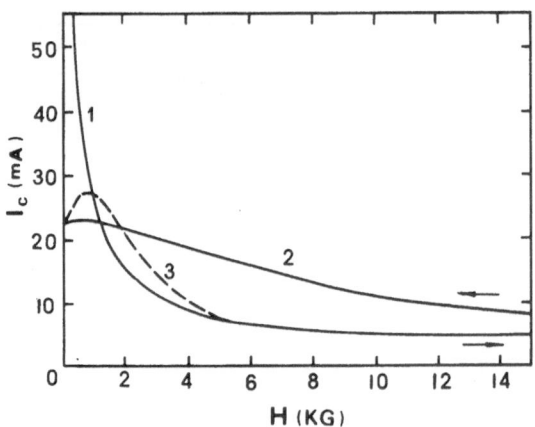

Fig. 3. The critical current Ic versus applied
magnetic field
(1) increasing applied field
(2) decreasing applied field
(3) increasing applied field

boundaries as well as the grain boundaries.

The applied magnetic field was supplied by a NbTi superconducting magnet outside the cryostat. A capacitance thermometer was used to control the temperature by using a temperature controller while applying a magnetic field.

Fig. 3 showed the critical current versus the applied
magnetic field at liquid helium temperature. It did not show
the original value of the critical current at zero field. The
critical current dropped down rapidly in small field region
and then decreased slowly with the increasing of the magnetic
fields (see curve 1 of Fig. 3). The critical current was
irreversible while decreasing the fields. The critical currents were higher than the initial values when the applied
magnetic fields were higher than 2 KG (see curve 2 of Fig 3)
and then crossed with curve 1. At zero applied field, the
critical current, which we defined as Ic1, was much lower than
the original value Ico. When the applied magnetic fields were
increased again, the critical current increased first and then
decreased as increasing the fields. A peak was displayed in
the Ic(H) diagram as shown in the curve 3 of Fig. 3. The
critical current Ic1 at zero applied field was depended on how
high the applied magnetic field had been reached. This was
also explained by trapped flux in the complex weak link network of Josephson coupling. When the applied field was
increased from zero in the first run the diamagnetic flux was
excluded from the whole sample. Then the flux compression in
the weak link of the grain boundaries will be associated with
the flux exclusion. It will steepen the initial Ic(H). As
the field was higher than Hc1 the flux would start to penetrate into the center of the most grains. When the field was
reduced, some of the trapped flux lines were no longer suppor-

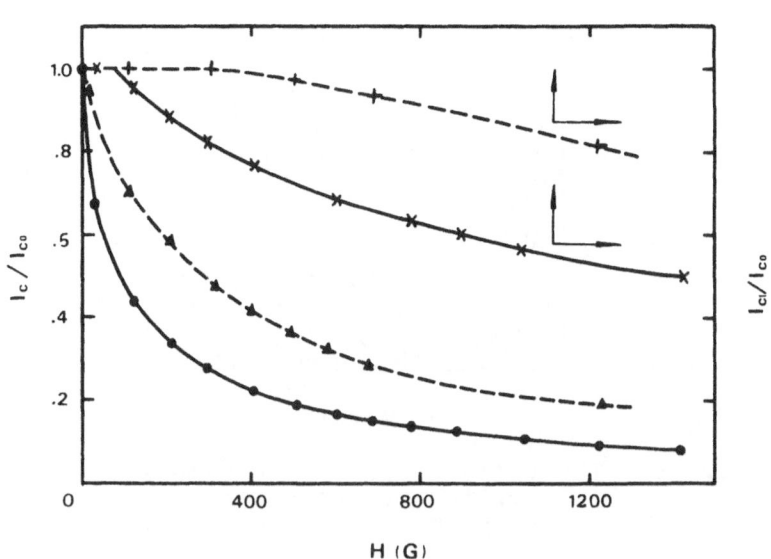

Fig. 4. The critical current Ic1 at zero field
versus magnetic field Ha, which has reached
before (curve 1 and 2) and critical current
Ic versus applied field H (curve 3 and 4).
(1) ---+--- Ic1 (Ha), H∥film,
(2) ---X--- Ic1 (Ha), H⊥film,
(3) ---▲--- Ic(H), H∥film,
(4) ---●--- Ic(H), H⊥film.

ted by the external field and the resulting moment would drive
a reverse flux component in the weakest junctions. The
average field within junctions became less than the applied
field so the critical currents in decreasing fields were
increased over the initial Ic(H) curve. After the flux was
trapped in the grains, the increasing applied field from zero
again would decrease the resulting moment in the weakest
junctions due to the opposite directions of both the trapped
flux and the applied field in the weak link. Fig. 4 showed
the critical currents Ic1 at zero field versus the fields Ha,
which had been reached, both for applied magnetic field parallel and perpendicular to the films. The Ic1 unchanged before
the Ha reached a certain value, which could be defined as
lower critical field Hc1 of the superconducting thin films.
In our case the Hc1 at liquid helium temperature was deduced
to be 300 G from the parallel field. This value was lower than
that Laborde et al. obtained from the magnetization measurements.[8] The reason was that the value of Hc1 was small and
hidden by the pinning of vortices as they mentioned. Fig. 4
also showed the variation of the critical currents with the
applied magnetic fields both for parallel and perpendicular
fields in a small field region.

Fig. 5 showed the dependence of the voltage of the sample
with the magnetic fields while the measuring current was kept
as a constant. Keeping the measuring current higher than the
zero field critical current Ic1, the sample was in the "resistance" state. With sweeping the applied magnetic fields the

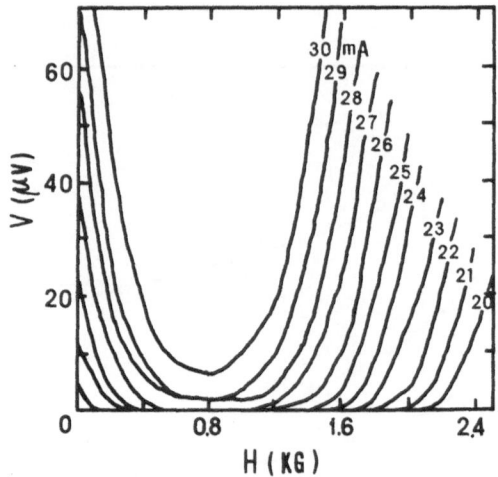

Fig. 5.   The voltage across the sample versus
the applied magnetic fields at
different measuring currents.

resistance of the sample was first decreased and "reenter"
into a superconducting state and showed a negative magnetoresistance.   We would like to emphasize that this phenomenon
occurred only after the applied magnetic field had been
increased to higher than Hcl and then the zero field critical
current Icl was lower than the initial critical current   Ico,
which was the value before applying any field.   The position
at the valley depended on how high the applied field had been
increased to.   The valley bottom position was a new symmetric
point.   It implied how high the flux was trapped in the film.

We measured the resistance as a function of applied
magnetic fields at different temperatures as shown in Fig. 6.
The field was perpendicular to the film.   The transition

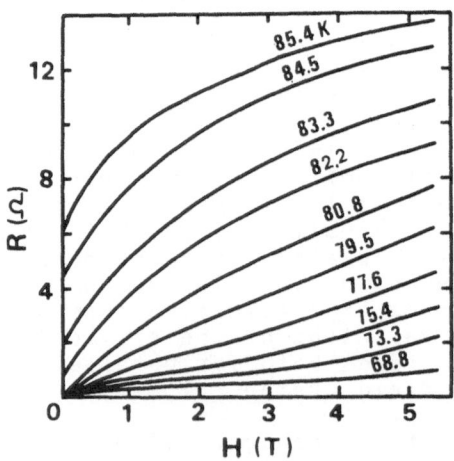

Fig. 6.   Resistive transition curves in the
applied magnetic fields at different
temperatures.

broadness increased as the magnetic field increased. Below zero resistance temperature of 80 K, a small resistance even occurred in a field around 1-2 KG. It was also caused by the flux creep in the weak link network as mentioned by Tinkham[9].

In summary, we have measured the temperature dependence of the critical current in YBaCuO thin films showing that Ico varies linearly with (Tc-T) below crossover temperature Tx. In the region close to Tc, a nonlinear tail can be fitted by $(1-T/Tc)^n$ with n=3/2. It indicates that the critical current is limited by a three dimensional percolating network of Josephson weak links. The irreversible behavior of the critical current in the magnetic fields have been observed and can also be explained by the trapped flux in the complex weak link network.

ACKNOWLEDGEMENTS

This work was supported by the National Natural Science Foundation of China (No. 1860860 and No.9587007) and Chinese Superconductivity Research Center.

References

1. J. E. Evetts and B. A. Glowacki, The relation of critical current irreversibility to trapped flux and microstructure in polycrystalline $YBa_2Cu_3O_7$, to be published in Cryogenics.
2. J. Aponte, H. C. Abache, and M. Octavio, Critical currents in the high Tc superconductors $Dy_1Ba_2Cu_3O_{7-y}$, $Y_1Ba_2Cu_3O_{7-y}$ and $Bi_2Ca_2Sr_2Cu_6O_y$, Preprint.
3. J. W. C. de Vries, M. A. M. Gijs, G. M. Stollman, T. S. Baller, and G. N. A. van Veen, Critical current as a function of temperature in thin $YBa_2Cu_3O_7$ films, J. Appl. Phys., 64:426 (1988).
4. D. S. Yee, R. J. Gambono, M. F. Chisholm, J. J. Cuomo, P. Madakson, J. Karasinski, Critical current and texture relationships in $YBa_2Cu_3O_7$ thin films, AIP Conf. Proc. 165:132 (1988).
5. Y. Enomoto, T. Murakami, M. Suzuki, and K. Moriwaki, Largely anisotropic superconducting critical current in epitaxially grown BaYCuO thin film, Jpn. J. Appl. Phys. 26:L1248 (1987).
6. J. Q. Zheng, G. G. Zheng, D. Q. Li, W. Wang, J. M. Xue, Y. D. Cui, D. Q. Xiao, L. Cao, W. Y. Guan, Study of high-Tc superconducting YBaCuO thin films, Solid State Comm., 65:59 (1988).
7. J. R. Clem, B. Bumble, S. I. Raider, W. J. Gallagher, and Y. C. Shih, Ambegaokar-Baratoff-Ginzberg-Landau crossover effects on the critical current density of granular superconductors, Phys. Rev., B35:6637 (1987).
8. O. Laborde, J. L. Tholence, P. Lejay, A. Sulpice, R. Tournier, J. J. Capponi, C. Michel, and J. Provost, Critical field Hc and critical current in $YBa_2Cu_3O_7$ up to 20 tesla, Solid State Commun., 63:877 (1987).
9. M. Tinkham, The resistance transition of high temperature superconductors, preprint.

MAGNETIZATION, FLUX CREEP AND TRANSPORT IN Tl-Ca-Ba-Cu-O THIN FILMS

E. L. Venturini, J. F. Kwak, D. S. Ginley, B. Morosin, and R. J. Baughman

Sandia National Laboratories
Albuquerque, NM 87185-5800

INTRODUCTION

The enormous experimental and theoretical effort on high-temperature copper-oxide-based superconductors began with the La-Ba-Cu-O system [1], shifted to the Y-Ba-Cu-O system [2], and recently has focused on the Bi-Ca-Sr-Cu-O [3] and Tl-Ca-Ba-Cu-O [4] systems. Tl-Ca-Ba-Cu-O ceramics have zero resistance and Meissner effect up to 125 K [5,6], and exhibit at least five structurally distinct superconducting phases.[7-10] Two common phases, $Tl_2Ca_2Ba_2Cu_3O_x$ (Tl-2223 with a c-axis of 35.9 Å) and $Tl_2CaBa_2Cu_2O_y$ (Tl-2122, c = 29.3 Å), contain double Tl-O planes sandwiched by Ba-O sheets and blocks of double (Tl-2122) or triple (Tl-2223) Cu-O planes separated by intermediate Ca layers.

Static magnetization, flux creep, and electrical transport data are presented for thin films in the high-temperature superconducting Tl-Ca-Ba-Cu-O system. Unoriented polycrystalline 0.7 μm-thick films of both Tl-2223 and Tl-2122 were prepared by sequential e-beam evaporation onto yttria-stabilized cubic zirconia and strontium titanate followed by careful post-deposition anneals. Low-field diamagnetic shielding shows very large demagnetization normal to the film, arising from current flow over macroscopic in-plane distances. Above 50 K there is negligible magnetization hysteresis consistent with easy flux motion. The magnetization exhibits a logarithmic time dependence due to activated flux creep with a small pinning energy of 0.083 eV.

The transport properties are very different from those of polycrystalline $YBa_2Cu_3O_{7-\delta}$ (Y-123) films. Tl-2223 films have shown critical current densities up to 240,000 A/cm² at 76 K (110,000 A/cm² for Tl-2122 films) with low sensitivity to in-plane magnetic fields and moderate sensitivity to perpendicular fields. This anisotropic behavior indicates that grain boundary Josephson junctions are not dominating the transport. The strong intergranular links and high critical current densities show the potential of these polycrystalline thin films for applications.

THIN FILM DEPOSITION AND STRUCTURE

Thin films were deposited in a 1.5×10^{-5} mbar oxygen atmosphere by sequential e-beam evaporation of Tl, Ca, Ba, and Cu metals onto substrates of $SrTiO_3$, polycrystalline MgO, sapphire, and yttria-stabilized cubic zirconia (YSZ).[11,12] The best films to date have been prepared on YSZ and $SrTiO_3$. The metals were deposited in a 25-layer sequence of Cu-Ba-Ca-Tl, starting and ending with Cu, to a total thickness of 0.7 μm with each metal layer thickness adjusted to achieve a net stoichiometry of either Tl-2223 or Tl-2122. The substrate temperature was kept below 50°C during deposition. The resulting films are very air and moisture sensitive.

The key step in thin film processing is proper control of the oxygen and thallium partial pressures during post-deposition sintering and annealing.[12] This control was accomplished by placing the substrate with the deposited metals face down on a sintered Tl-Ca-Ba-Cu-O pellet of the same stoichiometry as the film, air annealing at 850°C for 7-15 minutes, and then annealing in flowing oxygen at 750°C for 30 minutes with a furnace cool. The air anneal produces the morphology observed in the final films.

Scanning electron micrographs show a polycrystalline microstructure in all films consisting of plate-like growths up to 20 μm in size. Frequently, small crystallites of excess thallium oxide are observed on the surface of the films. X-ray diffraction for a number of the films on both YSZ and $SrTiO_3$ shows a random orientation of grains with the intended phase as the predominant one (>80%). Energy dispersive X-ray analysis indicates that the original metal deposition ratio (except for Tl) is maintained in the fully annealed films. Films that are a few atomic percent rich in Tl or Ca have shown the highest critical current densities, while films that are Tl deficient have low critical currents and pronounced weak link behavior.

STATIC MAGNETIZATION

The static magnetization was measured versus magnetic field and temperature in a SQUID susceptometer (Biomagnetic Technologies, San Diego, CA). Meissner effect data (flux expulsion by cooling in a magnetic field from above the superconducting transition) were recorded in 25 Oe to determine the magnetic onset temperature and the volume fraction of superconductor. The onset was 103 K for a Tl-2122 thin film and 109 K for a Tl-2223 film. Meissner fractions up to 75% were measured at 5 K with the field applied parallel to a Tl-2122 film on YSZ.

Diamagnetic shielding was determined by cooling a film in nearly zero field and then applying the measurement field. Fig. 1 compares the low-field shielding response at 5 K for an unoriented polycrystalline 0.7 μm-thick Tl-2122 film on YSZ for fields parallel and perpendicular to the film. Note that the vertical scale for the perpendicular data (solid triangles) is 200× greater than that for the parallel data (open triangles). This film has a Meissner effect onset near 103 K. The total film area was 3.0×3.5 mm^2 for a volume of 0.006 mm^3 using the estimated 75% dense coverage from scanning electron micrographs. The diamagnetic response of the YSZ substrate is negligible at the fields in Fig. 1.

The parallel shielding data correspond to a volume susceptibility $\chi_v = -0.87/4\pi$, while the perpendicular data give $\chi_v = -315/4\pi$. This difference is due to the demagnetization normal to a thin film. The enhancement in the perpendicular direction implies a demagnetization factor $D = 0.997$, requiring that the shielding currents flow through a region at least 0.35 mm in diameter (500x the film thickness). This demonstrates the ability of the film to transport screening currents over large continuous areas, confirming the negligible effects of possible weak links between

396

grains in agreement with the transport critical current results discussed below.

Fig. 2 compares high-field magnetization loops from 2 mT to 5 T and back to 2 mT on a semi-log scale at 5 and 77 K (note the vertical axis scale change) with the field applied normal to the same 0.7 μm Tl-2122 film on YSZ. The solid triangles represent increasing field strength (following zero field cooling) while the open triangles show decreasing fields. The data are corrected for the diamagnetism of the YSZ substrate and sample holder (measured directly). There is negligible hysteresis above 0.5 T at 77 K, while the hysteresis extends to 5 T at 5 K. This behavior is due to activated flux motion [13-14] resulting from weak flux pinning. Direct measurements of flux creep presented below confirm this interpretation.

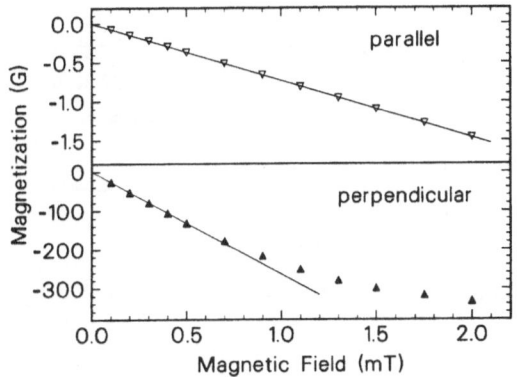

Fig. 1. Low-field diamagnetic shielding response at 5 K with the field applied parallel and perpendicular to a 0.7 μm unoriented polycrystalline Tl-2122 film on a YSZ substrate.

The minimum in the shielding signal for increasing field in Fig. 2 occurs near 10 mT at 77 K and 20 mT at 5 K, but this value is far above the lower critical field, H_{c1}. The data in Fig. 1 for fields parallel to the film are linear to 2 mT, suggesting that H_{c1} is greater than this value. Similar measurements on bulk ceramics cut into thin plates show that the shielding response deviates from linearity above 4 mT at 5 K and 2 mT at 77 K; these values are reasonable estimates of H_{c1} in polycrystalline bulk and thin film Tl-2223. However, they are an order of magnitude smaller than a recent measurement [15] of H_{c1} = 90 mT at 4.2 K in bulk ceramic Tl-2223.

To demonstrate the problems in applying the Bean critical state model [16] to a superconducting film with weak flux pinning, we can compare the inferred critical current densities J_{cm} from the magnetization hysteresis at 5 and 77 K with the direct transport values J_{ct}. The magnetization data were obtained on a Tl-2122 film which was scribed into smaller segments averaging 0.1 cm on a side for critical current studies. Hence at 77 K the remanent magnetization (low-field value following the loop in Fig. 2) is 3 G, and J_{cm} would be $\approx 60\times3/0.1 = 1800$ A/cm² compared to J_{ct} of 110,000 A/cm² from direct transport measurements [17]. This nearly two order of magnitude discrepancy demonstrates that the critical state model may not be applicable in these weakly pinned materials. The situation is considerably better at 5 K where flux motion is limited: the remanent magnetization of 750 G implies $J_{cm} \approx 450,000$ A/cm² compared to $J_{ct} = 2,000,000$ A/cm².

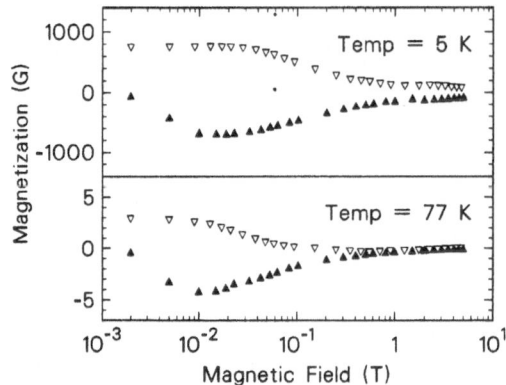

Fig. 2. High-field magnetization loops at 5 and 77 K with the field applied perpendicular to the same Tl-2122 thin film as in Fig. 1.

ACTIVATED FLUX CREEP

Flux creep in ceramic [18] and single crystal [14] samples of $YBa_2Cu_3O_{7-\delta}$ has been reported. Our measurements [19] on bulk ceramic Tl-2223 confirm flux creep in this material as well. Here we report measurements on a thin film with the magnetic field applied normal to the film. Fig. 3 shows magnetization versus time at 10 K illustrating flux creep for a Tl-2223 polycrystalline film on $SrTiO_3$. The solid triangles reflect flux motion out of the film following a step change from 1 T to nearly zero field. The open triangles show flux creep into the film after a step change from nearly zero field to 50 mT. These situations are quite different: for flux creep out of the film, the measured signal reflects critical current flow at nearly zero field arising from pinned flux remaining after a field sweep from the mixed state to zero field (top data in Fig. 3). For flux creep into the film, the shielding response reflects critical current flow to exclude the applied field (bottom data in 50 mT). Clearly, the flux motion is similar in these two physically distinct measurements.

The magnitude of the magnetization for the Tl-2223 thin film shown in Fig. 3 is substantially less than the remanent magnetization after the high-field loop for the Tl-2122 film in Fig. 2. The Tl-2223 film used for the flux creep measurements showed substantially less transport critical current than our "best" films, in agreement with the magnetic data.

398

However, the creep results are representative of flux motion in films since the magnetization probes the strongly linked regions.

The semi-log plot in Fig. 3 empasizes the logarithmic time dependence of the magnetization signal for times between 200 and 4000 seconds. These data are consistent with thermally activated flux motion.[13,14,19,20] The data for flux motion out of the film were extended to 7500 seconds with no change in behavior. Zero time for the magnetization data is defined as the time when the field sweep in our SQUID susceptometer reaches the measurement field. Due to the design of this instrument, there is a dead time of approximately 200 seconds after the field sweep stops before useful data can be recorded.

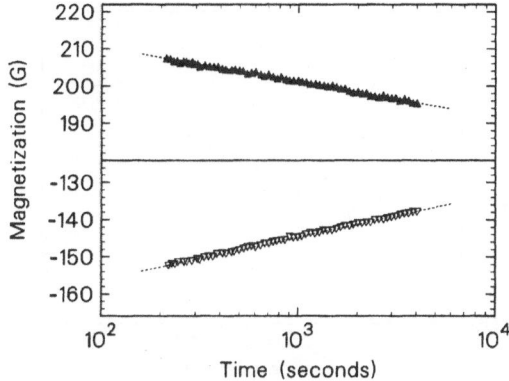

Fig. 3. Magnetization versus time at 10 K illustrating flux creep for a Tl-2223 polycrystalline film on $SrTiO_3$. The logarithmic time dependence is consistent with thermally activated flux motion. The solid triangles show creep out of the film following a step change from 1 T to nearly zero field at a nominal time of 0 seconds. The open triangles show flux creep into the film following a step change from nearly zero field to 50 mT. The field is applied normal to the film for both measurements.

The dashed lines in Fig. 3 represent a simple two-parameter linear fit to the logarithmic time dependence:

$$M(t) = M(t=1) + R \cdot \ln(t) \tag{1}$$

where $M(t)$ is the magnetization at time t, $M(t=1)$ is the extrapolated magnetization at $t = 1$ second, and $R = dM(t)/d\ln(t)$ is the creep rate. Similar measurements at temperatures between 5 and 40 K for flux creep into the film at 50 mT show a similar logarithmic time dependence. Substantial deviations from this simple dependence have been observed in single crystals of Tl-2122 between 20 and 40 K as discussed elsewhere.[21]

The two parameters in Eq. (1) can be used to extract an average pinning potential U by fitting the magnetization time dependence for data taken over a range of temperatures [14,19]:

$$R/M(t=1) = kT/U \qquad (2)$$

where R and M(t=1) are defined in Eq. (1) and k is the Boltzmann constant. Fig. 4 shows the ratio of creep rate out of the film to magnetization at t = 1 second versus temperature, demonstrating the linear relation according to Eq. (2). The dashed line corresponds to U = 0.083 eV which is a rather small pinning potential. This weak flux pinning with activated motion explains the hysteresis collapse at 77 K shown in Fig. 2. Eq. (2) predicts no flux creep at T=0 (no thermal energy), while the data in Fig. 4 do not extrapolate to zero at T=0. This suggests that the creep in these films may be driven by mechanisms more complex than simple thermal activation.

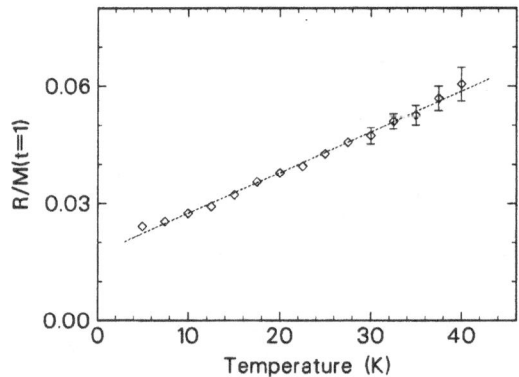

Fig. 4.   Ratio of logarithmic creep rate R out of the Tl-2223 film to the extrapolated magnetization M at t = 1 second versus temperature (parameters obtained from linear fits to data such as that shown by the open triangles in Fig. 3). The linear temperature dependence yields an activation energy barrier to flux motion of 0.083 eV.

TRANSPORT CRITICAL CURRENT DENSITY

The transport critical current density (J_{ct}) measurement procedure has been discussed elsewhere.[17] Here we concentrate on details relevant to the possibility of flux creep effects. Linear four-probe J_{ct} data were obtained on films scribed with large active areas (~ 0.1 mm^2) between the inner voltage contacts. It was usually necessary to use 200-1000 μs current pulses with a <1% duty cycle in order to minimize heating effects. Even so, it was not always possible to eliminate heating effects at the lowest temperatures and highest currents. J_{ct} was defined by the observation of a 2μV signal (the minimum quantifiable signal) on an oscilloscope with differential input. J_{ct} measurements at 76 K and H=0-0.08 Tesla were made with the films immersed directly in liquid nitrogen using a copper coil for field generation. For all other measurements, the films were mounted in an evacuable probe immersed in liquid helium in the bore of a superconducting Nb-Ti coil. The samples were routinely warmed above T_c after field cycles in order to eliminate trapped flux.

J_{ct} values up to 110,000 A/cm² have been observed for unoriented Tl-2122 films at 76 K, with J_{ct} varying as $(1-T/T_c)^2$ from 5 K to 90 K. Even higher values have been obtained on unoriented and oriented polycrystalline Tl-2223 films.[22] No systematic differences were observed among the various film morphologies and crystallographic phases. The temperature dependence is consistent with grain-boundary pinning as the dominant mechanism determining J_{ct}.

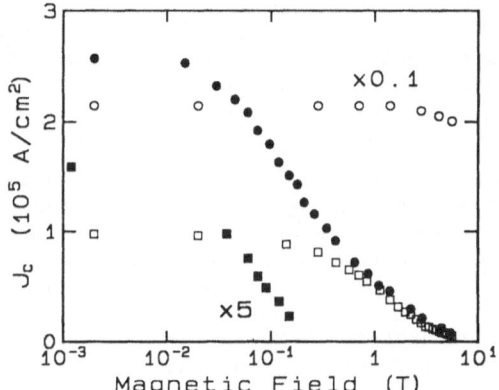

Fig. 5. Semi-log plot of critical current density versus magnetic field for a 0.7 μm unoriented polycrystalline film of Tl-2122 at 4 K (open circles) and 76 K (open squares) with the field in the film plane (but normal to the current), and for a 0.2 μm oriented polycrystalline film of Tl-2223 at 32 K (solid circles) and 80 K (solid squares) with the field normal to the film. The lowest field point for each data set was actually measured at H=0. Note that the 4 K data have been reduced by a factor of 10, and the 80 K data multiplied by 5, in order to show all the data together.

Fig. 5 shows the magnetic field dependence of J_{ct} at several temperatures. The data at 32 K and 80 K were taken in fields normal to the film plane, while those at 4K and 76 K were done with fields in the film plane, but normal to the nominal current direction. The fall-off is much stronger for normal fields, but still considerably slower than similar data for weakly linked bulk ceramic. The anisotropy <u>per se</u> is inconsistent with weak-link dominated critical current. The reason that grain boundaries should be so benign in these films remains a question, but may involve the existence of a multiplicity of closely related but distinct crystallographic phases which are all high-temperature superconductors.

Despite the favorable situation at the grain boundaries, the observed critical currents really are not all that high or remarkably well-behaved in applied magnetic fields. J_{ct}'s above 10^7 A/cm^2 are routinely obtained in granular NbN and Nb-Ti material, sometimes with J_{ct} initially <u>increasing</u> with applied field.[23] Nor do the present J_{ct} results compare with those [24] recently obtained on Bi(Pb)-Ca-Sr-Cu-O thin films at 77 K. It is only by comparison to other polycrystalline materials in which the grains are weakly linked that these films are impressive. The reason that J_{ct} is not higher, of course, is the very weak defect flux pinning in these materials due to their short coherence lengths. This problem is compounded by the high transition temperatures, resulting in the easy flux motion discussed above.

Comparison of Figs. 2 and 5 shows analogous behavior of the magnetization and transport critical currents with applied magnetic field normal to the film plane. However, the transport J_c is much larger than that indicated by the magnetization. This apparent discrepancy is understandable in the context of weak flux pinning. The critical state model used to derive critical current from magnetization assumes a uniform current density everywhere in the sample, with pinning forces just balanced so that the flux lines are stationary. In a weakly-pinned system, however, flux will begin to penetrate the sample just above H_{c1} and the screening currents will be confined to the vicinity of the sample surface, with a flux gradient determined by the strength of the pinning. This is the reason that the critical state model is strictly applicable only to highly hysteretic (hard) superconductors. These materials can in no way be regarded as hard.

Given weak flux pinning, it is not surprising that flux motion effects may be significant. The question here is whether such effects are important for the J_{ct} measurements. In order to discuss this, it is necessary to consider the flux creep model of Anderson and Kim.[13] One assumes that flux lines are weakly pinned at potential wells of energy U, and that the effective pinning strength is significantly reduced in the presence of a flowing current and a magnetic field. Then thermally activated flux motion will give rise to an electric field E. Following the formalism of Campbell and Evetts [25], but allowing for backward as well as forward flux creep:

$$E = B\Omega d \, \exp\left(-\frac{U}{kT}\right) \sinh\left(\frac{UJ}{kTJ_{c0}}\right). \tag{3}$$

In this equation, J_{c0} is the value J_c would take in the absence of flux creep, B is the magnetic induction, d is the spacing between pinning centers, and Ω is the oscillation frequency for a coupled flux bundle. The magnetization versus time data suggest that $U/kT_c < 10$ for a typical Tl-2223 film, an even lower ratio than that for Y-123.

Note that Eq. (3) predicts a nonzero E for any nonzero values of B and J. Thus, formally, J_c is always zero (since there is an induction B from the self-field of an applied J) in flux creep models. In fact, Eq. (3) predicts an exponential rise in voltage with current. We show below that this is not observed for these films suggesting, consistent with the magnetization data, that flux creep is not <u>thermally</u> activated. If the behavior were exponential only at lower voltages than those we are able to measure, we would still expect some type of fall-off from exponential behavior, whereas we actually see a clean cubic dependence (see below).

Practically, some finite electric field, E_c, is used to measure J_c. This gives rise to an effective critical current found by inserting E_c in Eq. (3), which is easily inverted to yield J_c when $UJ_c \gg kTJ_{c0}$:

$$J_c = J_{c0} \left(1 - \frac{kT}{U} \ln \frac{B\Omega d}{E_c} \right).$$ (4)

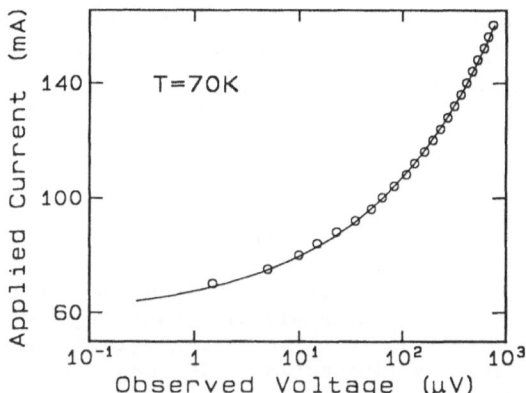

Fig. 6. Semi-log plot of applied current versus observed voltage for a polycrystalline epitaxial film of Tl-2223 at 70 K and zero field. The fit shown is $(i/i_{min} - 1)^3 = V/V_0$, as predicted by a model for non-uniform current density in the film, with $i_{min} = 58$ mA and $V_0 = 226$ μV. Note that our nominal i_c defined at $V = 2\mu$V would be 70 mA. Regardless of the model used to describe the observed behavior, it indicates that flux creep effects are not important for the critical current measurements.

Since the frequency Ω appears in Eq. (4), it is tempting to identify E_c/Bd with an experimental frequency. In the case of pulsed J_{ct} measurements, however, the appropriate time scale is not obvious, since pulse rise time, pulse width, and repetition rate are all involved, but the magnetic field is changed quite slowly. Moreover, since the application of a current pulse results in strong local flux gradients which could induce flux motion, one might even expect pulse-derived data to be too low. In fact, we have obtained the same J_{ct} results by pulse and dc measurements in cases where ohmic heating was negligible.

The true significance of Eq. (4) for J_{ct} measurements, whether pulsed or d.c., lies in the suppression of J_c below J_{c0} by flux creep. The effect of using a larger E_c is to reduce this suppression. Thus the measured J_{ct}

is a <u>lower</u> bound on the values which may be attained with more finely grained material or through the introduction of stronger pinning centers.

We wish to emphasize that the synthesis of films with high J_{ct} values is not trivial, since precise control of the processing protocols is required. Films that are nominally identical compositionally and structurally can show up to order-of-magnitude variations in J_{ct}, although the temperature and field dependences are similar. Films that had demonstrable differences from those reported here (e.g., with Tl deficiency) had much lower J_{ct}'s dominated by intergranular weak links. It is clear, therefore, that J_c within an individual film may well be markedly non-uniform due to variances in grain boundary quality. In fact, we have shown that the current versus voltage above 1 μV (Fig. 6) is completely consistent with a model of non-uniform J_{ct}, and that the nominal J_{ct} value defined at V=2μV is only fractionally higher than the minimum J_{ct} in the film.[17]

CONCLUDING REMARKS

Unoriented polycrystalline thin films of both Tl-2122 and Tl-2223 have been prepared by sequential e-beam evaporation of the metals followed by careful post-deposition anneals. The films were characterized by X-ray diffraction, energy dispersive X-ray analysis, static magnetization and electrical transport. Above 50 K there is negligible magnetization hysteresis consistent with easy flux motion. The magnetization exhibits a logarithmic time dependence between 5 and 40 K due to activated flux creep with a pinning barrier energy of 0.083 eV. The creep does not appear to be due to simple thermal activation. This weak pinning prevents use of the critical state model to infer critical current densities from magnetization data.

An increase in pinning should be beneficial for use of the films shielding applications, and might also enhance the critical current density. A recent calculation [26] suggests that the high temperature superconductors with their extremely short coherence lengths may show intrinsic flux pinning due to the discrete lattice. If this is correct, one expects that highly localized damage from ion implantation would also provide enhanced pinning. Various sample treatments are being explored in an attempt to increase pinning including rapid thermal annealing, slight compositional variations, and ion implantantion.

Transport critical current densities are large (over 10^5 A/cm^2 at 76 K and 10^6 A/cm^2 at 4 K) and exhibit only modest magnetic field dependence. This suggests that transport in these films is not dominated by weak links between grains, in contrast to the dramatic effects of weak links in bulk ceramic and polycrystalline films of Y-123. These promising results in the first few months of research on Tl-Ca-Ba-Cu-O thin films clearly demonstrate the potential for applications. Despite the undisputed occurrence of flux creep in the thallium-based superconductors under certain conditions, as shown by our magnetization versus time data, there is no clear evidence for significant flux creep effects on the transport critical current data. If flux creep is significant, our results represent a lower bound on the attainable critical current. We have also shown that the observed current-voltage behavior is consistent with non-uniform critical currents in the films.

ACKNOWLEDGEMENT

This work at Sandia National Laboratories was supported, in part, by the United States Department of Energy, Office of Basic Energy Sciences, under Contract No. DE-AC04-76DP00789. The technical assistance of T. Castillo, R. Hellmer, M. Mitchell, and G. Pannell, Jr. is appreciated.

REFERENCES

1. J. G. Bednorz and K. A. Muller, Z. Phys. B $\underline{64}$, 189 (1986).
2. M. K. Wu, J. R. Ashburn, C. J. Torng, P. H. Hor, R. L. Meng, L. Gao, Z. J. Huang, Y. Q. Wang and C. W. Chu, Phys. Rev. Lett. $\underline{58}$, 908 (1987).
3. H. Maeda, Y. Tanaka, M. Fukutomi and T. Asano, Japan. J. Appl. Phys. $\underline{27}$, L209 (1988).
4. Z. Z. Sheng, A. M. Hermann, A. El Ali, C. Almasan, J. Estrada, T. Datta and R. J. Matson, Phys. Rev. Lett. $\underline{60}$, 937 (1988); Z. Z. Sheng and A. M. Hermann, Nature $\underline{332}$, 138 (1988).
5. D. S. Ginley, E. L. Venturini, J. F. Kwak, R. J. Baughman, M. J. Carr, P. F. Hlava, J. E. Schirber and B. Morosin, Physica C $\underline{152}$, 217 (1988).
6. S. S. P. Parkin, V. Y. Lee, E. M. Engler, A. I. Nazzal, T. C. Huang, G. Gorman, R. Savoy and R. Beyers, Phys. Rev. Lett. $\underline{60}$, 2539 (1988).
7. R. M. Hazen, L. W. Finger, R. J. Angel, C. T. Prewitt, N. L. Ross, C. G. Hadidiacos, P. J. Heaney, D. R. Veblen, Z. Z. Sheng, A. El Ali and A. M. Hermann, Phys. Rev. Lett. $\underline{60}$, 1657 (1988).
8. B. Morosin, D. S. Ginley, P. F. Hlava, M. J. Carr, R. J. Baughman, J. E. Schirber, E. L. Venturini and J. F. Kwak, Physica C $\underline{152}$, 413 (1988).
9. S. S. P. Parkin, V. Y. Lee, A. I. Nazzal, R. Savoy, R. Beyers and S. J. LaPlaca, Phys. Rev. Lett. $\underline{61}$, 750 (1988).
10. B. Morosin, D. S. Ginley, J. E. Schirber and E. L. Venturini, Physica C $\underline{156}$, 587 (1988)
11. D. S. Ginley, J. F. Kwak, R. P. Hellmer, R. J. Baughman, E. L. Venturini and B. Morosin, Appl. Phys. Lett. $\underline{53}$, 406 (1988).
12. D. S. Ginley, J. F. Kwak, R. P. Hellmer, R. J. Baughman, E. L. Venturini, M. A. Mitchell and B. Morosin, Physica C $\underline{156}$, 592 (1988).
13. Y. B. Kim, C. F. Hempstead and A. R. Strnad, Phys. Rev. Lett. $\underline{9}$, 306 (1962); P. W. Anderson, Phys. Rev. Lett. $\underline{9}$, 309 (1962).
14. Y. Yeshurun and A. P. Malozemoff, Phys. Rev. Lett. $\underline{60}$, 1676 (1988).
15. W. Reith, P. Muller, C. Allgeier, R. Hoben, J. Heise, J. S. Schilling and K. Andres, Physica C $\underline{156}$, 319 (1988).
16. C. P. Bean, Phys. Rev. Lett. $\underline{8}$, 250 (1962); C. P. Bean, Rev. Mod. Phys. $\underline{36}$, 31 (1964).
17. J. F. Kwak, E. L. Venturini, R. J. Baughman, B. Morosin and D. S. Ginley, Physica C $\underline{156}$, 103 (1988).
18. C. Giovannella, G. Collin, P. Rouault and I. A. Campbell, Europhys. Lett. $\underline{4}$, 109 (1987); C. Giovannella, P. Rouault, I. A. Campbell and G. Collin, J. Appl. Phys. $\underline{63}$, 4173 (1988).
19. M. E. McHenry, M. P. Maley, E. L. Venturini and D. S. Ginley, submitted.
20. M. R. Beasley, R. Labusch and W. W. Webb, Phys. Rev. $\underline{181}$, 682 (1969).
21. E. L. Venturini, J. F. Kwak, D. S. Ginley, B. Morosin and R. J. Baughman, in preparation.
22. J. F. Kwak, E. L. Venturini, R. J. Baughman, B. Morosin and D. S. Ginley, Cryogenics (accepted).
23. D. Dew-Hughes, Phil. Mag. B $\underline{55}$, 459 (1987).
24. M. Klee, J. W. C. de Vries and W. Brand, Physica C $\underline{156}$, 641 (1988).
25. A. M. Campbell and J. E. Evetts, Adv. Phys. $\underline{21}$, 199 (1972).
26. L. Schimmele, H. Kronmuller and H. Teichler, Phys. Stat. Solidi (b) $\underline{147}$, 361 (1988).

COMPOSITION ANALYSIS OF HIGH T_c Bi-Sr-Ca-Cu-O THIN FILMS BY RBS AND EPMA

Neelkanth G. Dhere, John P. Goral, Alice R. Mason, Nalin R.
Parikh* and Bijoy K. Patnaik*

Solar Energy Research Institute
1617 Cole Boulevard, Golden, CO 80401, USA
*Dept. of Physics and Astronomy, University of North Carolina
Chapel Hill, NC 27599-3255, USA

INTRODUCTION

Precise knowledge of the variation of the composition is necessary for
deposition of thin films over large areas. Preparation of multi-component
thin films from single source has been important for many technological
applications. This is especially true for the development of high-T_c,
perovskite-oxide superconducting phases which may be formed with three or
more metallic elements in a stoichiometric proportion deposited on a multi-
component substrate[1-5]. Variation of film thickness usually shows a bell- or
´M´-shaped profile with respect to the axis of a circular, ring type or
cylindrical target (Fig. 1). The concentration of individual elements de-

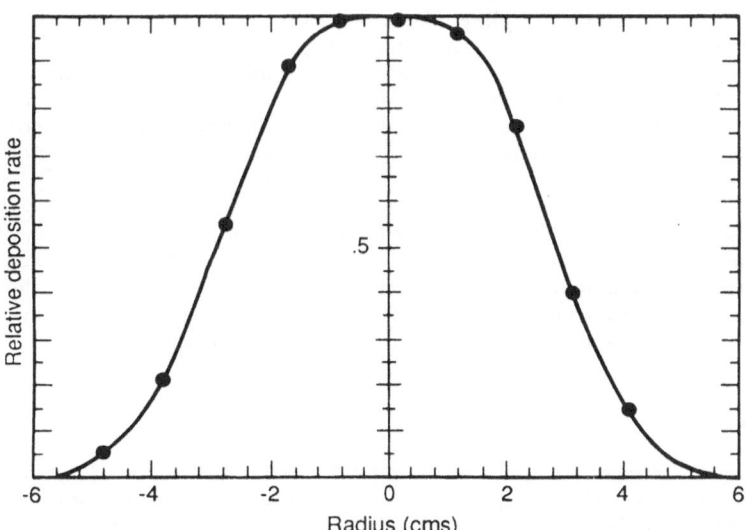

Fig. 1. Bell-shaped thickness profile of sputter-deposited thin
films, with respect to the axis of the ring target.

posited from a multi-component source may vary radially over substrate areas. Moreover, the deposition rates of the constituents may also vary over the duration of the deposition. It is, therefore, necessary to employ techniques which will measure accurately the variation of composition over the substrate area as well as over the film thickness.

A novel sputtering system (UNIFILM) with a computer-controlled rotation and scanning of the substrates is being utilized for the deposition of thin films. The program utilizes experimental thickness profiles for simulating the thickness variation under different substrate scanning routines. A more advanced program for thickness as well as composition uniformity is being developed independently of the program supplied by the manufacturer.

Electron probe X-ray microanalysis (EPMA) is routinely utilized in many laboratories for determination of composition of thin films with thicknesses of >10000 Å. Because of the small beam size, mapping of the variation of the composition along a line (line-scan) or over an area (area-scan) is possible with this technique. In EPMA, since the excitation for the emission of characteristic X-rays emitted by individual elements is provided not only by the primary electron beam but also by the secondary X-rays emitted by other elements, the measurements are affected by the presence of other elements (matrix effect). Standards for each element are utilized to obtain X-ray yields under the specific beam conditions. Computer programs are available for calculation of the matrix effect in a homogeneous and infinite thin film. At lower thicknesses, EPMA tends to be less accurate, especially because of the effect of the substrate which is more difficult to compensate.

Rutherford back-scattering (RBS) can measure the composition of a multi-component thin film, with or without the substrate, with good accuracy provided the RBS signals from individual elements do not overlap. It also provides information on the composition as a function of depth for thicker films. The incident ion is directly scattered by the target elements individually, without secondary matrix effects. The composition of the multi-element film, however, affects the stopping power of the material through which the ion traverses. This, in turn, affects the scattering yield per channel (energy). The stoichiometry of the of the film is calculated from the yields of all the elements, assuming linear additivity of stopping powers of the individual elements in proportion to their composition (Bragg´s rule). The stopping powers generally vary somewhat slowly with energy and small variations in the composition of the film or presence of small amounts of impurities not accounted for in the composition do not affect the yields appreciably. Secondary matrix effects are non-existent in RBS. Nor do the signals corresponding to scattering from the substrate pose any problem for thin films such as the one under study. The details of RBS techniques are available elsewhere[6].

RBS can provide area- or line- scan information only on a rougher scale (few mm) compared to EPMA because of the large area of the analyzing beam. Both techniques are non-destructive. The present paper compares the results of composition analysis with EPMA and RBS. These results will serve as a basis for the thickness and composition uniformity program.

2. EXPERIMENTAL

Thin films of Bi-Sr-Ca-Cu-O were deposited from single targets by RF magnetron sputtering. The ring targets consisting of pressed and sintered mixtures of oxide powders for the Sloan S310 type sputter gun were procured from KEMA. Base pressures of <1 x 10^{-7} Torr were obtained with cryo-pumped deposition system (UNIFILM). During RF magnetron-sputter deposition the total pressures of argon and oxygen were maintained in the range 3-5 x 10^{-3}

Torr. Ratio of argon to oxygen partial pressures was varied in the range 2 to 1. The deposition set-up and procedure have been described earlier[5,7].

The substrates consisted of (100) faces of single crystal MgO and fused quartz glass plates. The deposition rates and thicknesses were calibrated with a quartz crystal rate and thickness monitor, prior to the actual depositions. The actual thicknesses were measured at a step on a partially covered smooth quartz substrate with a stylus profilometer. In earlier work, difficulties were encountered in making mechanical contact to a conducting pad on an insulating substrate for DC biasing. Hence RF biasing was considered as more appropriate[7] and was employed in this work. The signal of the RF was derived from the sputtering RF power supply and was fed into the RF amplifier circuit of a trans-receiver. DC bias developed on the RF electrode backing the substrates was measured across a L-C filter. The compositions of the films deposited without and with different values of RF bias were analyzed by EPMA and RBS. The principle of EPMA are well known[8]. Energy dispersive and wavelength dispersive EPMA were carried out with 10 keV electrons in electron-probe X-ray microanalyzer (CAMECA). The standards utilized for Bi, Sr, Ca, Cu, Mg and O were metallic bismuth, strontium titanate, calcium fluoride, metallic copper and magnesium and silica respetively. RBS was carried out with 2 MeV He^+ ions back-scattered at an angle of 165°. The RBS set-up is shown in fig. 2. Calculations showed that the peaks from the metallic components Bi, Sr, Cu and Ca do not overlap for film thicknesses upto 1500 Å. The quantities of the metallic elements were determined from the areas under the separate peaks from Bi, Sr, Cu, and Ca. A computer program, WIN-SPEC developed at the Dept. of Physics and Astronomy, University of North Carolina, was utilized for simulating the RBS spectra and thus obtaining the atomic concentrations of the metallic components of the thin films[9].

3. RESULTS AND DISCUSSION

Two different target compositions were employed viz. Bi:Sr:Ca:Cu:O :: 2.2:2.0:0.8:2.0:8.1 and 2.0:2.0:1.8:2.8:9.6. The convenient film thicknesses for EPMA are >10000 Å whereas for RBS they are <1500 Å. Hence the films were deposited in both these ranges. It was found that the deposition process itself did not contribute additional impurities to the thin films. Figure 3 shows the EPMA line-scans of the variations of the composition across a thin film RF sputter-deposited with 1:1 ratio of Ar to O_2 partial

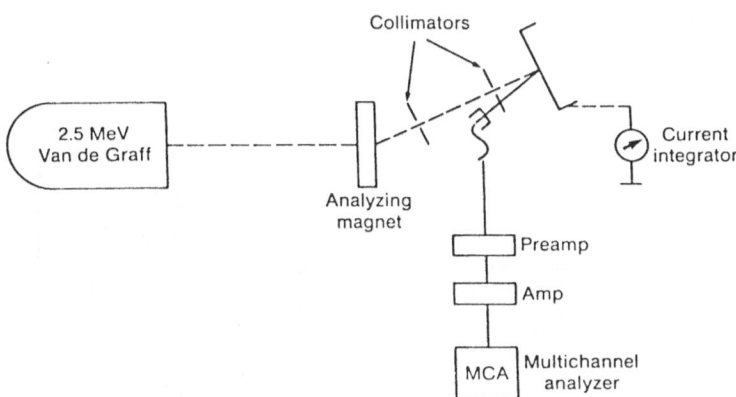

Fig. 2. Schematic of Rutherford back-scattering set-up.

pressures, on unheated and unbiased substrate utilizing the first target. As can be seen, the composition remains fairly constant across the sample. The average chemical composition of the thin film, as measured by EPMA, was Bi:Sr:Ca:Cu:O :: 3.0:2.0:0.8:1.9:13.4. As can be noted, the film became enriched 36 at.% in Bi while the composition of the other elements remained mostly constant.

Fig. 4 shows a RBS spectrum of a thin film, RF sputter-deposited with 2:1 ratio of Ar to O_2 partial pressures, on an unheated and unbiased MgO substrate, employing the second target. The back-scattering peaks, in the order of diminishing energy (channel number), are from Bi, Sr, Cu and Ca and are separate. Hence the results of the RBS analysis can be assumed to be fairly accurate, except for the concentration of Ca, the peak from which is small. The steps at lower energies are due to Mg and O signals. Simulated RBS spectra, shown as dashed lines in the fig. 4, obtained with the help of a computer program, WIN-SPEC, were matched with the experimental spectra to obtain the compositions of the thin films[9]. The stoichiometric composition of the metallic components of the film was found to be Bi:Sr:Ca:Cu :: 2.3:2.0:2.0:2.9. Table I shows the variation of the chemical compositions, measured by RBS, of Bi-Sr-Ca-Cu-O thin films with an increasing RF bias, maintaining the ratio of argon to oxygen partial pressures at 2:1. The values of negative DC voltage developed on the backing electrode have been indicated. As can be seen, in the case of an unbiased substrate, the concentration of Bi is ~15% higher in the film than in the target, and that the Bi concentration decreases monotonously with increasing RF bias. The copper concentration is slightly higher in the sputtered film than in the target, for the unbiased substrates. With increasing RF biasing it decreases slightly upto a bias of -80 V and then increases considerably. The accuracy of the measurements for Ca is not very good. It may be concluded that the concentrations of Sr and Ca do not change much as compared to the composition of the target. Table I also shows the compositions obtained by EPMA. There is considerable discrepancy in the concentrations measured by RBS

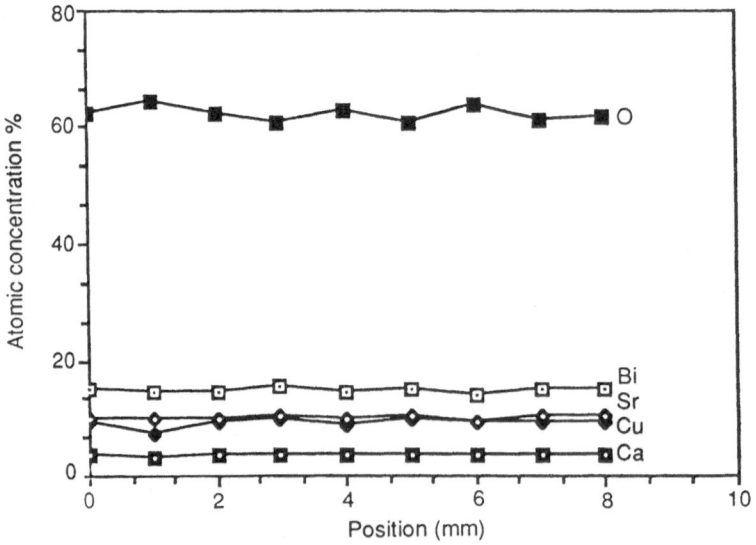

Fig. 3. EPMA line-scan of the variation along a thin film with an average composition Bi:Sr:Ca:Cu :: 3.0:2.0:0.8:1.9

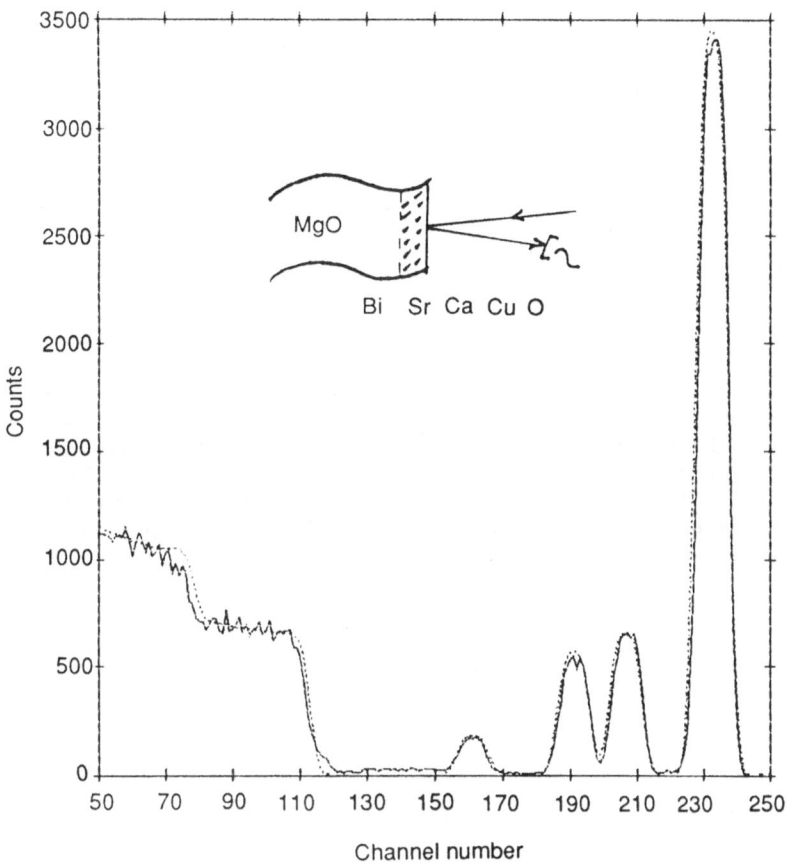

Fig. 4. Experimental and simulated (dashed lines) RBS spectra from thin film, <1500 Å, with an average stoichiometric composition Bi:Sr:Ca:Cu :: 2.3:2.0:2.0:2.9 deposited on unheated and unbiased MgO substrate.

and EPMA. Firstly film thicknesses of <1500 Å, are inadequate for EPMA and the beam sees a large portion of the magnesium oxide substrate. Attempts were made to compensate for the presence of the magnesium oxide substrate by assuming that the film contained magnesium. However, this did not improve the matching between the RBS and EPMA results. The insulating nature of the film as well as the substrate also contribute to the inaccuracy of the EPMA results.

Figure 5 shows experimental and simulated (dashed lines) RBS spectra from a film, >10000 Å thick, deposited by employing the second target on MgO substrates heated to 525° C and RF biased to -80 V. The higher film thickness would reduce the interference from the substrate in the EPMA measurements. Also at substrate temperatures above 450° C, the films are resistive and hence there is no charging-up of the sample by the electron beam. Table II shows the results of RBS and EPMA analysis of this film. For thicker films, the RBS signals overlap and hence the yields/channel rather than the areas under the peaks must be utilized for obtaining the elemental concentrations. Even though the accuracy is poorer, the results can still be used as a guideline. As can be seen, the compositions, measured by both the techniques match well for Bi and Sr. There is still considerable discrepancy

TABLE I. The chemical composition of metallic elements of
Bi-Sr-Ca-Cu-O thin films sputter-deposited on
unheated substrates under different RF bias
conditions.

	RBS ANALYSIS				EPMA ANALYSIS			
RF Bias	Bi	Sr	Ca	Cu	Bi	Sr	Ca	Cu
0 V	2.3	2.0	2.0	2.9	2.5	2.0	2.0	3.6
- 40 V	1.9	2.0	2.0	2.9	2.0	2.0	1.4	3.3
- 80 V	1.8	2.0	1.7	2.7	1.9	2.0	1.3	3.0
-120 V	1.2	2.0	1.8	3.8	1.3	2.0	1.3	3.7

between the results for Ca and Cu. As described above, the higher thickness
as well as the higher film conductivity would make EPMA analysis less erro-
neous. The EPMA results can still be seen to be affected by the matrix
effect and low film thickness. Still higher thicknesses in conjunction with
higher electron beam energies would also permit a more appropriate choice of
characteristic X-ray energies for EPMA analysis.

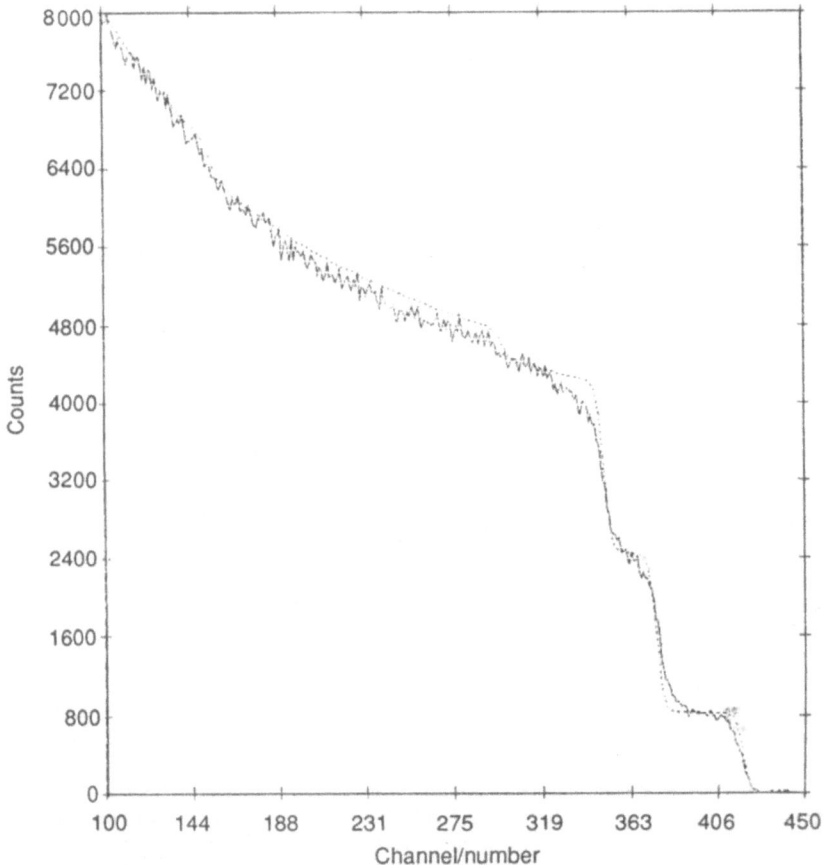

Fig. 5. Experimental and simulated (dashed lines) RBS
spectra from Bi-Sr-Ca-Cu-O thin film, >10000 Å,
deposited on MgO substrate heated to 525° C
and RF biased to -80 V.

TABLE II. The chemical composition of metallic
elements of Bi-Sr-Ca-Cu-O thin films
sputter-deposited on substrates
heated to 525° C and Rf biased to
-80 V.

| RBS ANALYSIS | | | | | EPMA ANALYSIS | | | |
Bi	Sr	Ca	Cu		Bi	Sr	Ca	Cu
0.2	2.0	1.3	3.9		0.2	2.0	0.8	2.8

It may be concluded that errors are introduced in EPMA measurements by
the matrix effect, the insulating nature of the thin films and the substrate
contribution for the thinner films. This is important because several
groups depend on EPMA measurements for the determination of composition of
multi-component thin films. The utilization of the standards which do not
match approximately the film composition can be considered as a major
drawback. In the future, attempts would be made to deposit conducting Bi-
based thin films of different thicknesses upto 15000 Å, on biased and unbiased substrates heated to temperatures >500° C and then utilize some of
the films analyzed by RBS as standards for EPMA analysis. The results will
be utilized for development of a program to obtain composition and thickness
uniformities over large areas.

4. CONCLUSIONS

Bi-Sr-Ca-Cu-O thin films, RF sputter-deposited from single-ring targets
on unheated and unbiased substrates, contain higher concentrations of Bi as
compared to the target, while the concentration of other elements is approximately the same. With increasing RF bias there is a continuous diminution
in Bi concentration. Cu shows considerable increase at high RF biases.
Errors are introduced in EPMA measurements due to the matrix effect, the
insulating nature of the thin films and from the substrate for small film
thicknesses. The RBS results are fairly accurate for films with thicknesses
below 1500 A, and it can serve as a convenient complementary technique for
improving the accuracy of EPMA.

ACKNOWLEDGEMENTS

This work was supported by the U.S. Department of Energy under
Contract No. DE-AC02-83CH10093. It was also partially supported by the
Brazilian Ministry of Education, through CAPES; and Brazilian National
Research Council (CNPq).

REFERENCES

1. K. Wasa, M. Kitanabe, H. Adachi, K. Setsune, and K. Hirochi, Superconducting Y-Ba-Cu-O and Er-Ba-Cu-O thin films prepared by sputtering deposition, Proc. AIP-AVS Topical Conf. on Thin Film
   Processing and Characterization of High T_c Superconductors,
   Anaheim, CA, (1988), 38.
2. M. Hong, J. Kwo, C. H. Chen, A. R. Kortan and D. D. Bacon, Single-
   phase high T_c superconducting $Tl_2Ba_2Ca_2Cu_3O_{10}$ films, To be
   published in Proc. AIP-AVS Topical Conference on Thin Film Pro-

cessing and Characterization of High-Temperature Superconductors, Atlanta, GA, (1989).

3. H. Liu, W. Zhang, L. Zhou, Z. Mao, B. Li, M. Yan, L. Cao, Z. Chen, Y. Ruan, D. Peng and Y. Zhang, Superconducting properties in $(Bi2-x-yPbxSby)Sr2Ca2Cu3Oz$ system (x=0, 0.1, 0.3, 0.5; y=0, 0.1), submitted to Physica C.

4. M. A. Subramanian, J. Gopalkrishnan, C. C. Torardi, P. L. Gal, E. D. Boyes, T. R. Askew, R. B. Flippen, W. E. Farneth and A. W. Sleight, Superconductivity near liquid nitrogen temperature in the Pb-Sr-Ca-R-Cu-O system (R=Y or rare earth), Technical Report, E. I. du Pont (1988).

5. N. G. Dhere, J. P. Goral, A. R. Mason, R. G. Dhere and R. H. Ono, Single-target magnetron sputter-deposition of high-T_c superconducting Bi-Sr-Ca-Cu-O thin films, J. Appl. Phys., $\underline{64}$, (1988), 5259.

6. W. K. Chu, J. W. Mayer and M. A. Nicolet, "Backscattering Spectrometry", Academic Press, New York, (1978).

7. N. G. Dhere, R. G. Dhere and J. Moreland, Effect of substrate temperature and biasing on the formation of 110 Bi-Sr-Ca-Cu-O superconducting single target sputtered thin films, To be published in Proc. AIP-AVS Topical Conference on Thin Film Processing and Characterization of High-Temperature Superconductors, Atlanta, GA, (1989).

8. L. C. Feldman and J. W. Mayer, "Fundamentals of Surface and Thin Film Analysis", North Holland, New York, (1986).

9. E. Frey, WIN-SPEC, Ph. D. Thesis, Dept. of Physics and Astronomy, University of North Carolina, Chapel Hill, NC, (1988).

DETERMINATION OF THE SUPERCONDUCTING TRANSITION

ONSET TEMPERATURE IN SMALL–VOLUME SPECIMENS

T. F. Ciszek and E. Tarsa[†]

Solar Energy Research Institute
1617 Cole Blvd., Golden, CO 80401, USA

ABSTRACT

A technique is described for characterizing the normal-to-superconducting transition in small-volume samples via the Meissner effect. The sample is placed in the inductor coil of an LC circuit that oscillates at a resonant frequency $f = 1/2\pi\sqrt{LC}$, sustained with the aid of a negative-conductance tunnel diode. As the sample temperature T is lowered through the transition, f and T are monitored. Magnetic flux exclusion from the superconductor (and hence from the coil) decreases L and hence increases f. A plot of f vs. T characterizes the transition. A variety of small samples were characterized by this method, and detection limits were determined for a 4.8-mm-diameter, 60-μH inductor coil. The transition onset temperature could be determined for bulk material with mass $\geq$ 5 mg, thin single crystals with dimensions $\geq$ $0.5 \times 0.5 \times 0.08$ mm^3, mesh #80 powders, and films with area $\geq$ 4 mm^2.

INTRODUCTION

The characterization of superconducting transition onsets can be problematic for some sample geometries such as powders, small single crystals, and small fragments of thin films or sintered pellets. We have successfully employed a modification of the tunnel diode oscillator (TDO) method[1] for such samples. For example, the transition onset could be determined in oxygen-annealed YBa$_2$Cu$_3$O$_{7-\delta}$ crystals as small as $0.5 \times 0.5 \times 0.08$ mm^3. The method is non-destructive and no lead or probe attachments are required. While the specimen temperature is ramped through the transition region using a suitable cryostat and control system, the resonant frequency f of an LC oscillator circuit with its coil encircling the specimen is monitored. The oscillations are sustained with the aid of a negative conductance tunnel diode. As flux exclusion from the test specimen takes place, flux in the coil consequently changes as well. The accompanying decrease in inductance L of the circuit causes f to increases and a plot of f vs T allows the onset temperature of superconductivity to be determined. The technique was successfully applied to a variety of YBa$_2$Cu$_3$O$_{7-\delta}$ and Bi-Sr-Ca-Cu-O-type superconductors including sintered materials, powders, single crystals we grew from eutectic solution and films we grew from the liquid phase. In this presentation, we describe our findings for lower limits of mass and volume that can be characterized

---

[†]Present address: Grad. School, Dept. of Phys., U. of CA, Santa Barbara, CA.

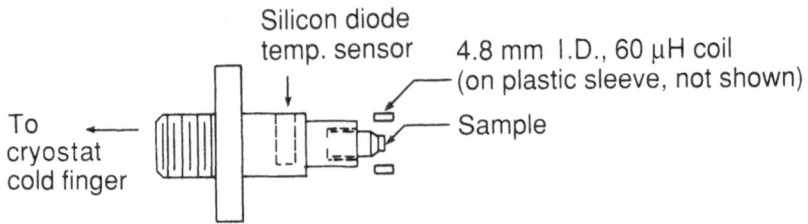

Fig. 1. Sample holder.

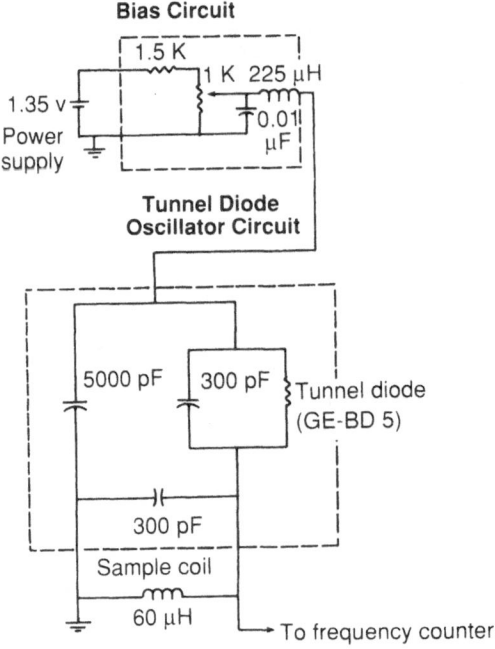

Fig. 2. Tunnel diode oscillator circuit

by the technique when a 4.8-mm-diameter inductor coil is employed. We also
present representative f vs. T curves for a variety of materials.

EXPERIMENTAL

Small samples are mounted on the end of a 2.4-mm-dia x 5-mm-long copper
rod with thermally-conducting epoxy. The rod is coated with thermal joining
compound (zinc oxide in a silicone base) before it is inserted in the sample
holder shown in Figure 1. The sample holder screws onto the cold finger of a
cryostat. A cylindrical Scientific Instruments Si-400NN silicon diode temperature sensor was also coated with this compound before it was placed in a
recessed hole in the sample holder. Larger samples are coated with thermal
joining compound and inserted directly in the sample holder (in place of the

copper rod). The inductor coil is wound on a plastic sleeve that is placed over the end of the sample holder. The sleeve is not shown in Figure 1.

The inductor coil is the only component of the tunnel diode oscillator circuit that resides in the cryostat. The rest of the circuitry shown in Figure 2 is contained in shielded electronics boxes for the bias circuit and the tunnel diode oscillator circuit proper. After placing the sample holder in the cryostat and applying 1.35 v to the bias circuit, the oscillations are monitored at the sample coil/frequency counter lead with an oscilloscope and the 1 KΩ potentiometer is adjusted to obtain stable oscillations and maximum amplitude. Then the scope is removed and temperature ramping is initiated.

We used an Air Products cryostat with a Displex expander module and a Scientific Instruments model 5500 temperature controller for temperature ramping and monitoring. A Phillips model 6669 universal frequency counter was used to measure f. A Hewlett Packard Series 80 computer was used for data acquisition. T and f were typically recorded at 0.5 K intervals between 200 K and 25 K. The measurements were repeated for increasing temperature.

RESULTS AND DISCUSSION

Bulk Sintered Material

A sintered pellet of $YBa_2Cu_3O_{7-\delta}$ was prepared by heating ground, mixed, and pressed oxide and carbonate powders. A 60.6 mg segment of the pellet was mounted on the end of a copper rod as previously described. Measurements were made of f vs T and then the segment size was reduced by grinding away a portion of the sintered $YBa_2Cu_3O_{7-\delta}$ material before each subsequent measurement. A series of such measurements is shown in Figure 3. The transition onset could be observed down to a segment weight of approximately 4 mg. Non-systematic error on the order of $\pm$ 2 K is observed in the transition points of the various f vs. T plots. This is thought to be due mainly to poor thermal contact reproducibility in mounting the copper rod on the cryostat cold finger extension which holds the silicon diode temperature sensor. A variable-temperature, liquid-helium research dewer with a larger uniform temperature zone may have advantages over the cold-finger type of cryostat in this respect, and would also reduce errors due to temperature gradients between the tip of the sample and the temperature sensor.

Single Crystals

A single crystal of $YBa_2Cu_3O_{7-\delta}$ was grown from solution in excess CuO. After annealing two days in O_2 at 550°C, it was mounted on the end of a 2.4-mm-dia copper rod with thermally conducting epoxy such that its flat face was perpendicular to the rod's axis and about 0.6 mm from the rod's end. Measurements were made of f vs T and then the crystal size was reduced before each subsequent measurement. The appearance of the 0.08-mm-thick crystal after reduction to 0.55 mm^2 area is shown in Figure 4, and the series of measurements appears in Figure 5. The transition onset could be observed down to a crystal area of about 0.2 mm^2. An area of 0.3 mm^2 is probably the lower limit for reasonably accurate onset temperature determination, since it is the smallest area that gives onset values consistent with those observed for larger crystals.

The frequency changes with T even when no sample is present, since the properties of the copper rod and inductor are also temperature dependent. This is a smooth, stepless, monotonic increase of f with T. However, it is particularly noticeable on the expanded frequency scale used for these small

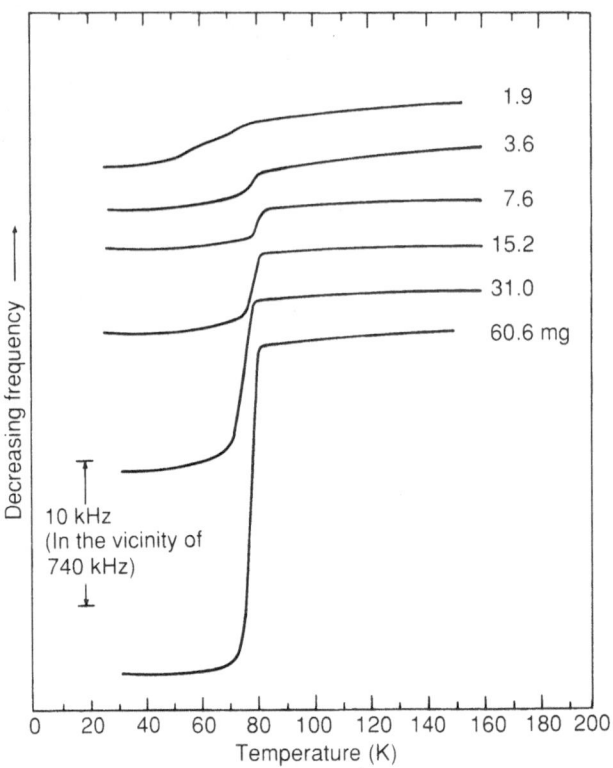

Fig. 3.   Superconductivity onset transitions for a sintered $YBa_2Cu_3O_{7-\delta}$
pellet fragment after progressive mass reductions

Fig. 4.   Natural face of an annealed, 80-μm-thick $YBa_2Cu_3O_{7-\delta}$ crystal.   Its
area has been mechanically reduced to 0.55 mm^2.

crystals, and steps should be taken to minimize the effect or at least standardize it so that its contribution could be subtracted from the f vs. T measurement of small samples. Once again, it would be advantageous to have a larger isothermal zone that does not rely on a cold finger for cooling. Then the copper rod could be eliminated.

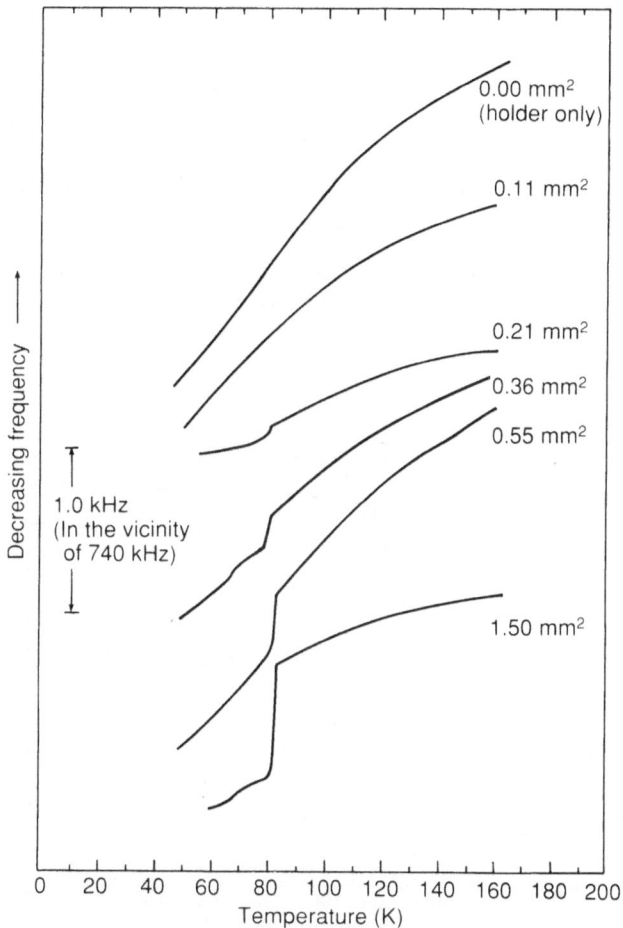

Fig. 5. A series of f vs. T measurements for the crystal shown in Figure 4. Each represents a different crystal area obtained by mechanical abrasion

Other Sample Geometries

The tunnel diode oscillator characterization technique has been applied to a variety of sample geometries. In addition to the single crystals and sintered bulk material discussed above, three other geometries that were successfully characterized include granular powders, melt-cast wires, and

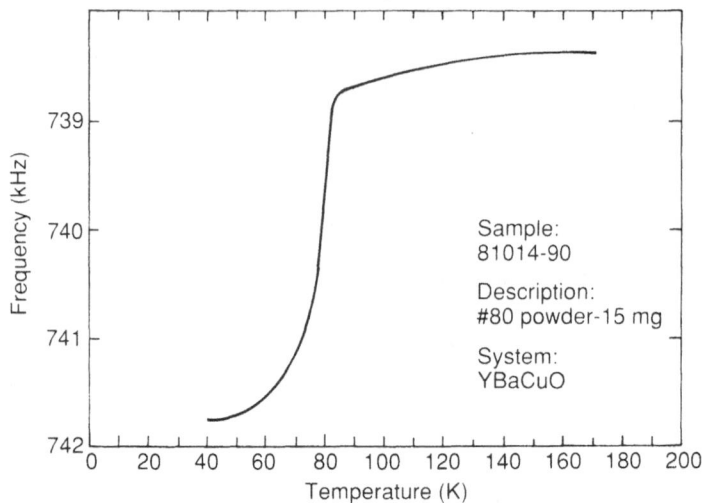

Fig. 6.   An f vs. T curve for a 15-mg sample of [#]80-mesh granular
$YBa_2Cu_3O_{7-\delta}$ powder fragments embedded in epoxy

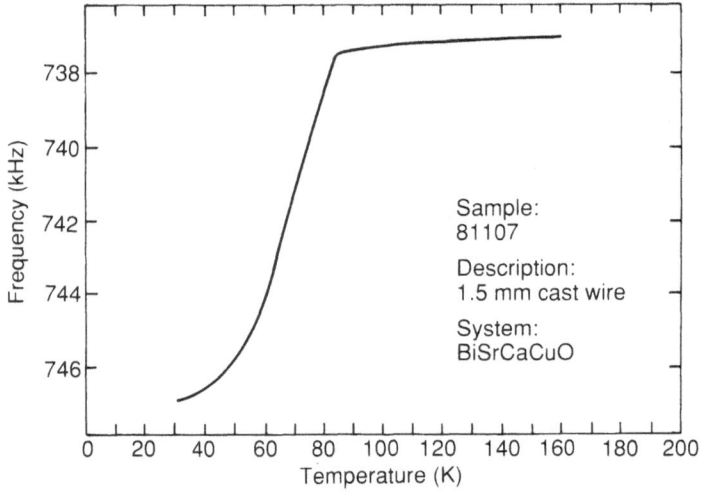

Fig. 7.   An f vs. T curve for a 1.5-mm diameter Bi-Sr-Ca-Pb-Cu-O
melt cast wire.

films deposited on foreign substrates. In Figure 6, an f vs. T curve is
shown for a 15-mg sample of #80-mesh granular $YBa_2Cu_3O_{7-\delta}$ powder fragments
embedded in thermally conducting epoxy at the end of a copper rod. This type
of transition characterization is particularly complimentary for powders that
have been studied by nuclear magnetic resonance techniques. The f vs. T
curve for a 1.5-mm diameter Bi-Sr-Ca-Pb-Cu-O wire that we melt cast in an
Al_2O_3 ceramic sleeve is shown in Figure 7. The f vs. T curve in Figure 8 is
for a Bi-Sr-Ca-Cu-O film grown from the liquid phase on a substrate and

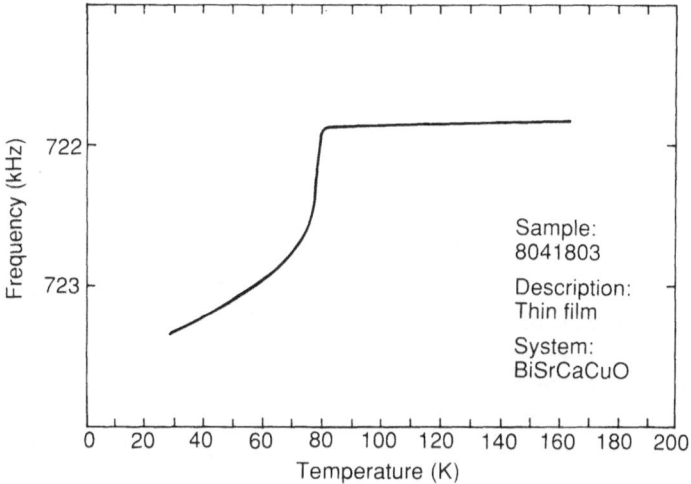

Fig. 8. An f vs. T curve for a Bi-Sr-Ca-Cu-O film grown from the liquid
phase on a substrate and subsequently removed

subsequently removed so as to be free-standing. Its thickness is several
hundred microns. Low-field ac magnetic susceptibility was measured on the
same sample at the National Institute for Standards and Technology (Boulder
laboratories). The two onset temperature determinations agreed within 2 K,
and the real component of the susceptibility vs. temperature curve had a
shape that is similar to that of the f vs. T curve shown in Figure 8.

SUMMARY AND CONCLUSIONS

A wide variety of superconducting samples have been characterized by the
tunnel diode oscillator technique. They include single crystals, sintered
pellet segments, films, melt-cast wires, and granular powders. The smallest
sample sizes that could be tested for superconductivity onset temperature
were about 0.5x0.5x0.1 mm^3 in the case of single crystals. Sample weights of
bulk sintered material as low as 5 mg were suitable for characterization. No
experimentation was done with inductor coil diameters smaller than 4.8 mm.
Larger ones have been used effectively for larger samples, and it is possible
that smaller samples yet could be measured with a smaller inductor. The
electrical circuit is simple and inexpensive to fabricate. Standard cryostat
technology can be used to cool the samples. However, care must be taken to
assure that the sample and temperature sensor are at the same temperature.
This is probably one of the largest sources of error in our cold-finger
cryostat system. Determinations of the onset temperature seem to be
reproducible to within about ±2 K, and they also agree with susceptibility
determinations within about the same amount (based on very limited testing).
No electrical contacts to the sample are needed in the TDO method.

The tunnel diode provides stable, undamped oscillations within its
characteristic conduction region — a bias range in which the diode displays a
negative conductance due to quantum mechanical tunneling of electrons across
its p-n junction. This negative conductance counteracts the internal

resistance of the circuit and allows stable oscillations for a given values of L and C in the tank circuit. The Meissner effect changes L and hence the oscillation frequency. L also changes due to temperature effects on the coil and sample holder. These effects on f vs T should be minimized or at least standardized and subtracted from the sample measurement curve when extremely small samples are characterized. They are negligible for sample weights above about 30 mg, with the experimental configuration used here.

ACKNOWLEDGMENTS

The authors are appreciative of the excellent technical assistance that was provided by Charles Evans and Terry Schuyler during this investigation. We also wish to thank Ron Goldfarb of NIST-Boulder for the low field ac magnetic susceptometry measurement. This work was partially supported by the Office of Energy Storage and Distribution of the US Department of Energy.

REFERENCES

1. John N. Fox and John U. Trefny, Am. J. Phys. **43** (1975) 622.

MEASUREMENT OF SUPERCONDUCTING TRANSITION TEMPERATURES OF THIN FILMS OF YBa$_2$Cu$_3$O$_7$ BY FOUR-TERMINAL RESISTANCE AND MAGNETIC SUSCEPTIBILITY

J. R. Matey

David Sarnoff Research Center
CN 5300
Princeton, NJ 08543-5300

INTRODUCTION

The superconducting transition temperature, T_c, is one of the most important parameters which we use to characterize superconducting materials. T_c can be determined by

1. Measuring the resistance of the sample, to detect the zero resistivity of the superconducting state.

2. Measuring the magnetic susceptibility of the sample, to detect the perfect diamagnetism of the superconducting state.

3. Measuring the heat capacity of the sample, to detect the second order transition.

Since the first method is the simplest of the three, it is the most frequently used. This is quite reasonable for homogeneous materials in which all parts of the sample undergo the superconducting transition at the same temperature. For such materials, the three methods agree very well — Soulen[1] has shown that T_c's determined by all three methods agree to within 0.1 mK for well annealed polycrystalline indium. In homogeneous superconductors it is not unusual to find transition widths of the order of mK and fractional widths, $\delta T/T_c$ of the order of 0.001. Indeed, Schooley[2] and his colleagues at the National Bureau of Standards developed a set of superconducting fixed points whose transitions are used as reference points for calibrating low temperature thermometers.

Non-homogeneous materials, on the other hand, can have broad transitions: different parts of the sample may have different transition temperatures. In such cases, the sample resistance goes to zero when a percolation path is established. According to the Bruggeman effective medium theory[3], superconducting percolation begins at 50% or 33% volume fraction of superconductor, depending on whether the sample morphology is two or three-dimensional. Since the sample resistance is zero below the percolation threshold, resistance measurements cannot provide information about changes below threshold. In many applications, such information is vital. For example, a superconducting thin film which has a 33% volume fraction of superconductor in a matrix of normal conductor would be a poor candidate for a microwave application[4].

Examination of any recent issue of *Applied Physics Letters* will show that the fractional transition widths of current high temperature superconductors are ~ 0.1, much larger than that of their low temperature counterparts [5,6,7]. This is strong evidence that the current materials are non-homogeneous. Hence, measurement of the transition temperature by resistive means alone is inadequate. We have developed a thin film susceptometer for measuring the magnetic properties of superconducting thin films. Together with four-terminal resistance measurements, it provides a more complete picture of the superconducting transition.

EXPERIMENTAL DETAILS

The basic idea of our thin film AC susceptometer is shown below.

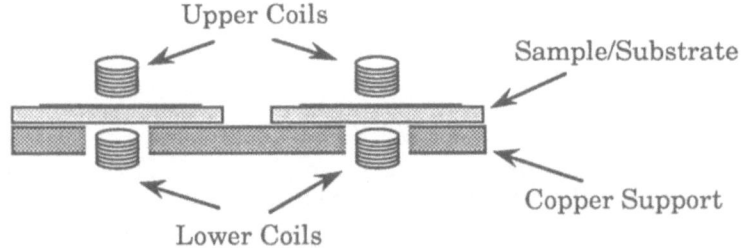

Fig. 1 Schematic view of the thin film AC susceptometer. Only two of the four coil sets are shown. The thermometer, leads and other details are also omitted.

Our samples are typically thin films of high temperature superconductors, ~ 1μ thick, coated on one surface of 0.76 x 1.07 x 0.011 cm dielectric substrates. They are mounted on an OFHC copper support which is attached to a cold finger within the sample chamber of the cryostat. The samples are attached with Apiezon N grease, to provide thermal contact. The cryostat temperature is measured using a calibrated diode thermometer attached to the copper support and controlled with a commercial temperature controller.[8] Electrical leads for the thermometer and the samples are in good thermal contact to the copper support. The cryostat is cooled by immersion in either liquid nitrogen or liquid helium. We find that we can reach 45K by pumping on the nitrogen, even though the nitrogen freezes at 63.1 K [9].

Concentric, cylindrical coils are placed above and below the sample as shown above. The coils are 3.2 mm in radius and consist of 120 turns of #40 copper wire wound on teflon forms. The mutual inductance of the coil pairs is measured as a function of temperature using a commercial LR analyzer[10] . When the sample becomes superconducting, its diamagnetism shields the upper coil from the lower, reducing the mutual inductance. The shielding would be complete for an ideal, large area sample, whose thickness is large compared with the London penetration depth. The mutual inductance would fall to zero at the transition and would be constant above and below the transition. In practice, there are several other effects which must be considered:

- Coupling between the leads connecting the coils to the inductance analyzer. This coupling can be of either sign. It can change if the leads are disturbed.

- Shielding of the coils by the copper which supports the samples. This will change with temperature, as the skin depth of the copper changes with temperature.

Both of these effects are apparent in the data presented below. Neither prevents us from determining the superconducting transition temperature of our samples.

The measurement system has been automated using a personal computer[11] and a relay interface[12] which can switch between four sets of mutual inductance coils. The same instrumentation can also be used to make four terminal resistance measurements on two other samples during the same run. Measurements can be made, either as a function of time, measuring the mutual inductance and temperature as the cryostat slowly warms up, or as a function of temperature, stabilizing the temperature at each point.

The coils are excited with a current of about 0.6 mA. The measured amplitude of the on-axis field of the excited coil is about 10 µT at the sample position. The experiments are typically run at 10 kHz.

EXPERIMENTAL RESULTS AND ANALYSIS

As a test of the system, we measured an 800 nm thick Pb film coated on one of our standard sapphire substrates and a 0.025 cm thick Nb sheet of similar dimensions. The results of these experiments can be seen in the figures below. The LR analyzer is capable of measuring both the in-phase and quadrature components of inductance. This can be modeled as a measurement of a lossless inductance, L, and an equivalent series resistance, L'. Both L and L' are plotted, with the data normalized to their 300K values. The 300K values are

Sample	L (µH)	L' (Ω)
Pb, 800 nm	5.51	0.10
Nb, bulk	4.93	0.88

There are a number of notable features:

- The 30K values are well below the 300K values. Both L and L' increase smoothly and monotonically from 30 K to 300K. L varies because the copper sample support partly shields the coils, and the effectiveness of the shielding increases as the the conductivity of the copper increases with decreasing temperature. The L' losses in the coil windings and in the copper support also decrease with decreasing temperature.

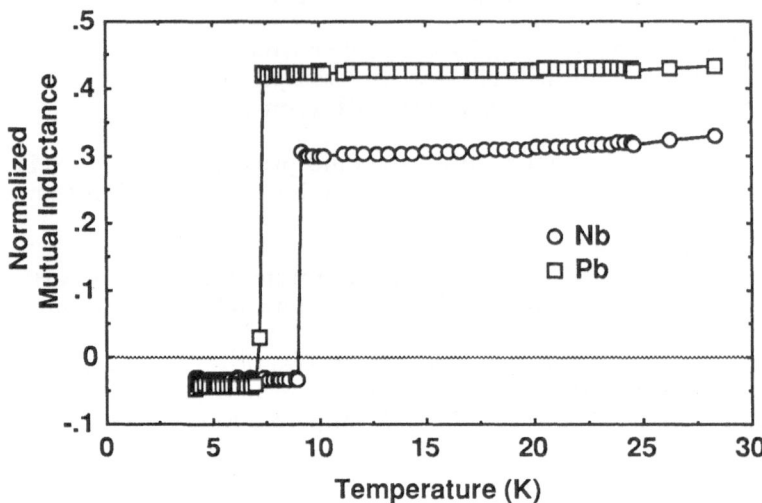

Fig. 2 Mutual inductance for the Nb and Pb test samples. The ordinates
have been normalized to the 300K value.

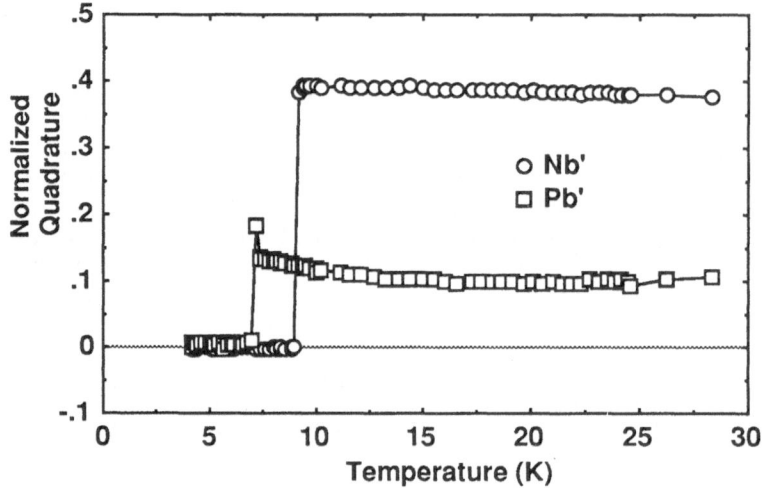

Fig. 3 The quadrature part of the mutual inductance for the Nb and Pb test
samples. The ordinates are normalized to 300K.

- The transitions occur at the accepted values for Pb and Nb[13], 7.23 and 9.25 K respectively.

- The width of the transition is smaller than the temperature grid on which the data was taken, ~ 0.1 K. The fractional width of the transitions is ~ 0.01 or less.

- There is an apparent glitch in the quadrature data for Pb at the transition. It is real.

- The mutual inductances become slightly negative below the transitions. This is due to coupling between the leads leading into the cryostat and could be reduced by using shielded or twisted pair leads.

Consider now some data from a reactively sputtered $YBa_2Cu_3O_7$ film on a ZrO_2 substrate. The details of the preparation of the film have been presented elsewhere.[14] Susceptometer and four terminal resistance measurements were made on the same film in two separate runs. The data have been normalized to their 300K values and plotted together in the figure below.

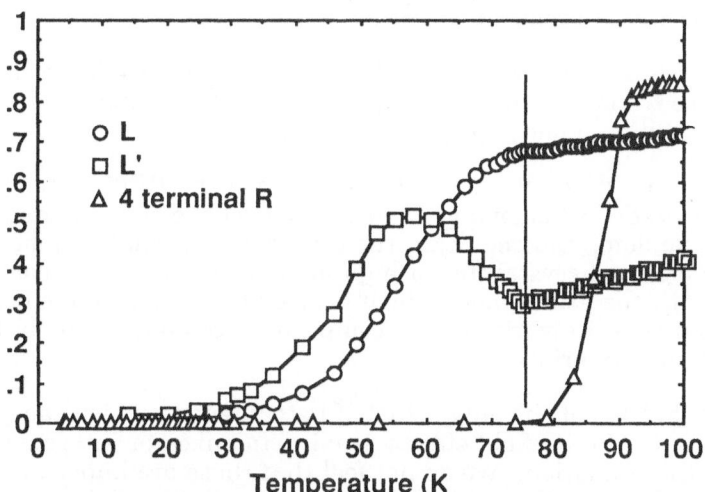

Fig. 4   Normalized susceptometer and 4 terminal resistance data for a 1μ thick, $YBa_2Cu_3O_7$ film on cubic ZrO_2.

The onset of the resistance transition for this sample is ~ 90 K, the resistance transition is complete at 75 K; the 10% and 90% points are at 83 and 88 K. There is no information in the resistance data below 75 K, since the resistance is zero. It is clear, however, from the susceptometer data that there is a lot going on below 75 K!

The inductance transition *begins* at 75 K, the temperature at which the resistance transition is just complete. The resistance goes to zero when there is a superconducting percolation path through the sample. Magnetic flux lines cannot cross a superconducting path[15]. Hence, the motion of magnetic flux lines through the sample is unimpeded until percolation paths become established, and motion becomes progressively more impeded as the superconducting percolation paths become dense. For a film in which the grain size of the superconducting and non-superconducting components is small in comparison with the film thickness, the three dimensional Bruggeman effective medium approximation (BEMA) [16] predicts superconducting percolation at a 33% volume fraction of superconductor.

We can also think about percolation paths through the non-superconducting part of the sample. The BEMA predicts the existance of non-superconducting percolation paths as long as at least 33% of the material is not in the superconducting state. Magnetic field lines can thread these paths. We can naively interpret the beginning of the transition in L as the point at which there is a 33% superconductor fraction, and the end as the point at which the fraction is 66%.

It can be energetically favorable for flux lines to break through superconducting paths at weak points in the path if the strength of the magnetic field exceeds a local critical field. Such events are typically not reversible and give rise to losses. These losses can be seen in the L' data. We see a peak which begins at the point where superconducting percolation paths are first established. As the temperature is decreased, additional paths become available for interaction with the magnetic flux, and the losses increase. At the same time, the amount of magnetic flux in the sample is decreasing and the strength of the weak links are increasing. The competition between these effects produces the peak.

It is important to note that the losses in this film are *larger* below the zero resistance transition than they are above ! Measurements of the resistance transition alone give no hint of this behavior.

We have discussed these results in terms of the BEMA. In the BEMA, the sample is made by removing small volumes of the superconducting host material, at random, and replacing them with the normal inclusion material. Unfavorably oriented crystallites of high anisotropy, chemical reactions of crystallites with the atmosphere which destroy the superconductor, and crystallite to crystallite variations in composition could all contribute such inclusions to the material.

The BEMA is not the only model of heterogeneity which can give rise to broad transitions. Finite field effects, leading to mixed or intermediate states, can broaden the transition. We do not feel that these are important in our measurements because the applied fields are small, 10 μT, in comparison with H_{c1}[17] , 0.2T at 0 K, and the demagnetizing fields are of similar strength for our geometry.

Another alternative is the weakly coupled grain model[18] in which the superconductor is composed of grains of superconductor coated with a non-superconducting interface material. On the basis of the data in this paper, it is not possible to distinguish this model from the BEMA. The actual state of affairs is probably some combination of inclusions and weakly coupled grains.

The behavior seen in figure 4 is typical of all films which we have tested to date, including films provided to us by other laboratories.

SUMMARY

A uniform, thin film, high temperature superconductor should have a sharp transition, and the four terminal resistance and the susceptometer transitions should occur at the same temperature. Experimentally, we find that the transitions are broad, and the susceptometer and resistance transitions occur at different temperatures. Until uniform films are available, comparison of resistance and thin film susceptometer data can be an important test of the quality of thin film superconductors.

ACKNOWLEDGEMENTS

My thanks to Y. Arie for the thin film samples used in these experiments, to B. Brycki for constructing the sample holder and coil assembly and to J. Gittleman and R. Williams for helpful discussions.

REFERENCES

1   R. J. Soulen, Jr. & J. H. Colwell, The Equivalence of the Superconducting Trasition Temperature of Pure Indium as Determined by Electrical Resistance, Magnetic Susceptibility, and Heat Capacity Measurements, J. Low Temp. Phys. 5:325 (1971).

2   J. F. Schooley, R. J. Soulen, Jr., & G. A. Evans, Jr., Preparation and Use of Superconductive Fixed Point Devices, SRM 767, NBS Special Publication 260-44, National Bureau of Standards , Washington, DC (1972).

3   We use the Bruggeman symmetric effective medium approximation as discussed in R. Landauer, Electrical Conductivity in Inhomogeneous Media, in: "Electrical Transport and Optical Properties of Inhomogeneous Media," J. C. Garland and D. B. Tanner, Ed., AIP Conference Proceedings No. 40, American Institute of Physics, New York (1978).

4   J. I. Gittleman and J. R. Matey, Modeling the Microwave Properties of the $YBa_2Cu_3O_7$ Superconductors, J. Appl. Phys. to be published January 1989.

5   J. Zhao, K. Dahmen, H. O. Marcy, L. M. Tonge, T. J. Marks, B. W. Wessels, & C. R. Kannewurf, Organometallic Chemical Vapor Depostiion of High Tc Superconducting Films using a Volatile Fluorocarbon based Precursor, Appl. Phys. Lett. 53:1750 (1988).

6   A. J. Panson, R. G. Charles, D. N. Schmidt, J. R. Szedon, G. J. Machiko & A. I. Braginski, Chemical Vapor Deposition of $YBa_2Cu_3O_7$ Using Metalorganic Chelate Precursors , Appl. Phys. Lett. 53:1756 (1988).

7   R. D. Lorentz & J. H. Sexton, Oriented High Temperature Superconducting BiSrCaCuO Thin Films Prepared by Ion Beam Deposition, Appl. Phys. Lett. 53:1654 (1988).

8   Lake Shore Cryotronics, Westerville, Ohio 43081. Model 805 controller and DT-470 diode temperature sensor.

9   Solid nitrogen has a substantial vapor pressure well below its freezing point, 63.1 K. Another important consideration is that solid nitrogen is more dense than the liquid, so that it will not rupture the cryostat as it freezes.

10 Hewlett-Packard, Palo Alto, CA. Model 4192.

11 An IBM PC/XT running DOS 3.3. Programs written in Borland TurboPascal 4.0.

12 ICS Electronics Corp, San Jose, CA 95112. Model4874.

13 B. W. Roberts, Properties of Selected Superconductive Materials, NBS Technical Note 724, National Bureau of Standards, Washington, DC. (1972), page 10.

[14] Y. Arie & J. R. Matey, Reactive Sputtering of Superconducting Thin Films, at: AVS Topical Conference on Thin Film Processing and Characterization of High Temperature Superconductors, October 1988. Proceedings to be published in the AVS series of the American Institute of Physics Conference Proceedings, February, 1989.

[15] Except at a weak link, as discussed below.

[16] Landauer, loc. cit.

[17] P. H. Hor, R. L. Meng, J. Z. Huang, C. W. Chu, and C. Y. Huang, Upper Critical Field of High Temperature $Y_{1.2}Ba_{0.8}CuO_{4-\delta}$ Superconductor, App. Phys. Comm. 7:129 (1987).

[18] J. R. Clem, Granular and Superconducting Glass Properties of the High Temperature Superconductors, Proceedings of the International Conference on High Temperature Superconductors and Materials and Mechanisms of Superconductivity, ed. J. Muller and J. L. Olsen, North Holland, Amsterdam (1988).

DETERMINATION OF OXIDATION STATES OF COPPER IN YBaCuO THIN FILMS BY ELECTROCHEMICAL ANALYSIS

S. Salkalachen, E. Salkalachen, P.K. John, H.R. Froelich, B.Y. Tong and S.K. Wong*

Department of Physics and Centre for Chemical Physics
*Department of Chemistry, The University of Western Ontario
London, Ontario N6A 3K7, Canada

INTRODUCTION

There have been numerous published reports on the nature of oxidation states of copper in high-Tc superconducting materials. The primary interest in the valence studies of copper stems from the fact that ionic states and electron distribution of copper play a crucial role in the superconducting mechanism. It is now well established that the crystal structure and transition temperature[1,2] are intrinsically correlated with the oxygen content and, in turn, the valency of copper. Oxygen is a highly mobile species in these compounds where changes in its composition are easily induced by annealing[3].

Techniques commonly used in the valence determination of bulk superconducting materials are x-ray photoelectron spectroscopy (XPS) and x-ray absorption spectroscopy (XANES). However, interpretation of the observed data is often ambiguous and non-quantitative due to the existence of more than one chemical species and also because of the difficulty in resolving them. XPS core electron studies have identified the dominant divalent Cu^{++} species in the $Y_1Ba_2Cu_3O_{7-x}$ superconductor by the presence of strong satellite peaks accompanying the Cu $2p_{3/2}$ and Cu $2p_{1/2}$ lines and also by the larger linewidth of the Cu 2p spectra[4]. Chemical shifts of 0.95 - 1.4 eV are observed[5,6] for model systems of CuO and Cu_2O, using which a rough estimation of the content of the monovalent Cu^+ state has been made. While most XPS studies have failed to provide direct evidence for Cu^{+++} states in the superconductor, through a careful analysis of the XPS Cu $2p_{3/2}$ core electron spectra Steiner et al[7] were able to show that small concentrations of the trivalent state ($\leq$ 10%) could exist in YBaCuO samples which were annealed and quenched below 500 $^\circ$C.

Probing the electronic configuration of the material using x-ray absorption spectroscopy has been more rewarding due to its inherent capability to differentiate the Cu^+ and Cu^{++} species. However, although copper model compounds such as Cu_2O(I), CuO(II) and $KCuO_2$(III) display excellent energy resolution in Cu K-edges, the XANES spectra of the YBaCuO compound do not directly yield well-resolved maxima and have to be reproduced by superposition techniques. These studies[8,9] also indicate that additional oxygen does not contribute Cu^{+++} ions but oxygen deficiency results in the formation of Cu^+ state as estimated from the peak of the derivative of XANES spectra. Lengler et al[10] suggested that the x-ray spectra of the Cu K-edge in $Y_1Ba_2Cu_3O_{6.9}$ material can be synthesized by superposition of 2 divalent Cu^{++} ions as in Y_2BaCuO_5, 0.84 chemically trivalent Cu^{+++} ions as in $KCuO_2$ and 0.16 monovalent Cu^+ ions as in Cu_2O.

Thus, while the presence of Cu^{++} and Cu^+ species in bulk $Y_1Ba_2Cu_3O_{7-x}$ material has been relatively well established, a really quantitative estimation of the oxidation states has not been easy. Moreover, recent observations suggest that the analysis becomes more difficult when the above valence techniques are applied to thin films. For example, when vacuum-based measurements employing ion, x-ray or electron beam are used to probe the material, the sample surface undergoes changes in oxygen stoichiometry and copper valency. It was recently established[6] that x-ray bombardment of the material in ultra high vacuum causes changes in the Cu valence state from Cu^{+++} via Cu^{++} to Cu^+. It was also reported[11,12] that the intensity of the Cu 2p satellite peak decreases after Ar^+ ion etching of thin films and that changes in oxygen stoichiometry were observed when the films were irradiated with electrons. The relatively unstable oxidation states and non-stoichiometric nature of thin film structures have made quantitative analysis very difficult.

We report here an electrochemical reduction technique to obtain quantitative information on the oxidation states of copper in YBaCuO thin films. In this method, chemical potentials of the various oxidized copper species (Cu^{++}, Cu^+) are established by successive reduction at constant current, which then provide a means to obtain their absolute abundances from the corresponding charge.

EXPERIMENTAL DETAILS

Thin films of YBaCuO material were deposited in an a.c. sputtering system system equipped with dual targets ($Y_1Ba_2Cu_3O_7$, Tc = 91 K) and with facilities to vary substrate temperature, target-to-target distance, gas composition and pressure. Films for the electrochemical experiments were deposited on Molybdenum foil using suitable masks. The rate of deposition was typically 20 $\overset{\circ}{A}$/minute at oxygen pressure of 200 mTorr.

In the electrochemical experiment, a known quantity of current is passed between a platinum electrode and the thin film sample, while measuring the chemical potential of the latter with reference to a calomel electrode as a function of time. The plateaus seen on the potential-time curve represent the quantity of charge consumed during the various reduction steps. The total number n of oxide species is directly proportional to the electric charge as

$$n \quad = \quad \frac{i\,t\,A}{NF\,a} \quad atoms/cm^2 \qquad (1)$$

where i = current in Ampere, t = time in second, A = Avogadro Number (6.023 $\times$ 10^{23}), N = number of charges required for reduction of 1 gram mole of the species (here N = 2), F = Faraday Number (96493 coulombs) and a = area of the reduced surface in cm^2.

The accuracy and resolution of the experiment are determined by the measurements of area (a), current (i) and time (t). The error in the measurement of area can be minimized by using a specified area for all the films. If the films are reduced too rapidly, considerable uncertainty is introduced in the measurement of time. For increased sensitivity, the current must be small and constant. A typical value of current density used in the present experiments is 0.17 mA/cm^2, the current being controlled to within $\pm0.1\%$. An aqueous solution of 0.1 N KCl in deionised water was used as the electrolyte which was deoxygenated by bubbling pure N_2 gas for 30 minutes before each set of experiments. It is reasonable to assume that thin films of YBaCuO are less sensitive to water than porous bulk materials and also, the short periods required for the reduction process would not cause decomposition.

There have been previous attempts to synthesize[13,14] and characterize[15,16] high-Tc superconducting oxides by electrochemical methods. Rosamilia et al[15] have reported the reduction steps associated with the transformation of Cu^{++} to Cu^+ and to Cu^0 in bulk materials of $Y_1Ba_2Cu_3O_7$ and $La_{1.8}Sr_{0.2}CuO_4$ but did not investigate it in detail so as to extract any quantitative information. In a study[16] of the cyclic voltammetry of bulk $Y_1Ba_2Cu_3O_7$ material, the observed cathodic current was attributed to the reduction of copper oxide components. This was inferred from the changes in surface morphology and composition. No quantitative data were, however, obtained on the oxidized species.

In order to ascertain and calibrate for the chemical potentials of copper and its oxides, we first performed reduction experiments using Cu_2O and CuO films. Values of 0.5 V, 0.8 V and 1.0 V have been established[17] for CuO(II), Cu_2O(I) and Cu^0 states respectively. The equivalent chemical reactions are,

$$CuO + 2e^- + H^+ \xrightarrow{\text{it}_1} Cu + OH^- \qquad (2) \quad \text{and,}$$

$$Cu_2O + 2e^- + H^+ \xrightarrow{\text{it}_2} 2\,Cu + OH^- \qquad (3)$$

where t_1 and t_2 are the respective reduction times.

RESULTS AND DISCUSSION

A typical electrochemical reduction curve for an as-sputtered YBaCuO film (deposition temperature T_{dep} = 400 ^{0}C, oxygen partial pressure PO_2 = 200 mTorr) is illustrated in Figure 1 along with those for evaporated and thermally oxidized copper films. The plateaus at 0.5 V, 0.8 V and 1.0 V correspond to cupric (Cu^{++}), cuprous (Cu^+) and neutral copper (Cu^0) states respectively. The concentration of Cu^{++} and Cu^+ species in the YBaCuO film as computed from the corresponding electrical charge are 1.14×10^{17} and 1.24×10^{17} atoms/cm^2 respectively.

We investigated the variation of oxidized copper content in the sputtered YBaCuO films as a function of various sputtering parameters such as substrate temperature, gas composition, pressure and subsrate-target geometry. We used the ratio Cu^+/Cu^{++}, or more simply t_2/t_1, to study the variation of the relative abundances of Cu^+ and Cu^{++} species in the deposited films. This was considered necessary because the films fabricated in the same sputtering cycle showed large inhomogeneities in thickness and composition. Figure 2 shows the resulting plot for room temperature and high temperature (400 ^{0}C) deposited films. It is significant to note that the ratio Cu^+/Cu^{++} in films deposited at 400 °C is generally lower and less variable than that for room temperature deposited films. This implies that high deposition temperatures favour growth of films containing relatively higher concentration of Cu^{++} species. An analysis of the different oxidation species in these films reveals that films deposited at high temperature (400 °C) contain nearly four times the amount of oxidized copper species (Cu^+ and Cu^{++}) than that found in room temperature deposited films (Table I). The spatial variation of the Cu^+/Cu^{++} ratio with substrate-target distance may be associated with temperature gradients present in the discharge region especially in the vicinity of the target area where the electric lines of force originate. This was more pronounced when the films were not externally heated.

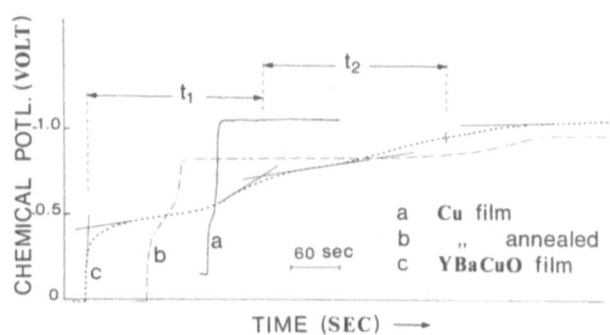

Fig.1. Typical electrochemical reduction curves of:
(a) evaporated Cu film, (b) Cu film annealed at 250 °C for 8 min. in air and, (c) as-sputtered YBaCuO film, deposited at 400 °C, in 200 mTorr oxygen for 4 hrs.

Films were also deposited at 400 °C under various partial pressures keeping the discharge power nearly constant. Owing to the observed dependence of composition, thickness and oxide content of the films on substrate-target geometry, we chose films from a specific location in the sputtering chamber for the electrochemical experiments. We employed the ratio $Cu^{++}/(Cu^+ + Cu^{++})$ to evaluate the relative content of oxidized copper species in each film which thus eliminated uncertainties due to thickness variations.

Table I Quantitative analysis of oxidized copper species in as-sputtered YBaCuO films.

Oxide Species	Concentration (atoms/cm^2)		
	T$_{dep}$:	room temp.	400 ^{o}C
Cu^{++}		3.00×10^{16}	1.14×10^{17}
Cu$^+$		2.13×10^{16}	1.24×10^{17}
Cu$^+$ + Cu^{++}		5.13×10^{16}	2.38×10^{17}

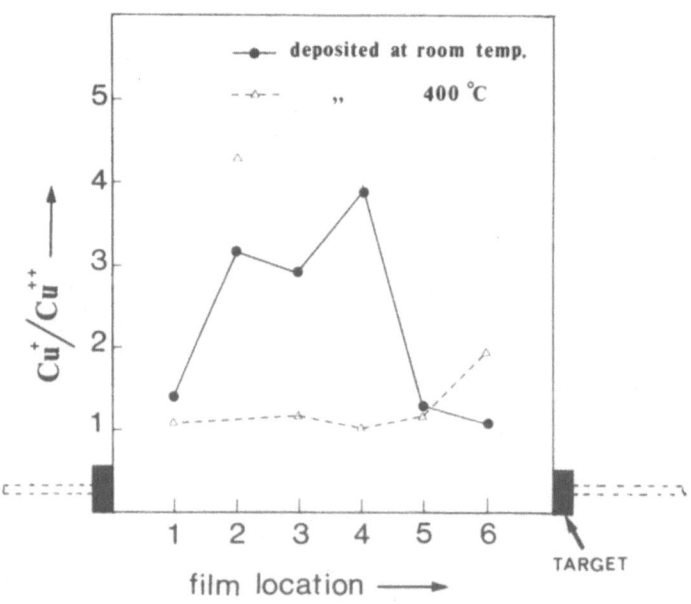

Fig.2. Variation of Cu$^+$/Cu^{++} ratio in as-sputtered YBaCuO films deposited at room temperature and at 400 ^{o}C.

This is shown in Figure 3. It is evident that the concentration of divalent copper species increases rather rapidly in the range 100 mTorr to 200 mTorr oxygen partial pressure, but rises slowly thereafter up to 1000 mTorr. This result is consistent with the growth kinetics and observed data on resistivity[18] of copper oxide films reactively sputtered under increasing oxygen pressures. The observed superconductivity[19] in as-sputtered films of YBaCuO at high oxygen pressure (2.5 - 3 Torr) and deposition temperature (550 $^{\circ}$C) may be associated with the degree and role of copper oxidation states particularly a large concentration of divalent species.

The sputtered films were subjected to oxygen ion (5 - 10 keV) bombardment in vacuum (3×10^{-4} Torr) both at room temperature and elevated temperature (400 $^{\circ}$C). A significant reduction in the oxygen concentration, by as much as 48 %, was noticed in Rutherford backscattering compositional analysis of films thus 'oxygenated' at high temperature. When these samples were analysed electrochemically, dramatic changes were observed in the potential-time curves. The plateaus corresponding to both cupric and cuprous oxidation states at 0.5 V and 0.8 V disappeared completely after the bombardment oxygenation process. This means that the copper oxides actually undergo reduction during a high temperature anneal in vacuum. On the other hand, the room temperature process did not markedly affect the oxidized species in the films. This is in accord with the hypothesis that annealing in a reducing atmosphere (or vacuum) above 400 $^{\circ}$C causes release of oxygen from these compounds.

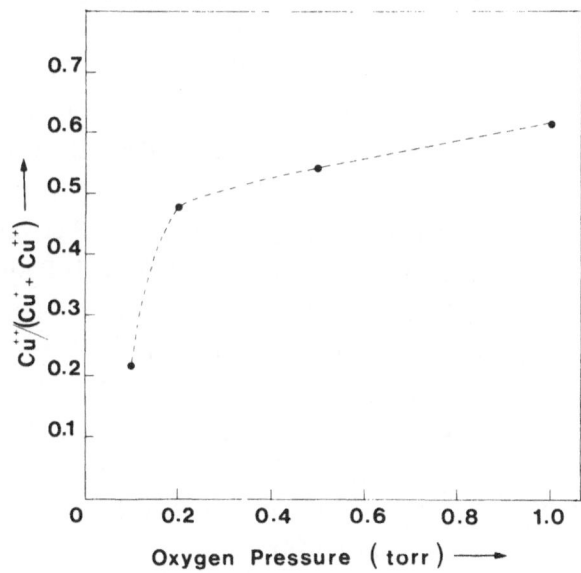

Fig.3. Variation of $Cu^{++}/(Cu^{+}+Cu^{++})$ ratio in as-sputtered YBaCuO films deposited at 400 $^{\circ}$C and under various oxygen partial pressures.

CONCLUSION

Quantitative data pertaining to cuprous (Cu^+) and cupric (Cu^{++}) states in as-sputtered YBaCuO films are determined by an electrochemical analysis. Results indicate that films deposited at 400 $^{\circ}$C contain nearly four times as much oxidized copper species as that found in films fabricated without substrate heating. When the oxygen partial pressure is raised during sputtering, the relative content of divalent copper species increases in the deposited films. When the films are subjected to ion bombardment at elevated temperature in vacuum, the oxide content is significantly reduced as evidenced by electrochemical and compositional data.

REFERENCES

1. W.E.Farneth, R.K. Bordia, E.M.McCarron III, M.K.Crawford and R.B.Flippen, Solid State Commun. 66, 953 (1988).

2. H.Ihara, H.Oyanagi, R.Sugise, E.Ohno, T.Matsubara, S.Ohashi, N.Terada, M.Jo, M.Hirabayashi, K.Murata, A.Negishi, Y.Kimura, E.Akiba, H.Hayakawa, and S.Shin, Physica C 153-155, 948 (1988).

3. A.Yoshida, H.Tamura, S.Morohashi and S.Hasuo, Appl. Phys. Lett. 53, 811 (1988).

4. H.Ihara, M.Jo, N.Terada, M.Hirabayashi, H.Oyanagi, K.Murata, Y.Kimura, R.Sugise, I.Hayashida, S.Ohashi and M.Akimoto, Physica C 153-155, 131 (1988).

5. P.Steiner, V.Kinsinger, I.Sander, B.Siegwart, S.Hufner, C.Politis, R.Hoppe and H.P.Muller, Z. Phys. B - Condensed Matter 67, 497 (1988).

6. W.Herzog, M.Schwarz, H.Sixl and R.Hoppe, Z. Phys. B - Condensed Matter 71, 19 (1988).

7. P.Steiner, S.Hufner, V.Kinsinger, I.Sander, B.Siegwart, H.Schmitt, R.Schulz, S.Junk, G.Schwitzgebel, A.Gold, C.Politis, H.P.Muller, R.Hoppe, S.Kemmler-Sack and C.Kunz, Z. Phys. B - Condensed Matter 69, 449 (1988)

8. A.Bianconi, A.Congiu Castellano, M.De Santis, P.Rudolf, P.Lagarde, A.M.Flank and A.Marcelli, Solid State Commun. 63, 1009 (1987).

9. T.Iwazumi, I.Nakai, M.Izumi, H.Oyanagi, H.Sawada, H.Ikeda, Y.Saito, Y.Abe, K.Takita and R.Yoshizaki, Solid State Commun. 65, 213 (1988).

10. B.Lengeler, M.Wilhelm, B.Jobst, W.Schwaen, B.Seebacher and U.Hillebrecht, Solid State Commun. 65, 1545 (1988).

11. D. van der Marel, J. van Elp, G.A.Sawatzky and D.Kietmann, Phys. Rev. B 37, 5136 (1988).

12. P.C.Healy, S.Myhra and A.M.Stewart, Phil. Mag. B 58, 257 (1988).

13. M.Schwartz, M.Rappaport, G.Hodes and D.Cahen, Physica C 153-155, 1457 (1988).

14. D.J.Zurawski, P.J.Kulesza and A.Wiechowski, J. Electrochem. Soc. 135, 1607 (1988).

15. J.M.Rosamilia, B.Miller, L.F.Schneemeyer, J.V.Waszczak and H.M.O'Bryan, Jr., J. Electrochem. Soc. 134, 1863 (1987).

16. H.Bachtler, W.J.Lorenz, W.Schindler and G.Saemann-Ischenko, J. Electrochem. Soc. 135, 2284 (1988).

17. S.Salkalachen, E.Salkalachen, P.K.John and H.R.Froelich, Appl. Phys.Lett. 53, 2207 (1988).

18. V.F.Drobny and D.L.Pulfrey, Proc. 13 th IEEE Photovoltaic Specialist's Conference, Washington, D.C., p. 180, 1978.

19. U.Poppe, J.Schubert, R.R.Arons, W.Evers, C.H.Freiburg, W.Reichert, K.Schmidt, W.Sybertz and K.Urban, Solid State Commun. 66, 661 (1988).

MIXED IONIC-ELECTRONIC CONDUCTION IN HT$_c$S

D.J. Vischjager, J. Schram, A. Mackor[*], and J. Schoonman
Laboratory for Inorganic Chemistry
Delft University of Technology
P.O. Box 5045, 2600 GA Delft, The Netherlands
[*] TNO Institute of Applied Chemistry
P.O. Box 108, 3700 AC Zeist, The Netherlands

INTRODUCTION

Application of HT$_c$S materials as thin-film electrodes in solid state electrochemical systems, like solid oxide fuel cells (SOFC) and sensors, at elevated temperatures is a distinct possibility, provided that sufficient stability and electronic conductivity can be maintained. In addition, for these applications, HT$_c$S materials need to exhibit also some oxygen ion conduction at the temperatures of operation. From the group of ceramic HT$_c$S we have selected $YBa_2Cu_3O_{7-x}$, 1, $EuBa_2Cu_3O_{7-x}$, 2, and $La_{1.85}Sr_{0.15}CuO_4$, 3, for detailed study. These materials have been shown to exhibit variable stoichiometry in the oxygen sublattice which can be related with ionic conduction at elevated temperatures[1-3].

For 1 oxygen stoichiometry and ordering have been the subject of thorough investigations. Oxygen ions in the CuO layer, common to the two barium-containing cubes in the 1 structure are the most loosely bound. These are the oxygen ions which disorder first on heating. Order-disorder phenomena in the oxygen sublattice are accompanied by loss of oxygen[4]. Schwartz et al.[2] have shown that the oxygen content of 1 is reduced at room temperature by electrochemical reduction in a propylene carbonate tetrabutylammonium perchlorate electrolyte. O'Sullivan and Chang[3] have used a solid-state oxygen concentration cell, comprising 1 and sputtered Au as electrodes, and yttria-stabilized zirconia (YSZ) as the solid electrolyte. Using this cell, it was possible to electrochemically drive oxygen into and out of the 1 electrode. The diffusivity of oxygen in 94% dense 1 was estimated to be ~ 5 x 10^{-8} cm^2 s^{-1} at 550 °C.

The effect of loss of lattice oxygen on the electrical properties of $YBa_2Cu_3O_{7-x}$ has been studied by Fiory et al.[5]. These authors demonstrate that this loss accounts for the sharp rise in electrical resistivity at T > 400 °C, since the increase of x with temperature causes the density of electronic states to decrease, and the elastic scattering rate due to oxygen vacancy order to increase. Similar observations were reported by Tu et al.[6]. They relate the resistivity increase, as measured in He as ambient, to effusion of oxygen. The rate of diffusion is assumed to be limited by reactions at the solid-gas interface.

In these studies, d.c. measurements are used with electrodes that are blocking to ionic currents, and ohmic for electronic charge carriers. Hence, the reported resistivity variations are to be related to the majority charge carriers in this oxide. To obtain the ionic conductivity in $\underset{\sim}{1}$ a different type of electrode is required.

Sofar, litte is known about the mixed ionic-electronic conductivity, the stability and the electrocatalytic properties of the new ceramic superconductors at high temperatures. These materials may replace, for instance, the current state-of-the-art cathode material $La_{0.85}Sr_{0.15}MnO_3$ in SOFC reactors. This cathode material is responsible for the largest polarisation loss in the Westinghouse SOFC design.

We have studied ionic transport in $\underset{\sim}{1}$-$\underset{\sim}{3}$, using a solid-state electrochemical cell, comprising ionically reversible $Pt(O_2)/YSZ$ electrodes which are blocking for electronic charge carriers, and the ceramic superconductor as a mixed-conducting solid electrolyte. Both a.c. impedance and d.c. polarization measurements have been performed. Part of the work on $\underset{\sim}{1}$ has already been published[1].

EXPERIMENTAL

Ceramic superconductors $\underset{\sim}{1}$-$\underset{\sim}{3}$ have been prepared by solid-state reactions, using as starting materials Y_2O_3 (Ventron, 99.99%), Eu_2O_3 (Ventron, 99.9%), La_2O_3 (Ventron, 99.9%), $SrCO_3$ (Merck, 99%), $BaCO_3$ (Merck 99%) and CuO (Ventron). Stoichiometric amounts of the reactants were thoroughly mixed, pressed into pellets (5 kbar) and subsequently fired. For $\underset{\sim}{1}$ firing conditions are 950 °C for 16 hours in air. After grinding the pellets of $\underset{\sim}{1}$ and pressing new pellets, these were sintered in oxygen at 950 °C for another 16 hours. The temperature was then

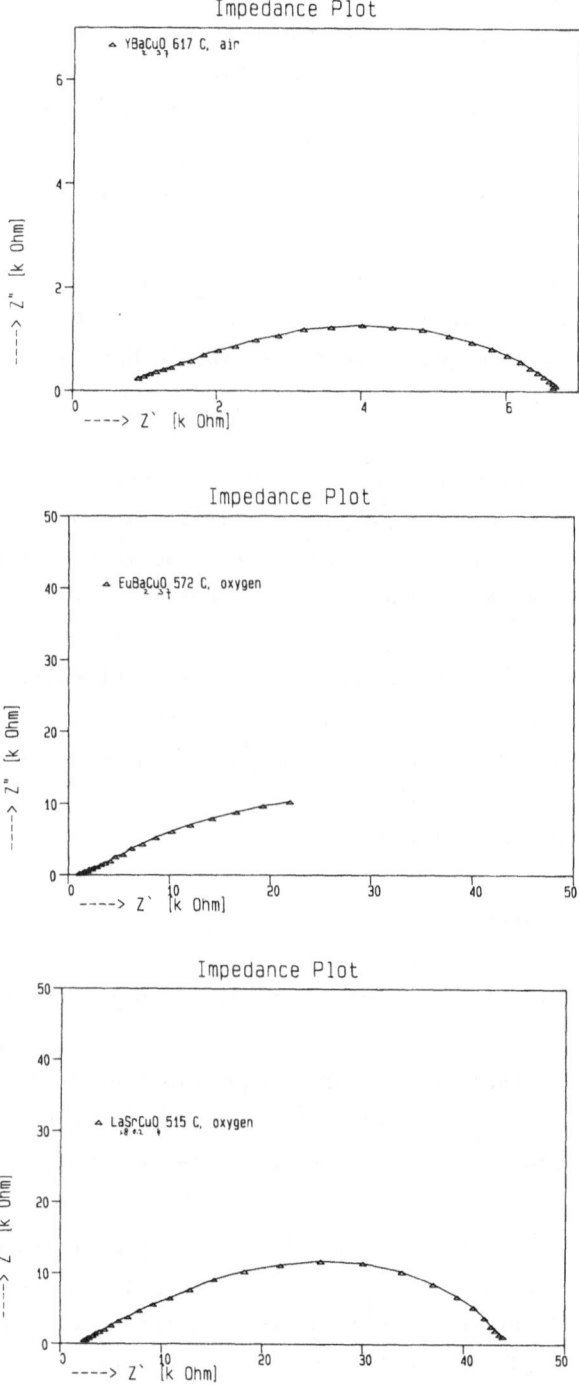

Figure 1. Impedance spectra of $YBa_2Cu_3O_{7-x}$, $EuBa_2Cu_3O_{7-x}$ and $La_{1.85}Sr_{0.15}CuO_4$ in different oxygen ambients. Frequency range 0.1 Hz to 65 kHz.

lowered to 400 °C, and the pellets were kept in oxygen at this temperature for 12 hours.

For $\underset{\sim}{2}$ firing conditions are 16 hours at 900 °C in air, regrinding and pressing, and subsequently sintering for 17 hours at 950 °C in oxygen. The sintered pellets were then kept at 450 °C for 10 hours. For $\underset{\sim}{3}$ solid state reaction was achieved at 1100 °C for 17 hours in air. Following the procedure described for $\underset{\sim}{1}$ and $\underset{\sim}{2}$, sintering took place at 1100 °C for 18 hours in oxygen.

X-ray diffraction revealed $\underset{\sim}{1}$ and $\underset{\sim}{2}$ to have the orthorhombic structure, and $\underset{\sim}{3}$ the tetragonal structure. The critical temperatures are 92 K ($\underset{\sim}{1}$), and 23 K ($\underset{\sim}{3}$). $EuBa_2Cu_3O_{7-x}$ did not exhibit a Meissner effect at liquid nitrogen temperature.

The ionic transference numbers were studied with the cell (air)Pt/YSZ/mixed conductor, HT_cS/YSZ/Pt(air). Zirconia, stabilized with 8 mole percent (m/0) yttria (Gimex) was used in combination with Pt as the ion-reversible electrode. Further details have been reported by Vischjager et al.[1]. The small-signal a.c. response of this cell arrangement was recorded using a computer-controlled Solartron 1250 Frequency Response Analyser, in combination with a Solartron 1286 Electrochemical Interface. A Marquardt non-linear least-squares parameter estimation for complex data has been used for the analysis of the admittance and impedance (immittance) spectra.

Results and Discussion

The small-signal a.c. response of $\underset{\sim}{1}$, $\underset{\sim}{2}$ and $\underset{\sim}{3}$ is presented in Figure 1.

The response data could be fitted to equivalent circuits $R_1W_1pR_2sW_2p$ for $\underset{\sim}{1}$, and $\underset{\sim}{2}$, and $R_2W_2pR_{dl}C_{dl}psR_bs$ for $\underset{\sim}{3}$. Here p stands for parallel and s for series. The equivalent circuits are presented in Figure 2.

A detailed analysis of the immittance spectra of $YBa_2Cu_3O_{7-x}$ has been reported before[1]. The equivalent circuit for $\underset{\sim}{1}$ and $\underset{\sim}{2}$ comprises the bulk ionic conductivity (R_2^{-1}), a grain boundary polarization effect of oxygen ions (R_1W_1p), while the parallel branch representing the electronic conductivity is manifested by a Warburg impedance only. The electronic resistivity in series with the Warburg impedance is extremely small, and does not show up in the immittance analysis, and hence the equivalent circuit.

The results for $\underset{\sim}{3}$ reveal grain-boundary polarization not to occur in this material. The part $R_{dl}C_{dl}pR_b s$ in the equivalent circuit for $\underset{\sim}{3}$ is to be related with the $(O_2)Pt/YSZ$ electronically blocking electrodes. Ideally, the small-signal a.c. response of the cell $(O_2)Pt/YSZ/Pt(O_2)$ can be modelled with the equivalent circuit $R_{dl}C_{dl}pR_b s^1$. R_{dl} and C_{dl} represent a Faraday impedance and a double layer capacitance,

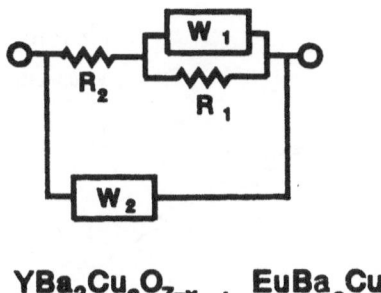

$$YBa_2Cu_3O_{7-x} \quad , \quad EuBa_2Cu_3O_{7-x}$$

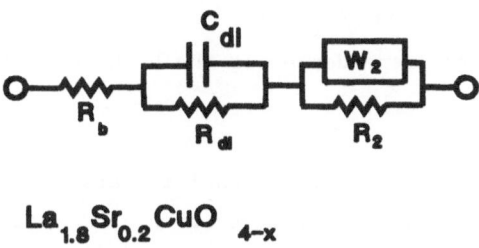

$$La_{1.8}Sr_{0.2}CuO_{4-x}$$

Figure 2. Equivalent circuits for the cells $(O_2)Pt/YSZ/HT_cS/YSZ/Pt(O_2)$

respectively, at the $(O_2)Pt/YSZ$ interface, while R_b represents the bulk ionic conductivity of YSZ. In several instances, C_{dl} needs to be replaced with a constant-phase element ϕ having an admittance of the form $Y_\phi = k(i\omega)^\alpha$ with $0.5 < \alpha < 1$. This can often be related with current inhomogeneities, which occur in polycrystalline samples. The ionic conductance (R_2^{-1}) of $\underset{\sim}{1}$, $\underset{\sim}{2}$ and $\underset{\sim}{3}$ is plotted as specific ionic conductivities in an Arrhenius plot of log σT vs T^{-1} (Figure 3).

The ionic conductivities are about a factor of 10^2 to 10 smaller than that of YSZ (8 m/o) in the temperature region 415 to 780 °C, indicating that oxygen ion conduction in these HT_cS materials is remarkably high.

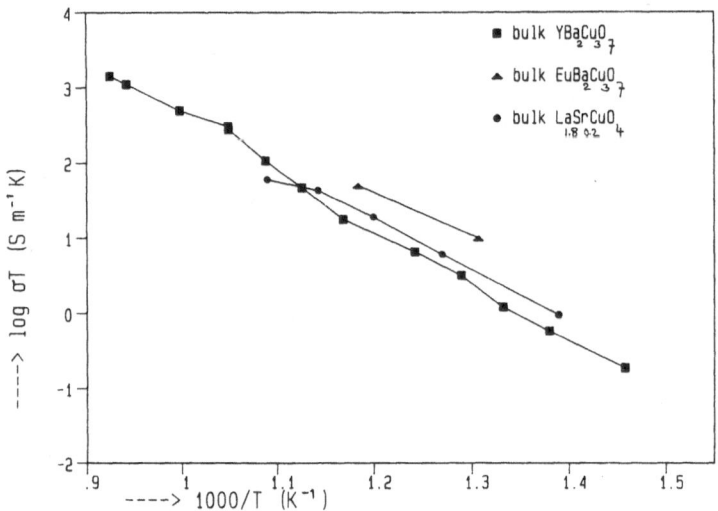

Figure 3. Temperature dependence of the bulk ionic conductivity of $\underset{\sim}{1}$, $\underset{\sim}{2}$ and $\underset{\sim}{3}$.

The conductivity activation enthalpies are 1.51 eV ($\underset{\sim}{1}$), 0.93 eV ($\underset{\sim}{2}$) and 1.27 eV ($\underset{\sim}{3}$). The numerical values for R_2 as obtained from fitting the impedance spectra in Figure 1 are presented in Table 1 along with the values for the specific electronic resistivity of these materials. From the respective ionic conductivities, the oxygen ion transference numbers t_i have been calculated, and included in Table 1.

Table 1

HT_cS	T(°C)	R_2(kΩ)	R_{el}(Ωm)	t_i
$YBa_2Cu_3O_{7-x}$ (air)	617	1	1.3×10^{-3}	7.28×10^{-7}
$EuBa_2Cu_3O_{7-x}$ (O_2)	572	2	-	5.6×10^{-7}
$La_{1.85}Sr_{0.15}CuO_4(O_2)$	515	30	2.4×10^{-3}	2.0×10^{-7}

In the case of $\underset{\sim}{2}$ we have estimated the transference number by assuming R_{el} $(\underset{\sim}{1})$ = R_{el} $(\underset{\sim}{2})$.

For $\underset{\sim}{1}$ the transference numbers vary from 2×10^{-7} to 8×10^{-6} in the temperature range 500 to 625 °C. The present data for $\underset{\sim}{1}$ have been used to calculate the oxygen ion diffusion coefficient. At 503 °C $\sigma(V_O^{\cdot\cdot})$ of $\underset{\sim}{1}$ has the value 4×10^{-5} S cm^{-1} in air. The oxygen ion vacancy concentration at 500 °C is 5×10^{21}/cm^3 in air. For the mobility of the oxygen ion vacancies we then obtain $\mu(V_O^{\cdot\cdot}) = 2.5 \times 10^{-8}$ cm^2/Vs. With the Einstein relation $\mu/D = q/kT$ results for $D(V_O^{\cdot\cdot})$ the value 1.67×10^{-9} cm^2/s in air. For the tracer diffusion coefficient the value 10^{-10} cm^2/s ($PO_2 = 1$ atm, T = 500 °C) was reported. With regard to chemical diffusion coefficient data the present value is in between these and the tracer diffusion coefficient.

In Figure 4 we have plotted the temperature dependence of the non-stoichiometry parameter x as obtained from the literature[7]. The variation of x in the orthorhombic phase reveals an activation energy of 0.22 eV, which is far below the conductivity activation enthalpy of 1.51 eV for $\underset{\sim}{1}$. We are, therefore, inclined to believe, that the present ionic conductivity measurements are not dominated by the loss of oxygen of $\underset{\sim}{1}$, and hence can be used to calculate diffusion coefficients.

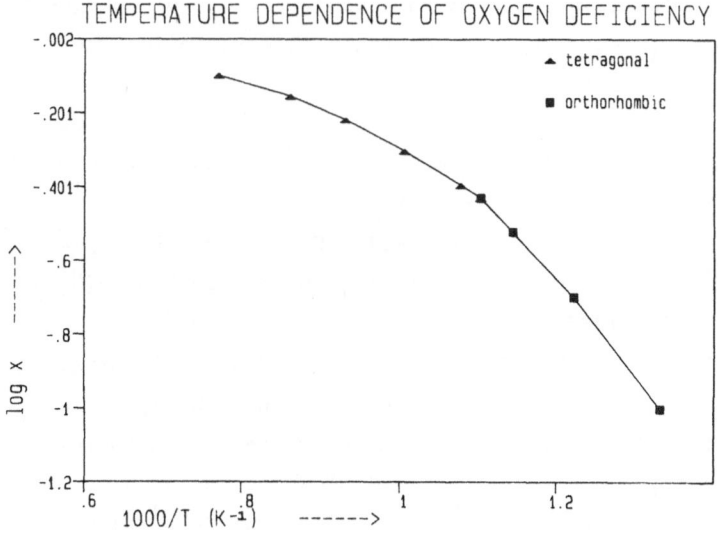

Figure 4. Temperature dependence of x as calculated from ref. 7.

Usually, chemical diffusion coefficients are obtained by recording conductivity changes upon changing the ambient from air to oxygen. These diffusion coefficients are apparent diffusion coefficients which should be analysed using ambipolar diffusion models. In general, such apparent diffusion coefficients are larger than diffusion coefficients calculated from conductivity data.

The present study indicates that $(O_2)Pt/YSZ$ electrodes effectively block electronic currents at the YSZ/HT$_c$S interface, and hence can be used to determine the ionic transference numbers in HT$_c$S materials. Further research is being carried out on $La_{2-x}M_xCuO_4$ (M = Sr,Ba).

ACKNOWLEDGEMENT

D.J.V. acknowledges financial support by the Netherlands Organization for Applied Scientific Research TNO.

REFERENCES

1. D.J. Vischjager, P.J. van der Put, J. Schram and J. Schoonman, Oxygen diffusion in $YBa_2Cu_3O_{7-x}$; an impedance spectroscopy study, <u>Solid State Ionics</u> 27: 199 (1988).

2. M. Schwartz, M. Rappaport, G. Hodes and D. Cahen, Electrochemical preparation and properties of oxygen deficient $YBa_2Cu_3O_{7-x}$, <u>Physica C</u> 153-155: 1457 (1988).

3. E.J.M. O'Sullivan and B.P. Chang, Study of oxygen transport in $Ba_2YCu_3O_{7-\delta}$ using a solid state electrochemical cell, <u>Appl. Phys. Letters</u>, in press.

4. G. van Tendeloo and S. Amelinckx, Direct observation of the order-disorder transformation of the oxygen sublattice of $YBa_2Cu_3O_{7-\delta}$, <u>Phys. Stat. Sol. A</u> 103: K1 (1987).

5. A.T. Fiory, M. Gurvitch, R.J. Cava and G.P. Espinosa, Effect of oxygen desorption on electrical transport in $YBa_2Cu_3O_{7-\delta}$, <u>Phys. Rev. B</u>. 36: 7262 (1987).

6. K.N. Tu, C.C. Tseui, S.I. Park and A. Levi, Oxygen diffusion in superconducting $YBa_2Cu_3O_{7-\delta}$ oxides in ambient helium and oxygen, <u>Phys. Rev. B</u> 38: 772 (1988) and references cited therein.

7. P.K. Gallagher, Characterization of $Ba_2YCu_3O_x$ as a function of oxygen partial pressure. Part I: Thermoanalytical Measurements, <u>Adv. Ceramic Mater.</u> 2: 632 (1987).

8. K. Kitazawa, H. Takagi, K. Kishio, T. Hasegawa, S. Uchida, S. Tajima, S. Tanaka and K. Fueki, Electronic properties of cuprate superconductors, <u>Physica C</u> 153-155: 9 (1988).

PHOTODETECTION WITH HIGH-T_c SUPERCONDUCTING FILMS

J. Talvacchio, M. G. Forrester, and A. I. Braginski

Westinghouse R&D Center
Pittsburgh, Pennsylvania 15235

INTRODUCTION

One of the promising applications of high-T_c superconducting thin films is the detection of infrared radiation. The purpose of this paper is to clarify which measurements are needed to identify the basic mechanism responsible for a photo-response, and to summarize the measurements made to date.

The motivation for studying high-T_c photodetectors stems largely from the results shown in Fig. 1. Figure 1 is a comparison of the detectivity, D^*, of two different superconducting Josephson junction detectors[1,2] with semiconductor detectors and D^* calculated for a 300 K background-limited photoconductor with a 180° field-of-view.[3] The shaded regions include data from 19 different semiconductor detectors.[3] Each semiconductor detector has a high D^* in a narrow wavelength range compared to the potentially broadband sensitivity of Josephson junction detectors.

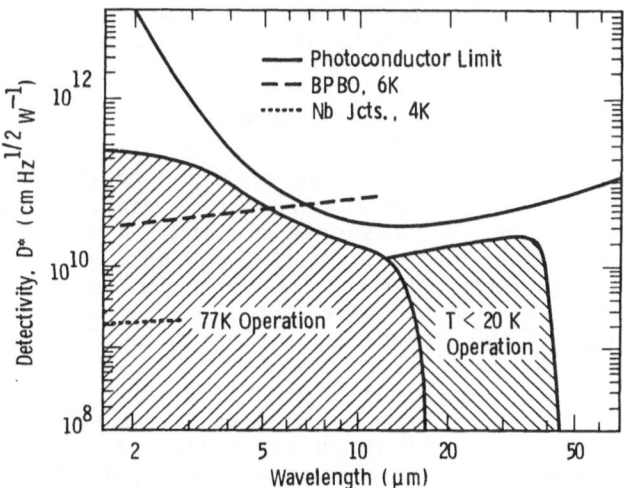

Fig. 1 - Detectivity of a granular $BaPb_{0.7}Bi_{0.3}O_3$ film measured at 6 K,[1] and discrete Nb tunnel junctions measured at 4.2 K,[2] compared with semiconductor detectors.[3]

In superconducting detectors exposed to radiation with energy greater than the gap energy, $\Delta(T)$, the absorbed radiation reduces the gap energy, modifying the current-voltage characteristics of Josephson junctions. Two possible mechanisms for gap-energy reduction will be considered here. The first is an equilibrium mechanism in which the film acts as a bolometer and the response is proportional to the temperature coefficient of resistance of a Josephson junction. The disadvantage of a bolometer is that the response is slow unless the amplitude of the response is traded off for efficient cooling. The second mechanism is dependent on the creation of quasiparticle pairs by the incident radiation. This non-equilibrium quasiparticle density is measured in Josephson junction I-V curves as a reduced gap energy. The response time of the non-equilibrium mechanism is determined by the quasiparticle recombination time – as short as 10^{-10} s in a strong-coupling superconductor at $T \simeq T_c$.[4]

The data from superconducting detectors shown in Fig. 1 were measured in detectors composed of different types of Josephson junctions. The lower D^* data is from a $Nb/Al_2O_3/Nb$ Josephson tunnel junction.[2] The response at 4.2 K was attributed to a bolometric effect since it decreased by approximately three orders of magnitude when the sample was measured in superfluid helium at 2 K. The $BaPb_{0.7}Bi_{0.3}O_3$ (BPBO) detectors were granular films which had grain-boundary Josephson junctions with I-V characteristics similar to point-contact or microbridge junctions.[1] These detectors not only had very high D^* values, but had response times on the order of 1 ns – attributed to a non-equilibrium response mechanism. Three of the important properties of BPBO which contributed to the detector results are also present in high-T_c oxide superconductors, typified by $YBa_2Cu_3O_7$ (YBCO).

1. Both BPBO[1] and YBCO[5] have higher absorption and lower reflection coefficients than transition metal superconductors due to their lower density of electronic states.
2. The decrease in gap energy due to an excess quasiparticle density is inversely related to the density of states and is, therefore, greater in the oxide superconductors.
3. Grain-boundary Josephson junctions are readily formed in the oxide superconductors. A higher density of junctions can be realized in granular films than with discrete microbridge junctions lithographically defined. Tunnel junctions of these materials have not been made due to their short coherence lengths, unstable surfaces, and high processing temperatures.

SIMULATED RESPONSES TO RADIATION

This section will present the results of simple models for the bolometric and non-equilibrium responsivities, r_B and r_{NE}, of granular films modeled as microbridge Josephson junctions connected in series. The responsivity will be considered as a function of bias current, I_B, and reduced temperature, $t = T/T_c$. The important results – those which are neither obvious without invoking a model nor overly model-dependent – should clarify which characteristic forms of the responsivity are useful in identifying equilibrium and non-equilibrium response mechanisms.

The series arrays used in this model each consisted of 100 junctions with critical currents at $T = 0$ which had a Gaussian distribution centered at I_{c-avg} and standard deviation, σ, expressed as a percentage of I_{c-avg}. No junctions were permitted to have a critical current in the "tails" of the distribution, that is, less than zero or greater than $2I_{c-avg}$. The resistance at currents much greater than the critical current, R_n, was assumed to be the same for each junction and independent of temperature. The particular distribution chosen for R_n was not an important assumption. However, the

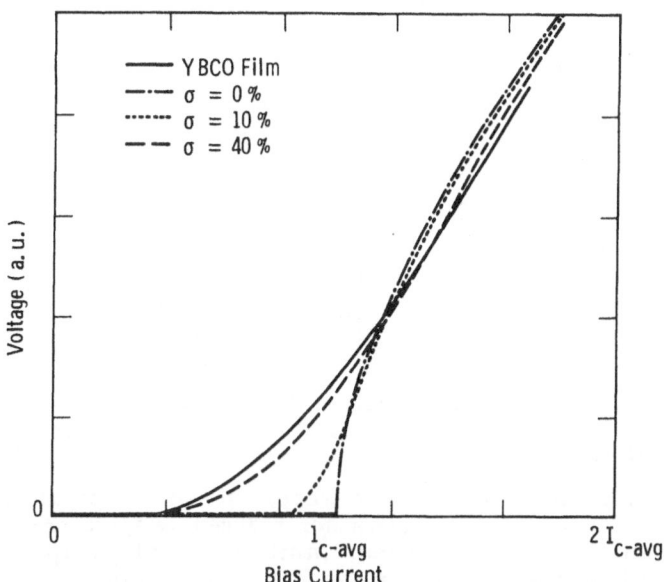

Fig. 2 – Calculated I-V curves for three series arrays with $\sigma = 0$, 10, and 40%, and the I-V curve of a bridge patterned in a YBCO film. The critical current of the bridge at T = 30 K was 0.25 mA and its resistance at 1 mA was 45 Ω.

assumption of temperature independence for R_n will give a poor simulation of those superconducting films which exhibit strongly semiconducting behavior of resistivity for $T > T_c$. The t = 0 I-V curves for these series arrays were calculated in the RSJ model[6] and are shown in Fig. 2. The I-V curve of a 10 μm wide by 90 μm long bridge patterned in a sputtered YBCO film is also shown in Fig. 2 for comparison of the model I-V curves with that of a film for which photo-response data has been published.[7]

The general form of the bolometric and non-equilibrium responsivities is given below for a single current-biased junction where Φ is the incident photon flux in W/cm^2, I_B is the bias current, I_c is the temperature-dependent Josephson critical current, and n_q is the quasiparticle density. The only approximation made in Equs. 1 and 2 is that R_n is independent of temperature. Note that the responsivity is expressed in units of V-cm^2/W – in agreement with Ref. 2 – since the response, δV,* is independent of detector area for a fixed photon flux for either mechanism.*

$$r_B = \frac{\delta V}{\delta \Phi} = I_B \left.\frac{\partial R}{\partial T}\right|_{I_B} \frac{\delta T}{\delta \Phi} \simeq \left.\frac{\partial V}{\partial I_c}\right|_{I_B} \frac{dI_c}{dT} \frac{\delta T}{\delta \Phi} \qquad (1)$$

$$r_{NE} = \frac{\delta V}{\delta \Phi} \simeq \left.\frac{\partial V}{\partial I_c}\right|_{I_B} \left.\frac{\partial I_c}{\partial \Delta}\right|_{T} \left.\frac{\partial \Delta}{\partial n_q}\right|_{T} \frac{\delta n_q}{\delta \Phi} \qquad (2)$$

---

* The bolometric response, $\delta V = r_B \, \delta \Phi = I_B \, \partial R/\partial T \, \delta T$, where bridge resistance, R, is independent of area for a fixed aspect ratio, δT is independent of area for a fixed ratio of cooling area to detection area, and I_B is assumed to be adjusted for maximum response.

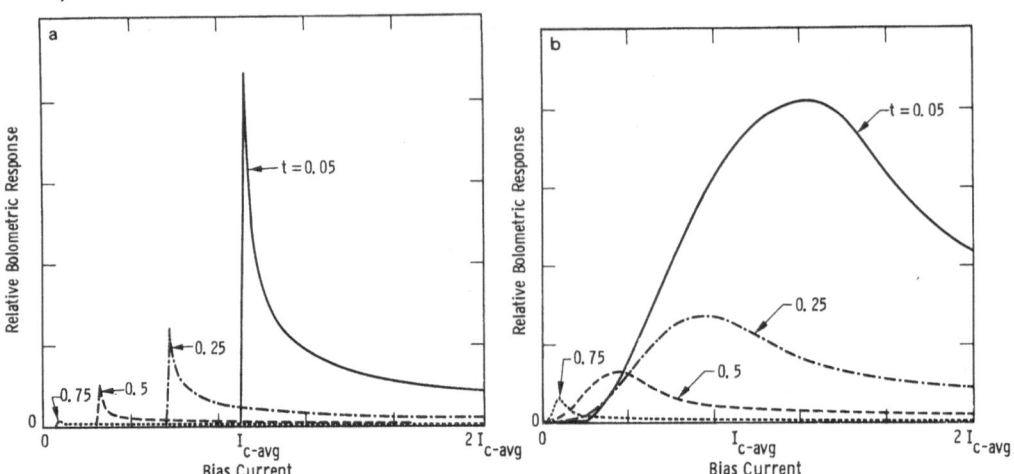

Fig. 3 - The simulated bolometric response of junction arrays with (a) $\sigma = 0$, and (b) $\sigma = 40\%$ for reduced temperatures of 0.05, 0.25, 0.5, and 0.75. The discontinuities in (a) at $I_B = I_c(T)$ were rounded by the discrete nature of the calculation.

The only factor in Equs. 1 and 2 that depends on I_B is $\partial V/\partial I_c$, and it appears in each equation. Therefore, the general shape of measured photo-response plotted as a function of bias current cannot be used to distinguish which response mechanism is dominant. Simulated curves of response versus bias current are plotted in Figs. 3(a) and (b) for series junctions with $\sigma = 0$ and 40%, respectively, using $\partial V/\partial I_c$ calculated from the RSJ model. The qualitative features of $\delta V(I_B)$ data published in Refs. 1, 8, and 9, are present in the curves in Fig. 3.

In contrast to the effects of changing bias current, the effect of changing temperature is completely different in Equs. 1 and 2. The temperature dependence of $\partial V/\partial I_c$ is described by $I_c(T)$. The following relationships were used in the simulations for $I_c(T,\Delta)$ and δn_q:

$$I_c \propto (1-t)^2 \qquad \text{(Ref. 10)} \qquad (3)$$

$$\frac{\partial I_c}{\partial \Delta} \propto (1-t)^{1/2} \qquad \text{(Ref. 11)} \qquad (4)$$

$$\delta n_q(T) \propto \tau_{eff}(T) \propto t^{-1/2} e^{\Delta(0)/kT} \simeq t^{-1/2} e^{1.76/t} \qquad (5)$$

where τ_{eff} is an effective quasiparticle recombination time.[12] The effective lifetime was assumed to have the same dependence as the simple recombination time,[4] and a weak coupling BCS relationship was assumed for $\Delta(0)/kT_c$. Following Ref. 1, $\partial \Delta/\partial n_q$ was taken as a constant — an appropriate assumption for δn_q small compared to the equilibrium quasiparticle density.

The only temperature dependence which has not been specified at this point is that of the change in temperature of a bolometer, $\delta T = P/(KA/d)$, where — for a particular bridge of area, $A = 10 \times 90 \ \mu m^2$ — $P \simeq 0.6 \ \mu W$ was the laser power incident on the bridge,[14] and $K(T)$ was the thermal

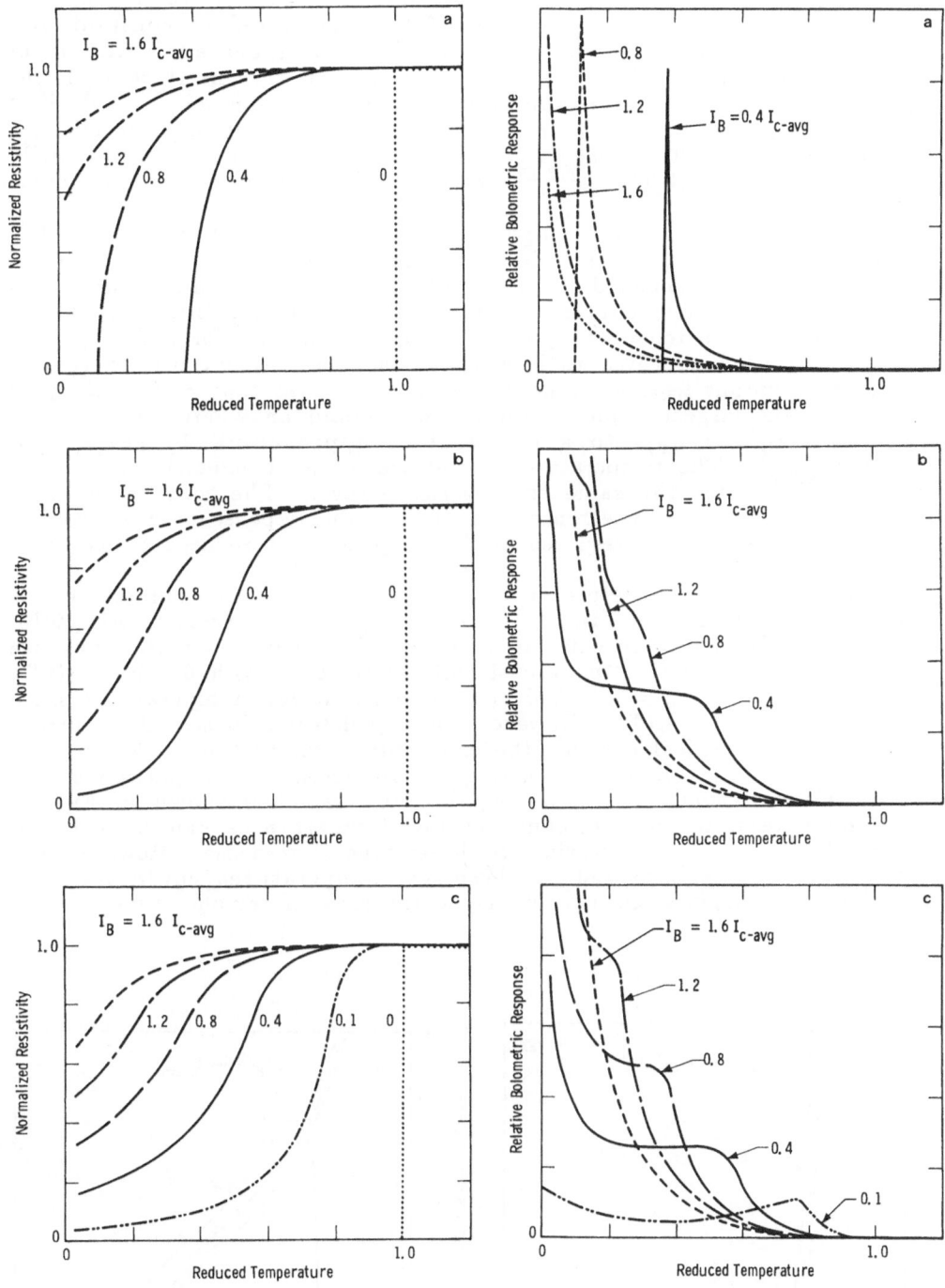

Fig. 4 - Calculations of resistance plotted for various values of the bias current as a function of temperature for series arrays with σ = (a) 0, (b) 40, and (c) 90%.

Fig. 5 - Calculations of bolometric response plotted for various values of the bias current as a function of temperature for series arrays with σ = (a) 0, (b) 40, and (c) 90%.

conductivity of the thermal barrier of thickness, d. The temperature rise was estimated to be 10 mK from the slope of V(T) measured at constant bias current. The thermal barrier was assumed to be a degraded film layer ~100 nm thick which is known to exist adjacent to the interface with the substrate.[15] The thermal conductivity calculated from δT and P, K $\simeq$ 10^{-4} W/cm/K, is the same order of magnitude as that of oxygen-deficient, tetragonal YBCO.[13] The temperature dependence of the thermal conductivity of tetragonal YBCO, K(T) α $t^{1/2}$,[13] was used in our model of photoresponse.

Since K varies slowly at high temperature (t > 0.5), the bolometric response follows the temperature derivative of $R(I_B,T)$ as T → T_c. Figure 4 contains a series of calculated $R(I_B,T)$ curves for three series arrays with σ = 0, 40, and 90%. The curves in Fig. 4 show that an infinitely sharp transition to the superconducting state has been assumed (for small I_B) and that the transition appeared to broaden as the bias current was increased. The most significant feature of the curves in Fig. 4 was that — depending on σ and I_B — the largest value of dR/dT was found anywhere from T = 0 (Fig. 4(a), $I_B \geq I_{c-avg}$) to a temperature approaching T_c (Fig. 4(c), I_B = 0.1 I_{c-avg}). The temperature dependence of the bolometric response is shown in Fig. 5 for the same three series arrays. The most significant feature in Fig. 5 is the appearance of a peak in the response near T_c coupled with a rising value of the response as the temperature is reduced to <0.1T_c.

The calculated temperature dependence of the non-equilibrium response is shown on a linear scale in Fig. 6(a) and on a logarithmic scale in Fig. 6(b). The results in Fig. 6 were calculated for junctions with σ = 10% but are representative of simulations performed with distributions with 0 $\leq$ σ < 100%. The curve for I_B/I_{c-avg} = 0.4 had zero response at low temperatures since $\partial V/\partial I_c$ = 0 for $I_B < I_c(T)$. However, for any detector biased at a current greater than the critical current, the exponential temperature dependence of effective quasiparticle lifetime dominated the response. A comparison of the calculations for bolometric and non-equilibrium responses shown in Figs. 5 and 6 indicate that measurements at low temperatures can be used to increase the possibility of observing the latter type of response. However, the large values of τ_{eff} expected at these low temperatures could make it impossible to distinguish the two effects on the basis of response time.

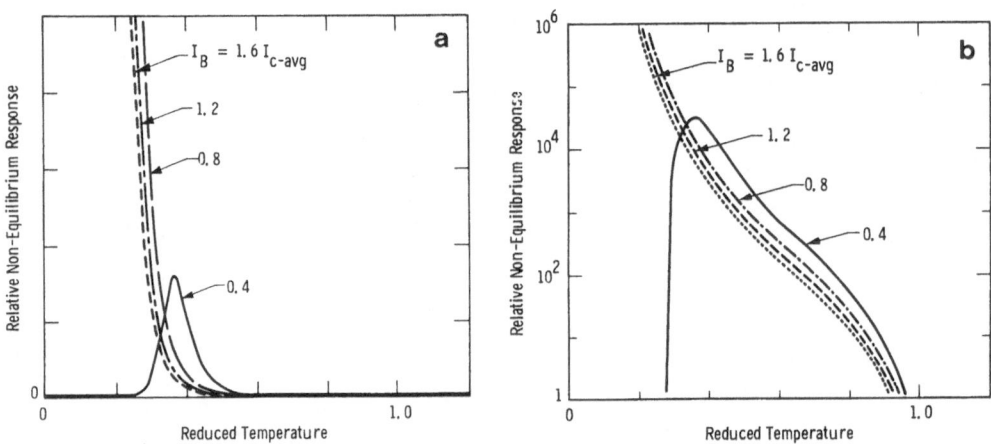

Fig. 6 - The calculated non-equilibrium response plotted on (a) a linear and (b) a logarithmic scale for various values of the bias current for a series of junctions with σ = 10%.

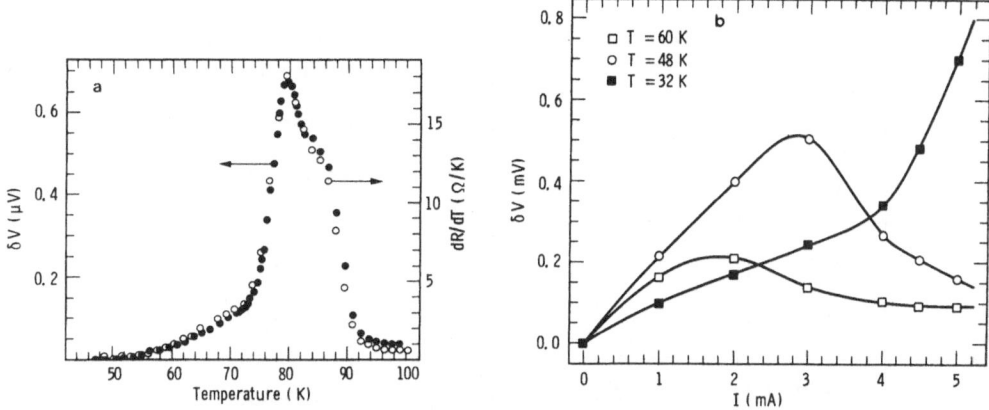

Fig. 7 - (a) A comparison of measurements of dR/dT and the response of a YBCO detector to 10.6 μm radiation which indicates that the response is purely bolometric. (b) Measured photo-response plotted as a function of bias current for a YBCO detector held at various temperatures.

DISCUSSION OF EXPERIMENTAL RESULTS

With the exception of one report on the response of $LaSr_{1-x}Cu_xO_4$,[16] YBCO films have been used for all photodetection measurements with high-T_c films. Figure 7 shows response data which is typical of YBCO samples thought to have a purely bolometric response. The data in Fig. 7(a) is from a granular YBCO film sputtered onto a BaF_2 buffer layer as described in Ref. 14, and is a comparison showing close agreement between the temperature dependence of dR/dT and photo-response. A comparison of Fig. 7(a) with Fig. 5 shows that the model series array with the broadest distribution and biased at low I_B (Fig. 5(c); $\sigma = 90\%$) had a peak in the response at high reduced temperatures in agreement with the experimental results. However, the correlation between measured response and dR/dT was better than indicated by Equ. 1 or Figs. 4 and 5. The difference is that the thermal conductivity was modeled as $K(T) \propto t^{1/2}$ and the thermal conductance of the measured detector apparently varied more slowly than that in the vicinity of T_c. The interface layer between the high-T_c film and the substrate was modeled as tetragonal $YBa_2Cu_3O_6$, but certainly consisted of other oxides as well – especially BaO.[15] The important aspect of the temperature dependence of K is not the detailed dependence near T_c, but the fact that, in simulations, $K(T)$ led to a rising response as the temperature was decreased $<< T_c$. Equally important, although not considered in the steady-state model, is that $K(T)$ affects the time response of a bolometric δV measured at various temperatures.

Fig. 7(b) shows the dependence of photo-response on bias current for a detector which behaved as a bolometer based on the measurements of response versus temperature and time presented in Ref. 7. In contrast to the conclusions of Ref. 9, the presence of a peak in $\delta V(I_B)$ is not inconsistent with a bolometric response. The peak is consistent with the calculated curves of Fig. 3(b) and the discussion of Equs. 1 and 2.

A summary of published photo-response measurements is presented in Table 1. Since the dependence of response on bias current does not help to identify the detection mechanism, only the temperature and temporal dependences are included. The temperature dependence is indicated by a sketch of $\delta V(T)$. The temporal dependence is indicated by the measured

Table 1 - Summary of reported infrared responses of YBCO films.

Reference	Temperature dependence	Temporal response	Possible mechanism, sensitivity
Forrester et al. (14)	ΔV [graph] T	$\tau \approx \mu sec$– 0.1 sec	Bolometric $r \approx 4 \times 10^3$ V/W $D^* \approx 10^8$ cm√Hz/W
Strom et al. (19)	ΔV [graph] T_c T	$\sim \omega^{-1/2}$	Bolometric + "Optically induced flux flow" $D^* \approx 10^7$
Osterman et al. (9)	ΔV [graph] T	$\tau < 0.25$ msec	Bolometric + nonequilibrium? at 1.8 K $r \approx 0.01$ V/(W/cm^2)
Wilson et al. (8)	ΔV [graph] T	$\tau < 0.1$ msec	Bolometric + nonequilibrium? $r \approx 3000$ V/W $D^* \approx 5 \times 10^7$
Enomoto et al. (20)	?	$\sim \omega^{-1/2}$	Bolometric $r \approx 10$ V/W (@ 10 kHz)
Brocklesby et al. (21)	ΔV [graph] T	"On" $\approx$ 10 –100 ns	Bolometric $r \sim 0.1$ V/W

response time, the maximum response time with a limit determined by measurement apparatus, or the dependence of δV on the frequency at which incident radiation was chopped. Some reports of bolometric responses[17,18] were excluded from Table 1 since such data is adequately represented and typical data are shown in Fig. 7.

The most interesting data in Table 1 are the points measured at low temperature by Osterman et al.[9] and Wilson et al.[8] Although the time response is unknown for this data, even a slow response ($>>\mu sec$) would not rule out the non-equilibrium detection mechanism since recombination times should be long as $T \rightarrow 0$. With the temperature dependence as the only remaining guide, the low-temperature data of Ref. 8 can be explained as the effect of $K(T)$ on a bolometric response. The change of response with temperature is not exponential as predicted by the non-equilibrium model. The single low-temperature data point of Ref. 9 was measured with the sample immersed in superfluid helium to increase $K(T)$ and minimize a bolometric response. Additional measurements made with small changes in the helium bath temperature could be sufficient to identify the detection mechanism.

The response data from Ref. 19 is noteworthy in that a second peak in the response was measured at a temperature ($\sim25K$) where the slope of $R(T)$

was not a maximum. The response was relatively slow and decreased as $\sim\omega^{-1/2}$ for kHz frequencies. Measurements of the response time as a function of temperature could be used to identify the non-equilibrium mechanism considered in this paper.

SUMMARY AND CONCLUSIONS

The issue of which response mechanism is active in the detection of infrared radiation by high-T_c films does not have a direct bearing on their potential usefulness. The response of any detector simply needs to be large enough to raise the level of the detector noise higher than that of other noise sources in the system, and fast enough for the particular application at hand. For most applications, a bolometric (equilibrium) response is inadequate to meet these two needs simultaneously, but a non-equilibrium detection mechanism has not been unambiguously observed in high-T_c superconducting films. Measurements of peaks in δV versus bias current and increases in δV at low temperatures are not sufficient indicators of a non-equilibrium response. Both the temperature and time dependence of the response of a non-equilibrium detector are expected to be dominated by the exponential temperature dependence of effective recombination times for quasiparticles. Therefore, the two most convincing tests for establishing that a non-equilibrium detection mechanism is active are response measurements which show:

1. An exponentially increasing magnitude of response as the temperature is decreased.
2. An exponentially increasing response time as the temperature is decreased.

ACKNOWLEDGMENTS

This work was supported by Air Force Office of Scientific Research Contract No. F49620-88-C-0030.

REFERENCES

1. Y. Enomoto and T. Murakami, "Optical Detector Using Superconducting $BaPb_{0.7}Bi_{0.3}O_3$," J. Appl. Phys. <u>59</u>, 3808 (1986), and Y. Enomoto, T. Murakami, and M. Suzuki, "Infrared Optical Detector Using Superconducting Oxide Thin Film," Physica C <u>153-155</u>, 1592 (1988).
2. D. P. Osterman, M. Radparvar, and S. M. Faris, "Optical Detection by Suppression of the Gap Voltage in Niobium Junctions," to be published in IEEE Trans. Magn. <u>MAG-25</u> (1989).
3. W. L. Wolfe and G. J. Zissis, eds., <u>The Infrared Handbook</u> (Office of Naval Research, Washington, 1978), p.-11-87.
4. S. B. Kaplan, C. C. Chi, D. N. Langenberg, J. J. Chang, S. Jafarey, and D. J. Scalapino, "Quasiparticle and Phonon Lifetimes in Superconductors," Phys. Rev. B <u>14</u>, 4854 (1976).
5. I. Bozovic, K. Char, S. J. B. Yoo, A. Kapitulnik, M. R. Beasley, T. H. Geballe, Z. Z. Wang, S. Hagen, N. P. Ong, D. E. Aspnes, and M. K. Kelly, "Optical Anisotropy of $YBa_2Cu_3O_{7-x}$," Phys. Rev. B <u>38</u>(7), 5077 (1988).
6. L. G. Aslamazoff and A. I. Larkin, "Josephson effect in Superconducting Point Contacts," JETP Lett. <u>9</u>, 87 (1969).
7. M. G. Forrester, M. Gottlieb, J. R. Gavaler, and A. I. Braginski, "Optical Response of Epitaxial Films of $YBa_2Cu_3O_{7-\delta}$," Appl. Phys. Lett. <u>53</u>(14), 1332 (1988).
8. J. A. Wilson, J. M. Myrosznyk, R. E. Kvass, M. D. Jack, P. R. Norton, A. H. Hamdi, J. V. Mantese, A. L. Micheli, J. Y. Josefowicz,

A. T. Hunter, and D.B. Rensch, "IR Detection in Superconducting YBaCuO," paper A2, presented Mtg. IRIS Specialty Group on Infrared Detectors, Baltimore, August 1988.

9. D. P. Osterman, R. Drake, R. Patt, E. K. Track, M. Radparvar, and S. M. Faris, "Optical Response of YBCO Thin Films and Weak-Links," to be published in IEEE Trans. Magn. MAG-25 (1989).

10. K. Moriwaki, Y. Enomoto, and T. Murakami, "Josephson Junctions Observed in $La_{1.8}Sr_{0.2}CuO_4$ Superconducting Polycrystalline-Films," Jpn. J. Appl. Phys. 26(4), 521 (1987).

11. This is a simplification based on K. K. Likharev, "Superconducting Weak Links," Rev. Mod. Phys. 51, 116 (1979).

12. A. Rothwarf and B. N. Taylor, "Measurement of Recombination Lifetimes in Superconductors," Phys. Rev. Lett. 19, 27 (1966).

13. J. Heremans, D. T. Morelli, G. W. Smith, and S. C. Strite III, "Thermal and Electronic Properties of Rare-Earth $Ba_2Cu_3O_x$ Superconductors," Phys. Rev. B 37(4), 1604-1610 (1988).

14. M. G. Forrester, M. Gottlieb, J. R. Gavaler, and A. I. Braginski, "Optical Response of Epitaxial and Granular Films of $YBa_2Cu_3O_{7-\delta}$ at Temperatures from 25K to 100K," to be published in IEEE Trans. Magn. MAG-25, (1989).

15. J. Talvacchio, J. R. Gavaler, and J. Greggi, "Comparison of $YBa_2Cu_3O_7$ Films Grown by Solid-State and Vapor-Phase Epitaxy," to be published in IEEE Trans. Magn. MAG-25, (1989).

16. K. Moriwaki, Y. Enomoto, and T. Murakami, "I-V Characteristics of $(La_{1-x}Sr_x)_2CuO_4$ Superconducting Films and Infrared Irradiation Effects," Jpn. J. Appl. Phys. Suppl. 26-3, 1147 (1987).

17. P. L. Richards, S. Verghese, T. H. Geballe, and S. R. Spielman, "The High T_c Superconducting Bolometer," to be published in IEEE Trans. Magn. MAG-25, (1989).

18. F. H. Garzon, "Fabrication of Infrared Detectors Based on Granular Thin Films of $YBa_2Cu_3O_7$," presented 174th Electrochem. Soc. Mtg., Chicago, 1988.

19. U. Strom, E. S. Snow, R. L. Henry, P. R. Broussard, J. H. Claassen, and S. A. Wolf, "Photoconductive Response of Granular Superconducting Films," to be published in IEEE Trans. Magn. MAG-25, (1989), and U. Strom, E. S. Snow, M. Leung, P. R. Broussard, J. H. Claassen, and S. A. Wolf, "Optical Response of Granular Superconducting Films of Y-Ba-Cu-O and NbN/BN," SPIE Proceedings Vol. 948, High-T_c Superconductivity: Thin Films and Devices, edited by R. B. van Dover and C. C. Chi (SPIE, Bellingham, Washington, 1988), p. 10. Temporal dependence data obtained from M. Leung, private communication.

20. Y. Enomoto, T. Murakami, and M. Suzuki, "Infrared Optical Detector Using Superconducting Oxide Thin Film," Proc. 5th Intl. Workshop on Future Electron Devices (Miyagi-Zao, 1988), p. 325.

21. W. S. Brocklesby, D. Monroe, A. F. J. Levi, M. Hong, S. H. Liou, J. Kwo, and C. E. Rice, "Electrical Response of Superconducting $YBa_2Cu_3O_{7-\delta}$ to Light," preprint, 1988.

THE FABRICATION OF JOSEPHSON JUNCTIONS ON Y-Ba-Cu-O THIN FILMS

Jiro Yoshida, Koichi Mizushima, Masayuki Sagoi,
Yoshiaki Terashima and Tadao Miura

Research and Development Center, Toshiba Corporation
1, Komukai Toshibacho, Saiwaiku, Kawasaki 210, Japan

ABSTRACT

Electrical contact properties between noble metals and Y-Ba-Cu-O (YBCO) thin films prepared by multi-target rf sputtering method on (100) $SrTiO_3$ and (100) MgO substrates were investigated. The temperature dependence of the contact resistance below the critical temperature of YBCO films suggested the existence of an insulating layer at the Au/YBCO interface. On the other hand, heat treated Ag/YBCO interface gave a very low contact resistance which could not be detected within the resolution limit of our apparatus. Based on this finding, we tried to fabricate Josephson junctions using lead as a counter electrode on the YBCO thin films covered by a thin silver layer expecting that the proximity effect worked between silver and YBCO. A dc "superconducting current" with undetectable resistance and a highly hysteretic current voltage (I-V) characteristic were observed below the transition temperature of lead.

I. INTRODUCTION

Since the discovery of high-Tc superconductors, various efforts have been directed towards the understanding of the physics of high temperature superconductivity as well as a possible application of these materials to various fields of industry. The preparation of a high quality thin film is one of the key technologies for the application to electronics. Various methods have been utilized to fabricate thin films, including sputtering, electron-beam evaporation, laser ablation and chemical vapor deposition. High quality thin films with zero resistivity above 80 K were reported without a high temperature post annealing process.[1-3]

Another requirement for electronics application is to combine these thin films with other materials, including metals, insulators or semiconductors, in order to realize certain kinds of device functions. In spite of a large volume of reports concerning thin film formation technology, little work has been reported about the combination of high Tc superconductors with other materials.[4,5] A very serious problem of high Tc superconductors is the existence of a degraded layer on the surface, which is believed to be formed by the escape of oxygen or chemical reaction with water vapor. This degraded layer, together with the inherently short coherence length, makes it difficult to realize a junction structure with high Tc materials.

This paper reports our efforts on investigating the interface properties between YBCO and metals aimed at deriving superconductivity on the very surface of the YBCO thin film. Also, a preliminary result of Josephson junctions fabricated on YBCO thin films using lead as a counter electrode will be described. In Sec. II. our thin film preparation technology on which the following junction experiments rely will be described. Section III deals with the electric properties at the YBCO/metal interfaces. In Sec. IV, the fabrication process of the Josephson junction and the junction characteristics will be discussed.

II. THIN FILM PREPARATION

The multi-target rf magnetron sputtering technique was used to fabricate the YBCO thin films. Yttrium metal, copper metal and sintered Ba_2CuO_3 ceramic target were simultaneously sputtered in an argon-oxygen mixture gas. The rf power to each target was regulated to minimize the stoichiometric deviation of the film from an ideal composition. The sputtering gas pressure was around 0.65 Pa with the oxygen partial pressure of 0.32 Pa. $SrTiO_3$ and MgO single crystals both with a (100) surface were used as substrates. The substrates were heated to temperatures ranging from 530 °C to 730 °C during film growth in order to realize a superconducting film without a post annealing process. The film thickness was from 500 nm to 700 nm and the deposition rate was about 3 nm/min. After the deposition, the films were cooled to about 200 °C under the sputtering gas condition and then 1 atmospheric pressure oxygen was introduced into the chamber. More details about the thin film preparation technology will be published elsewhere.[6]

Figure 1 shows a SEM micrograph and a RHEED pattern obtained for a film grown on a (100) $SrTiO_3$ substrate. Although a small grain structure with around 0.3 μm diameter was seen in the SEM micrograph, indicating a polycrystalline nature of the film, the RHEED pattern confirmed a strong c-axis orientation perpendicular to the film plane. An X-ray diffraction pattern also confirmed this orientation.

Figure 2 shows the resistivity versus temperature relation for an as-grown film. Zero resistivity was obtained at around 80 K. Temperature dependence of the critical current density is shown in Fig. 3.

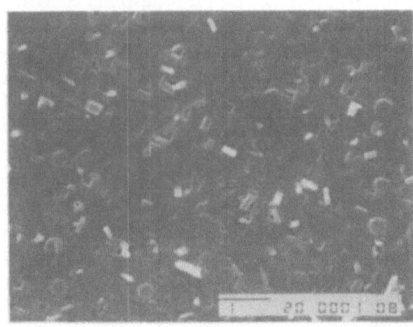

Fig. 1   SEM micrograph and RHEED pattern of YBCO thin film on $SrTiO_3$ (100) substrate

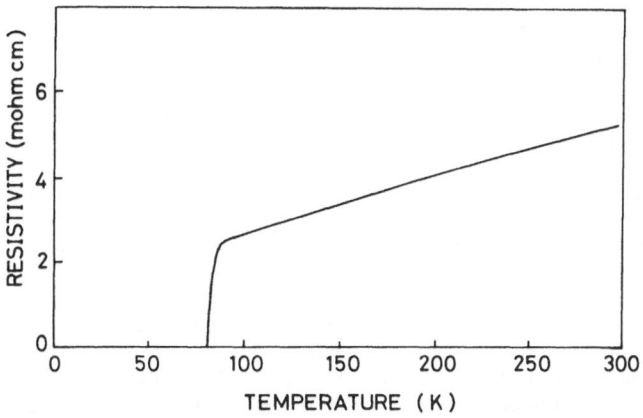

Fig. 2   Resistivity versus temperature for an as-grown film

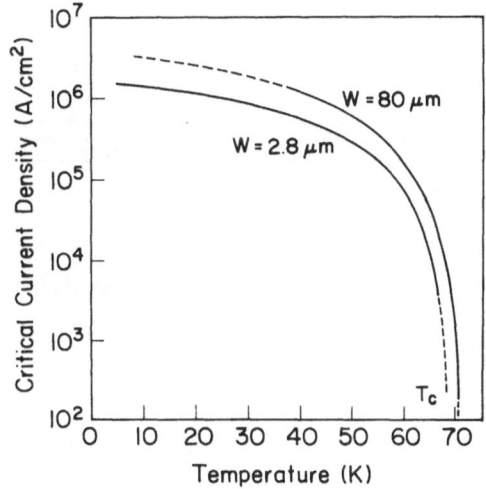

Fig.3   Critical current densities versus temperature for stripe pattern
with different width W

The measurements were performed with a narrow stripe pattern fabricated
by a wet etching process which will be described in Sec. IV. The
critical current density exceeds $1x10^6$ A/cm^2 at about 25 K and 45 K below
the transition temperature for the 80 μm wide and 2.8 μm wide stripe
patterns, respectively. Slight decreases in the critical temperature and
the critical current density were observed with the decrease in the
stripe width. This is attributed to the polycrystalline nature of the
film.

III. ELECTRIC PROPERTIES AT METAL-YBCO INTERFACE

The electric properties at the metal-YBCO interface were investigated by comparing the temperature dependence of resistance obtained from conventional four-probe measurements and that from two-probe measurements using gold and silver as electrodes. The sample structures are shown in Fig.4. Gold and silver electrodes with 0.5 mm diameters were prepared on the film surface by vacuum evaporation. Electrical leads were attached to the electrode using indium, one lead per electrode for the four-probe measurements and two leads per electrode for the two-probe measurements.

Figure 5 shows an example of the temperature dependence of the resistance, where the solid and broken curves are for the two-probe and four-probe measurements, respectively. The two-probe and four-probe results show a resembling temperature dependence, except for their absolute values. The two-probe resistance was about three times larger than the four-probe resistance above the critical temperature. This difference in the apparent resistance value is attributed to the difference in the current distribution, that is, the spreading resistance near the electrodes.

The small residual resistance below the critical temperature was examined from the current-voltage (I-V) characteristic on an oscilloscope.

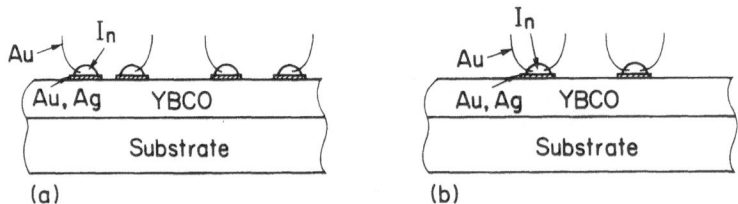

Fig.4 Sample structures for (a) four-probe measurement and (b) two-probe measurement

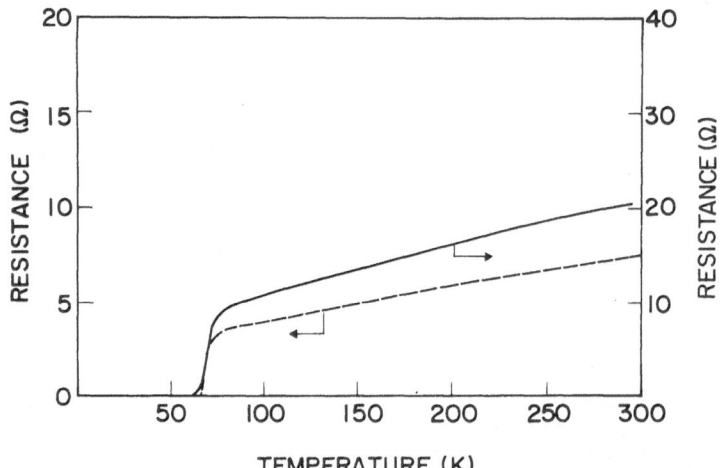

Fig.5 Resistance versus temperature curves obtained by two-probe measurement (solid curve) and by four-probe measurement (broken line)

An example of the I-V characteristic below the critical temperature of YBCO is shown in Fig. 6. The residual resistance was derived from the inclination of the I-V characteristic below the critical current of the film. As shown in Fig. 7 with black and open circles, detectable resistances were observed for the Au/YBCO interface. The annealing process between 500°C and 850°C in an oxygen atmosphere after electrode formation could not reduce the resistance value to below the detection limit of our apparatus. An almost temperature independent ((a),(b) in Fig. 7) behavior was observed for the residual resistance at the Au/YBCO interface. These results indicate the existence of a thin insulating layer through which electrons can tunnel at the interface. On the other hand, a very low residual resistance was observed for the silver electrode specimen annealed at 550°C in an oxygen atmosphere, as shown by triangles in Fig. 7. The shaded region in Fig. 7 indicates the region where the resistance of the indium contact could not be neglected. This result drived us to the hypothesis that superconductivity may come out to the surface of YBCO and that the proximity effect may be expected between YBCO and silver. Unfortunately, however, a ball-up phenomenon occurred for silver during the annealing process. This made the silver surface very inhomogeneous.

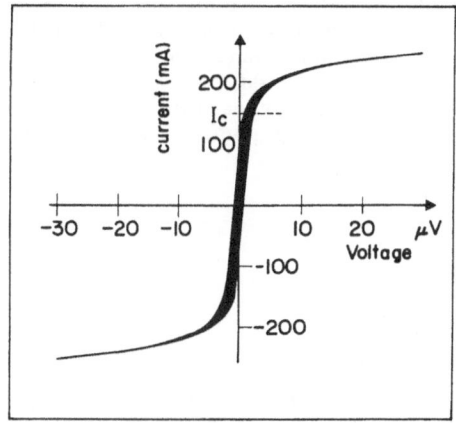

Fig. 6 Current-voltage characteristic obtained for two-probe measurement.

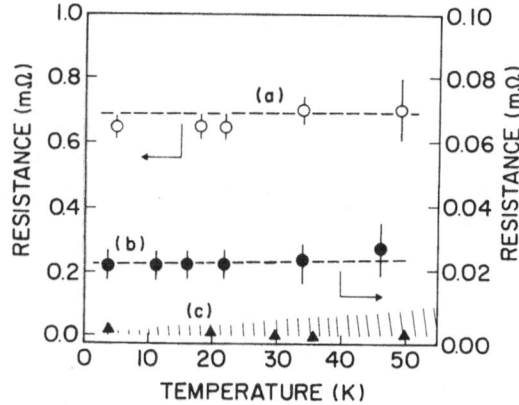

Fig. 7 Contact resistance between gold and YBCO ((a)-(b)) and between silver and YBCO (c) at low temperatures

In order to test the hypothesis, tunnel junctions with lead-indium alloy as a counter electrode was fabricated on the Ag/YBCO multilayers. This will be described in the next section.

IV. FABRICATION OF SMALL JUNCTIONS

Small junctions ranging from 20x20 μm to 100x100 μm were fabricated with a photolithography process to investigate the feasibility of Josephson junctions on YBCO thin films with a Pb-In alloy as the counter electrode. Fig. 8 is a schematic representation of the fabricated junction. The fabrication process started with a wet etching of YBCO thin films using dilute nitric acid. After the etching, silver was evaporated onto the places where the junction and the bonding pads were to be made. The patterning of the silver film was performed using the conventional lift-off process. The silver film thicknesses were 1000 Å and 3000 Å for the junction and the bonding pads, respectively. The film was annealed at 550 °C in oxygen flow for two hours. The junction area was defined with a negative resist and the resist film was made to remain so as to work as an interlayer isolation dielectric. Then, a Pb-In alloy was evaporated and patterned with the lift-off process to provide a counter electrode. An artificial barrier of thermally oxidized aluminum was introduced between the Pb-In alloy and the Ag/YBCO multilayer. No substantial difference in the junction characteristics was observed with the variation in the oxidation condition. This is probably due to the inhomogenity of the junction caused by the ball-up of the silver during the heat treatment. The fabrication process completed with the deposition of gold on the bonding pads. Figure 9 shows a microphotograph of the fabricated junction.

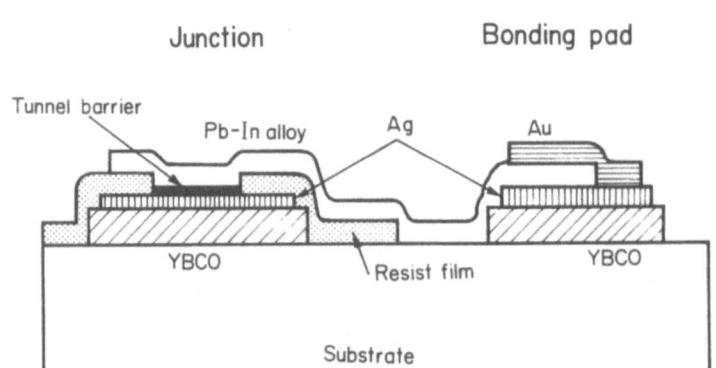

Fig. 8 Schematic representation of small junction fabricated by photolithography process

Figure 10 shows the I-V characteristic observed for a 40x40 μm junction at 4.2 K. A dc "superconducting current" with an undetectable voltage drop (<1 μV) and a hysteretic I-V characteristic with a voltage jump amounting to 20 mV were observed. The critical current Ic and the minimum finite voltage current Im at which the I-V curve returned to the

Fig. 9 Microphotograph of fabricated small junction

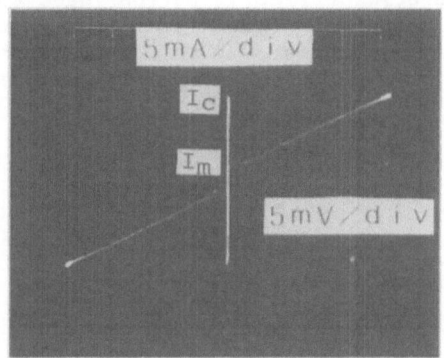

Fig. 10 I-V characteristic observed for 40x40 μm junction at 4.2 K

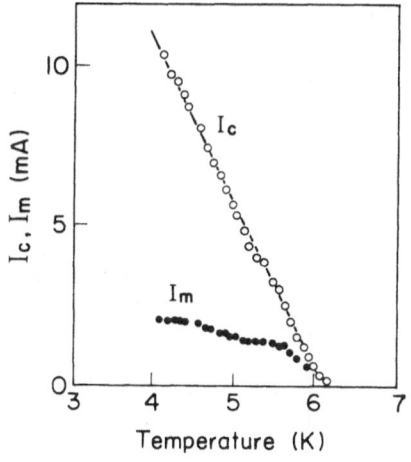

Fig. 11 Temperature dependences of critical current Ic and
minimum finite voltage current Im

"superconducting current" state are plotted in Fig. 11. The critical current showed a linear dependence on Tc-T and no saturation was observed down to 4.2 K. This is in contrast to the Ambegaokar and Baratoff theory for tunnel junctions provided that the tunnel junction is formed between lead and either YBCO or the Ag/YBCO multi-structure with a large gap energy. A plausible explanation is that the junction has a weak link structure or is a mixture of a fine weak-link and a tunnel junction. If this is the case, the hysteretic behavior is the result of a finite capacitance value associated with the junction, and the junction characteristic may be well described by a simple resistively shunted junction (RSJ) model. As a test of this probability, the McCumber's admittance ratio $\beta = 2eIcC/hG$ was calculated as a function of the observed hysteresis parameter $\alpha c = Im/Ic$ by assuming a constant conductance and capacitance value, and the result was compared with the theoretical one[8]. In this calculation, the conductance value G at around 20 mV, which was 0.53 mho, was used and the capacitance value C was treated as a fitting parameter. Best fitting was obtained for C=0.33 pF, as shown in Fig.12, where the solid line shows the McCumber's theoretical curve and circles represent the experimental values. Although a very good agreement was obtained between the theory and the experiment, care must be taken to make a fit to the McCumber's theory as discussed by Skocpol et al[9]. They showed that the self-heating effect in microbridges caused hysteresis in the I-V characteristic and that the hysteresis parameter depended on $Ic^{-1/2}$. This dependence is shown in Fig. 12 as a broken line. Because of the resemblance of the two theoretical curves, it is very difficult to distinguish to which curve the experimental values fit better.

Although uncertainty remains on the origin of the hysteresis, observation of the dc "superconducting current" with undetectable resistance ($<1 \times 10^{-8}\ \Omega cm^2$) is very encouraging. It seems to ensure the superconducting contact between the Pb-In alloy and YBCO. This fact also

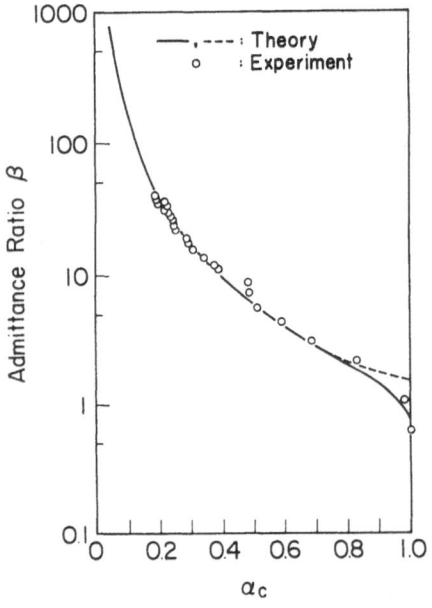

Fig. 12
Comparison of β versus αc relation between experiment (open circles) and McCumber's theory (solid line). The broken line shows the result of the self-heating hotspot model.

suggests that the thin silver layer on the YBCO can prevent the pronounced reaction between the YBCO and the Pb-In alloy which is often observed between YBCO and easily oxidized metals. The development of a homogeneous junction structure is needed to confirm the possibility of a superconducting contact between YBCO and other superconductors. Plasma oxidation of the YBCO surface prior to metal deposition may provide a way to realize a homogeneous Josephson junction with a metal buffer layer which acts as a barrier to supress the reaction between YBCO and other materials.

V. CONCLUSION

Electrical contact properties between noble metals and YBCO were investigated using thin films prepared by rf magnetron sputtering. It was revealed that an insulating layer was formed between gold and YBCO, whereas the contact resistance between silver and YBCO was less than the detection limit ($<1 \times 10^{-8} \Omega \, cm^2$). Based on the result, tunnel junctions were fabricated on the silver and YBCO multi-layer with lead as a counter electrode. Small junctions fabricated with a photolithography process exhibited a dc "superconducting current" with undetectable resistance below the critical temperature of lead, and a hysteretic behavior was observed in the I-V characteristic. These results demonstrate the feasibility of Josephson junctions on YBCO thin films.

REFERENCES

1. H. Adachi, K. Hirochi, K. Setsune, M. Kitabatake and K. Wasa, Low-temperature process for the preparation of high Tc superconducting thin films, Appl. Phys. Lett. 51, 2263 (1987)
2. T. Terashima, K. Iizima, K. Yamamoto, Y. Bando and H. Mazaki, Single-crystal $YBa_2Cu_3O_{7-x}$ thin films by activated reactive evaporation, Jpn. J. Appl. Phys. 27, L91 (1988)
3. X. D. Wu, A. Inam, T. Venkatesan, C. C. hang, E. W. Chase, P. Barboux, J. M. Tarascon and B. Wilkens, Low-temperature preparation of high Tc superconducting thin films, Appl. Phys. Lett. 52, 754 (1988)
4. A. Nakayama, A. Inoue, K. Takeuchi and Y. Okabe, Y-Ba-Cu-O/AlOx/Nb Josephson tunnel junctions, Jpn. J. Appl. Phys. 26, L2055 (1987)
5. H. Akoh, F. Shinoki, M. Takahashi and S. Takada, S-N-S Josephson junction consisting of Y-Ba-Cu-O/Au/Nb thin films, Jpn. J. Appl. Phys. 27, L519 (1988)
6. T. Miura, Y. Terashima, M. Sagoi, K. Kubo, J. Yoshida and K. Mizushima, Effects of oxygen partial pressure on properties of Y-Ba-Cu-O films prepared by magnetron sputtering, presented at the American Vacuum Society 35th National Symposium, Atlanta (1988)
7. V. Ambegaokar and A. Baratoff, Tunneling between superconductors, Phys. Rev. Lett., 10, 486 (1963)
8. D. E. McCumber, Effect of ac impedance on dc voltage-current characteristics of superconductor weak-link junctions, J. Appl. Phys., 39, 3113 (1968)
9. W. J. Skocpol, M. R. Beasley and M. Tinkham, Self-heating hotspots in superconducting thin-film microbridges, J. Appl. Phys., 45, 4054 (1974)

NON-AQUEOUS ONE-MICRON FEATURE SIZE LITHOGRAPHY

OF SUPERCONDUCTING FILMS

R. Boerstler, C. Carey, and A. Poirier

Sanders Associates, A Lockheed Company
95 Canal Street
Nashua, New Hampshire 03061

Y. Huang, J. Ryu, and C. Vittoria

Center for Electromagnetics Research
Northeastern University
Boston, Massachusetts 02115

Conventional lithography was used to pattern $Y Ba_2 Cu_3 O_{6+x}$ thin films. The effort is focused on application of equipment and materials commonly used in industrial microelectronic facilities. Techniques for fabricating medium scale integrated superconducting circuits with minimum feature sizes of one micron have been successfully developed under this program. Attention has been turned toward building devices (SQUID) and microstrip structures to evaluate the effect of film properties on device performance. Test patterns currently used to guide process development consist of a one micron weak link SQUID and coplanar microwave resonators.

A totally non-aqueous process using PMMA (496k molecular weight), deep uv (220nm) exposure and ion beam milling has been developed. An approximately 1.3 micron thick PMMA layer is required to etch a one micron thick $YBa_2Cu_3O_{6+x}$ film sufficiently far into the ZrO_2 substrate to remove the conducting interface/substrate layer. The exposed PMMA is patterned by dissolving the irradiated PMMA with methyl isobuytl ketone.

INTRODUCTION

Methods for fabricating patterns in thin film superconductors are needed to implement electronic devices. Fine line patterns down to the micron feature size level are needed to obtain improved performance from high temperature superconducting films relative to the metallic line currently used. The short superconduction coherence length, together with the reactivity of the material and the sensitivity of grain boundaries in the high temperature superconducting materials requires that processing chemicals and techniques be carefully chosen.

A conventional resist, optically exposed, developed, and etched with an aqueous acid solution[1,2] is the fastest and most inexpensive way to proceed. However, water tends to react with the superconductor[3,4]. This can be particularly deleterious to microwave applications where the first 1000 A of superconducting material is important. A thin layer of low Tc material can prevent the realization of low surface resistance superconducting lines. The high Tc superconductors etch rapidly in acid etches, which makes fine line resolution difficult. A number of direct writing processes have been explored which track a focused beam over the sample to create a pattern in high-Tc superconducting film. This includes laser[5,6,7] as well as ion beam[8] techniques. Direct writing techniques are slow and mechanically complex. Some damage is always induced in superconductors adjacent to the edges of the lines drawn.

Intermediate in speed and cost between direct writing and conventional wet chemistry is ion beam milling. Recently at Sanders and elsewhere[9,10] ion beam patterning of high Tc superconducting films has been carried out. Conventional resist systems can be optically exposed and developed with non-aqueous solvents. The superconductor is then milled away in the unprotected areas. The ion beam milling approach described here has been taken down to under one micron feature size. Other investigations have demonstrated 5 micron[10] and 2 micron[9] feature sizes.

The advantages of the ion beam etching are:

- Moderate speed
- Nonaqueous processing
- Excellent linewidth control on submicron patterns
- Dry etching
- Excellent vertical definition
- Amenable to large area circuits (2" wafers currently)

The effort reported below details an ion beam milling technique for performing one micron level lithography on 1 cm^2 superconducting film.

PATTERNING

The major thrust of the present effort was to employ commonly used microfabrication techniques which might serve as the basis of commercial processes. Initially, the major concern was to develop a low temperature non-aqueous process. However as the work proceeded, it became clear that certain steps were critical in order to pattern the film without destroying its superconducting properties. The film had to be protected from direct impact of the ion beam on the bare superconducting surfaces. In addition, any resist remaining after processing had to be removed with a process which did not degrade the superconducting lines. A low temperature oxygen plasma ash procedure proved to be successful for the final resist removal step.

The process flow is presented in Table 1. The development of the patterns in the superconducting films is accomplished with an ion beam etch process. Subsequent metalization is accomplished with lift off process. A separate mask is required for each of the two patterning steps. The poly-methyl methacylate spun on to the sample is used for all steps requiring a resist.

A two level PMMA-coplanar system is used for the ion beam milling resist to overcome problems associated with the surface roughness characteristics of the superconducting films. A SEM photomicrograph of the as received superconducting films is shown in Figure 1(a). The film has a platalete structure with a surface roughness of about ± 0.25 microns as illustrated by the surface profilometry trace in Figure 1(b). The average thickness of the samples ranged between 0.8 - 1.4 microns. The combined effect of roughness and average thickness variation across a one cm chip tend to expose the high portions of the superconductor to ion beam. This problem was solved using a planarizing layer of copolymer (PMMA-MAA) under the final soft baked PMMA resist layer.

TABLE 1. Fabrication Sequence for Patterning
$YBa_2Cu_3O_{6+x}$ Superconducting Films.

1. Ion Beam Patterning
    - Surface profile sample
    - Spin coat P(MMA-MAA) Copolymer
    - Air convection bake 160 degrees C 30 minutes
    - Spin coat 9% PMMA (Polymethylmethacrylate)
    - Air convection bake 160 degrees C 30 minutes
    - Image resist by vacuum contact. Exposure wavelength 220-240 nm.
    - Develop PMMA; Methyl Isobutyl Ketone:Isopropanol
    - Develop P(MMA-MAA); Ethoxyethyl Acetate:Ethanal
    - Ion mill 30 minutes, 1000 Volts, 0.5 mA beam current
    - Oxygen ash remaining resist

2. Liftoff Metalization
    a. Double-layer Resist
        - Spin coat P(MMA-MAA) Copolymer
        - Air convection bake 160 degrees C 30 minutes
        - Spin coat 9% PMMA
        - Air convection bake 160 degrees C 30 minutes
        - Image resist by vacuum contact. Exposure wavelength 220-250 nm.
        - Develop PMMA; Methyl Isobutyl Ketone:Isopronopanol
        - Develop P(MMA-MAA); Ethyoxyethyl Acetate:Ethanol

    b. Contact Formation
        - Planar oxygen ash
        - Electron beam/sputter evaporate one micron of gold
        - Solvent liftoff
        - Planar oxygen ash
        - Sinter gold contacts: 300 degrees C, 1 hour, oxygen atmosphere
        - Measure contact resistance

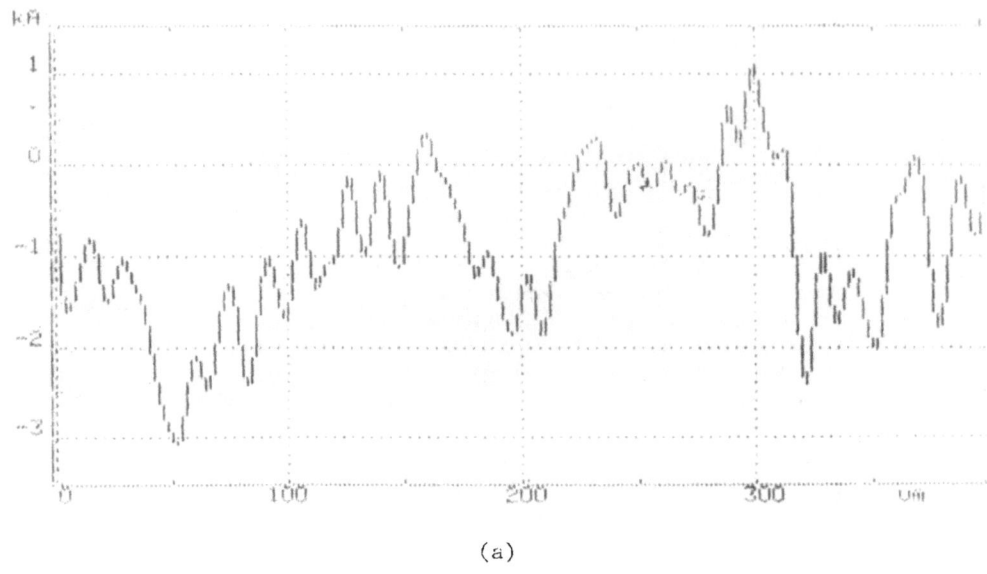

(a)

Distance

(b)

Figure 1. Surface Profile Data and SEM Micrograph
Depicting Surface Roughness of $YBa_2Cu_3O_{6+x}$
Superconducting Thin Film

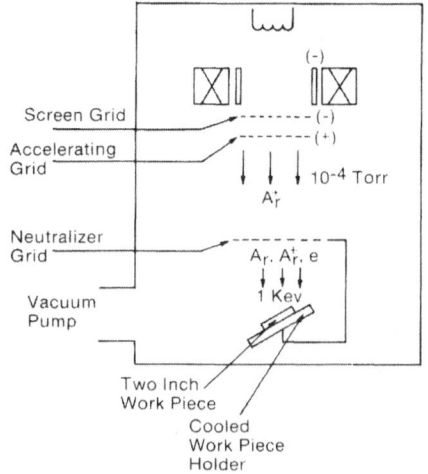

Screen Grid

Accelerating Grid

Neutralizer Grid

Vacuum Pump

(-)

(-)

(+)

10^{-4} Torr

A_r^+

A_r, A_r^+, e

1 Kev

Two Inch Work Piece

Cooled Work Piece Holder

Figure 2. Ion Beam Milling System for
Patterning Superconducting Films

Figure 3. Microwave Circuit Patterned
in $YBa_2Cu_3O_{6+x}$ Superconductor

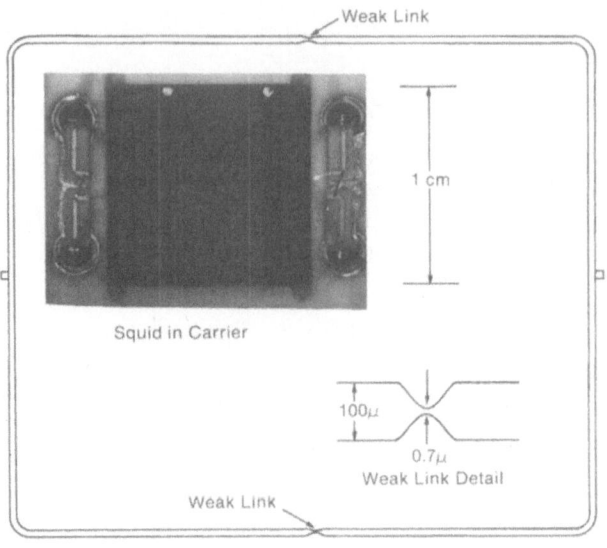

Figure 4. High Temperature Superconducting
SQUID Device Mounted in Test Fixture

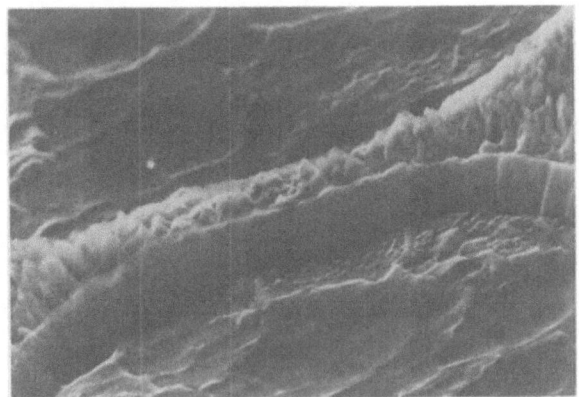

Figure 5. SEM Micrograph of
Weak Link Details

The PMMA system allowed uv exposure for sharp definition (220-250nM) and non-aqueous (Methyl Isobytul Ketone:Ethanol) pattern development. The result of the pattern development is a positive image coating of PMMA-MAA over the portions of the superconductor to be saved.

The unwanted superconductor was sputtered away in an ion beam mill apparatus shown in Figure 2. A Kaufman source generates a flux of argon ions in a low pressure chamber. A neutralizer grid is used to discourage substrate charging. The substrate is mounted on a cooled substrate holder. A 90° angle of incidence was employed in all the work to date.

Etch rate experiments were performed on the 9% PMMA, P(MMA-MAA), and $YBa_2Cu_3O_{6+x}$ superconductor. The ion mill conditions for this experiment listed in Figure 2. The beam current density is 0.5 mA/cm^2. The samples were mounted on a water cooled substrate holder which maintained the substrate temperature below 100°C. The sample was rotated about an axis normal to the beam axis at two revolutions per minute. The component material etch rate measured under these conditions are:

PMMA, Polymer resist	275 A/min
PMMA P(MMA-MAA) copolymer	300 A/min
$YBa_2Cu_3O_{6+x}$	300 A/min

Sufficient copolymer thickness is applied under the PMMA resist to allow the potentially conducting diffusion layer formed during film growth to be removed without exposing the top surface of the superconducting pattern.

FABRICATION RESULTS

Test patterns which demonstrate image area patterns and small linewidths were created in superconducting films using the ion beam based dry etch approach. A microwave FET metalization mask was available which allowed a wide variety of linewidths and shapes to be milled into the superconducting film. The results of ion beam milling on a one centimeter square superconducting film is shown in Figure 3. The mask pattern was faithfully reproduced over the entire chip down to 0.7 micron feature size.

A weak link was fabricated to make a minimum feature size element with SQUID-like behavior. The details of this pattern are shown in Figure 4. A roughly one centimeter loop (100 microns wide) of superconductor is symmetrically interrupted with a 10 micron long neck. The minimum width at the neck is 0.7 microns.

A detail of the neck region is shown in the SEM photomicrograph in Figure 5. The 0.25 micron surface roughness is clearly visible on the top surface of the superconducting line. The sharp vertical definitions should be noted in this photograph. In addition, the roughness in the original surface is replicated to some extent in the base substrate.

CONCLUSIONS

Microfabrication processes for patterning and metallizing superconducting films without destroying the superconducting properties of the films has been demonstrated. High resolution (0.2 micron limit of resolution) uv defined polymethyl methacrylate resist can be used to pattern large area (2") superconducting films. Residual resist can be cleaned off with a low temperature oxygen plasma ash technique. A copolymer planarization and O_2 plasma can be used to do fine lithography in the presence of large surface roughness in as grown high Tc superconducting films. Minimum feature sizes of the order of one micron have been demonstrated. However, linewidth down to the order of 0.2 microns should be possible using the ion beam milling approach to film patterning.

REFERENCES

1. H. Yamada, U. Kawabe, DC Squid with High-Critical-Temperature Oxide-Superconductor Film, Jap. Jour. of App. Phys. 26:L1925 (1987)

2. M. Tonouchi, Y. Sakaguchi, T. Kobayshi, Chemical Etching of High-Tc Superconducting Films in Felcox-115 solution, Jap. Jour. of App. Phys. 27:L98 (1988)

3. N.P. Bangal, A.L. Sandkuhl, Chemical Durability of High-Temperature Super-conductor $YBa_2Cu_3O_6$-x in Aqueous Environment, App. Phys. lett. 52:323 (1988)

4. M.F. Yan, R.L. Barns, H.M. O'Bryan, Jr., P.K. Galligher, R.C. Sherwood, S. Jin, Water Interaction with the Superconducting $YBa_2Cu_3O_7$, App. Phys. lett. 51:532 (1987)

5. J. Mannhart, M. Scheverman, C.C. Tsuei, M.M.Oprysko, C.C. Chir, C.P. Umbaeh, R.H. Koch, C. Miller, Micropatterning of High Tc Films with and Excimer laser, App. Phys. Lett. A 46:331 (1988)

6. G. Liberts, M. Eyett, D. Bauerle, Direct Laser Writing of Superconducting Patterns into Semiconducting Ceramic Y-Ba-Cu-O, App. Phys. Lett. A. 46:331 (1988)

7. H.C. Pandey, Y.K Jain, S.K. Bhatnagar, B.R. Singh, W.S. Khokle, Direct Laser Beam Writing on YBaCuO Film for Superconducting Microelectri Devices, Jap. Jour. App. Phys. 27:L1517 (1988)

8. S. Matsui, Y. Ochai, Y. Kojima, H. Tsuge, N. Tsuge, N. Takado, K. Asakawa, H. Matsuteva, J. Fujita, T. Yoshitake, Y. Kubo, Focused Ion Beam Processes for High Tc Superconductors, J. Vac. Sci. Technical, B6:900, (1988)

9. A. Enokihara, H. Higashino, K. Setsune, T. Mitsuyu, K. Wasa, Superconductivity in a 2 micron Wide Strip line of Gd-Ba-Cu-O Thin Film Fabricated by Low Temperature Process, Jap. Jour. App. Phys. 27:L1521 (1988)

10. G.C. Hilton, E.B. Harris, D.J. Van Harlingen, Growth, Patterning, and Weak-link Fabrication of Superconducting $YBa_2Cu_3O_{6-x}$ Thin Films, Appl. Phys. lett. 53:1107 (1988)

LIMITATIONS DUE TO THE CRITICAL CURRENT DENSITY IN SUPERCONDUCTING LINES FOR FUTURE VLSI PACKAGING

Bernard Flechet and Jean Chilo

BULL and LEMO/INPG/CNRS

LEMO, 23 Avenue des Martyrs 38031 Grenoble Cedex, France

INTRODUCTION

At hardware level the speed of complex high performance electronic system is limited by two basic componenents:
- Integrated circuit chips.
- Interconnections between chips.

Today, the constraint of an increasing speed of chip to chip communication has driven to the improvement of interconnection and packaging techniques towards high density [1,2]. This goal can be achieved by scaling down sizes and pitches of the lines. Using conventional lines (Al or Cu), several limitations [3] arise from the increasing resistance and from the stronger coupling between adjacent lines (inducting distorsion and noise immunity degradation). Thus, the possibilities at 77 K of new high critical temperature superconductors [4,5] are explored because superconducting lines present a very low series resistance [6] without incurring the penalties of the scaling rule [7]. The cross-section of the lines will be reduced down to another limitation arising from the critical current density (Jc). It is the maximal current density that these films can carry without destroying their superconductive state.

Therefore, it is important to investigate the current density distribution and the influence of edges and proximity effects. In this paper, the principal propagating characteristics (77 K) for superconducting (YBaCuO) and normal (Al or Cu) lines are compared when their dimensions are scaling down. Next, the consequences on the current density distribution are pointed out and the limitations due to Jc are shown (assuming $Jc=10^6$ A/cm^2).

In order to make some kind of comparisons among various interconnection approaches, a standard line geometry is defined (Fig. 1). This particular structure results in a characteristic impedance of about 50 Ω. The principal parameters of the structure are the microstrip width (W) and the spacing between lines (S).

THEORETICAL BACKGROUND

Electrical Modeling

In the simplified model of transmission lines (Fig. 1), we assume parallel microstrip lines with an infinite length in the z-direction. In this condition, the quasi-TEM analysis can be used if the wavelength of fields traveling along the structure is more important than the cross-section of the lines . Thus, the current density is in the z-direction and depending only of x and y directions.

The Potential Method[8] is used in order to calculate series impedance (Z=R+jωL) and shunt admittance, reduced to the capacitive admittance (Y=jωC) assuming a lossless dielectric. These quantities are matrices when coupling phenomena exist and matrix equations are solved with a Moment Method [9]. The Potential Method takes into account proximity effects and fringing effects at the edges of the lines. In addition, the

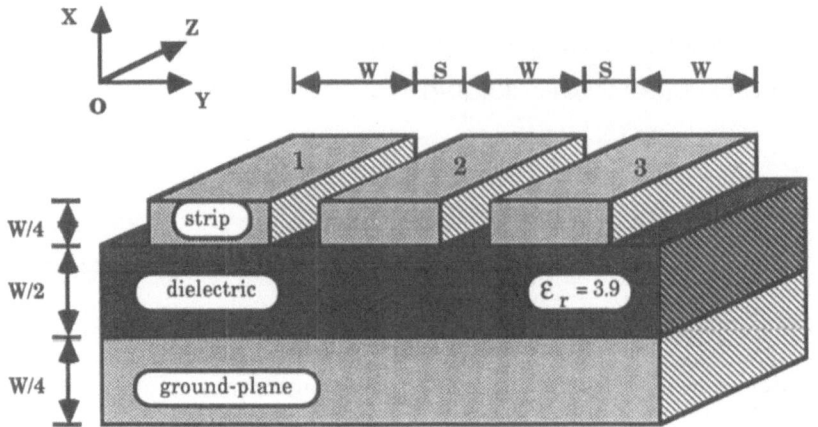

Fig. 1 Standard line structure.

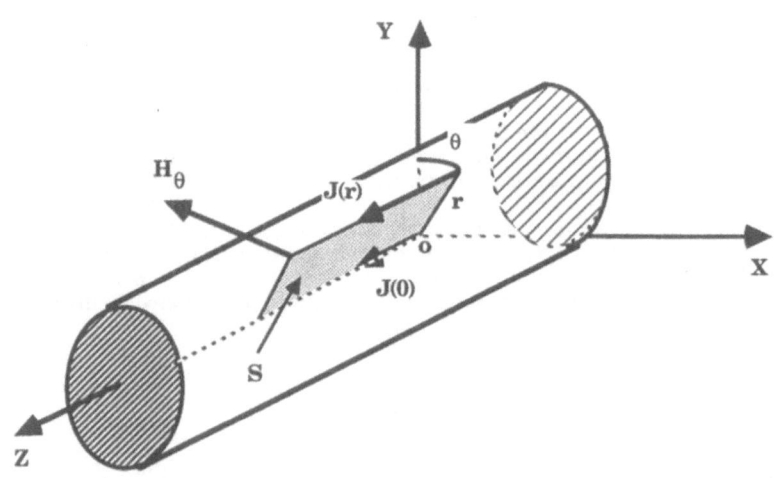

Fig. 2 Simplified model of transmission line.

current density distribution can be determined in the cross-section of the lines by imposing the driving conditions.

Superconductor characteristics

The difference between conventional conductor and superconductor is introduced at the conductivity level, i.e in the series impedance Z. We suppose that the new high Tc superconductors can be characterized with the theory of London and two-fluid model [10]. In addition, we don't take into account the high Tc superconductor anisotropy which may appear, for example, in the London's penetration depth and in the critical current density. [11,12]. Thus, the conductivity is complex and given by:

$$\sigma^* = \sigma_n + j\sigma_{sc}$$

The real part of σ^* is the conductivity of normal electrons, related to the losses in superconducting film. At temperature T, σ_n is given by:

$$\sigma_n = \sigma_{nc} \, (T/Tc)^4$$

where σ_{nc} is the normal state conductivity just above the transition temperature (Tc). The imaginary part of the complex conductivity arises from London's equations:

$$\sigma_{sc} = [1-(T/Tc)^4] \, / \, \omega\mu_0\lambda_L^2(T=0 \text{ K})$$

where λ_L is the penetration depth of London for superconductor:

$$\lambda_L (T) = \lambda_L (T=0 \text{ K}) \, [1-(T/Tc)^4]^{-1/2}$$

This imaginary conductivity accounts for energy stored in the film[13]. The characteristic relation of the two-fluid model is given by the total current density:

$$J_{total} = J_n + J_{sc} = \sigma^* E$$

As J is in the z-direction (Fig. 2) and using London's equations, the integration of the Maxwell's equation on a rectangular area S(r) brings to:

$$J_{sc}(r) - J_{sc}(0) = - (\frac{1}{\lambda_L})^2 \int_0^r H_\theta dr$$

Thus, the superconductive current density is allocated to an inhomogeneous distribution in the microstrip line. It is very important to note that this current distribution is identical with dc and ac current. As shown in this equation, the distribution shape is imposed by:

- The material parameters (λ).
- The sizes of superconducting lines (integral boundary).
- The surrounding parameters (Hθ). This magnetic field component arises from the nature, the shape and the distance of the other lines and from their current distribution.

The same calculation for normal conductor can be made with Ohm's equation. In this case, the $\partial/\partial t$ operator appears in addition to the influencial parameters σ, r and Hθ. This operator traduces the time-frequency dependence of the normal current distribution (skin effect).
The two-fluid model is accurate for frequencies much below the gap-frequency [14]. For this work, as in very high speed digital system, frequencies are limited to 100 GHz which is far below the $\approx$6.8 THz gap-frequency of YBaCuO [15].

Table 1. Used parameters in examples of transmission lines.

type	model	strip	dielectric	ground plane	Tc K	λ (77K) μm	ρₙ (77K) μΩm
super-conductor	1	YBaCuO		YBaCuO	92.5	0.1941	4.16
	2	YBaCuO	$\varepsilon_r = 3.9$	YBaCuO	92.5	0.1941	4.16
	3	YBaCuO		YBaCuO	92.5	0.1941	4.16
normal conductor	1	Al		Al	-	-	0.00667
	2	Al	$\varepsilon_r = 3.9$	Al	-	-	0.00667
	3	Cu		Cu	-	-	0.00192

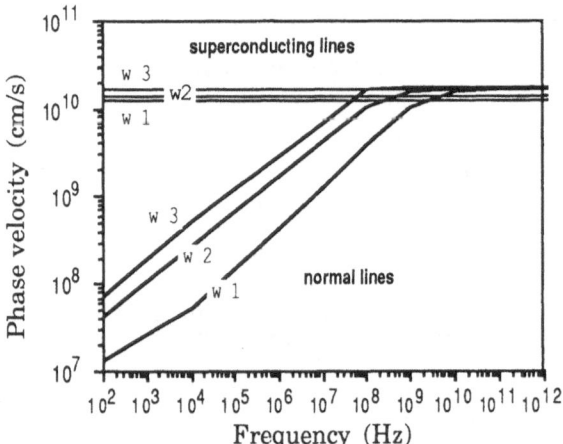

Fig. 3   Phase velocity at 77 K as a function of frequency for 3 models (w1=2 μm, w2=7 μm, w3=20 μm).

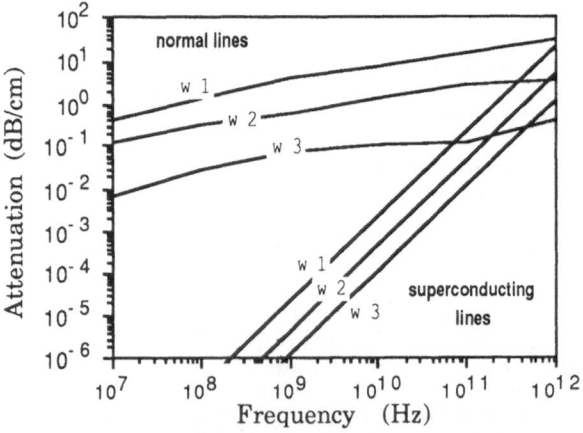

Fig. 4   Attenuation at 77 K as a function of frequency for 3 models (w1=2 μm, w2=7 μm, w3=20 μm).

Material parameters listed in table 1 are used in our calculations[15,16] and three line structures are investigated :

- Model 1 with W1= 2 µm.
- Model 2 with W2 = 7 µm.
- Model 3 with W3 = 20 µm.

The propagation characteristics (phase velocity vø and attenuation α) at 77 K are ploted in Fig.3 and in Fig.4 respectively. Fig.3 shows that the phase velocity is independent of frequency up to 1 THz for superconducting lines. Thus, these lines are non dispersive. The phase velocity of normal conductor lines is much less than that of a superconducting line for frequencies below 100 MHz and the very large dispersive character up to this frequency range is pointed out.

We should note the effect of the increasing kinetic inductance for superconducting interconnects[17]. When the dimensions are scaling down, the phase velocity is reduced because of the increasing total inductance. The charcteristic impedance (Zo) is affected in the same way:

- Model 1: L_{total} = 0.374 µH/m and Z_0= 48.8 Ω.
- Model 2: L_{total} = 0.296 µH/m and Z_0 = 50.1 Ω.
- Model 3: L_{total} = 0.282 µH/m and Z_0 = 56.3 Ω.

This fact appears when the penetration depth becomes no negligible in front of the dielectric and superconductive layers.

The attenuation of superconducting lines increases as the increasing frequency square (Fig. 4). However, this attenuation remains small, below 10^{-3} dB/cm, at frequency up to 10 GHz for all models. The best compromise between attenuation and high density possibilities is obtained with the smallest superconductive structure. In this case, dispersion and loss are negligible for length and frequency ranges of very high speed systems. Thus, superconducting interconnects with cross-section of 2 x 0.5 $µm^2$ should be able to operate with low loss and low dispersion[18].

CURRENT DENSITY DISTRIBUTION STUDY

The current density flowing down the line increases as we reduce their transversal dimensions. In addition, the fringing fields effects at the edges of the wires are increasing when sizes are scaling down. This is due to the London's penetration depth[8].Thus, it is important to investigate the current density distribution in the lines in order to determine the influence of dimensions and proximity effects. Using the Potential Method , a complex current density is obtained. The imaginary part arises from the normal conductivity existence. It should be noted that real and imaginary parts of J are not exactly fitted with superconductive and the normal current given by the two-fluid model. But the ratio:

Q (f) = imaginary (J) / real (J)

is representative of the normal current density in the line.

Single line (response at 1mA)

The current density distribution have been determined for single driving lines (Fig.1 with I2=1 mA and I1=I3=0). According to equation (1), the frequency is not influencial on the superconductive current distribution. Up to 100 GHz, Q (f) is smaller than 3 x 10^{-2} and the normal current is negligible in front of the superconductive current. The greater part of the current is located near the lower microstrip plane and strongly increasing with x and y directions near the edges.

In Fig. 5 and 6 are shown the real and imaginary current distribution in the case of model 2 (W2 = 7 µm). These curves represent the current density variations for three x-planes in the microstrip and from edge to edge in the y-direction:

- (1): near the lower plane, x = 9.72 x 10^{-2} µm.

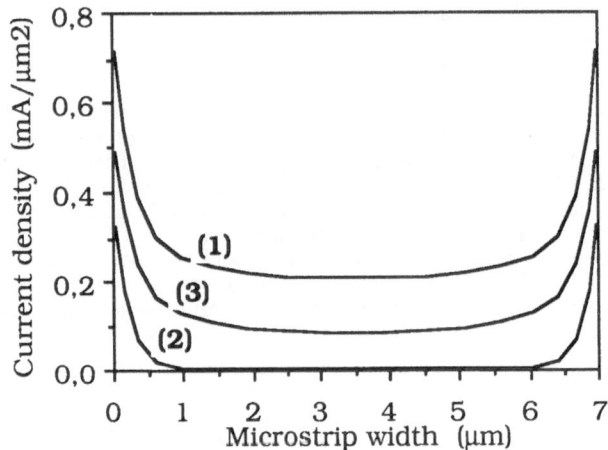

Fig. 5 Real part of the current density in the microstrip (W=7 μm)
Response at 1 mA.
(1) near the lower plane,
(2) near the middle plane,
(3) near the upper plane.

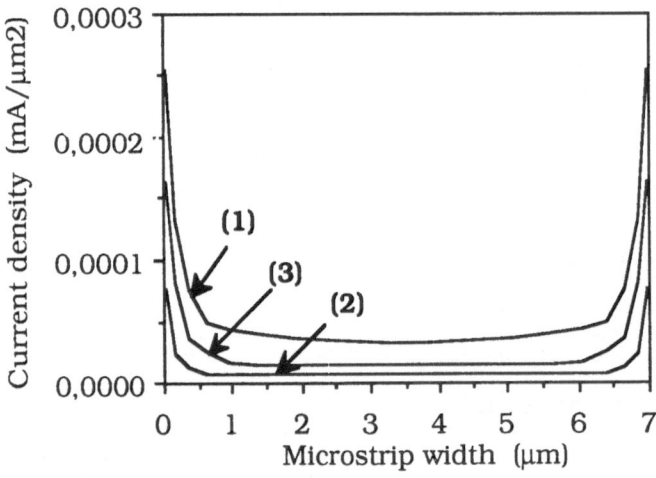

Fig. 6 Imaginary part of the current density in the microstrip (W=7 μm)
Response at 1 mA.
(1) near the lower plane,
(2) near the middle plane,
(3) near the upper plane.

- (2): near the middle plane, x = 0.875 µm.
- (3): near the upper plane, x = 1.652 µm.

where x=0 is the microstrip lower plane.
The maximal values of the current density are critical parameters in the use of superconducting lines. They are plotted in Fig. 7 as a function of the sizes of the standard line structure. If another driving current are excited instead of 1 mA, the maximal value will be propotional to those obtained with 1 mA.

Coupling effects (response at 1 mA)

The current density will be modified by the magnetic crosstalk due to the eventual other lines. In order to estimate this effect, we have imposed I1=1 mA and I3=-1 mA in the lines on both sides of the previously studied line. The new current density in this middle line is given by:

$$J_{total} = J_{single} + J_{additional}$$

where:

- J_{single} is the current density in the middle line without current in other lines.
- $J_{additional}$ is the current density arising from the coupling effects.

The additional current density in the middle line is drawn in Fig. 8 in the case of model 2 with a relative spacing S=W=7 µm. It shoud be noted that the total additional current is zero and symetrically distributed because of the proximity effects.
As before, the additional current distribution of superconducting lines is independent on the frequency and the maximal value is located near the edges. Thus, the current density and the maximal value can be determined in the case of single line or coupled lines as a function of the geometric parameters (W and S).

LIMITATIONS DUE TO THE CRITICAL CURRENT DENSITY

The reduction of superconducting wire sizes and wire pitches is stated because superconductors are only capable of supporting a limited current density. Actually Jc is in the range of 10^5 - 10^7 A/cm^2 for dc current [12,19,20]. As the current is independently distributed of the frequency, a limit to the reduction can be given by the dc critical current density.
The maximal current density in the interconnection lines is imposed by the logic driving conditions. The peak current flowing into the line, for the kind of high frequency operation that we have been considered, is simply:

$$I_{peak} = V_{signal} / Z_0$$

where:

- V_{signal} is the voltage amplitude of the logic signal.
- Z_0 is the characteristic impedance.

The value of the voltage amplitude is a very technology dependent issue. For example, CMOS circuits use 5 volts logic level and high speed bipolar circuit, like ECL, currently use 800 mV signals.

Single line

Assuming Jc=10^6 A/cm^2 (i.e 10 mA / µm^2) and a 50 Ω characteristic impedance, the smallest width Wc of the standard line geometry is 8.2 µm by using an ECL technology (Ipeak = 16 mA) . With indentical conditions, the critical wire size increases up to Wc = 50 µm for a CMOS technology. In Fig. 9 are shown the relationship between the critical sizes of a single line and the driving voltage of logic technologies. It is important to note that the critical sizes are proportional in the inverse ratio to Jc. Note also that the restrains arising from the low critical magnetic field Hc1 are less stringent than the dc critical current density one. Therefore, in assuming Hc1=500 Oe

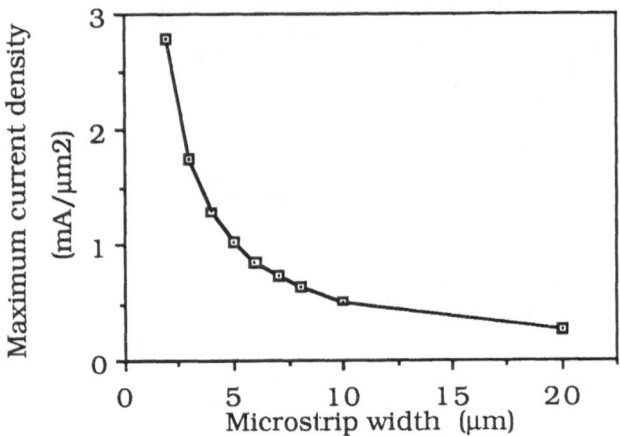

Fig. 7   Maximal values of the current density in the microstrip versus the microstrip width of the standard structure. Response at 1 mA.

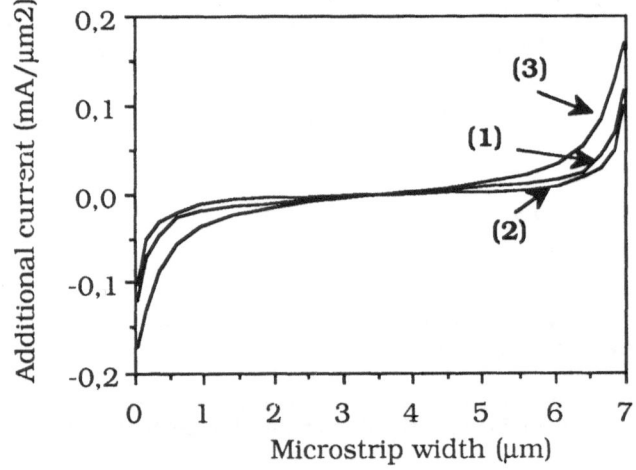

Fig. 8   Additional current in the microstrip (W=7 μm, S=7 μm).
Response at 1 mA.
(1) near the lower plane,
(2) near the middle plane,
(3) near the upper plane.

16 (linear surface current density of about 60 mA / μm^2), a line width of 10 μm should carry a surface current of 600 mA instead of about 20mA.

<u>Coupled lines</u>

The magnetic crosstalk compels to extend the wire sizes since the additional current density appears particulary near the edges where the driving current density is maximal. A critical spacing (Sc) is associated with the critical sizes (Wc). Sc and Wc can be determined in order to obtain a minimal wire pitch.

For example, the critical current density constrains to use standard line structure with Wc=14 μm and Sc=6 μm for ECL technology and Wc= 88 μm with Sc=32 μm for CMOS technology.

In the case of minimal wire pitch, Wc and Sc are plotted in Fig.9 as a function of the driving voltage (assuming Jc=10^6 A/cm^2). Thus, a high wiring density compels to use low driving voltage, below 1 V. If the driving voltage is chosen more important, the cross-section of superconducting interconnections must be greater and they do not present a significant advantage over the conventional one[7].

It is important to keep in mind that these limits are peculiar to the critical current density and do not take into account the crosstalk noise peculiar to the logic system.

SUMMARY

The critical current density may be the most important factor for small dimension superconducting interconnects. The analysis of the current density distribution compels to introduce critical wire sizes as a function of:
- The critical current density.
- The logic voltage level of the considered technoloy.
In the case of a single line, the critical sizes are proportional in the inverse ratio of the critical current density. If a coupled lines network is used, the magnetic crosstalk must be token into account in order to determine the new critical dimensions (width and spacing). A minimal wire pitch can be obtained with each driving voltage.

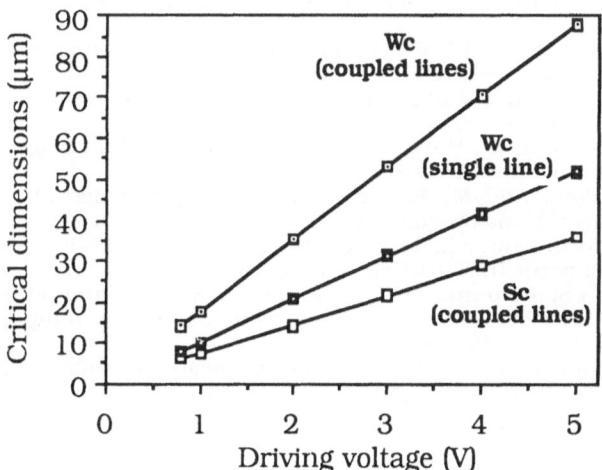

Fig. 9   Critical sizes and spacings versus the driving voltage.
Assuming Jc = 10^6 A/cm^2.
Single line: V1=V3=0.
Coupled lines: V1=V2=-V3.

In this conditions, the superconductive state is preserved. Thence, the lines are nondispersive and attenuation remains below 10^{-2} dB/cm for frequencies up to 100 GHz. But, the analysis of the current density distribution shows that superconductive interconnects become attractive only for system operating at lower voltage (up to 1 V). Making the assumptions that the thin film processing and the critical parameters requirements of high Tc superconductors can be met, superconducting interconnections could be used with low loss, high bandwidth and high wiring density.

REFERENCES

1. M.R. Pinnel and W.H. Knausenberg, Interconnection system requirements and modeling, AT&T Technical Journal, vol.66, July/August, (1987).
2. R.J. Jensen, J.P. Cummings and H. Vora, Copper polymide materials system for high performance packaging, IEEE CHMT-7, (1984).
3. R.G. Saenz and E.M. Fulcher, An approach to logic circuit noise problems in computer design, Computer Design, April (1969).
4. J.G. Bednorz and K.A Muller, Possible high Tc superconductivity in the BaLaCuOsystem, Z. Phys, B64, (1986).
5. M.K. Wu and al., Superconductivity at 93 K in a new mixed-phase YBaCuO compound system, Phys. Rev. Lett. 58, (1987).
6. R.E Mattick, "Transmission Lines for Digital and Communication Networks", MacGraw Hill, N.Y, (1969).
7. R. Ruby, A comparison of copper transmission lines and superconducting transmission lines at 77K for second level packaging, Superconductors in Electronics Commercialisation Workshop, San Fransisco, (1987).
8. J. Chilo and C. Monllor, Magnetic field and current distribution in a system of microstrip lines, IEEE MAG 19 (3), (1983).
9. R.F. Harrington, "Field Computation by Moment Method", MacMillan,N.Y, (1968).
10. M. Tinkam, "Superconductivity", Gordon and Breach, N.Y, (1965).
11. Y. Enomoto and al., Largely anisotropic superconductivity critical current in epitaxially grown BaYCuO thin film, Jap. J. Appl. Phys. , Vol.26, L1248, (1987).
12. R. Hammond, R. Barton, Recent progress in superconducting thin film technology, Superconductors in Electronics Commercialisation Workshop, San Fransisco, (1987).
13. J.M. Pond and al., Mesurements and modeling of kinetic inductance, IEEE MTT35 vol.12, (1987).
14. T. Van Duzer and C.W. Turner, Principles of superconductive devices and circuits, New York, (1981).
15. B. Young and T. Itoh, Loss reduction in superconducting microstrip like transmission lines, IEEE MTT-s Digest, (1988).
16. R.J Cava and al., Bulk Superconductivity at 91 K in single phase oxigen..., Phy. Rev. Lett. vol.58 (16), (1987).
17. J.M. Pond and M. Krowne, Slow-vawe properties of superconducting microstrip transmission lines, IEEE MTT-s Digest, (1988).
18. O.K. Kwon and al., Superconductors as very high speed system level interconnects, IEEE EDL vol. 8 (12), (1987).
19. A.M. DeSantolo and al., Preparation of High Tc and Jc films of YBaCuO using laser evaporation of a composite target containing BaF, Appl. Phys. Lett. 52 (23), (1988).
20. P. Chaudhari and al., Critical current meausurements in epitaxial films of YBaCuo compounds, Phys. Rev. Lett. 58 (25), (1987).

COEVAPORATED Bi-Sr-Ca-Cu OXIDE FILMS AND THEIR PATTERNING

FILM FABRICATION OF ARTIFICIAL (BiO)/(SrCaCuO) LAYERED STRUCTURE

T.S. Kalkur, R.Y. Kwor, S. Jernigan and R. Smith*

Dept. of Electrical and Computer Engineering
University of Colorado at Colorado Springs
Colorado Springs, CO 80933

* Kaman Sciences Corporation, Colorado Springs, CO 80933

INTRODUCTION

Initially, after the discovery of high T_c superconducting materials, most of the studies were confined to yttrium based 1–2–3 compounds. However, Bi and Tl based compounds are attracting the attention of many investigators recently because of their relatively lower sensitivity to oxygen and easy annealing[1]. In addition to a semiconducting phase[2], these compounds have multiple superconducting phases with transition temperatures between 85 and 110 K. Various approaches to obtaining high T_c superconducting BiCaSrCu oxide films have been reported. These include electron beam coevaporation[3], ion and reactive ion beam deposition[4-5], multisource magnetron sputtering[6], spin–on coating[7] and laser ablation[8]. For applications in electronics, another important aspect is the patterning of thin superconducting films. Numerous papers have been published regarding the patterning of yttrium based 1–2–3 films by wet chemical etching[9], ion beam etching[10] and laser ablation[11], but there are only a few papers on the patterning of Bi based compounds. In this paper we report the results of the fabrication of BiSrCaCu oxide superconducting thin films by coevaporation of CaF_2, SrF_2, Bi and Cu followed by a post–deposition annealing in wet oxygen; and their patterning by wet chemical etching.

SAMPLE PREPARATION

The experimental set up used for the coevaporation of Bi–Cu–CaF_2–SrF_2 films is shown in Fig. 1. CaF_2 and SrF_2 mixtures in the form of broken crystals are loaded in a single tungsten boat. Since the boiling points of these two compounds are very close (2489 C for SrF_2 and 2500 C for CaF_2) the composition of Sr and Ca in the deposited film can be controlled by adjusting their ratio in the tungsten boat. Bi is evaporated using an electron beam and Cu is evaporated from a tungsten boat. The composition of the constituents was monitored independently by crystal monitors mounted on a tube and directed towards respective sources. The typical evaporation rates used were: Cu–0.8 A/sec, Bi–1.2 A/sec, and CaF_2–SrF_2 mixture–3.2 A/sec. The film thickness used for the study was 0.52 micron (Cu 800 A, Bi 1200 A, and CaF_2 + SrF_2 3200 A). The expected composition of the film after annealing was $Bi_1Ca_{0.7}Sr_{1.3}Cu_2O_x$. The films were deposited on yttrium saturated (100) cubic zirconia or MgO and annealed in a wet oxygen ambient. A two step annealing cycle was used. In the first step, the wafers were annealed at a temperature of 725 C for 15 to 60 minutes; in the second step, the wafers were annealed at a temperature of 825 C for 5 minutes and then slowly pulled out of the furnace. A typical annealing cycle is shown in Fig. 2.

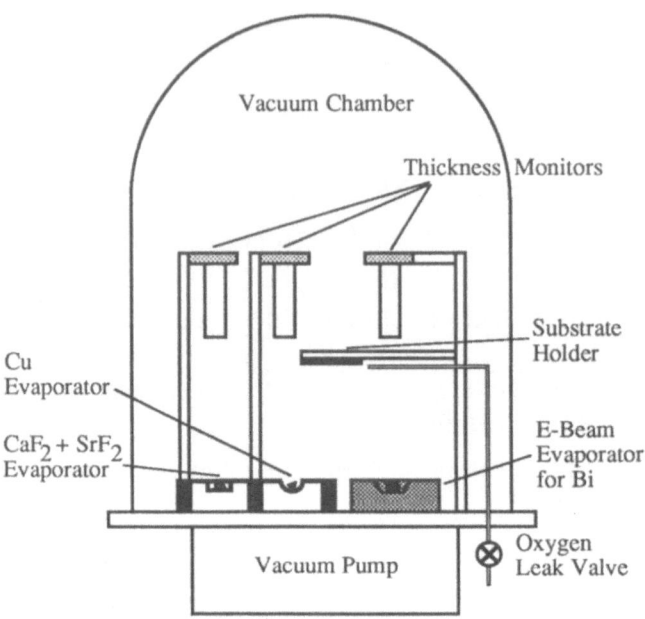

Fig 1. Set Up for the Coevaporation of Bi, CaF_2 + SrF_2 and Cu.

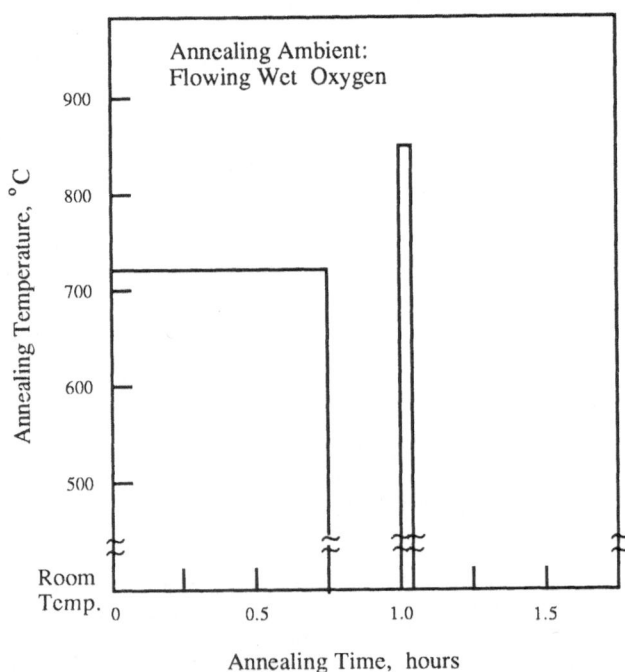

Fig 2. Typical Thermal Cycle used for the Annealing of CaSrBiCuO films.

The photopatterning of the BiSrCaCu films has been performed using AZ 1400 photoresist. The films were etched in dilute HNO_3 or HCl. For some wafers a thin layer (200 A) of Cu was first evaporated on the films in order to minimize the surface degradation during photoprocessing.

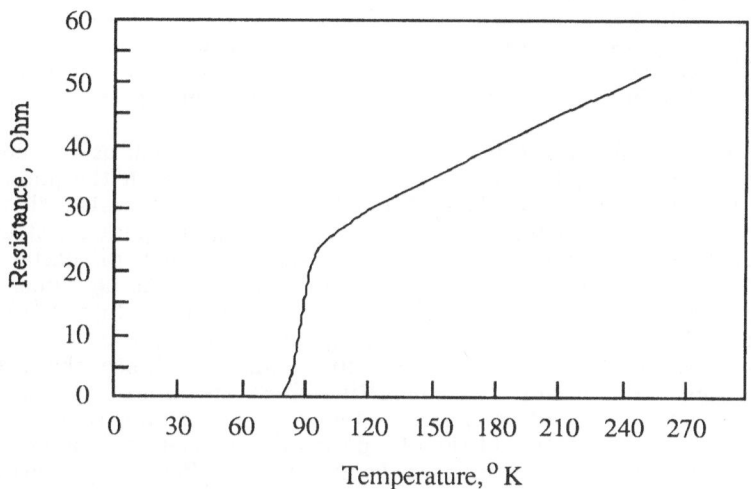

Fig 3. Resistance vs Temperature Characteristics of CaSrBiCuO film on Zirconia.

RESULTS AND DISCUSSION

The as—deposited films on zirconia and MgO were conducting but did not show any superconducting transition. The typical time—temperature curve for BiSrCaCu oxide film on zirconia substrate after annealing is shown in Fig. 3. The resistance starts dropping rapidly at 100 K and zero resistance was achieved at 78 K. Similar transitions were observed for films on MgO substrate. In the case of films on zirconia substrate, the annealing time was found to be critical. Increasing annealing time increases the resistivity of the film drastically, and some areas of the film were found to become insulating. This is due to the interaction of BiSrCaCu oxide film with zirconia substrates, consistent with the observations by Sullivan et al[6]. They deposited Bi—Sr—Ca—Cu oxide films by DC magnetron sputtering on zirconia, and after annealing, the films became insulating. This may be due to their prolonged high temperature treatment (860 C for 30 minutes) during the annealing cycle. For films on MgO substrates, no change in resistivity was observed when the annealing time was increased to 10 minutes at 850 C. Films on sapphire and SiO_2 substrates were found to be highly resistive after annealing. This shows that MgO is a good choics as substrate for BiSrCaCu oxide superconducting films.

The microstructure of the BiSrCaCu oxide films were investigated by X—ray diffraction using Cu K$_\alpha$ radiation. Fig. 4 shows the X—ray diffraction pattern of an annealed superconducting film on MgO substrate. A series of well—defined peaks that fall into two groups, were observed. Calculations performed on one set of peaks showed the formation of an 84 K superconducting phase with a c—axis spacing of 30.5 A. This value is slightly lower than the observations of Rice et al[3] but in agreement with the observations of Hetherington et al[12]. Satoh et al[2] observed some dependence of c axis spacing on the calcium composition in the film. The other set of peaks belongs to the semiconducting phase with a c—axis spacing of 24.5 A. The surface morphology of the film (Fig. 5) is found to be rough but the connectivity between various regions is clearly observed. Fig. 6 shows the X—ray diffraction spectra of a superconducting film on the zirconia substrate. Compared to Fig. 4 additional peaks can be found. This might be due to the interaction of ZrO_2 with the superconducting film, giving rise to various intermediate compounds. The surface morphology of the film (Fig.7) was found to be rough and the interconnectivity between various regions is not clearly seen, but the film was found to be superconducting.

The patterning of the BiSrCaCu oxide film was performed using the AZ 1400 photoresist. Slight interaction of the superconducting film with the photoresist was observed. However, EDAX analysis did not show any observable change in the composition of the film. After exposure and developing, the photoresist on the film was removed using acetone. Fig. 8 shows the formation of patterns on the superconducting film without any etching. This shows that the developer (MF—312) for positive photoresist partially dissolves the superconducting film. EDAX analysis of the film did not show any significant change in the composition of the film. This shows that the developer dissolves the film uniformly. Since the color of the photoresist and the color of superconducting films are almost the same, it was difficult to prevent the overdevelopment of the pattern. For films covered with a thin layer of copper, the dissolution of the film during development of the photoresist was found significantly reduced. In this case, because of the difference in contrast for film covered with copper and the photoresist, the developing time could be controlled more accurately.

The BiSrCaCu films with the patterned photoresist were post baked at a temperature of 140 C for 20 minutes. They were cooled down to room temperature and etched in various concentrations of dilute nitric acid. Nitric acid with various concentrations were used as etchants and the etching rates were characterized with the help of a Dektak profilometer. The etching for 10% nitric acid (10% nitric acid and 90% deionized water) was found to be too fast. Reasonably slow rates for bare films (not covered by Cu) can be obtained by using 2% nitric acid (etching rate = 1.5 μm/min.) or 1% nitric acid (etching rate = 0.8 μm/min.). The films covered with a thin layer of copper took an additional 5—10 seconds to dissolve the copper. After the etching, the pattern formed on the film was examined in a scanning electron microscope and a typical pattern formed is shown in Fig. 9. The pattern edge was found to be very sharp, meeting the requirement for the fabrication of devices with fine features. The relative composition of the various constituents of the film was once again analyzed by EDAX and no observable change due to etching was found.

Etching experiments were also performed in 1% and 2% hydrochloric acid. Their film etching rate was almost the same as that for nitric acid but the etching was highly non—uniform. The pattern edges were very rough and were not of device quality.

CONCLUSIONS

BiSrCaCu oxide superconducting films have been successfully fabricated by coevaporation of Bi, Cu, CaF_2 and SrF_2; followed by wet oxidation. The degradation of the film during the patterning process could be reduced by depositing a thin overlayer of copper. Dilute nitric acid was found to be a better etchant than dilute HCl in defining fine patterns.

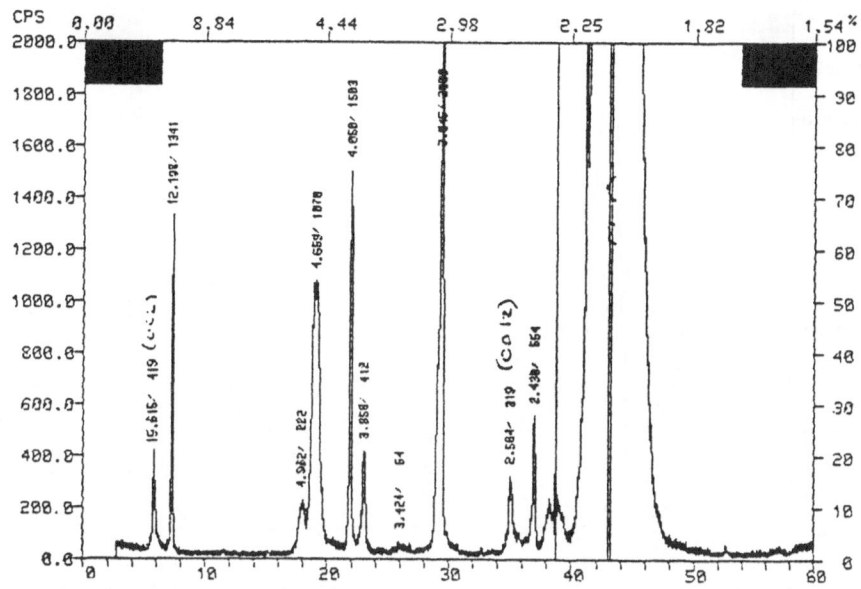

Fig 4. X—ray Diffraction Pattern of CaSrBiCu Oxide Film on MgO substrate.

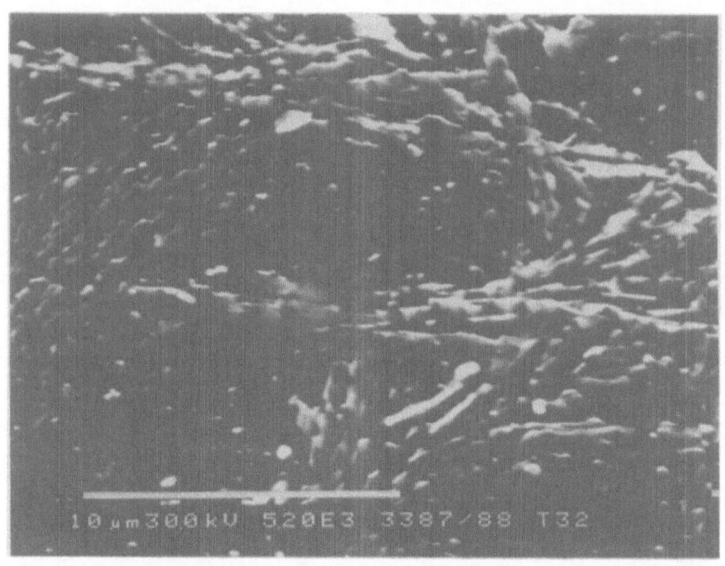

Fig 5. Surface Morphology of CaSrBiCu Oxide Film on MgO Substrate.

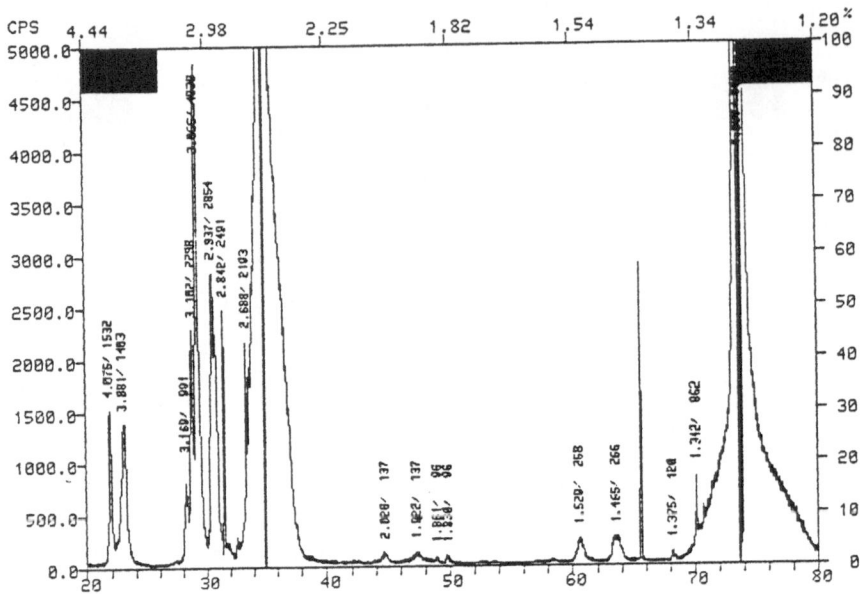

Fig 6. X—ray Diffraction Pattern of CaSrBiCu Oxide Film on ZrO$_2$ Substrate.

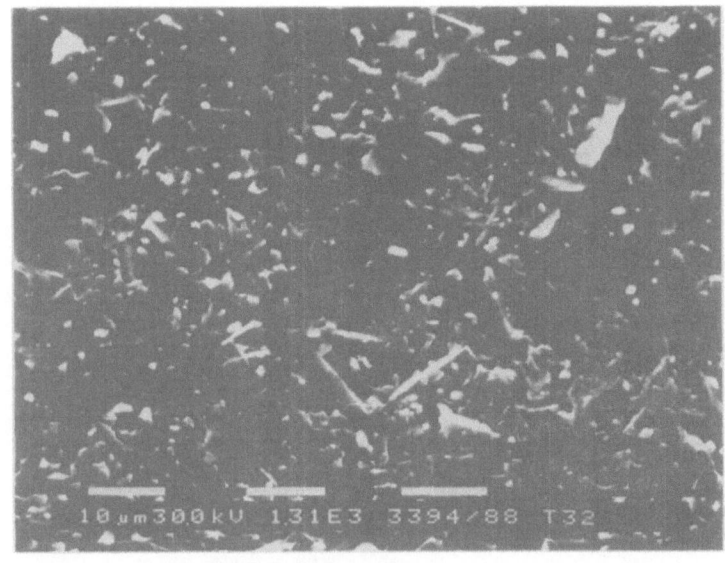

Fig 7. Surface Morphology of CaSrBiCu Oxide Film on ZrO$_2$ Substrate.

Fig 8. Pattern Formed on the CaSrBiCu Film After Developing of Photoresist.

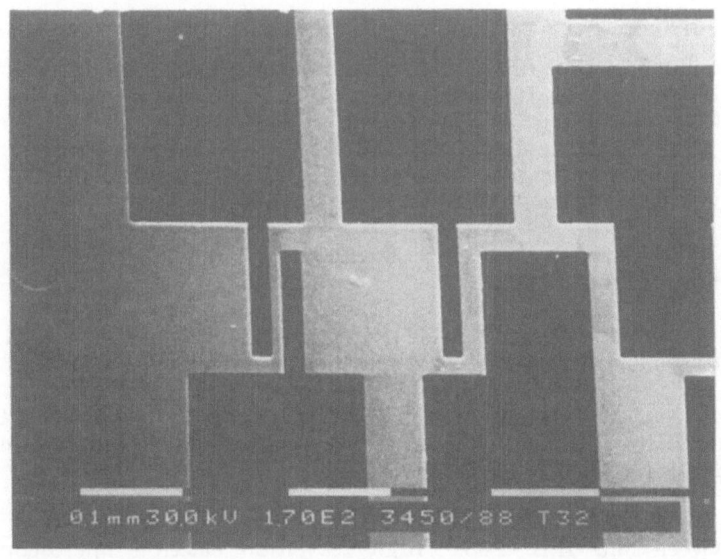

Fig 9. Etched Pattern of a CaSrBiCu Oxide Film with 200 A of Cu Overlayer.

ACKNOWLEDGEMENT

The authors are grateful to Kaman Sciences Corporation of Colorado Springs, for the support of this project.

REFERENCES

1. H. Maeda, Y. Tanaka, M. Fukutomi, and T. Asano, "A new high T_c oxide superconductor without a rare earth element", Japan J. Appl. Phys. 27, L 209 (1988).

2. T. Satoh, T. Yoshitake, Y. Kubo, and H. Igarashi, "Composition dependence of superconducting properties in $Bi_2(Sr_{1-x}Ca_x)_{n+1}$ Cu_nO_y(n=2,3) thin films, Appl. Phys. Lett. 53, 1213 (1988).

3. C.E. Rice, A.F. Levi, R.M. Fleming, P. Marsh, K.W. Baldwin, M. Anzlower, A.E. White, K.T. Short, S. Nakahara, and H.L. Stromer, "Preparation of superconducting thin films of calcium strontium bismuth copper oxides by coevaporation", Appl. Phys. Lett. 52, 1828 (1988).

4. R.D. Lorentz and J.H. Sexton, "Oriented high temperature Bi–Sr–Ca–Cu oxide thin films prepared by ion–deposition", Appl. Phys. Lett., 53, 1654 (1988).

5. A.B. Harker, P.H. Kobrin, P.E.D.Morgan, J.F. DeNatale, J.J. Ratto, I.S. Gergis, and D.G. Howitt, "Reactive ion deposition of thin films in the bismuth – calcium – strontium – copper oxide ceramic superconductor system", Appl. Phys. Lett., 52, 2186, (1988).

6. B.T. Sullivan, N.R. Osborne, W.H. Hardy, J.F. Cardon, B.X. Yang, P.J. Michael and R.R. Parsons, "Bi–Sr–Ca–Cu–O superconducting thin films deposited by dc magnetron sputtering", Appl. Phys. Lett., 52, 1992, (1988).

7. S.L. Furcone and Y.M. Chiang, "Spin–on Bi_4 Sr_3 Ca_3 Cu_4 O_{16+x} superconducting thin films from citrate precursors", Appl. Phys. Lett., 52, 2180 (1988).

8. M.K. Jaggi, M. Meskoob, and S.F. Wahid and C.J. Rollins, "Superconductivity in thin films of Bi–Sr–Ca–Cu oxide deposited via laser ablation of oxide pellets", Appl. Phys. Lett., 53, 1551 (1988).

9. I.Shih and C.X. Qui, "Chemical etching of Y–Ba–Cu–O thin films", Appl. Phys. Lett., 52, 1523 (1988).

10. J.W.C. de Vries, B. Dam, M.G.J. Heijman, G.M. Stollman, and M.A.M. Gijis, C.W. Hagen and R.P. Griessen, Appl. Phys. Lett., 52, 1904, (1988).

11. J. Mannhart, M. Scheuermann, C.C. Tsui, M.M. Oprysko, C.C. Chi, C.D. Umbach, R.H. Koch and C. Melton, "Micropatterning of high T_c films with an excimer laser", 52, 1271 (1988).

12. C.J.D. Hetherington, R. Ramesh, M.A. O'Keefe, R. Kilas and G. Thomas, S.M. Green and H.L. Luo, "High resolution electron microscopy of the C = 30.5 A and C = 38.2 A polytypoids in the Bi–Ca–Sr–Cu–O superconductor", Appl. Phys. Lett., 53, 12, (1988).

dV/dI DOUBLE PEAK STRUCTURES IN SUPERLATTICE-BASED TUNNEL JUNCTIONS

L. Maritato*, A.M.Cucolo†, R. Vaglio†, C. Noce‡,
J. L. Makous# and C. M. Falco#

*INFN-LNF, C.P. 13, 00044 Frascati, Italy
†Dipartimento di Fisica, Università di Salerno, Baronissi (SA), Italy
‡Dipartimento di Fisica Teorica, Università di Salerno, Baronissi (SA), Italy
#Department of Physics, Optical Sciences Center and Arizona Research Laboratories, University of Arizona, Tucson, AZ, USA

ABSTRACT

We realized high quality tunnel junctions using bcc-bcc Mo-Ta superlattices in the modulation wavelength range 16-450 Å as the first electrode. The dV/dI *vs.* V characteristics showed a double peak structure in the wavelength range 50-250 Å. We fitted the temperature behavior of these structures using both a two band model and the proximity effect theory. Preliminary calculations of a possible microscopic explanation of this data are presented.

EXPERIMENTAL PROCEDURE

The fabrication by sputtering of bcc-bcc Mo-Ta superlattices with structural coherence even in the monolayer limit with alternating individual atomic planes of Mo and Ta,[1,2] allowed several studies of the structural, electronic and superconducting properties of this kind of superlattice.[3-4]

For the present work we realized high quality Mo-Ta superlattice-based tunnel junctions using two procedures. In the first, after a chemical definition of the Mo-Ta base electrode geometry the superlattices were sputter-cleaned removing ≈150 Å of the topmost layer of ≈200 Å of Ta. The remainder Ta was then oxidized by heating the samples at 200 °C for 30 sec. In the second procedure the samples were masked during the deposition. After a thermal oxidation of several days at room temperature the geometrical definition of the junctions was obtained by a collodion technique. The second electrode was in both cases a thermally evaporated Pb film ≈5000 Å thick. Both procedures resulted in tunnel junctions with very leakage currents, as shown in Fig. 1.

Using standard tunneling techniques[5] we measured the dV/dI *vs.* V

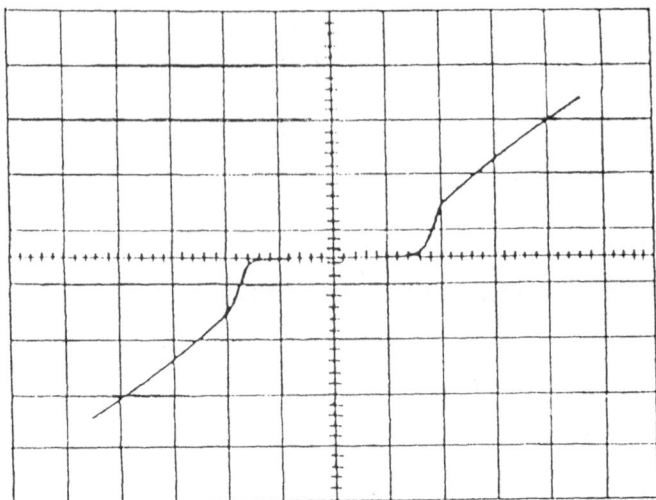

Fig. 1 - I-V characteristic at 1.5 K of a Mo-Ta/Ta$_x$O$_y$/Pb tunnel junction based on a superlattice with λ = 29 Å. X-axis scale = 1 mV/div. Y-axis scale = 20 μA/div.

characteristics of our junctions. In the superlattice modulation wavelength range 50-250 Å, the dV/dI *vs.* V curves showed a double peak structure,[2] as illustrated in Fig. 2.

The temperature behavior of these peaks was equally well fitted by the MacMillan proximity effect theory[6,7] and the Suhl, Matthias and Walker (SMW) two-band theory.[8]

In Fig. 3 is shown the temperature behavior of the two peaks for a junction based on a superlattice with the modulation wavelength Λ = 175 Å. The energy gap of Pb at T = 0, $\Delta_{Pb}(0)$ = 1.32 meV, has been subtracted from the y-scale. The solid lines represent the best theoretical fit obtained using a proximity effect theory,[7] taking the parameters Γ_N and Γ_s equal to 0.035 and 0.14 respectively and the molybdenum

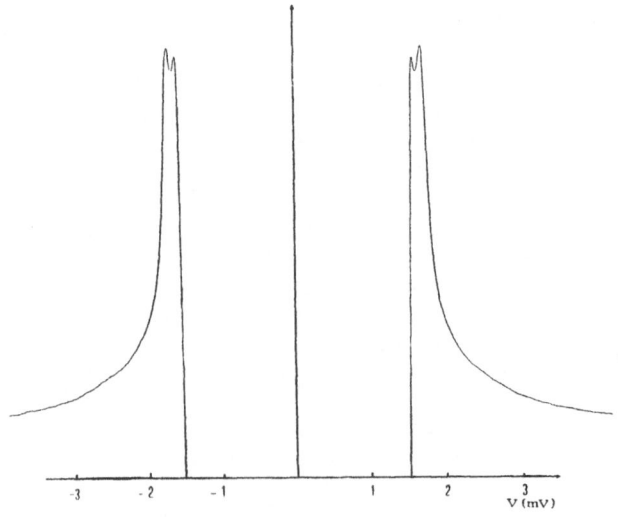

Fig. 2 - -dV/dI *vs.* V characteristic of a Mo-Ta/TaxO$_y$/Pb tunnel junction based on a superlattice with λ = 114 Å

and tantalum electron-phonon coupling constants equal to 0.1971 and 0.22. In Fig. 4, is shown the temperature behavior of the peaks of the junction based on a superlattice

$$H = H_O + H_T \tag{1}$$

where $H_O = H_{BCS}^{(1)} + H_{BCS}^{(2)}$ describes the BCS behavior of the two electrodes, and H_T is the term responsible for the tunneling of electrons from one side of the barrier to the other.

The temperature dependence of the gap has been determined in the framework of Thermo-Field Dynamics (TFD) theory. In order to derive the equation for the gap we have calculated the two-point Green function

$$G_i^{ab}(x_1 - x_2) = <O(\beta)| T[\phi_i^a(x_1)\phi_i^{b+}(x_2)]| O(\beta> \tag{2}$$

where ϕ_i is the electron Heisenberg field and $|0(\beta)>$ is the vacuum in TFD. The requirement that the energy should be observable gives

$$G_i^{ab} = S_i^{ab} \tag{3}$$

where S_i^{ab} is the two-point free Green function. Using this condition in the equation of the two-point Green function in the Random Phase Approximation, we derive the following equation for the gap on the i side of the junction

$$\Delta_i = -\lambda_i \hbar \Delta_i \int d^3k/(2\pi)^3 \, 1/(2\varepsilon_{ik}) \, [1 - 2f_F(\varepsilon_{ik})]$$

$$- \Delta_j T^2 \int d^3k/(2\pi)^3 \, 1/(2\varepsilon_{jk}) \, [1-2f_F(\varepsilon_{jk})] \tag{4}$$

where

$$f_F(\varepsilon_{ik}) = 1/\{1 + \exp[\beta(\varepsilon_{ik}^2 + \Delta_i^2)^{1/2}]\} \tag{5}$$

and where λ_i is the electron-phonon coupling constant in the "i" side, T is the tunneling matrix element and f_F is the Fermi function.

Since the integrations are cut off by the Debye energies ω_{Di}, Eq. (4) can be rewritten as

$$\Delta_i[1+ \lambda_i \hbar N_i(O)F_i(E)] = - \hbar \Delta_j T^2 N(O) F_j(E) \tag{6}$$

where $N_i(O)$ is the density of states of the Fermi level and $F_i(E)$ is

$$F_i(E) = \int_E^{\omega_{Di}} d\varepsilon_i^i \, [1 - 2f_F(E)] \tag{7}$$

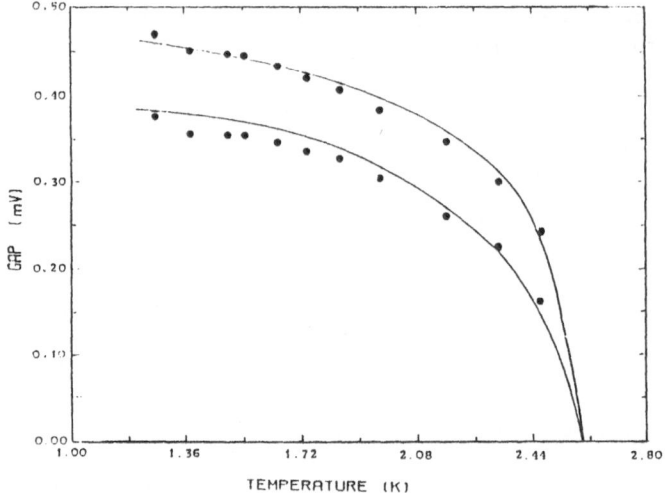

Fig. 3 - Double peak voltages as a function of the temperature for a junction based on a superlattice with $\lambda = 175$ Å. The solid lines are the best theoretical fit obtained with proximity effect theory[7].

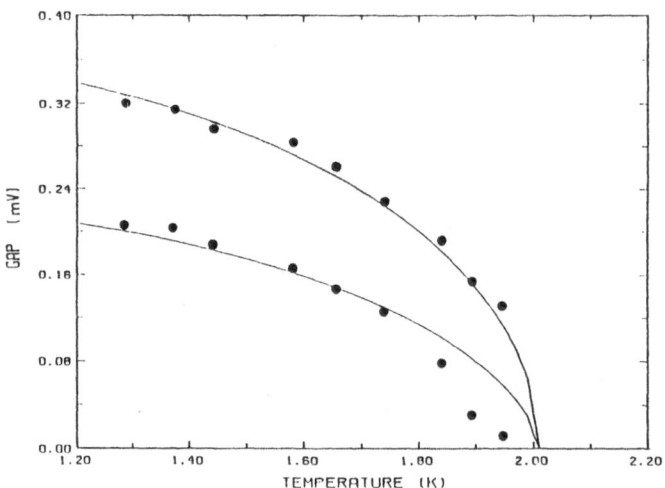

Fig. 4 - Double peak voltages as a function of the temperature for a junction based on a superlattice with $\lambda = 50$ Å. The solid lines are the best theoretical fit obtained with SMW theory[8].

with $\Lambda = 50$ Å. In this case the solid lines are the best fits obtained by the SMW two-band theory with the two parameters V_A and V_B respectively equal to 0.83 and 10.[2]

DISCUSSION

The ability to fit our data by using both the McMillan proximity effect theory and the SMW two-band theory is related to the microscopic equivalence of these two models. This can be shown by considering the Hamiltonian used by McMillan in his theory

$$E = (\varepsilon_i^2 + \Delta_2^2)^{1/2} \tag{8}$$

By putting $V_i = -\hbar\lambda_i$ and $V_j = -\hbar T^2$ we deduce immediately the gap equation of SMW[2,8]. We stress that this result holds if the tunneling Hamiltonian is treated in the second order approximation.

ACKNOWLEDGEMENT

The work at the University of Arizona was supported by the U.S. DOE, under contract No. DE-FG02-87ER45297.

REFERENCES

1. J. L. Makous et al., Japn. J. Appl. Phys., 26:1467 (1987).
2. L. Maritato et al., to be published in Phys. Rev. B, (15 Dec. 1988).
3. A. M. Cucolo et al., in "Progress in High Temperature Superconductivity," Vol, 4, Proceedings of the II Soviet-Italian Symposium on Weak Superconductivity, Naples, Italy, A. Barone and A. Larkin, eds., p. 283 (1987)
4. J. L. Makous et al., Proceedings of the Materials Research Society (1987), in press.
5. J. M. Rowell, in "Tunneling Phenomena in Solids," E. Burstein. ed., Plenum, New York (1969)
6. W. L. McMillan, Phys. Rev. 175:537, (1968)
7. J. Vrba and S. B. Woods, Phys. Rev., B3, 2243 (1971).
8. H. Suhl, B. T. Matthias and L. R. Walker, Phys. Rev. Lett., 3, 552 (1959)

PROCESSING AND PATTERNING TECHNIQUES FOR THIN FILMS OF $YBa_2Cu_3O_7$

C.M. Mombourquette, J.G. Beery, D.R. Brown
R.A. Lemons, and I.D. Raistrick

Electronics Research Group
Los Alamos National Laboratory
Los Alamos, NM 87545

ABSTRACT

Superconducting thin films of $YBa_2Cu_3O_7$ were prepared by co-evaporation of copper, yttrium, and barium fluoride onto single-crystal strontium titanate substrates. E-beam evaporation was used for the yttrium and copper, and thermal evaporation was used for the BaF_2. The as-deposited films were converted to the superconducting oxides by a post-evaporation anneal in dry and wet oxygen. We discuss here the patterning and processing of the films into a current constriction device geometry.

INTRODUCTION

This paper discusses the deposition, processing, and patterning of thin films of high-temperature superconductors for electronic applications. A companion paper in these proceedings discusses the details of the phases present, the morphology and crystalline texture of the films.[1] We concentrate here upon the processing aspects involved in converting the thin films into electronic devices.

FILM PREPARATION

The films are vacuum deposited and then annealed to produce the superconductive crystal structure. The typical films discussed in this paper consist of a 5000 Å thick layer of $YBa_2Cu_3O_7$ grown on single crystal strontium titanate, orientation <100>. The evaporation system has two electron-beam sources for the copper metal and the yttrium metal and a thermal evaporation source for the barium fluoride. The substrate was not heated and a witness sample was evaporated on a sapphire substrate simultaneously.

The as-deposited films are relatively smooth and are amorphous mixtures of copper, yttrium, and barium fluoride. The superconducting perovskite structure is formed during the post deposition annealing procedure. That procedure typically consists of heating to 850°C in dry oxygen, adding water to the oxygen stream and holding for 60 minutes at

850°C. The water is then removed and, under a flow of dry oxygen, the film is held at 750°C for three hours and then at 450°C for three hours followed by a furnace cool.

CHARACTERIZATION

Figure 1(a) is a scanning electron micrograph of an as-deposited film; Fig. 1(b) is a film after the 850° anneal. X-ray diffraction texture analysis indicate that the needles are oriented with their c-axes parallel to the plane of the substrates. At higher temperatures there are fewer needle-like structures and the grains become more plate-like. The plate-like structures are aligned such that their c-axes are perpendicular to the plane of the substrate. This crystal-line texture is discussed more fully in the companion paper.[1] A conventional four-point dc probe is used to measure the resistivity of the material. Figure 2 is a set of R vs T plots for a patterned film. (The four curves are for different active areas on the same device; two 30 μm square and two 100 μm square bridges). The presence of a superconducting transition gives little information regarding the fraction of 1-2-3 phase in the film; only that there is enough to form a conducting path.

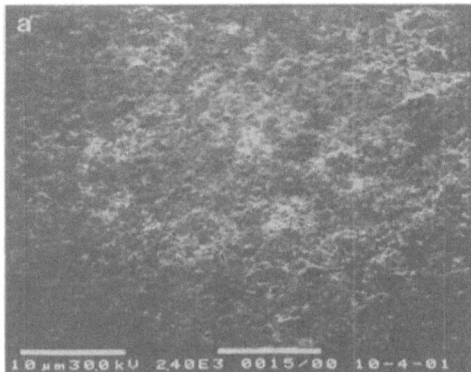 

Fig. 1. Scanning Electron Microscope (SEM) micrograph of (a) as-deposited amorphous film and (b) post-annealed superconducting film.

One of the techniques we used to determine the stoichiometry of the films is Rutherford backscattering spectrometry (RBS). Figure 3 is a portion of a RBS spectrum for a witness film deposited on a sapphire substrate. This witness sample allows an accurate determination of the metallic ratios by Rutherford backscattering; when the substrate is strontium titanate the ion scattering edges from the strontium and titanium complicate the analysis. Since the aluminum in the substrate has a much lower mass than Cu, Y, and Ba, a thin film gives three well separated peaks from which the metal stoichiometry of the film may be determined. The area under the peaks is integrated and weighted by the scattering cross section for each metal to determine the atomic ratios. Since, at the relatively high beam energies needed to separate the peaks, the scattering may not be totally Rutherford, experimentally determined scattering cross sections are used. These techniques were developed by Martin et al.[2]

502

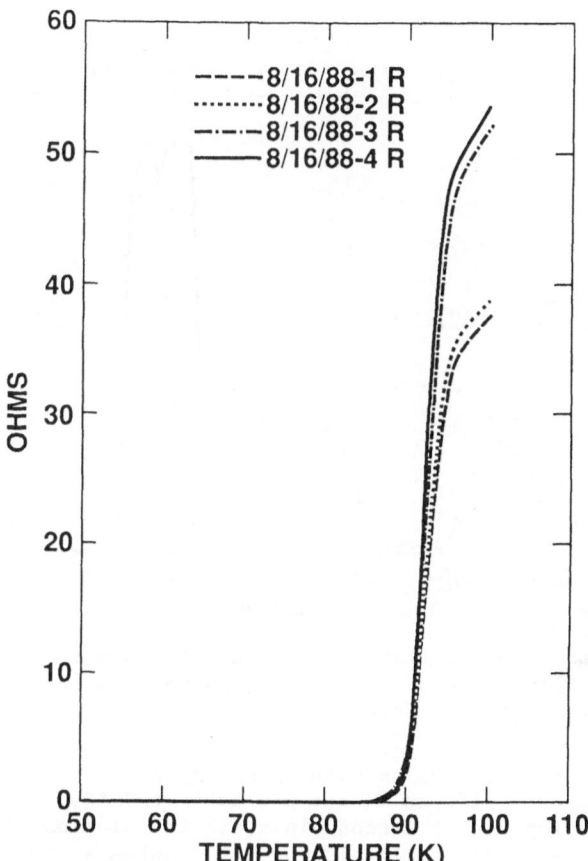

Fig. 2.   DC resistivity measurements for a 5000 Å superconducting film on single crystal <100> SrTiO$_3$. Upper curves are for 30 μm constrictions, lower curves for 100 μm constrictions.

Table I gives a list of the processing steps we are using to pattern the films. Most of the work has been carried out with a current constriction test pattern mask. This pattern consists of a matrix of nine superconducting areas with a central constriction ranging in size from 10 μm x 10 μm to 100 μm x 100 μm. We initially attempted to pattern the films with a wet etching technique, but observed severe undercutting due to the open morphology of the films. Presently we use RF argon sputter etching for our films. Most of the experiments have been performed on films patterned after post deposition anneal.

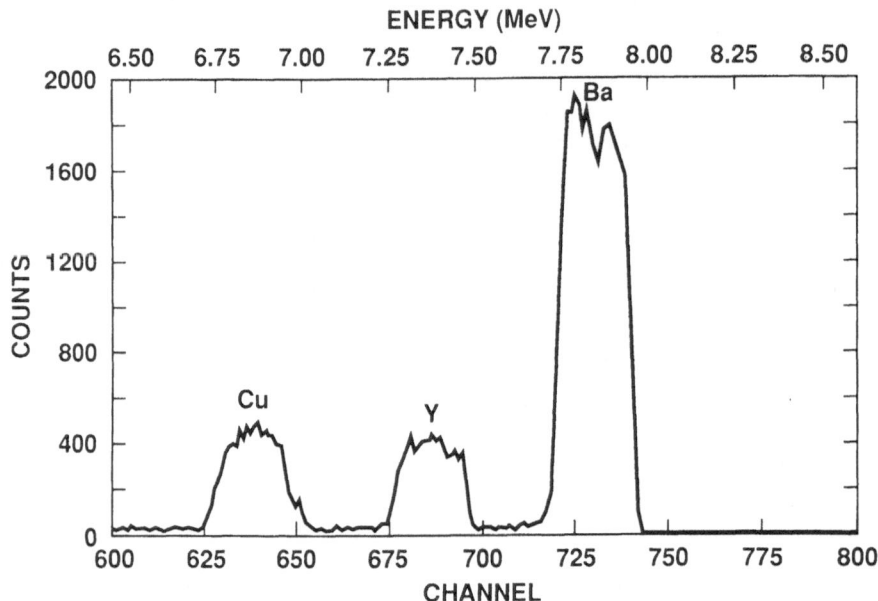

Fig. 3. Rutherford Backscattering spectrum of 8.8 MeV α-particles for a thin film of $YBa_2CU_3O_7$ on a sapphire substrate. The chemical composition can be determined from the peak areas.

The oxide patterning begins with spinning Z1350J photoresist at 5000 rpm followed by a fifteen minute pre-bake at 85°C. We expose with 56 mJ/cm^2 and develop for 30 seconds in a dilute solution of Shipley Microposit developer. Sputtering is for 120 minutes at 60 W in an argon pressure of 9 μm. The sputtering is done in two stages of 60 minutes each with an intermediate repatterning. EMT130 stripper is heated to 90°C; stripping requires 5 minutes.

The gold metallization begins with spinning KTI9000 photoresist, exposure is 84 mJ/cm^2, developer is again Shipley Microposit, not diluted. Developing time is 45 seconds. An oxygen sputter etch is performed for 10 minutes, at 60 W and in 9 μm of oxygen. This etch cleans the film surface and improves metal adhesion. A 2500 Å layer of gold is sputtered followed by stripping with EMT130 for 5 minutes.

After oxide patterning and metallization the devices are annealed for 3 hours in dry oxygen at 450°C, which restores oxygen removed in the processing. An ultrasonic wire bonder attaches 1 mil gold wire; the device is heated to 125°C during wire bonding. Contact resistances are less than 5 μΩ cm^2. The completed device is secured to a package with silver expoxy.

In addition to our standard processing steps presented in Table I, we have investigated two other techniques: patterning in the as-deposited state prior to anneal and using a chlorobenzene liftoff process. Figure 4 presents micrographs of a 30 μm x 30 μm active area, which was patterned before anneal. This process gives somewhat better edge definition than the standard post-anneal patterning. We are presently evaluating these alternative routes with regard to the effect of the properties of the films.

TABLE I

PROCESS STEPS FOR PATTERNING AFTER ANNEAL

1. OXIDE PATTERNING

- 5000 Å HTSC EVAPORATION
- SPIN ON POSITIVE PHOTORESIST
- EXPOSE; DEVELOP
- RF Ar SPUTTER ETCH
- RESIST STRIP

2. METALLIZATION

- SPIN ON POSITIVE PHOTORESIST
- EXPOSE; DEVELOP
- OXYGEN RF SPUTTER ETCH
- SPUTTER DEPOSIT METAL (2500 Å)
- RESIST STRIP

3. POST PROCESS LOW TEMPERATURE ANNEAL AT 450°C

The best edge definition has been obtained by a chlorobenzene lift-off process. After the film is coated with photoresist, it is soaked in chlorobenzene prior to exposure. Figure 5 shows micrographs of a different test pattern using the chlorobenzene liftoff process. Preliminary electrical tests on these films are encouraging.

The sputter etch and metallization processes reduce the T_c substantially, but a subsequent 450°C anneal in oxygen restores the films to nearly their original prepatterned T_c (Fig. 6). We have performed a series of tests to determine the source of the degradation in T_c. In addition, processing also causes a small (approximately 4 K) irrecoverable decrease in T_c. We attribute this degradation to the presence of water in the developer, stripper, etc.

The reductions in T_c are not drastic; patterned devices still have T_c's in the 85-89 K range compared to 90-92 K range for the prepatterned films.

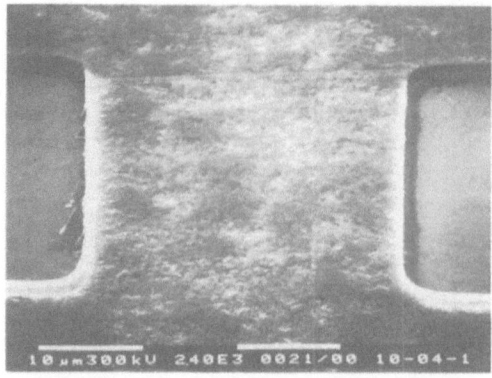

BEFORE HIGH TEMPERATURE ANNEAL

Fig. 4. SEM micrographs of 30–μm x 30–μm bridge. Patterned before 850°C anneal.

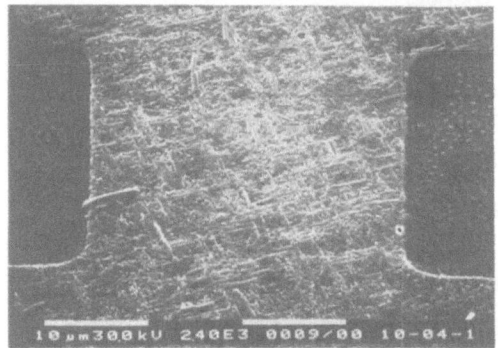

AFTER HIGH TEMPERATURE ANNEAL

BEFORE HIGH TEMPERATURE ANNEAL

Fig. 5. 10–μm line patterned by liftoff, using chlorobenzene-treated photoresist.

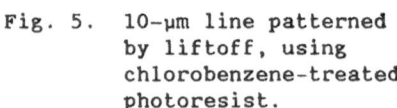

AFTER HIGH TEMPERATURE ANNEAL

EFFECT OF SPUTTER ETCHING

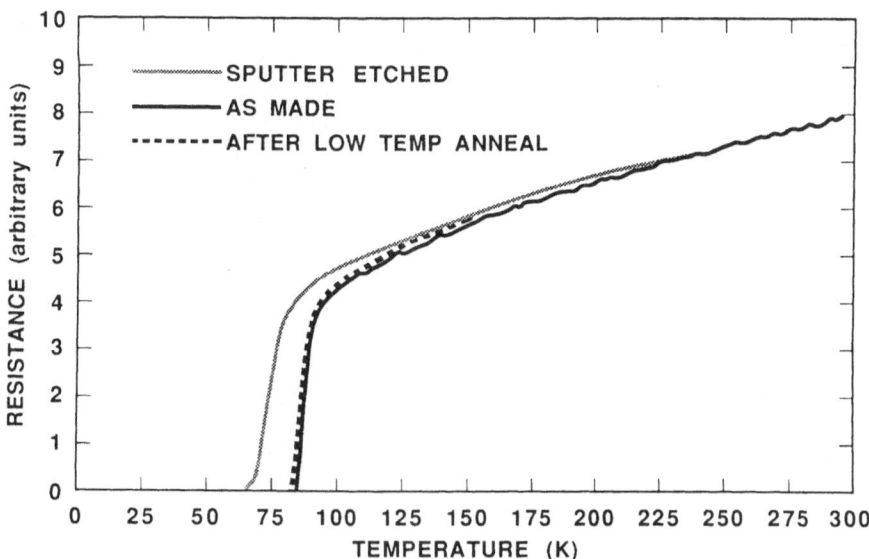

Fig. 6. Effect of sputter etching and subsequent low-temperature anneal on DC resistance of unpatterned film.

SUMMARY

We have developed processes for growing and patterning thin films of $YBa_2Cu_3O_7$ films, which retain their superconducting properties. Even though the films are not fully dense, we obtain features down to 10 μm in size.

ACKNOWLEDGMENTS

We gratefully acknowledge the assistance of Ruth Sherman in annealing the films and Joe Tesmer, Caleb Evans, and Mark Hollander for help with the RBS data and the four-point-probe resistance measurements.

REFERENCES

1. I. D. Raistrick et al., "$YBa_2Cu_3O_7$ Thin Films Prepared by Evaporation from Y, Cu, and BaF_2," these proceedings.

2. J. A. Martin, M. Nastasi, J. R. Tesmer, and C. J. Maggiore, "High-Energy Elastic Backscattering of Helium Ions for Compositional Analysis of High-Temperature Superconductor Thin Films," App. Phys. Lett. 52 (25), 2177–2179 (1988).

HIGH MAGNETIC FIELD SHIELDING

WITH SUPERCONDUCTING NbTi-Cu MULTILAYER FILMS

Sochi Ogawa, Masaaki Yoshitake, Kazu Nishigaki*,
Takao Sugioka**, Masaru Inoue**, and Yoshiro Saji**

Osaka Pref. Industrial Research Institute
Kobe University of Merchantile Marine*
Koatsu Gas Kogyo Co., Ltd.**

Abstract

We have studied the magnetic shielding effects (MSC) of NdTi single layer and NbTiCu multilayer films of various thickness in vertical field, as well as of designed hollow superconducting cylinders in transversal magnetic field to the axis of the cylinder. The total thickness (nd) dependence of MSC per unit thickness $\Delta B_m/nd$ (ΔB_m: maximum MSC, d: film thickness, n: number of NbTi films in a multilayer film) in the multilayer films increased exponentially with decreasing nd and increasing n in the region of nd $\gtrless$ 3 µm and in any case of n=1,2,3. The relation between $\Delta B_m/nd$ and nd can be expressed well by the following formula on the basis of these results;

$$\ell n(\Delta B_m/nd)=\alpha-\beta\ell n(nd), \quad \alpha,\beta \text{ (constant)}.$$

These studies have developed for high magnetic shielding devices of hollow cylinders with superconducting NbTi-Cu multilayer films. The device provides a space with free from high magnetic field up to 0.8 T.

From these results, we conclude that it is possible to completely seal off or form a field free space against flux density of one tesla or more by laminating NbTiCu multilayer films of reasonable thickness.

INTRODUCTION

In many superconducting devices and applications such as SQUID and EMT using strong magnetic fields, it is necessary to shield the magnetic field efficiently. For this purpose, superconducting magnetic shield is required powerful mean. High magnetic shielding above a few teslas, however, has not been applied to the superconducting devices, and reports on this problem have scarcely been presented previously(1)(2).

In recent year, it was shown that $\Delta B_m/d$ (ΔB_m:max. magnetic shielding effects(MSE), d: film thickness) increased by decreasing d and obtained maximum value at d=0.7µm. The value of MSE was more than 20 times of bulk NbTi plate on our study of magnetic shielding effects of superconducting NbTi thin film deposited by sputtering method on Al foil.(3)

In this work, We carry out basic study on MSE of NbTi single layer films and NbTi-Cu multilayer films of various thickness in vertical field, and of designed hollow superconducting cylinders in transversal magnetic field to the axis of the cylinder.

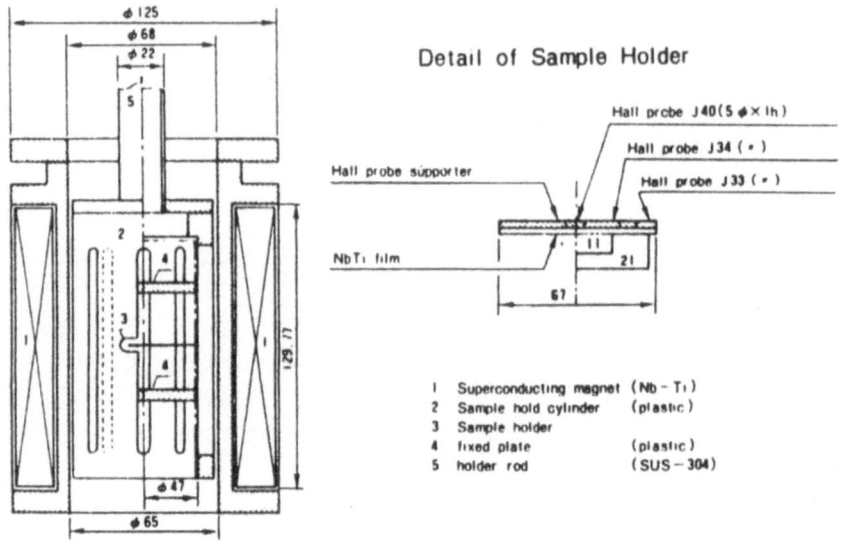

Detail of Sample Holder

1 Superconducting magnet (Nb - Ti)
2 Sample hold cylinder   (plastic)
3 Sample holder
4 fixed plate   (plastic)
5 holder rod   (SUS - 304)

Fig.1. Cross-sectional view of core of apparatus

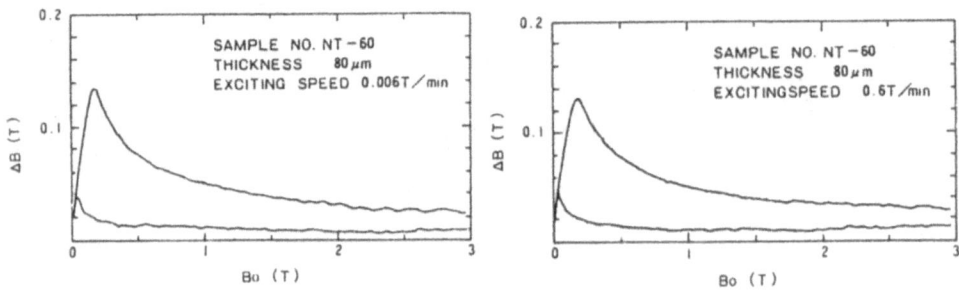

Fig.2. Shielding magnetic field versus magnetic flux density

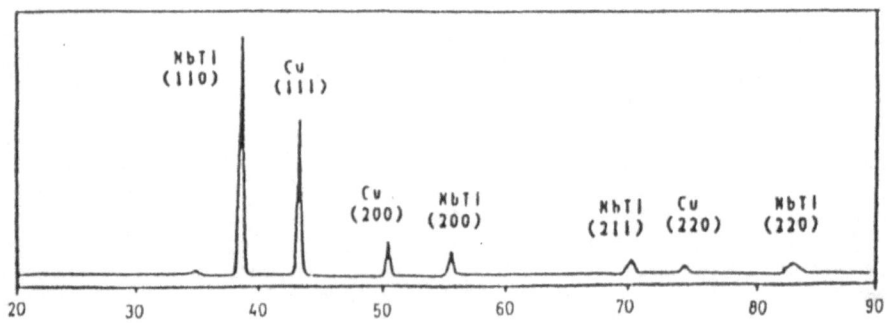

Fig.3. X-ray diffraction pattern of laminated superconducting
NbTi-Cu films (NbTi 2.9μm X 5 layers)

EXPERIMENTAL PROCEDURE

Sample Preparation

NbTi (50at%Ti, Tc=7.8K) films were prepared by RF magnetron sputtering method. It is possible to sputter with relatively high deposition rate and to obtain high purity NbTi films. The film were deposited on Al foil(100 x 100 x 0.015mm^t) at Ar gas pressure in 10^{-3} Torr range with sputtering power of 300W while the substrate temperature was kept at R.T.. The film thickness ranged from 0.1 to 80μm with a deposition rate of 0.1μm/min. The crystal structure of NbTi films as defined by X-ray diffractmeter was cubic structure of the same as that of the NbTi target used for sputtering. The crystalline size of the films was a few tens of angstroms.

The hollow superconducting cylinders were constructed by piling disks of NbTi-Cu multilayer films and Al plates.

Measurement system

As shown in Fig.1 the core of the apparatus is composed of superconducting solenoidal coil and cylindrical sample holder. A test film is inserted between pressing plates with three Hall probes and supported in the center of the coil so that the natural direction of magnetic shielding of coil and the surface of the film is in a right angle. The core of the apparatus is set in the cryostat by which test temperature is freely obtained between 4.2K and 1K. Measurements and data treatments are carried out automatically by a computer system.

EXPERIMENTAL RESULTS AND DISCUSSIONS

Experimental data are put in order based on shielding magnetic field by superconducting planer film. We defined by eq.(1), because the effects of screened magnetic field is made remarkably as much as possible.

$$\Delta B = Bo - Br \qquad (1)$$

Where Bo is natural magnetic flux density without screening NbTi film and Br is real magnetic flux density with the film.

Magnetic shielding effects of NbTi-Cu films

Figure 2 shows an example of measured data. Increasing the external magnetic field, the flux can not intrude into the shielding film. This means that the screened magnetic field ΔB is equal to the natural magnetic flux density Bo. Increasing the magnetic field from 0.1T, magnetic flux begins to intrude into the NbTi films. At 0.19T, the maximum shielding magnetic flux density ΔB of 0.136T is obtained. Increasing the magnetic field, the ΔB decreases gradually, but shielding effects of 0.015T is still remained at Bo=3T. Therefore, it is possible to shield high magnetic field of more than 3T by multilayer NbTi thin films. As magnetic fluxes intrude gradually from the edge rim of the NbTi film, the ΔB near the edge is much smaller than the one at the center of the film. In the case of the NbTi film, the value of WBm is independent of the exciting speed for the superconducting magnet.

Figure 3 shows the representative x-ray diffraction pattern of NbTi-Cu multilayers film(NbTi:2.9 μm x 5, Cu:2μm x 5). As being seen the diffraction pattern,individual reflections from NbTi and Cu can be clearly detected. The facts show that alloying of each layer film does not occur.

Figure 4 shows relation of maximum MSC ($\Delta Bm/nd$) with total thickness of NbTi layer(nd) using the number of NbTi layer(n) of NbTi-Cu multilayers film as a parameter.

In the region of nd>3μm, $\Delta Bm/nd$ shows linear increase with decrease of nd

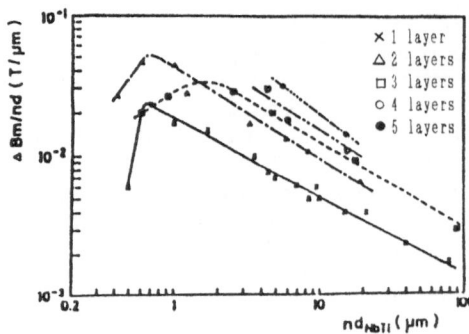

Fig.4.a. The thickness dependence of $\wedge Bm/nd$
with the parameter of the number
of NbTi layers for multi-layer
sheets.

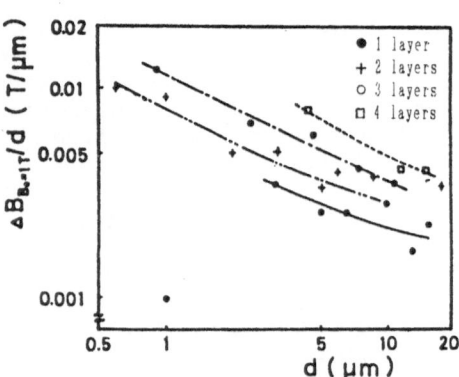

Fig.4.b. The thickness dependence of $\wedge Bm/nd$
in Bo of 1T.

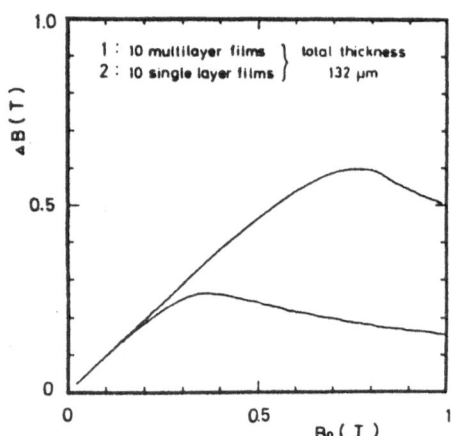

Fig.5. Magnetic shielding effect of multi-layer
films or 10 single layer films with the
same total thickness of NbTi layers.

Table.1. Magnetic shielding effect of superconducting
disks with a hole of different size.

O.D.(mm)	I.D.(mm)	$\triangle Bm$ (Tesla)
46	–	0.101
46	15	0.040
46	20	0.031
46	23	0.026
46	28	0.020

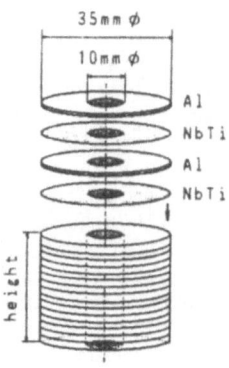

Fig.6. MFS device

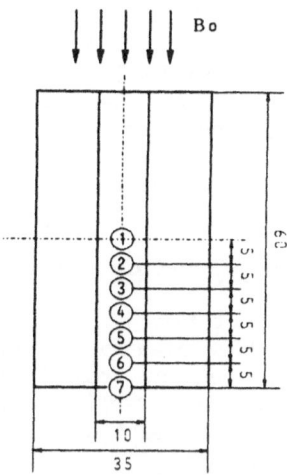

Fig.7. A cross section of the cylinder marked
with measurement points.
(sample No.5 in Table 2)

Table.2. Magnetic shielding effect of hollow superconducting cylinder.

NO	NbTi disk number	Aluminum disk (mm) thickness	number	cylinder (mm) O.D.	I.D.	length	maximum shielding effect $\triangle Bm$ (Tesla)
1	30	1.0	31	35	10	32	0.308
2	60	0.5	61	35	10	32	0.448
3	90	0.5	61	35	10	32	0.569
4	180	0.5	31	35	10	20	0.883
5	60	1.0	61	35	10	62	0.333
6	90	1.0	91	35	10	93	0.338

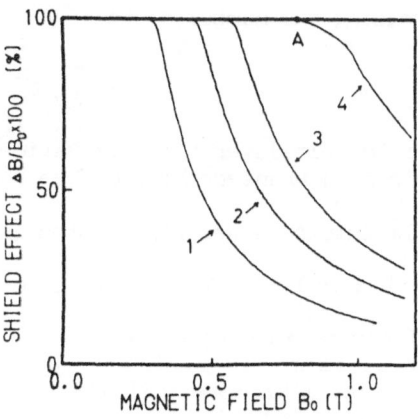

Fig.8. Magnetic shielding effect in the
center of cylinder.

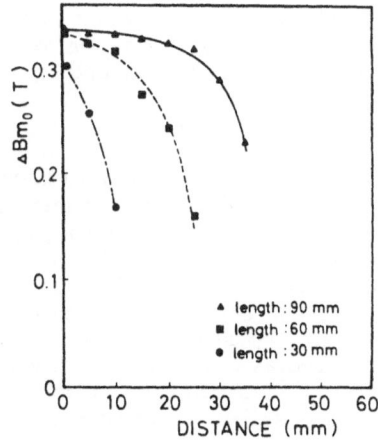

Fig.9. Magnetic shielding effect in superconducting cylinder of 30,60,90mm
in height.(sample No.1,5 and 6)

in any case of n=1,2,3, and its relation conforms to the formula(2).

$$\ln(\Delta Bm/nd) = \alpha - \beta \ln(nd) \qquad (2)$$

α, β : constant

Figure 5 shows results of high magnetic field shielded by ten multilayer films and ten single layer films. In two cases, each total thickness of NbTi is 132 μm. From this figure, as ΔBm is 0.6T, it is known that a high magnetic field of 0.6T is shielded and as ΔB is 0.4T where Bo=1T, a high magnetic field of 1T can be shielded by three layers of this film having less than 0.3mm thickness.

Application to magnetic field shielding device (MFS device)

Table 1 shows the magnetic shielding effects of five hollow superconducting NbTi-Cu multilayer films. The hollow disk has an outer diameter(O.D.) of 46mm and an inner diameter(I.D.) ranging from 0 to 28mm by using the multilayer films which were composed of two NbTi layers and one Cu layer of 2μm thickness respectively. As the increasing I.D. of the disk from 0 to 28mm ΔBm decreases from 0.10 to 0.02 T.

Table 2 shows the values of ΔBm at the center (point 1 in Fig.7) of MFS device, forming by piling the hollow superconducting disks and Al disks alternatively(see Fig.6). As the increase of number of superconducting disks from 30 to 90 ΔBm increases from 0.31 to 0.45 T.
The height of the superconducting cylinder remains 32mm. As the number of the superconducting disks in a unit increases, ΔBm becomes greater. As shown in table 2, the device of No.4 includes 180 of the superconducting disks (i.e. the NbTi layers are measured 0.72mm in thickness) providing more disks in one unit. The relation between a rate of magnetic shielding effects ($\Delta B/Bo$) and Bo of the devices No.1 to No.4 are shown in Fig.8. The No.4 can completely shielded up to 0.8T(correspond to point A).

Figure 9 shows the result of shielding effects in which the height of the superconducting cylinder is increased from 30 to 60 and 90 mm while the number of the superconducting disks in one unit remains unchanged. The ΔBmo(muximum complete magnetic shielding effects) of ambient magnetic field which can be interrupted by shielding completely at each of the positions on the cylinder is stated. As for a shielding area, an area in the cylinder having a height of 30mm covering from its center to a distance of 5mm can completely be shielded from the magnetic field of about 0.25 T. In the cylinder of 90mm in height, an area shielded from the magnetic field of 0.25 T extends from the center of the cylinder to a distance of 35mm.

In the results, a large area can completely be shielded from the magnetic field of a tesla or more, by increasing the number of the superconducting disks in one unit and also.

CONCLUSION

The magnetic shielding effects of NbTi-Cu multilayer films of various thickness in vertical field and of MFS devices in transversal field to the axis are summarized as follow,
1. The relation between $\Delta Bm/nd$ and NbTi film thickness was represented as eq.(2). $\Delta Bm/nd$ was raised with increasing n.
2. ΔBm of the sheet laminated multilayer films in total NbTi thickness of 132μm could be shielded the magnetic field of 0.6T.
3. The effective shield for high magnetic field is made possible by laminating superconducting NbTi-Cu multilayers film.
4. It is possible to completely shield up to 0.8 T or more with the MFS devise.

REFERENCES

1. Simizu, A. and Inoue, M. 'A Test of the superconducting shielding tube made of V3Ga type' IEEE. transaction on magnetics, vol.MAG-17, No.5 September (1981) pp.2146-2149.
2. Sato, S., Ikeuchi, M., Iwata, A., Saji, Y. and Kado, S., 'The Magnetic field screening with NbTi' proc. ICEC-9, (1982) pp.115-119.
3. Ogawa, S., Tada, E., Toda, H., Yositake, M., Sinpo, M. and Saji, Y. 'The Study of Superconducting NbTi films for Magnetic Shield' Proc. ICEC-11, (1986) pp.484-488.

INTRODUCTION TO SUPERCONDUCTIVITY

John U. Trefny

Physics Department
Colorado School of Mines
Golden, CO 80401

INTRODUCTION

We are all aware of the fact that a revolution is taking
place in the field of superconductivity. Properties, that for
75 years and despite the efforts of numerous scientists were
restricted to extremely low temperatures, were extended
dramatically in 1986 by Bednorz, Mueller [1] and others to new
classes of materials and to more easily-accessible conditions.
Because of the scope of these developments, it is probable
that new mechanisms for the phenomenon will need to be
identified and understood. Fortunately, there is a remarkable
number of properties and behaviors which the high-temperature
oxides share in common with conventional, low-temperature
superconducting materials. The purpose of this introduction
is to review those general features of superconductivity,
mostly at the phenomenological level, which are common to both
low- and high-temperature superconductors and which support
our understanding and application of both.

BASIC IDEAS

Perfect conductivity and superconductivity are two
different concepts. The distinction was discovered in 1933 by
Meissner and Ochsenfeld [2]. If a perfect conductor is
exposed to an increasing magnetic field, currents will be
induced to maintain constant flux within. If the field is
increased from zero, the final induction inside the material
will be zero as it was initially. However, if a material is
cooled to the perfectly conducting state while in a field, the
final induction within will be identical to the non-zero
starting value. The final state of the sample thus depends on
its history. The case of the ideal superconductor is quite
different. What Meissner and Ochsenfeld discovered is that
the magnetic induction is always zero, no matter what the
order of events has been. For example, when a superconductor
is cooled in an external field, shielding currents will arise
spontaneously to expel the magnetic flux from within.

The spontaneous expulsion of magnetic fields from the interior of ideal superconductors requires energy. Below the transition temperature, the superconducting state must have a lower free energy than the normal state. However, the additional energy required of the superconductor to expel the magnetic flux must be less than this energy difference if the material is to remain superconducting in a field. Qualitatively, this can only be true in fields less than some thermodynamic critical field, Hc, which should depend on temperature.

So far then we have two conditions required for the existence of superconductivity in a material: low temperatures and low magnetic fields. In fact, there is a third condition, first identified by Silsbee [3], which is that the electrical current density must not exceed a certain critical value. One way to see this is to recall that a current through any wire generates a magnetic field with its maximum value at the surface of the conductor. Silsbee's criterion is that the self-generated field at the wire's surface cannot exceed the critical field, and that the current must be correspondingly small if the superconducting state is to persist.

The result is to set three interrelated conditions for superconductivity. Generally speaking, critical temperatures, fields, and currents scale with one another so that materials with high critical temperatures will also have large values of the other properties too.

The central organizing idea of the modern view of superconductivity is that, below Tc, large numbers of electrons participate in a common, macroscopic quantum state. This apparent violation of the Pauli exclusion principle is made possible by the rearrangement of electrons (actually holes in the high-temperature oxides) into Cooper pairs with net zero spin and Bose statistics. Leon Cooper first demonstrated the instability of the metallic Fermi surface to such a rearrangement whenever the net interaction between electrons is attractive, no matter how weak [4]. Cooper's insight led soon afterward to the microscopic theory of Bardeen, Cooper and Schrieffer (BCS) which appeared in 1957 [5].

In the standard BCS model, the net attraction is caused by electrons interacting with the vibrations of the crystal lattice which contains them. Since lattice-vibrational energy is quantized in units called phonons, the mechanism is called the electron-phonon interaction. The basic idea is that one electron can polarize a local region of the positively-charged crystal lattice. If the strength and dynamics of the polarization are right, then a second electron may be attracted toward the same region.

From detailed calculations based upon this mechanism, Bardeen, Cooper, and Schrieffer were able to develop a formula which relates the superconducting critical temperature to other material properties:

$$T_c = T_D \exp(-1/NV). \tag{1}$$

Here T_D is the Debye temperature which characterizes the lattice-vibrational energies, N is the density of electronic states at the Fermi level, and V is a measure of the strength of the attractive electron-electron interaction. For typical materials, $T_D \cong 300$ K which leads to $T_c \cong 1$-10 K.

An important indication of the electron-phonon mechanism is the predicted dependence of T_c upon the mass of the lattice ions. This dependence, called the "isotope effect", is a simple consequence of the fact that lattice energies, and hence T_D, scale with the frequencies of the lattice vibrations. These frequencies are characterized by the familiar mass-spring formula $\omega = \sqrt{k/m}$. Thus, the BCS formula predicts that T_c would vary as $1/\sqrt{m}$ for samples of the same element but of different isotopic composition. The discovery of this effect in 1950 was an important precursor of the microscopic theory [6].

High transition temperatures can be explained within the BCS framework and in the context of Equation 1 only by extraordinarily large values of the parameters T_D, N, or V. Many experiments to test the validity of the BCS model and of the electron-phonon mechanism have been performed on both conventional and oxide superconductors.

A great deal of attention has been given to the possibility of an isotope effect in these systems. Substitution of the rare earths for yttrium in Y-Ba-Cu-O with little effect upon Tc already indicates that phonons are relatively unimportant. Similarly, early reports indicated that the substitution of oxygen-18 for oxygen-16 had no influence upon transition temperatures. The evidence now shows a positive isotope effect in some samples which means that phonons in fact do play some role, albeit probably a secondary one [7].

A second class of experiments relates to the question of whether the product NV is large (strong coupling) or small (weak coupling). Marsiglio et al. have recently extended the theory to the extreme strong-coupling case and have predicted the dependence of several dimensionless ratios upon coupling strength [8]. While the experiments are not yet decisive, the evidence is beginning to mount in favor of weak coupling. The implied small values of the product NV would require a large (i.e. non-phonon) prefactor in Equation 1 in order to explain the high transition temperatures.

PHENOMENOLOGY

The remainder of this introduction will be concerned with the phenomenology of superconductivity as it derives from the concept of a macroscopic quantum state. The superconducting quantum state may be characterized by a complex wave-function $\Psi = \sqrt{n}\, \exp(i\phi)$ where n is the condensate density and ϕ is its phase. In the presence of a magnetic vector potential, $\hat{A}$, the quantum-mechanical expression for the probability current density is:

$$\vec{J}(\vec{r},t) = (1/m^*)\ \mathrm{Re}\ \{\Psi^*[-i\hbar\, \vec{\nabla} - q\hat{A}]\Psi\}. \tag{2}$$

The electrical current density is therefore,

$$\vec{J}_e = -2e \; \vec{J} = - (e \; n \; \hbar/m) \vec{\nabla}\phi - (2 \; e^2 \; n/m) \; \vec{A} \qquad (3)$$

where we have assumed that $m^* = 2m$ and $q = -2e$.

An insight into the physics of this constitutive relationship can be found in the limit of $\vec{\nabla}\phi = 0$. In this case taking a time derivative leads to

$$\dot{\vec{J}}_e = - (2 \; e^2 \; n/m) \; \dot{\vec{A}} \qquad (4)$$

or, equivalently,

$$m \; \dot{\vec{v}} = - e \; \vec{E} \qquad (5)$$

which is the equation of a freely-accelerating charge.

The new constitutive expression, Eq. 3, also leads directly to the Meissner effect. Consider the Maxwell equation $\vec{\nabla} \times \vec{B} = \mu_o \vec{J}_e$. For a superconductor of uniform phase, which occupies the half-space $x > 0$, Eq. 3 becomes:

$$- \partial^2 A_y/\partial x^2 = - (2 \; e^2 \; n \; \mu_o/m) \; A_y. \qquad (6)$$

This has solutions $A_y = A_y(o) \; \exp \pm(x/\lambda)$ where $\lambda = \sqrt{(m/2e^2 n \mu_o)}$ Only the decaying solution is acceptable. Thus $\vec{A}$ will disappear in a depth λ which is called the penetration depth. Since $\vec{A}$ goes to zero on this scale, so too will $\vec{B}$. For most superconductors, λ is on the order of one micron or less.

A second consequence of the equation for $\vec{J}_e$ involves a phenomenon called flux quantization. Consider the integral of $\vec{J}_e$ around any closed loop. Since ϕ is a phase, $\oint \vec{\nabla}\phi \cdot \vec{dl} = 2\pi n'$ where n' is an integer. Also $\oint \vec{A} \cdot \vec{dl} = \oint (\vec{\nabla} \times \vec{A}) \cdot \vec{ds} = \oint \vec{B} \cdot \vec{ds}$ is the magnetic flux enclosed by the loop. Therefore,

$$\oint \vec{J}_e \cdot \vec{dl} = - (e \; n \; \hbar/m) \; 2\pi n' - (2 \; e^2 n/m) \Phi \qquad (7)$$

or

$$\Phi + (m/2 \; e^2 \; n) \oint \vec{J}_e \cdot \vec{dl} = - (h/2e) \; n'. \qquad (8)$$

For a thick superconducting loop, the integral can be taken deep within the material where $\vec{J}_e = 0$. In this case, Φ is quantized in units of $(h/2e) = \phi_o$, the flux quantum. This is equavalent to just 0.2 microgauss·cm² which is a very small unit indeed.

In zero magnetic field (and for some materials even in finite external fields), the normal-to-superconducting transition is second order -- that is it exhibits no latent heat. In this sense, superconductors act as prototypes for the modern theory of second-order phase transitions and critical phenomena. A pioneering theory was that of Ginzburg and Landau (1951) [9]. Although it was originally phenomenological, this theory has since been rederived from a microscopic approach to be valid near Tc and for slow spatial variations of Ψ and $\vec{A}$.

The low-temperature, "ordered" phase can be characterized by an order parameter which grows somehow from zero at Tc to larger values at lower temperatures. In the simplest case, the superconducting free energy can be written as

$$G_s = G_m + \alpha(T) |\Psi|^2 + 1/2 \beta(T) |\Psi|^4. \tag{9}$$

G_s will have a minimum for finite $|\Psi|^2$ if $\alpha(T) < 0$ and for $|\Psi|^2 = 0$ if $\alpha(T) > 0$. $\beta(T)$ may be taken as a positive constant, while the choice of $\alpha(T) = \alpha'(T-T_c)$ satisfies the conditions just expressed. Appropriate expressions for the parameters $\alpha(T)$ and β have since been derived from microscopic theory.

The minimum of G_s is obtained when $|\Psi|^2 = -\alpha(T)/\beta$. In this case, the free energy difference is given by $-\alpha^2/2\beta$. Since this same difference can be identified with $\mu_0 H_c^2/2$, we can predict the temperature dependence of H_c itself:

$$H_c \propto (T - T_c). \tag{10}$$

This in fact agrees quite well with the observed temperature dependence near Tc.

The real success of the theory comes when the spatial variation of Ψ is taken into account. The idea is to add a term of the form $p^2/2m$ to represent kinetic energy. Quantum-mechanically, p introduces gradients and, to preserve gauge invariance, the vector potential, $\vec{A}$. Thus, in the presence of a field, the full Ginzburg-Landau free energy becomes:

$$G_s = G_m + \alpha(T) |\Psi|^2 + 1/2 \beta |\Psi|^4 + B^2/(2\mu_0)$$
$$+ (1/(2m^*) |-i\hbar\vec{\nabla}\Psi - q\vec{A}\Psi|^2 \tag{11}$$

where $m^* = 2m$ and $q = -2e$.

One next generates two equations by minimizing the integral of this energy function over all space with respect to Ψ and $\vec{A}$. This results in the famous Ginzburg-Landau equations:

$$[-\vec{\nabla} \times (\vec{\nabla} \times \vec{A})]/\mu_0 - (ie\hbar/m^*) [\Psi^*\vec{\nabla}\Psi - (\vec{\nabla}\Psi^*)\Psi]$$
$$- (4e^2/m^*) \vec{A} |\Psi|^2 = 0. \tag{12}$$

and

$$\alpha\Psi + \beta |\Psi|^2 \Psi + (1/2m^*)(-i\hbar\vec{\nabla} + 2e\vec{A})^2 \Psi = 0. \tag{13}$$

Several important observations may be made. For the spacial case of $\vec{A} = 0$ and $\vec{J}_e = 0$, we can write:

$$\alpha\Psi + \beta\Psi^3 - (\hbar^2/2m^*) d^2\Psi/dx^2 = 0 \tag{14}$$

in one dimension. It follows that the quantity $\sqrt{-\hbar^2/2m^*\alpha}$ has dimensions of a length, $\xi(T)$. This is called the "coherence length" and is a measure of the shortest distance over which the order parameter may vary without excessive cost in energy.

A second natural length, the penetration depth, is also apparent. If the spatial variations of Ψ are ignored, then we discover:

$$\vec{\nabla} \times (\vec{\nabla} \times \vec{A}) = - (4 \mu_o e^2 |\Psi|^2/m^*) \vec{A} \qquad (15)$$

which contains the length

$$\lambda(T) = \sqrt{m^*/(4 \mu_o e^2 |\Psi|^2)} \qquad (16)$$

This is essentially equivalent to the form introduced earlier.

The ratio of the two lengths $\lambda(T)/\xi(T)$ is an important dimensionless parameter called the Ginzburg-Landau κ.

Calculations of the critical magnetic field will serve to illustrate the usefulness of the Ginzburg-Landau theory as well as provide additional insight into the nature of superconductivity itself.

Imagine a superconductor in a large, uniform magnetic field. As the field is reduced, the superconducting phase will eventually nucleate. At this point, Ψ will be sufficiently small to allow linearization of the Ginzburg-Landau equations:

$$\alpha \Psi + (1/2m^*) (-i \hbar \vec{\nabla} + 2 e \vec{A})^2 \Psi = 0. \qquad (17)$$

Also, the vector potential $\vec{A}$ can be taken to describe just the external field since any corrections due to superconducting screening currents will be small. The resulting equations are formally identical to the case of a particle of mass m* and charge −2e in a uniform field. Thus, we immediately find:

$$- \alpha = 1/2 m^* v_y^2 + (n + 1/2) \hbar \omega_c \qquad (18)$$

where the cyclotron frequency $\omega_c = 2eB/m^*$. The lowest "energy" corresponds to the highest possible nucleation field:

$$B_{c2} = - \alpha m^*/(\hbar e). \qquad (19)$$

In terms of H_c, we obtain the important result $H_{c2} = \sqrt{2} \kappa H_c$. When κ exceeds $1/\sqrt{2}$, superconductivity will first appear above the thermodynamic field, Hc. This cannot be the ordinary Meissner phase since that would be energetically unfavorable. In fact, it is a new phase, called the "mixed state" or vortex phase, of the type-II superconductor. For κ greater than $1/\sqrt{2}$, the superconductor will be flux-excluding or type-I.

Alternatively, one can examine the energetics of a normal-line inclusion in a superconducting matrix exposed to a uniform magnetic field. As one moves away from the line, the magnetic field will diminish in a characteristic distance λ while the order-parameter will grow in its characteristic distance ξ. Energetically, one gains an energy $\mu_o H^2 \lambda^2/2$ by allowing the field to penetrate. However, one will lose an energy $\mu_o H^2 \xi^2/2$ by destroying some volume of condensate. The important consideration is the ratio of energy gain to energy loss. This will equal unity when $H \cong H_c/\kappa$. For large κ the gain will outweigh the loss, and flux penetration will be favorable, as long as $H > H_c/\kappa$ which is below the ordinary critical field. This defines a lower critical field H_{c1}. For small κ, the gain will only outweigh the loss for $H > H_c$ which is academic.

The high-temperature, oxide superconductors are anisotropic with extremely large values of κ. The anisotropy is well-illustrated by the unit cell of $Y_1Ba_2Cu_3O_7$ which is slightly orthorhombic with a length-to-width ratio of about three. It is now well-established that the superconducting critical-current densities are significantly larger for currents flowing in the a-b planes than for currents along the c-axis.

The large values of κ are due primarily to unusually small values of the coherence length in these materials. Furthermore, both characteristic lengths are themselves anisotropic. Thus, the magnetic penetration depth of $Y_1Ba_2Cu_3O_7$ will vary from approximately 0.1 to 1.0 microns depending upon the direction of the field with respect to the crystal axes. The coherence length is estimated to be from 22 to 31 Å in the a-b plane but only about 5 Å along the c-axis [10]. The consequent values of κ, roughly 35 and 230 respectively, mean that this material is extreme type-II.

Direct evidence for the existence of a mixed state in this material has been obtained by Gammel et al. [11] using a high-resolution Bitter pattern technique. The fluxoid lattice is strikingly similar to those seen in conventional, low-temperature superconductors. Moreover, by comparing the observed fluxoid density to the known field in which the samples were cooled, it was possible to verify that the flux-quantum involves the charge 2e in agreement with the standard concept of Cooper pairs.

The fluxoids of the mixed-state of Type-II superconductors have important consequences for transport properties, especially critical currents. In the presence of a transverse current, fluxoids experience a Lorentz force and have a tendency to move and dissipate energy. This can only be avoided if fluxoid motion is prevented by "pinning centers". Unfortunately, the extremely small coherence lengths which characterize the high-temperature superconductors complicate the meeting of this requirement. Tinkham [12] has recently suggested that zero resistance may not be achieveable at room temperature no matter how high T_c and H_{c2} might be, because of the inevitablility of flux creep.

The last important aspect of superconductivity which will be discussed here is the Josephson Effect discovered in 1962 [13]. Like many of the other properties which have been described, this too depends essentially upon the macroscopic quantum nature of superconductivity. The starting point is to consider two superconductors in close proximity -- a "Josephson junction". If the wave-functions on either side are assumed to be coupled, then it is a simple matter to derive a set of equations which relate the superconductive tunneling current, I, the quantum-mechanical phase difference across the junction, δ, and the potential difference, V [14]:

$$I = I_o \sin (\delta) \tag{20}$$

and

$$\dot{\delta} = (2\pi/\phi_o) V \tag{21}$$

where I_o is proportional to the coupling strength.

If a potential difference is applied to a Josephson junction, ac supercurrents at the frequency V/ϕ_o will result. Conversely, a junction exposed to microwave radiation of frequency f will exhibit steps in its current-voltage characteristics at applied potential differences equal to multiples of $\phi_o f$. The latter phenomenon has been observed in $Y_1 Ba_2 Cu_3 O_7$ by Tsai et al. [15]. Careful analysis of the data has confirmed the factor 2e in ϕ_o just as in the case of the fluxoid-lattice experiments described above.

CONCLUSIONS

It is impossible in a brief review to adequately cover all of the essential aspects of superconductivity. The main points which have been emphasized here are that the conventional phenomenology of low-temperature superconductivity provides a useful framework for characterizing many of the properties of the high-temperature oxide materials. Although the electron-phonon mechanism cannot account for the highest transition temperatures, the basic framework of the BCS theory, including the concept of Cooper pairs, appears still to have relevance.

The new superconductors present numerous challenges to the materials engineer because of their anisotropy and small coherence lengths as discussed above. Additional problems which have not been described include the need for precise control of both stoichiometry and morphology. These difficulties are no less challenging than those of the theorist who has yet to discover the essential source of superconductivity in these systems or even to understand many of their normal-state properties. In view of these deficiencies in our current knowledge, it is clear that much work remains to be done.

ACKNOWLEDGEMENTS

I would like to recognize the contributions of my colleagues and students, especially Professor Baki Yarar of the Department of Metallurgical and Materials Engineering at CSM. Our recent work has been supported by the Engineered Materials Processing Center of the Colorado Advanced Materials Institute and also by the Colorado Center for Advanced Ceramics.

REFERENCES

1. J. Bednorz and K. A. Mueller, <u>Z. Phys.</u> B64:189 (1986).

2. W. Meissner and R. Ochsenfeld, <u>Naturwissenschaften</u> 21:787 (1933).

3. F. B. Silsbee, <u>J. Wash. Acad. Sci.</u> 6:597 (1916).

4. L. N. Cooper, <u>Phys. Rev.</u> 104:1189 (1956).

5. J. Bardeen, L. N. Cooper, and J. R. Schrieffer, <u>Phys. Rev.</u> 106:162 (1957).

6. C. A. Reynolds, B. Serin, W. H. Wright, and L. B. Nesbitt, <u>Phys. Rev.</u> 78:487 (1950); W. D. Allen, R. H. Dawton, M. Bar, K. Mendelssohn, and J. L. Olsen, <u>Nature</u> 166:1071 (1950).

7. B. Batlogg, G. Kourouklis, W. Weber, R. J. Cava, A. Jayaraman, A. E. White, K. T. Short, L. W. Rupp, and E. A. Rietman, <u>Phys. Rev. Lett.</u> 59:912 (1987); T. A. Faltens, W. K. Ham, S. W. Keller, K. J. Leary, J. N. Michaels, A. M. Stacy, H.-C. zur Loye, D. E. Morris, T. W. Barbee, L. C. Bourne, M. L. Cohen, S. Hoen, and A. Zettl, <u>Phys. Rev. Lett.</u> 59:915 (1987); K. J. Leary, H.-C. zur Loye, S. W. Keller, T. A. Faltens, W. K. Ham, J. N. Michaels, and A. M. Stacy, <u>Phys. Rev. Lett.</u> 59:1236 (1987); H. Katayama-Yoshida, T. Hirooka, A. J. Mascarenhas, Y. Okabe, T. Takahashi, T. Sasaki, A. Ochiai, T. Suzuki, J. I. Pankove, T. Ciszek, and S. K. Deb, <u>Japanese J. Appl. Phys.</u> 26:L2085 (1987).

8. F. Marsiglio, R. Akis, and J. P. Carbotte, <u>Phys. Rev.</u> 36:5245 (1987).

9. V. L. Ginzburg and L. D. Landau, <u>Zh. Eksp. Teor. Fiz.</u> 20: 1064 (1950).

10. G. W. Crabtree, W. K. Kwok, and A. Umezawa, Basic Properties of Copper Oxide Superconductors, <u>in</u>: "Quantum Field Theory as an Interdisciplinary Basis," F. C. Khanna, H. Umezawa, G. Kunstatter, and H. C. Lee, eds., World Science Publ. Co. Ltd., London (1988).

11. P. L. Gammel, D. J. Bishop, G. J. Dolan, J. R. Kwo, C. A. Murray, L. F. Schneemeyer, and J. V. Waszczak, <u>Phys. Rev. Lett.</u> 59:2592 (1987).

12. M. Tinkham, <u>Phys. Rev. Lett.</u> 61:1658 (1988).

13. B. D. Josephson, <u>Phys. Lett.</u> 1:251 (1962).

14. R. P. Feynman, R. B. Leighton, M. Sands, "The Feynman Lectures on Physics," Addison-Wesley Publishing Company, Reading (1965).

15. J. S. Tsai, Y. Kubo, and J. Tabuchi, <u>Phys. Rev. Lett.</u> 58:1979 (1987).

FERMIOLOGY OF HIGH T_C FILMS AND THE PROXIMITY EFFECT

Vladimir Z. Kresin

Stuart A. Wolf

Lawrence Berkeley Laboratory
1 Cyclotron Road
Berkeley, California 94720

Naval Research Laboratory
Washington, D. C. 20375

ABSTRACT

Thin superconducting films can be used to determine the major parameters of the new high T_c oxides. Magneto-acoustic experiments look
particularly promising. A noticeable decrease in T_c which is important
for experimental Fermiology, can be obtained with the use of an N-S-N
proximity system. The evolution of the Fermi surface with carrier
concentration can also be studied.

INTRODUCTION

As is known, the proximity effect in an S-N system allows one to
induce the superconducting state in the N film. On the other hand, the
presence of the N film depresses superconductivity in S film, so that
$T_c < T_c^S$ and $H_{cs} < H_{c2}^S$, where T_c^S and H_{c2}^S correspond to an isolated S
film. This decrease in T_c and H_{c2}, which usually is considered a
negative side of the proximity effect, can be used to analyze the normal
properties of the new high T_c materials. Indeed, the usual methods
(see below) frustrated by the high values of T_c and H_{c2}. It will be
shown in this paper that the use of thin and N-S-N proximity systems
will allow one to overcome this problem.

Fermiology: Analysis of Thin Films

The analysis of the normal and superconducting properties of
cuprates based on the Fermiology was proposed in our papers [1,2] and
then developed by G. Deutscher and the authors in [3]. The analysis is
carried out in momentum space; the anisotropy of the system is manifested in the shape of the Fermi surface (FS). FS can be reconstructed
with the aid of special experiments (see, e.g., ref. 4).

The major parmeters of the La-Sr-Cu-O and Y-Ba-Cu-O system were
evaluated in [1-3] with the use of heat capacity [5] and Hall effect
[6] data. We think that the small values of ϵ_F and v_F are the key

features of the oxides. Note that positron annihilation results [7] show the presence of the cylindrical part of FS in accordance with [1].

It would be interesting to employ the usual Fermiology techniques (see, e.g., ref. 4); then a detail reconstruction of FS is possible. In this connection, we think that the use of thin films as parts of proximity junctions is very promising.

Fermiology of thin films can be studied by several methods (see, e.g., ref. 8). For example, the cyclotron resonance can be used to determine m^*. The most promising method is ultrasound attenuation in an external magnetic field H. The attenuation oscillates as a function of H^{-1}, and the period is directly related to the cross section of FS: $\Delta \sim 2p_{\perp}^* H^{-1}$, [8,9] where $2p_{\perp}$ is the external diameter of the cross section of FS in the direction of sound wave propagation. By changing the direction of the ultrasonic wave and the orientation of the film relative to the interface, one can reconstruct FS in an effective way.

The main problem with this method is that it is necessary to carry out the experiment at low temperatures because the condition $q\ell \gg 1$ must be satisfied (q is the wave vector $\ell \simeq V_F \tau$ is the mean free path, τ is the relaxation time; ℓ decreases with increasing T). At low temperatures relaxation is not essential, and the electron's motion reflects the shape of FS. Note that electron-phonon scattering is the major mechanism for the attenuation of sound (see, e.g., ref. 10).

The normal properties of the high T_C oxides can be studied at T > T_c (or at T < T_c, if the superconducting state is destroyed by the critical field). Because of the large value of T_c it is difficult to meet the criterion $q\ell \gg 1$.

Fermiology and the Proximity Effect

Consider an N-S-N proximity system, where N is a normal film (e.g., Cu or Sb), and S is a high T_c film. For concreteness we consider La-Sr-Cu-O. Of course, one can also use an usual S-N junction, but the effect T_c depression will be stronger for the N-S-N system. It is desirable to have a thin S film with $d_s \lesssim 4.10^2$ Å; this is important because of the short coherence length in La-Sr-Cu-O, see, refs. 1 and 2, but, nevertheless, a theoretical analysis and the experimental data for dirty conventional films (see, e.g., refs. 11 and 12; the "dirty" limit is also characterized by a short coherence length) show that one can achieve a noticeable decrease of T_c relative to T_c^S. In addition, one should note that $H_{c2} < H_{c2}^S$; this follows directly from the expression[13] $H_{c2} = H_{c2}^S (1+n \ d_n/d_s)^{-1}$, where $n \simeq \sigma_n/\sigma_s$ (d_s, d_n are the thicknesses of the films and σ_s, σ_n are their normal conductivities). As a result, the value of $T_c \simeq 20-25$ K is perfectly realistic. Note also that ultra-sound attenuation in conventional proximity systems have been studied in several papers (see, e.g., ref. 14).

Consider the criterion $\alpha \equiv q\ell \gg 1$ for a La-Sr-Cu-O film. The relaxation time can be estimated from normal conductivity data because

$\sigma = ne^2 \tau/m*$. Thus we have $\alpha \simeq 2\pi f\sigma v_F (ne_2 u)^{-1}$, where n is the carrier concentration, and u and f are the sound velocity and frequency, respectively. Using the values $n \simeq 3 \times 10^{21}$ cm^{-1} [6], $v_F/u \simeq 10^2$ [1,2,15] σ(at $T \simeq 3 \times 10^2$ K) $\simeq 5 \times 10^3$ ohm^{-1} cm^{-1},[16] and $T \simeq 20$ K (such a low temperature can be obtained with the use of the proximity effect), we obtain $\alpha \simeq 3$. We have assumed linear dependence of the resistivity on T: $\sigma \simeq (\sigma_0/T)$. As a matter of fact, such a dependence has been observed in the region $T > T_c^S$ (see, e.g., ref. 16 and the theoretical analysis in ref. 3). It is possible that the decrease of T will lead to a sharper temperature dependence of σ at $T \lesssim 20$ K. In addition one should also take into account the residual term. These factors lead to a still larger value of α; hence the criterion $\alpha \gg 1$ can be met.

Dynamics of Fermiology and Doping

It is known that T_c is strongly affected by doping. For example, for La-Sr-Cu-O, $T_c \simeq 10$ K at $n \simeq 10^{21}$ cm^{-3} [17] (a similar effect has been observed in Y-Ba-Cu-O). In this connection, it would be interesting to measure the major parameters for different values of n. For example, one can study the superconducting of La-Sr-Cu-O at low temperatures ($T \simeq 10$ K) if $n \simeq 10^{21}$ cm^{-1} $< n_{max}$ (n_{max} corresponds to $T_c^{max} \simeq 36$ K). The proximity effect will allows one to decrease further even this low value of T_c. Such an analysis is of definite interest, and can be carried out by the usual methods of Fermiology.

CONCLUSION

The use of N-S-N proximity systems with high T_c films, is of definite interest, because it allows one to decrease T_c and to develop the Fermiology of thin films. In addition, it would be interesting to study the Fermiology of the high T_c materials with different degrees of doping. Magneto-acoustic experiments will allow one to reconstruct Fermi surface of the new materials. Recent progress in thin film technology makes such an experiment perfectly realistic, and it will be an important step in the understanding of the origin of high T_c.

ACKNOWLEDGMENT

The research of VZK was supported by the Office of Naval Research under Contract No. N00014-87-F0015 and carried out at Lawrence Berkeley Laboratory under Contract No. DE-AC03-76SF00098.

REFERENCES

1. V. Kresin and S. Wolf, Solid State Comm. 63:1141 (1987); J. Superconductivity, 1:143 (1988).

2. V. Kresin and S. Wolf, Novel Superconductivity, S. Wolf and V. Kresin, eds., Plenum, New York (1987), p.287.

3. V. Kresin, G. Deutscher, and S. Wolf, J. Superconductivity, in press; Proc. Latin-Amer. Conf. on High T_c (Rio de Janeiro, 1987) World, Singapore, in press.

4. A. Gracknell and K. Wong, The Fermi Surface, Clarendon Press, Oxford (1973).

5. N. Phillips, et al., ref. 2; R. Fisher, J. Gordon, and N. Phillips (review), J. Superconductivity, in press.

6. A. Braginskii, ref. 2; W. Kwok, et al., Phys. Rev B. 36:5343 (1987); N. Ong, et al., Phys. Rev. B 35:8807 (1987).

7. L. Smedskjaer, et al., Physica C, in press.

8. V. Kresin and B. Kokotov, Sov. Phys.-JETP 48:537 (1978).

9. V. L. Gurevich, Sov. Phys.-JETP 10:1190 (1960).

10. B. Geilikman and V. Kresin, Kinetic and Nonsteady-State Effects in Superconductors, Wiley, New York (1974).

11. G. Deutscher and P. G. de Gennes, Superconductivity, R. Parks, ed., Marced Dekker, New York (1969), p. 1005.

12. G. Deutscher, S. Hsieh, P. Lindenfeld, and S. Wolf, Phys. Rev. B 8:5055 (1979).

13. K. Biagi, V. Kogan, and J. Clem, Phys. Rev. B 32:7165 (1985).

14. Yu. Kolesnickensko, Sov. J. Low Temp. Phys. 11:386 (1985).

15. D. Bishop, et al., ref. 2, p. 659.

16. M. Gurvitch, et al., Physica C 153:1369 (1988).

17. J. Torrance, et al., Phys. Rev. Lett. 61, 1127 (1988).

THE HIGH Tc TL-BA-CA-CU-O SUPERCONDUCTING SYSTEM

A.M.Hermann and Z.Z.Sheng

Department of Physics
University of Arkansas
Fayetteville, AR 72701

INTRODUCTION

Discoveries of 30-K La-Ba-Cu-O superconductor [1] and 90-K Y-Ba-Cu-O superconductor [2] have stimulated a worldwide race for new and even higher temperature superconductors. Breakthroughs were made by the discoveries of the 90-K Tl-Ba-Cu-O system [3,4], 110-K Bi-Sr-Ca-Cu-O system [5,6] and 120-K Tl-Ba-Ca-Cu-O system [7-9]. Recently, high temperature superconductivity was also observed in the Tl-Sr-Ca-Cu-O system [10-12], and in the M-Tl-Sr-Ca-Cu-O with M = Pb [13,14] and rare earths [15]. In this paper, we present preparation procedures, structure, and some properties of the 120-K Tl-Ba-Ca-Cu-O superconductors. We discuss an unusual levitation phenomenon of the Tl-Ba-Ca-Cu-O superconductor due to flux pinning [16]. Finally, we present a new Tl_2O_3-vapor-process [17] which allows the highest temperature Tl-Ba-Ca-Cu-O superconductors to be easily made in the forms of complex bulk components, wires and fibers, and thick and thin films, and minimizes problems caused by toxicity and volatility of Tl starting compounds. Recent results on Tl203 vapoer-processing of thin film Ba-Ca-Cu-O precursors are included.

PREPARATION

Tl-Ba-Ca-Cu-O superconductive compounds form easily; there are many ways to make good-quality superconducting samples. One of the typical procedures in preparing the Tl-Ba-Ca-Cu-O samples which we use is the following. Ba-Ca-Cu-oxides are first prepared using the method similar to that we previously developed for preparation of Ba-Cu-oxides [18,19]. Appropriate amounts of $BaCO_3$, CaO (or $CaCO_3$), and CuO are mixed, ground, and heated at 925-950 $^{\circ}$C in air for 24-48 hour with several intermediate grindings. The resulting uniform black material is ground and served as master material. Appropriate amounts of Tl_2O_3 and Ba-Ca-Cu oxide (depending on the desired stoichiometry) are completely mixed and ground, and pressed into a pellet with a diameter of 7 mm and a thickness of 1-2 mm. The pellet is then put into a tube furnace which had been heated to 880-910 $^{\circ}$C, and is heated for 2-5 minutes in flowing oxygen, followed by furnace cooling to below 200 $^{\circ}$C.

STRUCTURE

The Tl-Ba-Ca-Cu-O system can form a number of superconducting phases.

Two phases, $Tl_2Ba_2Ca_2Cu_3O_{10+x}$ (2223) and $Tl_2Ba_2Ca_1Cu_2O_{8+x}$ (2212), were first identified [20]. The 2223 superconductor has a 3.85 x 3.85 x 36.25 Å tetragonal unit cell. The 2212 superconductor has a 3.85 x 3.85 x 29.55 Å tetragonal unit cell [20,21]. The 2223 phase is related to 2212 by addition of extra calcium and copper layers. In addition, the superconducting phase in the Ca-free Tl-Ba-Cu-O system is $Tl_2Ba_2CuO_{6+x}$ (2201) [20,22]. Fig. 1 shows schematically the arrangements of metal atom planes in these three Tl-based superconducting phases. The 2201 phase has a zero-resistance temperature of about 80 K, whereas the 2212 and 2223 phases have zero-resistance temperatures 108 K and 125 K, respectively [20-25]. It appears that the addition of each Ca and Cu layer increases the transition temperature about 20 K. If this trend continues linearly, it might be expected that 2234 phase will have a transition temperature at 140-150 K.

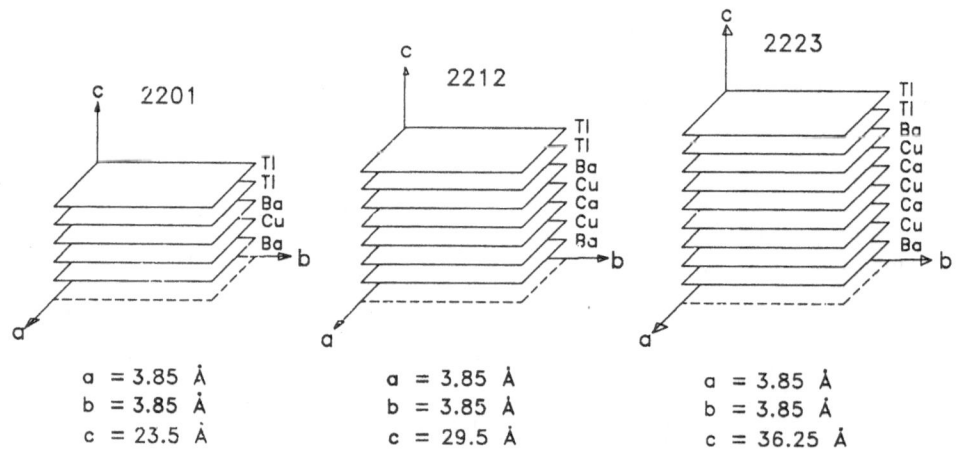

Figure 1   Schematic arrangements of the Tl-based
superconducting phases 2201, 2212 and 2223.

A new series of superconducting compounds with a single Tl-O layer, which we denote by $TlBa_2Ca_{n-1}Cu_nO_{2n+2.5}$, were recently also reported [26,27]. The Tl-Ba-Ca-Cu-O superconducting series should be represented

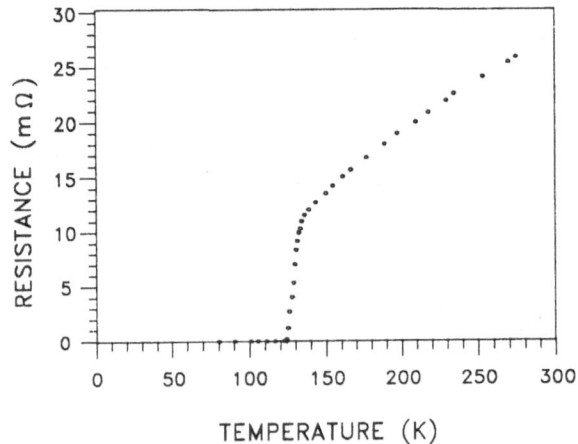

Figure 2   Resistance-temperature dependence of
a nominal $Tl_{2.2}Ba_2Ca_2Cu_3O_{10.3+x}$ sample.

using a general formula of $Tl_mBa_2Ca_{n-1}Cu_nO_{1.5m+2n+1}$ with m = 1 and 2, and n = 1, 2, 3, and 4. The Tc of the single Tl-O layer compounds also increases with the number of Cu-Ca layers, and is slightly lower than that of the corresponding double Tl-O layer compounds. Therefore, an increase of Tc might be achieved not only by increasing the number of Ca-Cu layers, but also by increasing the number of Tl-O layers.

TRANSITION TEMPERATURE

Fig. 2 shows resistance-temperature variation for a nominal $Tl_{2.2}Ba_2Ca_2Cu_3O_{10.3+x}$ sample. This sample has an onset temperature near 140 K, midpoint of 127 K, and zero resistance temperature at 122 K.

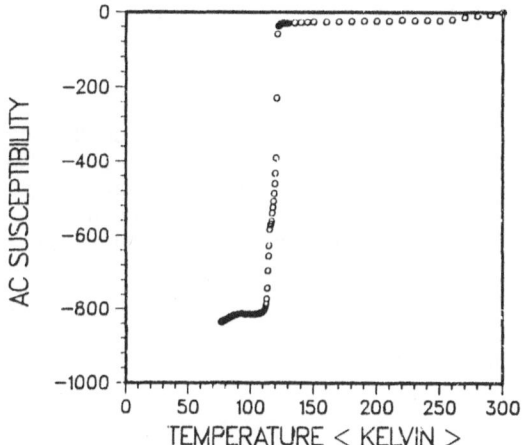

Figure 3   Temperature dependence of AC susceptibility for a nominal $Tl_2Ba_2Ca_2Cu_3O_{10+x}$ sample.

Fig. 3 shows a similar transition temperature for the 2223 phase by 5 kHz AC susceptibility measurements. The onset of the transition is 123 K.

LEVITATION

Properly prepared Tl-Ba-Ca-Cu-O samples can easily levitate over a magnet. Careful observations have shown that for some Tl-based samples, the force between the sample and the magnet is complicated. In particular, the Tl-Ba-Ca-Cu-O samples specially prepared can be suspended above, below, or to the side of a magnet. Fig. 4a shows two Tl-based superconductors suspended horizontally above and <u>below</u> a ring-shaped magnet. A similar magnetic effect showing the coexistence of repulsive and attractive forces was reported for some Y-Ba-Cu-O/AgO samples [28]. Fig. 4b shows two Tl-based superconductors suspended vertically near a ring-shaped magnet. The downward flowing nitrogen vapor is evident in figures 4a and 4b. The levitation beneath or at the side of the magnet clearly involves flux penetration into the non-superconducting regions and corresponding attractive supercurrents which are pinned. Corresponding large residual positive magnetic susceptibility following application of a magnetic field has been observed experimentally for the samples showing unusual levitation [29].

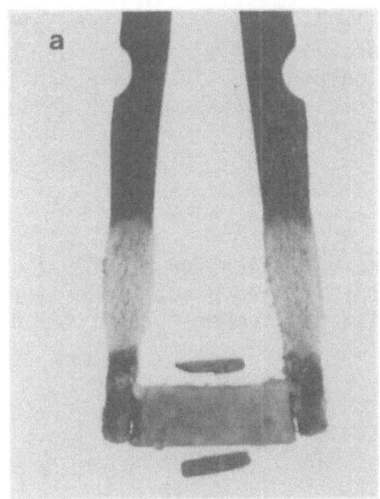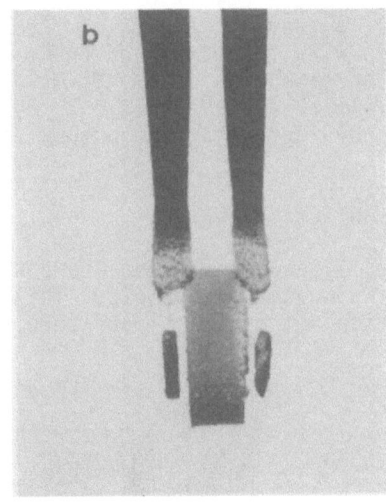

Figure 4 (a) two Tl-based superconductors
suspended horizontally above and
below a ring shaped magnet, and
(b) two Tl-based superconductors
suspended vertically near a ring
shaped magnet.

TL_2O_3-VAPOR PROCESS

In extensive preparation experiments on Tl-based superconductors, we
have found that Tl_2O_3 evaporates above its 717°C melting point, and the
vapor reacts with solid Ba-Ca-Cu-oxides, forming high-quality Tl-Ba-Ca-Cu-
O superconductors. This vapor-solid reaction has simplified the making of
Tl-Ba-Ca-Cu-O superconductors to the making of Ba-Ca-Cu-oxides, and this
minimizes the toxicity problem of Tl starting compounds. In particular,
this Tl_2O_3-vapor-process allows the Tl-Ba-Ca-Cu-O superconductors to be
easily made in the forms of complex bulk components, wires and fibers, and
thick and thin films by fabrication of the precursor Ba-Ca-Cu-oxides and
subsequent introduction of Tl [17].

An appropriate amount of the Ba-Ca-Cu-oxide powder was pressed into a
pellet, and the pellet was heated at 925-950 °C in flowing oxygen for 5-10
minutes and then was air-cooled. A small platinum boat was put in a
quartz boat, and a small amount of Tl_2O_3 (typically 0.1-0.2 gram) was put
into the platinum boat. The heated Ba-Ca-Cu-O pellet was put above the
platinum boat. The quartz boat with the contents was put into a tube
furnace, which had been heated to 900-925 °C, and was heated for about 3
minutes in flowing oxygen followed by furnace-cooling.

After the above heating treatment, the Tl_2O_3 in the platinum boat had
completely evaporated, and the Tl_2O_3-vapor-processed Ba-Ca-Cu-O pellet
formed a layer of Tl-Ba-Ca-Cu-O superconducting compound(s) on its bottom
surface. Figure 5 shows temperature dependences of resistance for some
Tl_2O_3-vapor-processed Ba-Ca-Cu-O precursors with the following
compositions: (A) $BaCa_3Cu_3O_7$, (B) $Ba_2CaCu_2O_5$, and (C) $Ba_2Ca_2Cu_3O_7$. Their
zero resistance temperatures are 110, 96 and 105 K, respectively, which
are comparable to those of corresponding sintered samples.

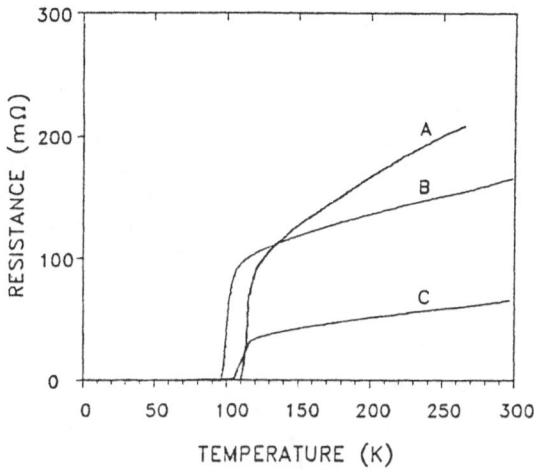

Figure 5   Temperature dependences of resistance
for Tl_2O_3-vapor-processed Ba-Ca-Cu-O
precursors: (A) $BaCa_3Cu_3O_7$, (B) $Ba_2CaCu_2O_5$,
and (C) $Ba_2Ca_2Cu_3O_7$.

In principle, this technique has simplified the preparation of Tl-Ba-
Ca-Cu-O to the preparation of Ba-Ca-Cu-O. Therefore, this technique
allows Tl-Ba-Ca-Cu-O superconducting components, such as bulk components,
wires and fibers, thick and thin films, to be easily made. In particular,
Tl_2O_3-vapor-processed molten Ba-Ca-Cu-oxides are also superconducting, and
thus this technique allows superconducting components to be easily made in
arbitrary shape. Figure 6 shows resistance-temperature dependences for a
Tl_2O_3-vapor-processed $Ba_2Ca_2Cu_3O_7$ thick wire (A) and for a Tl_2O_3-vapor-
processed $Ba_2Ca_2Cu_3O_7$ thick film (B). The thick film was prepared by
first melting $Ba_2Ca_2Cu_3O_7$ on a platinum substrate (heating at 980 ^oC in
flowing oxygen for 5 minutes) and then subjecting to Tl_2O_3-vapor-
processing. It reached zero resistance at 111 K. It must be emphasized
that this technique is of particular importance in making superconducting
Tl-Ba-Ca-Cu-O thin films [30,31], and recent thin film results are
described below.

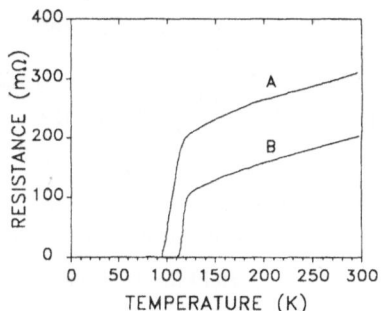

Figure 6   Resistance-temperature dependences for
a Tl_2O_3-vapor-processed $Ba_2Ca_2Cu_3O_7$ thick
wire (A) and a Tl_2O_3-vapor-processed
$Ba_2Ca_2Cu_3O_7$ thick film (B).

Experiments have also shown that some elements, although they cannot completely replace any component element in the Tl-Ba-Ca-Cu-O superconducting system, do not influence or influence slightly the superconducting behavior of the Tl-Ba-Ca-Cu-O system. We found that addition of these elements to the Ba-Ca-Cu-O precursors, followed by Tl-vapor-processing, can form various Tl-based superconductors which may satisfy various special requirements for practical applications. Fig. 7 shows resistance versus temperature for a Tl_2O_3-vapor-processed $In_2Ba_2Ca_2Cu_3O_{10}$ sample. Note that the $In_2Ba_2Ca_2Cu_3O_{10}$ itself is not superconducting with our preparation conditions.

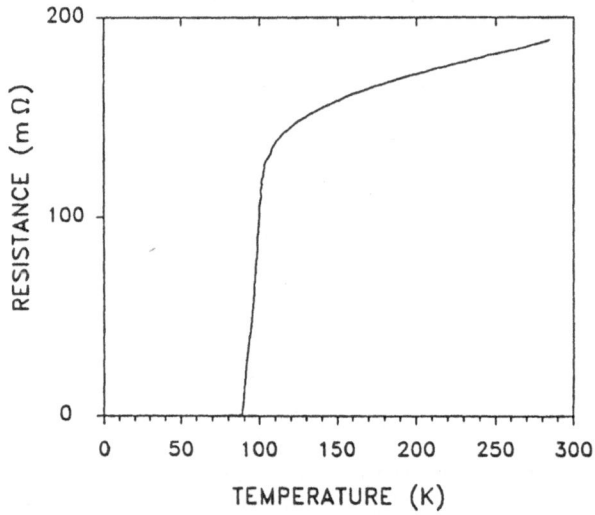

Figure 7   Resistance-temperature dependence
            for a Tl_2O_3-vapor-processed
            $In_2Ba_2Ca_2Cu_3O_{10}$.

THIN FILM

Thin film (2-3 micron) Ba-Ca-Cu-O precursors were deposited by N.J. Ianno, J.A. Woollam, and co-workers of the University of Nebraska by

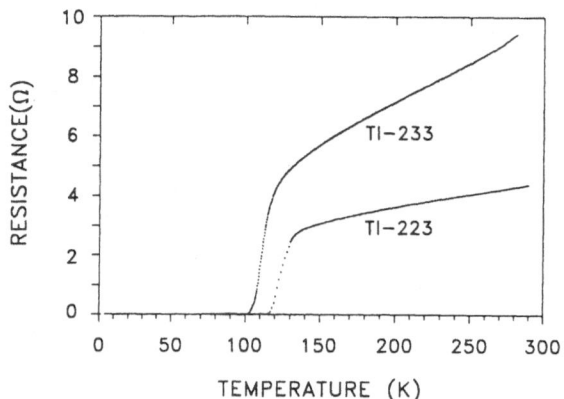

Fig. 8   Resistance-temperature dependences for
          a Tl_2O_3-vapor-processed $Ba_2Ca_2Cu_3O_7$ thin
          film precursor (Tl-223) and for a Tl_2O_3-
          vapor-processed $Ba_2Ca_3Cu_3O_8$ thin film
          precursor (Tl-233)

laser deposition onto Y-stabilized ZrO_2 substrates using a frequency-doubled Nd-YAG laser operating at 523 nm. These films were then treated at 900 $^\circ$C for 3 minutes in flowing oxygen and Tl_2O_3 vapor as described previously. Fig. 8 shows the resistivity-temperature dependences for a Tl_2O_3-vapor-processed $Ba_2Ca_2Cu_3O_7$ precursor and a Tl_2O_3-vapor-processed $Ba_2Ca_3Cu_3O_8$ precursor. Their zero-resistance temperatures are 115 K and 104 K, which are comparable the those of sentered Tl-based samples. The electronic properties of these films are highly process-dependent. Critical current-density measurements are in progress.

BEST PROPERTIES OF TL-BASED SUPERCONDUCTORS - A REVIEW

In an effort to help readers compare rare-earth-oxide-based and bismuth-based high temperature superconductors with thallium-based superconductors, we have prepared the following table (table 1) showing the best properties of Tl-based superconductors reported by a wide variety of researches.

Table 1. Properties of Tl-Ba-Ca-Cu-O Superconductors
----

* $T_c(R=0)$:

125 K,	bulk	IBM [25]
120 K,	thick film	Arkansas
120 K,	thin film	IBM [35]

* $J_c(77K, B=0)$:

22,000 A/cm^2,	bulk	Sandia [33]
300 A/cm^2,	intergranular	Oak Ridge [34]
>100,000 A/cm^2,	intragranular	Oak Ridge [34]
6,400 A/cm^2,	wire	Hitachi [35]
240,000 A/cm^2,	poly-thin film	Sandia [36]
3,200,000 A/cm^2,	poly/epi. thin film	Sumitomo [35]

* B dependence of $J_c(77K)$:

< 10-fold drop at 1 T ($B \perp J$, $B /\!/$ film plane)	Sandia [36]
10-fold drop at 1 T ($B /\!/$ film plane)	Sumitomo [37]

* $H_{c1}(T=0)$ 0.05 T   Tohoku University

* $H_{c2}(T=0)$ >100 T   Houston, Arkansas

* Thermal and chemical stability:

- No degradation in Tc at 800 $^\circ$C in air for several hours
- Oxygen stability - excelent
(no crystallographic transformation linked to oxygen loss)
- Sensitive to H_2O - yes   for sentered samples,
                - modest for melt samples (Arkansas)

* Weak Links:

films can be made with or without weak links (Sandia [39])

* Surface superconducting: ??

Many deductions can be made from the table:

1. The high Tc of the thallium compounds is extremely important for electronic devices where the Tc should be close to twice the operating temperature.

2. The rich variety of high Tc phases is also important for fabrication of wire and devices where often off-stoichiometry and less than ideal processing are achieved.

3. The critical currents are already high enough for many electronic devices.

4. The ability to form Tl-compounds in strong-linked form with only weak magnetic field dependence of Jc is important.

5. The Tl is bound tightly in superconductor once the compounds is formed.

6. Oxygen is tightly bound; there is no orthorhombic to tetragonal insulator transition.

An important question yet to be answered involves surface characterization of thin films (are the surface insulating as is the case for rare earth oxide superconductors - a severe difficulty for Josephson junction device fabrication?). An important problem to be solved involves the mechanical properties of Tl-based superconducting wire: the fabrication of wire with good mechanical strength.

These and other questions and problems are being addressed in many laboratories now.

REFERENCES

1) J.G.Bednorz and K.A.Muller, Z.Phys.B 64, 189 (1986).
2) M.K.Wu, J.R.Ashburn, C.T.Torng, P.H.Hor, R.L.Meng, L.Gao, Z.J.Huang, Y.Q.Wang, and C.W.Chu, Phys.Rev.Lett. 58, 908 (1987).
3) Z.Z.Sheng and A.M.Hermann, Nature 332, 55 (1988).
4) Z.Z.Sheng, A.M.Hermann, A.El Ali, C.Almason, J.Estrada, T.Datta, and R.J.Matson, Phys.Rev.Lett. 60, 937 (1988).
5) H.Maeda, Y.Tanaka, M.Fukutomi, and T.Asano, Jpn.J.Appl.Phys.Lett. 27, L207 (1988).
6) C.W.Chu, J.Bechtold, L.Gao, P.H.Hor, Z.J.Huang, R.L.Meng, Y.Y.Sun, Y.Q.Wang, and Y.Y.Xue, Phys.Rev.Lett. 60, 941 (1988).
7) Z.Z.Sheng and A.M.Hermann, Nature 332, 138 (1988).
8) Z.Z.Sheng, W.Kiehl, J.Bennett, A.El Ali, D.Marsh, G.D.Mooney, F.Arammash, J.Smith, D.Viar, and A.M.Hermann, Appl.Phys.Lett. 52, 1738 (1988).
9) A.M.Hermann, Z.Z.Sheng, D.C.Vier, S.Schultz, and S.B.Oseroff, Phys.Rev.B 37, 9742 (1988).
10) Z.Z.Sheng, A.M.Hermann, D.C.Vier, S.Schultz, S.B.Oseroff, D.J.George, and R.M.Hazen, Phys.Rev.B (to be published).
11) W.L.Lechter, M.S.Osofsky, R.J.Soulen,Jr., V.M.LeTourneau, E.F.Skelton, S.B.Qadri, W.T.Elam, H.A.Hein, L.Humphreys, C.Skowronek, A.K.Singh, J.V. Gilfrich, L.R.Toth, and S.A.Wolf (submitted).
12) S.Matsuda, S.Takeuchi, A.Soeta, T.Suzuki, K.Aihara, and T.Kamo (submitted).
13) M.A.Subramanian, C.C.Torardi, J.Gopalakrishnan, P.L.Gai, J.C.Calabrese, T.R.Askew, R.B.Flippen, and A.M.Sleight (submitted).
14) Z.Z.Sheng and A.M.Hermann (unpublished).
15) Z.Z.Sheng, L.Sheng, X.Fei, and A.M.Hermann (submitted).
16) W.G.Harter, A.M.Hermann, and Z.Z.Sheng, Appl.Phys.lett. 53, 1119 (1988).
17) Z.Z.Sheng, L.Sheng, H.M.Su, and A.M.Hermann, Appl.Phys.Lett. (accepted).

18) A.M.Hermann and Z.Z.Sheng, Appl.Phys.Lett. 51, 1854 (1987).
19) A.M.Hermann, Z.Z.Sheng, W.Kiehl, D.Marsh, F.Arammash, A.El Ali,
    G.D.Mooney, L.Sheng, J.A.Woolam and A.Ahmed, Appl.Phys.Comm. 7, 275
    (1987).
20) R.M.Hazen, L.W.Finger, R.J.Angel, C.T.Prewitt, N.L.Ross,
    C.G.Hadidiacos, P.J.Heaney, D.R.Veblen, Z.Z.Sheng, A.El Ali, and
    A.M.Hermann, Phys.Rev.Lett. 60, 1657 (1988).
21) L.Gao, Z.J.Huang, R.L.Meng, P.H.Hor, J.Bechtold, Y.Y.Sun, C.W.Chu,
    Z.Z.Sheng, and A.M.Hermann, Nature 332, 623 (1988).
22) C.C.Torardi, M.A.Subramanian, J.C.Calabrese, J.Gopalakrishnan,
    E,M.McCarron, K.J.Morrissey, T.R.Askew, R.B.Flippen, U.Chowdhry, and
    A.M.Sleight, Phys.Rev.B 38, 225 (1988).
23) M.A.Subramanian, J.C.Calabrese, C.C.Torardi, J.Gopalakrishnan,
    T.R.Askew, R.B.Flippen, K.J.Morrissey, U.Chowdhry, and A.M.Sleight,
    Nature 332, 420 (1988).
24) C.C.Torardi, M.A.Subramanian, J.C.Calabrese, J.Gopalakrishnan,
    K.J.Morrissey, T.R.Askew, R.B.Flippen, U.Chowdhry, and A.M.Sleight,
    Science 240, 631 (1988).
25) S.S.P.Parkin, V.Y.Lee, E.M.Engler, A.I.Nazzal, T.C.Huang, G.Gorman,
    R.Savoy, and R.Beyers, Phys.Rev.Lett. 60, 2539 (1988).
26) Y.Luo, Y.L.Zhang, J.K.Liang, and K.K.Fung (submitted).
27) S.S.P.Parkin, V.Y.Lee, A.I.Nazzal, R.Savoy, R.Beyers, and S.J.La
    Placa, Phys.Rev.Lett. 61, 750 (1988).
28) P.N.Peters, R.C.Sisk, E.W.Urban, C.Y.Huang, and M.K.Wu,
    Appl.Phys.lett. 52, 2066 (1988).
29) Z.Z.Sheng, Y.H.Liu, X.Fei, L.Sheng, C.Dong, W.G.Harter, A.M.Hermann,
    D.C.Vier, S.Schultz, and S.B.Oseroff (submitted).
30) C.X.Qiu and I.Shih, Appl.Phys.lett. 53, 523 (1988).
31) C.X.Qiu and I.Shih, Appl.Phys.lett. 53, 1122 (1988).
32) S.H.Lion, N.J.Ianno, B.Johns, D.Thompson, D.Meyer, and J.A.Woollam,
    Conference on Science and Technology of Thin Film Superconductors,
    Colorado Springs, Nov. 14-18, 1988 (to be published by Plenum
    Publishing Corporation).
33) D.S.Dinley, E.L.Venturini, J.F.Kwark, R.J.Baughman, M.J.Carr,
    P.F.Hlava, J.E.Schirber, and B.Morosin, Physica C 152, 217 (1988).
34) J.R.Thompsom, J.Brynestad, D.M.Kroeger, Y.C.Kim, S.T.Sekula,
    D.K.Christen, and E.D.Specht, (submitted to Phys.Rev.B).
35) T.Nakahara, International Symposium on Superconductivity, August 28-
    31, 1988, Nagoya, Japan.
36) J.F.Kwark, E.L.Venturini, R.J.Baughman, B.Morosin, and D.S. Ginley
    (preprint).
37) S.Yazu, International Symposium on Superconductivity, August 28-31,
    1988, Nagoya, Japan.
38) A.Iwasaki, N.Kobayashi, Y.koije, M.Kikuchi, Y.Syono, K.Noto, and
    Y.Muto (preprint).
39) D.S.Dinley, J.F.Kwark, E.L.Venturini, R.J.Baughman, R.P.Hellmer,
    M.A.Mitchell, and B.Morosin, Conference on Sience and Technology of
    Thin Film Superconductors, Colorado Springs, Nov. 14-18, 1988 (to be
    published by Plenum Publishing Corporation).

PARTICIPANTS

B. Abdelhak
University of Houston
Space Vacuum Epitaxy Center
Houston, TX 7704-5507

K. Agarwal
General Dynamics Space Systems Div
MS 23-8560 PO Box 85990
San Diego, CA 92138

A. Agarwala
Eastman Kodak Company
1669 Lake Avenue
Rochester, NY 14650-2036

J. Agostinelli
Eastman Kodak Company
Research Laboratories
Rochester, NY 14650

R. K. Ahrenkiel
Solar Energy Research Institute
1617 Cole Blvd
Golden, CO 80401

D. Albin
Solar Energy Research Institute
1617 Cole Blvd.
Golden, CO 80401

T. Aponick
Foster-Miller Inc.
350 Second Avenue
Waltham, MA 02154-1198

S. Arora
Mechanical Eng & Aerospace
Orlando, FL 32816

D. Arvizu
Ssndia National Laboratory
Box 5800
Albuquerque, NM 87185

J. Banner
Solar Energy Research Institute
1617 Cole Blvd.
Golden, CO 80401

R. Barkley
University of Colorado
Box 215 Dept of Chem & Biochem
Boulder, CO 80309-0215

J. Barnes
General Dynamics Space Systems
PO Box 85990
San Diego, CA 92138

T. S. Basso
Solar Energy Research Institute
1617 Cole Blvd
Golden, CO 80401

J. A. Beall
Nat'l Institute of Standards &
   Tech
325 Broadway 724.03 1-2137
Boulder, CO 80303

J. G. Beery
Los Alamos National Laboratory
PO Box 1663 MS D429
Los Alamos, MN 87545

E. Belohoubek
David Sarnoff Research Center
CN5300
Princeton, NJ 08543

R. F. Belt
Airtron-Litton
200 E. Hanover Ave
Morris Plains, NJ 07950

P. Berdahl
Lawrence Berkeley Laboratory
Bldg 90 Rm 2024
Berkeley, CA 94720

C. Berendt
USAF
Bolling AFB
Washington, DC 20332

J. Berzins
Superconductive Technologies Inc
4630 Indiana Street
Golden, CO 80403

R. Birkmire
Institute of Energy Conversion
University of Delaware
Newark, DE 19711

P. Bly
Litton Airtron
200 East Hanover Avenue
Morris Plains, NJ 07950-2496

R. Boerstler
Sanders Associates
Daniel Webster Hwy S NHQ6-1517
Nashua, NH 03061

B. Boone
Johns Hopkins University
Johns Hopkins Rd
Laurel, MD 20707

P. Broussard
Naval Research Laboratory
Code 6344
Washington, DC 20375

G. Burton
SAES Getters
1122 E Cheyenne Mountain Blvd
Colorado Springs, CO 80906

D. P. Byrne
Kaman Sciences Corporation
1500 Garden of the Gods Road
Colorado Springs, CO 80933

C. Caley
Stanford University
Chemistry Dept
Stanford, CA 94305

P. H. Carr
Rome Air Dev Ctr
Attn RADC/EEA
Hanscom AFB, MA 01731

A. Cavalleri
Centro ITC-CNR
Pante Di Povo(Universita Di
Fisica)
Trento,   ITALY

D. Chambonnet
Laboratoires de Marcoussis
Route de Nazey
Marcoussis Essonne, 91460 FRANCE

J.-C. Chang
Furukawa Electric Co
2-4-3 Okano, Nishi-ku
Yokohama,   JAPAN

K. C. Chen
General Atomics
10955 John Jay Hopkins Dr
San Diego, CA 92138-5608

Z. Chen
Stanford University
Department of Applied Physics
Stanford, CA 94305

J. M. Chesnutt
U.S. Air Force Academy
POB 5282
Colorado Springs, CO 80841

P. Chiang
Amdahl Corporation MS197
1250 E Arques
Sunnyvale, CA 94088

W. Childs
Teledyne Brown Engineering
300 Sparkman Drive, MS 108
Huntsville, AL 35807

K. Chopra
Indian Institute of Technology
Director Indian Inst Technology
Kharagpur W Bengal,   721302 INDIA

L. Chow
University of Central Florida
Physics Department
Orlando, FL 32816

D. Christen
Oak Ridge National Laboratory
PO Box 2008
Oak Ridge, TN 37831

T. Ciszek
Solar Energy Research Institute
1617 Cole Blvd.
Golden, CO 80401

J. Clarke
Lawrence Berkeley Laboratory
Dept of Physics
Berkeley, CA 94720

P. Clarke
Sputtered Films Inc
Box 4700
Santa Barbara, CA 93103

B. Clemens
General Motors Research
3011 Malibu Canyon Road MS RL63
Malibu, CA 90265

K. Craft
SAES Getters
1122 E Cheyenne Mountain Blvd
Colorado Springs, CO 80906

C. Cunningham
Stanford University
Physics Department
Stanford, CA 94305

A. Czanderna
Solar Energy Research Institute
1617 Cole Blvd.
Golden, CO 80401

A. DasGupta
U.S. Department of Energy
9800 S Cass Ave
Argonne, IL 60439

M. Davis
University of Houston
4800 Calhoun
Houston, TX 77004

A. DeLozanne
University of Texas
Department of Physics
Austin, TX 78712

S. Deb
Solar Energy Research Institute
1617 Cole Blvd
Golden, CO 80401

J. Debsikdar
EG&G Idaho Inc
PO Box 1625
Idaho Falls, ID 83415

P. C. Dent
APD Cryogenics Inc
1919 Vultee St
Allentown, PA 18103

V. Desai
Univ. of Central Florida
Mech Engineering & Aerospace
Orlando, FL 32816

N. G. Dhere
Solar Energy Research Institute
1617 Cole Blvd
Golden, CO 80401

D. Dunlavy
Solar Energy Research Institute
1617 Cole Blvd
Golden, CO 80401

E. Durbin
1403 N Hartford St
Arlington, VA 22201

R. Eaton
U.S. Department of Energy
1000 Independence Ave. SW
Washington, DC 20585

P. Eklund
University of Kentucky
Dept of Physics and Astronomy
Lexington, KY 40506

A. Erbil
Georgia Tech
School of Physics
Atlanta, GA 30332

R. Feenstra
Oak Ridge National Laboratory
PO Box 2008, Bldg 3137
Oak Ridge, TN 37831

S. Fiedziuszko
Ford Aerospace
3825 Fabian Way
Palo Alto, CA 94303

J.-I. Fujita
NEC Corporation
4-1-1 Miyazaki, Miyamae-ku
Kawasaki, 213 JAPAN

F. Garzon
Los Alamos National Laboratory
LANL MS D429
Los Alamos, NM 87545

J. Geerk
Kernforschungszentrum Karesnake
Stutensee, 7513
WEST GERMANY

A. Gelb
Hittman Mat & Med Components Inc
9190 Red Branch Road
Columbia, MD 21045

T. Geyer
China Lake Naval Weapons Center
Naval Weapons Center Code 3851
China Lake, CA 93555

R. F. Giese
Argonne National Laboratory
9700 South Cass Ave.
Argonne, IL 60439

D. Ginley
Sandia National Laboratories
PO Box 5800 Org 1144
Albuquerque, NM 87185

R. E. Glover
University of Maryland
Dept of Physics
College Park, MD 20742

J.-S. Goela
Morton Thiokal Inc/CVD Inc
185 New Boston Street
Woburn, MA 01801

K. E. Gray
Argonne National Laboratory
9700 S. Cass Avenue Bldg 223-MSC
Argonne, IL 60439

A. Greenwald
Spire Corporation
Patriots Park
Bedford, MA 01730

R. Greer
Kaman Sciences Corp.
1500 Garden of the Gods Road
Colorado Springs, CO 80907

H.-U. Habermeier
Max-Planck-Institute
Heisenbergstr 1
D 7000 Stuttgart-80,   WEST
GERMANY

R. Hammond
Superconductor Technologies, Inc
460 Ward Drive, Suite F
Santa Barbara, CA 93111

R. H. Hammond
Stanford University
Hansen Physics Lab
Stanford, CA 94305

R. E. Harris
National Inst of Standards & Tech
325 Broadway 724.03 1-2137
Boulder, CO 80303-3328

A. M. Hermann
University of Arkansas
104 Physics Bldg. Physics Dept
Fayetteville, AR 72701

B. Hichwa
Optical Coating Laboratory, Inc.
2789 Northpoint Parkway,M.S. 121-1
Santa Rosa, CA 95407-7397

G. Hoffer
Kaman Sciences Corporation
1500 Garden of the Gods Rd
Colorado Springs, CO 80933-7463

Y. R. Hsiao
Boeing Electronics
PO Box 24969, MS 9Z-80
Seattle, WA 98124-6269

E. Hu
Univ of California/Santa Barbara
University of California
Santa Barbara, CA 93117

A. T. Hunter
Hughes Research Laboratories
3011 Malibu Canyon Road, MS RL63
Malibu, CA 90265

N. Ianno
University of Nebraska
209 N WSEC
Lincoln, NE 68588

H. Jaeger
Max Planck Inst fur
Metallforschung
Heisenbergstr 5
7000 Stuttgart 80,   WEST GERMANY

D. Jedamzik
GEC plc
East Lane
Wembley Middlesex
HA97PP UNITED KINGDOM

D. Jennings
National Semiconductor C-2465
2900 Semiconductor Drive
Santa Clara, CA 94087

N. M. Jisrawi
Rutgers University
718 Bevier Rd
Piscatawany, NJ 08903

B. C. Johnson
E I Dupont de-Nemours Co
Central R&D Exper Station
304/C129
Wilmington, DE 19880-0304

B. Johs
University of Nebraska
209 N WSEC
Lincoln, NE 68585-0511

A. Kahan
RADC/ES
Hanscom AFB
Bedford, MA 01731

H. Kajikawa
Kobe Steel Ltd
1-5-5 Takatsukadai Nishi-ku
Kobe Hyogo, 673-02 JAPAN

R. Kampwirth
Argonne National Lab
9700 S. Cass Ave.
Argonne, IL 60439

T. Kawai
ISIR Osaka University
Mihogaoka
Ibaraki   Osaka, 567 JAPAN

G. Kerber
Hughes Aircraft
6155 El Camino Real
Carlsbad, CA 92009

B. F. Kim
Johns Hopkins University
Johns Hoplins Road
Laurel, MD 20707

H. Koinuma
Tokyo Institute of Technology
4259 Nagatsuta, Midori-ku
Yokohama Kanagawa, 227 JAPAN

V. Kresin
Lawrence Berkeley Laboratory
582 Westfield Way
Oakland, CA 94619

N. Kumar
MCC
12100 Technology Blvd
Austin, TX 78727

J. R. Kwo
AT&T Bell Laboratories
1D232 600 Mountain Ave
Murray Hill, NJ 07974

H. S. Kwok
State University of New York
214 Bonner Hall
Buffalo, NY 14260

D. Lathrop
Cornell University
Applied & Engineering Physics
Ithaca, NY 14850

A. Lee
TRW, Inc.
R1-2154 1 Space Park
Redondo Beach, CA 90278

F. Li
Chinese Academy of Science
Zhong Guan Cun Institute of
    Physics
Beijing,   CHINA

K. Li
Boeing
PO Box 24969, MS 97-80
Seattle, WA 98124-6269

C. Lichtenberg
University of Houston/NASA
CODE: EE36
Houston, TX 77058

R.-J. Lin
Industrial Tech Research Institute
195-5 Chung-hsing Rd Sec 4 Chutung
Hsinchu   Taiwan,   31015 CHINA

J. Lindquist
Aerojet Electro Systems Company
PO Box 296, MS 170-8244
Azusa, CA 91702

S.-H. Liou
University of Nebraska
260 Behlen Laboratory of Physics
Lincoln, NE 68588-0111

E. Logothetis
Ford Motor Company
PO Box 2053
Dearborn, MI 48009

S. Ludvik
Teledyne
1274 Terra Bella Ave
MountainView, CA 94043

A. Mackor
Institute of Applied Chemistry TNO
PO Box 108
3700 AC 2E1ST, THE NETHERLANDS

K. Makoto
Central Res Labs Matsushita
    Elec Co
3-15 Yagumonakamachi
Moriguchi 570,   JAPAN

V. Marchenko
Inst. for Prob. of
Microelectronics
142432 Chernogolovka
Moscow District,   RUSSIA

L. Maritato
INFN via E Fermi 40
Frascati, 00044 ITALY

J. Matey
David Sarnoff Research Center
CN 5300
Princeton, NJ 08543

O. K. Mawardi
Collaborative Planners Inc
15 Mornington Lane
Cleveland Heights, OH 44106

B. McAvoy
Westinghouse R&D
1310 Brulah Rd
Pittsburgh, PA 15235

B. W. McConnell
Oak Ridge National Lab
Bldg 3147, MS-6070
Oak Ridge, TN 37831-6070

R. D. McConnell
Solar Energy Research Institute
1617 Cole Blvd
Golden, CO 80401

J. McMenamin
Custom Engineered Materials Inc
4039 Avenida de La Plata
Oceanside, CA 92056

J. McNally
U.S. Air Force Academy
Department of Physics
Colorado Springs, CO 80840

T. E. McNeil
U.S. Air Force Academy
USAFA/DFP
Colorado Springs, CO 80918

M. Melich
Naval Postgraduate School
4637 N 24th Street
Arlington, VA 22207

E. Meloni
Strem Chemicals Inc
PO Box 108
Newburyport, MA 01950

Y. Miki
Stanford University
GP-B Hansen Labs
Stanford, CA 94025

D. E. Miller
Lawrence Livermore Laboratory
7000 East Avenue MS L-156
Livermore, CA 94550

T. A Miller
Ames Lab, ISU
Phsics 12
Ames, IA 50011

N. Mitra
Solar Energy Research Institute
1617 Cole Blvd
Golden, CO 80401

C. Mombourquette
Los Alamos National Laboratory
PO Box 1663 MS D429
Los Alamos, NM 87545

J. Moreland
Nat'l Institute of Tech &
    Standards
325 Broadway, Mail Code: 724.05
Boulder, CO 80303

J. G. Morse
Advanced Materials Institute
Colorado School of Mines
Golden, CO 80401

M. Moske
University of Goettingen
I Physical Institute
Bunsenstrasse9
D-3400 Goettingen,   WEST
GERMANY   0

R. Motes
U.S. Air Force Academy
4408B Qtrs USAFA
Colorado Springs, CO 80840

D. Mueller
Nat'l Institute of Stand & Tech
Gaithersburg
Gaithersburg, MD 20899

R. Muenchausen
Los Alamos National Lab
MS E549
Los Alamos, NM 87545

S. Nam
Wright State University
7735 Peters Pike
Dayton, OH 45414

R. Neifeld
U.S. Army Elect Tech & Devices Lab
Fort Monmouth
Fort Monmouth, NJ 07703

R. N. Nelson
Ford Aerospace Corporation
PO Box A Ford Road
Newport Beach, CA 92658-9983

M. (Max) Nofziger
City of Austin-City Council
PO Box 1088
Austin, TX 78767

P. Norris
EMCORE Corp
35 Elizabeth Ave
Somerset, NJ 08873

R. Noufi
Solar Energy Research Institute
1617 Cole Blvd
Golden, CO 80401

M. O'Connell
Commonwealth Scientific Corp
500 Pendleton St
Alexandria, VA 22314

K. Ogawa
Nat'l Res. Institute for Metals
Tsukuba Labs 1-2-1 Sengen
Tsukuba Ibaraki-Ken, 305 JAPAN

R. H. Ono
Nat'l Institute of Standards &
    Tech
325 Broadway 724.03 1-2137
Boulder, CO 80303

V. Palmieri
Infn. Labor. Nazionali Di Legnaro
via Romea 4
Legnaro (Padova), I-35020 ITALY

F. J. Pern
Solar Energy Research Institute
1617 Cole Blvd
Golden, CO 80401

J. Phillips
Coors Ceramics Co
17750 W. 32nd Ave
Golden, CO 80401

N. Phillips
Lawrence Berkeley Laboratory
Berkeley, CA 94720

S. Pike
Superconductive Technologies Inc
4630 Indiana St
Golden, CO 80403

B. Politt
PTB Berlin
Abbestr. 2-12 1000 Berlin 10
Berlin,   WEST GERMANY

D. Proffitt
Superconductive Components
1145 Chesapeake Ave
Columbus, OH 43212

J. Przybysz
Westinghouse R&D Center
401-3A12
Pittsburgh, PA 15235

X. Qiu
McGill University
3480 University St
Montreal, PO H3A2A7 CANADA

P. G. Quincey
National Physical Laboratory
Queens Road
Teddington Middlesex,   TW110LW
UNITED KINGDOM

F. Radpour
University of Oklahoma
202 West Boyd CEC 219
Norman, OK 73019

I. D. Raistrick
Los Alamos National Laboratory
MS D429
Los Alamos, NM 87545

R. Ralston
MIT Lincoln Laboratory
244 Wood Street, PO Box 73
Lexington, MA 02173-0073

K. Rebis
David Taylor Research Center
Code 2812
Annapolis, MD 21403

C. Reece
Continuous Electron Beam Accel Fac
12000 Jefferson Ave
Newport News, VA 23606

D. Resendes
Physical Sciences Inc
Research Park PO Box 3100
Andover, MA 02145

P. L. Richards
Lawrence Berkeley Lab
One Cyclotron Road
Berkeley, CA 94720

M. Robinson
Superconductor Technologies, Inc
460 Ward Drive   Suite F
Santa Barbara, CA 93111-2310

B. Rosenblum
Allied-Signal Aerospace Company
PO Box 419159
Kansas City, MO 64141-6159

548

P. Roth
Sandia National Laboratory
Div 6224 PO Box 5800
Albuquerque, NM 87185

L. Roybal
Solar Energy Research Institute
1617 Cole Blvd
Golden, CO 80401

J. Ryu
Northeastern University
235 FR 360 Huntington Ave
Boston, MA 02115

S. Salkalachen
University of Western Ontario
Dept of Physics
London Ontario,   N6A 3K7 CANADA

A. C. Schaffhauser
Oak Ridge National Laboratory
PO Box 2008 Bldg 4500S
Oak Ridge, TN 37831-6140

D. G. Schlom
Stanford University
226 McCullough
Stanford, CA 94305-4055

T. Schuyler
Solar Energy Research Institute
1617 Cole Blvd.
Golden, CO 80401

R. Seery
CEC Inc
12950 Rush Lane
Black Forest, CO 80908

R. Sega
University of Colorado
Dept. of Electrical Engineering
Colorado Springs, CO 80933

C. Segre
Illinois Inst of Technology
3301 S Dearborn
Chicago, IL 60616

I. Shih
McGill University
3480 University St
Montreal, PO H3A2A7 CANADA

S. Shinozaki
Ford Motor Company
10507 Bassett Drive
Livonia, MI 48150

J. Siebert
Ball Aerospace
PO Box 1062
Boulder, CO 80306

D. E. Siegfried
ION Tech Inc
2330 E Prospect
Fort Collins, CO 80525

R. E. Sievers
University of Colorado
Campus Box 215
Boulder, CO 80309

R. Simon
TRW, Inc
1 Space Park
Redondo Beach, CA 90278

C. Simpson
Naval Avionics Center   Code 813
6000 E 21st St
Indianapolis, IN 46219

D. R. Smith
Kaman Sciences Corporation
1500 Garden of the Gods Rd
Colorado Springs, CO 80933-3393

R. Sobolewski
University of Rochester
Dept of Electrical Engineering
Rochester, NY 14627

R. J. Soulen
Naval Research Lab
Code 6344
Washington, DC 20375-5000

J. Spargo
Hughes Aircraft
6155 El Camino Real
Carlsbad, CA 92009

M. Sparpaglione
ENICHEM
via Medici del Vascello 26
Milano,   20138 ITALY

D. Spaulding
University of Denver
2455 S Gaylord
Denver, CO 80210

K. Springer
Los Alamos National Laboratory
MS D429
Los Alamos, NM 87545

B. Stafford
Solar Energy Research Institute
1617 Cole Blvd.
Golden, CO 80401

J. L. Stone
Solar Energy Research Institute
1617 Cole Blvd.
Golden, CO 80401

J. Talvacchio
Westinghouse R&D Center
1310 Beulah Rd
Pittsburgh, PA 15235

J. Tate
Technical University of Munich
Physik Department E10
D-8046 Garching,   WEST GERMANY

G. Tennyson
U.S. Department of Energy
2505 Punta de Vista NE
Albuquerque, NM 87112

D. Theirlich
Universitat Wuppertal
Dipl Physics
Wuppertal,   WEST GERMANY

D. Thompson
University of Nebraska
209 N WSEC, UNL
Lincoln, NE 68588-0511

G. S. Tompa
EMCORE Corporation
35 Elizabeth Ave
Somerset, NJ 08873

J. Trefny
Colorado School of Mines
Physics Department
Golden, CO 80401

Y. Tzeng
Auburn University
200 Brown Hall
Auburn, AL 36849

G. Valco
Ohio State University
2015 Neil Ave (Dept. of EE)
Columbus, OH 43220

T. VanDuzer
University of California
EECS Dept
Berkeley, CA 94720

L. VanRoekel
Leybold Inficon Inc
6500 Fly Road
East Syracuse, NY 13057

T. Venkatesan
Bellcore
331 Newman Springs Road
Red Bank, NJ 07701

E. L. Venturini
Sandia National Laboratories
Division 1131
Albuquerque, NM 87185

M. Vichr
Air Products
2450 Autumnvale Drive
San Jose, CA 95131

J. Vrba
CTF Systems Inc
15-1750 McLean Ave
Port Coquitlam, BC V3C 1M9 CANADA

T. Walsh
National Inst. of Standards & Tech
325 Broadway - 1/2137 724.03
Boulder, CO 80303-3328

X.-K. Wang
Northwestern University
2215 Wesley Avenue
Evanston, IL 60208

K. Wasa
Matsushita Electric Co
3-15 Yagumonakamachi
Moriguchi 570,   JAPAN

R. Welty
University of Colorado
4279 C Monroe
Boulder, CO 80303

B. Wessels
Northwestern University
Technical Institute
Evanston, IL 60208

C. Whitman
CVC Products Inc.
525 Lee Rd Box 1886
Rochester, NY 14475

D. Wilde
Los Alamos National Laboratory
MD D429
Los Alamos, NM 87545

D. Williams
Ball Communications Systems Div
PO Box 1235
Broomfield, CO 80020-8235

S. Witanachchi
SUNY at Buffalo
330 Bonner Hall
Amherst, NY 14260

S. A. Wolf
Naval Research Laboratory
Code 6340
Washington, DC 20375

L. S. Wright
Royal Military College of Canada
Department of Physics
Kingston Ontario, K7K 5L0 CANADA

G.-C. Xiong
Peking University
Dept. of Physics
Beijing,   CHINA

H.-C. Yen
TRW
One Space Park R6/1575
Redondo Beach, CA 90278

J. Yoshida
Toshiba R&D Center
1, Komukai Toshibacho, Saiwaiku
Kawasaki Kanagawa, 210 JAPAN

C. B. Zarowin
Perkin Elmer Corp
100 Wooster Hgt Rd
Danburg, CT 06810

A. Zehnder
Paul Scherrer Institute
5324 Villigen,   SWITZERLAND

B.-R. Zhao
Chinese Acad. of Sciences
   (Physics)
Zhong Guan Cun
Beijing,   CHINA

G.-G. Zheng
Chinese Academy of Sciences
PO Box 603-46 Inst of Physics
Beijing, 100080 CHINA

INDEX